The Observer's Guide to Astronomy
Volume 2

PRACTICAL ASTRONOMY HANDBOOK SERIES

The Practical Astronomy Handbooks are a new concept in publishing for amateur and leisure astronomy. These books are for active amateurs who want to get the very best out of their telescopes and who want to make productive observations and new discoveries. The emphasis is strongly practical: what equipment is needed, how to use it, what to observe, and how to record observations in a way that will be useful to others. Each title in the series will be devoted either to the techniques used for a particular class of object, for example observing the Moon or variable stars, or to the application of a technique, for example the use of a new detector, to amateur astronomy in general. The series will build into an indispensible library of practical information for all active observers.

Titles available in this series

How can you find new novae and supernovae? How can you use photoelectric detectors to derive the temperatures of stars? And how can you calculate the orbits of meteors? The questions asked by serious amateur astronomers are answered in this authoritative and wide-ranging guide

Topics range from spectroscopy of meteors to visual and photographic observations of aurora, meteors, double stars and deep-sky objects. For each topic, sound practical methods of observation and the scientific background are given to lead you to better observations. Guidelines also show you how to record and catalogue your observations using the recognised professional terminology and classification schemes.

From t impl est visual observations of variable stars to observations of the most distant ga es with state-of-the-art CCD cameras and photoelectric photometers, this guide acked with practical tips for all types of amateur observations. It will develop tl servational skills of the keen novice and satisfy the more demanding needs of t xperienced amateur astronomer.

at your telephone call is cha
call the numbers as set out
c r code

The Observer's Guide to Astronomy
Volume 2

Edited by
PATRICK MARTINEZ

Translator
STORM DUNLOP

Published by the Press Syndicate of the University of Cambridge
The Pitt Building, Trumpington Street, Cambridge CB2 1RP
40 West 20th Street, New York, NY 10011-4211, USA
10 Stamford Road, Oakleigh, Melbourne 3166, Australia

First published as Patrick Martinez, *Astronomie, le Guide de l'Observateur* by
Edition Société d'Astronomique Populaire, and © Patrick Martinez 1987

This English language edition © Cambridge University Press 1994

English language edition first published 1994

Printed in Great Britain at the University Press, Cambridge

A catalogue record of this book is available from the British Library

Library of Congress cataloguing in publication data

Astronomie, le guide de l'observateur. English.
The observer's guide to astronomy/edited by Patrick Martinez:
translator Storm Dunlop. – English language ed.
p.   cm. – (Practical astronomy handbooks)
Includes bibliographical references.
ISBN 0 521 37068 X (v. 1) (hc). – ISBN 0 521 37945 8 (pbk.: v.1).
ISBN 0 521 45265 1 (v. 2) (hc). – ISBN 0 521 45898 6 (pbk.: v.2).
1. Astronomy–Amateurs' manuals. 2. Astronomy–Observers' manuals.
I. Martinez Patrick. II. Title.
III. Series: Practical astronomy handbook series.
QB63.A7813   1994
520–dc20   93-29830   CIP

ISBN 0 521 45265 1 hardback
ISBN 0 521 45898 6 paperback

The contributors

Michel Blanc

Roland Boninsegna

René Boyer

Christian Buil

Roger Chanal

Serge Chevrel

François Costard

Jean Dijon

Jean Dragesco

Michel Dumont

Pierre Durand

Robert Futaully

Claude Grégory

Alain Grycan

Jean Gunther

Serge Koutchmy

Jean Lecacheux

Michel Legrand

Jean-Louis Leroy

Jari Mäkinen

Jean-Marie Malherbe

Patrick Martinez

Marie-Josèphe Martres

Jean-Claude Merlin

Thierry Midavaine

Bruno Morando

Régis Néel

Gualtiero Olivieri

Olivier Saint-Pé

Jean Schwaenen

Jean-Pierre Tafforin

Jacques Barthès (drawings)

Summary

Contents

Preface

Amateur astronomers

Active amateur astronomers engage in their favourite activity as a hobby, in their spare time. They are able to devote a limited amount of time to it, being restricted by their jobs, whose hours are rarely arranged to suit people who want to spend the night looking through a telescope!

Amateurs are limited financially, because they either have to provide the money themselves or obtain it from an astronomical group, most of which do not generally have large amounts of funds.

Many instruments bought commercially by amateurs are small refractors or reflectors, and the majority of these are probably gathering dust in attics, either because their owners were disappointed in their performance, or because they were purchased in a burst of short-lived enthusiasm or, finally, because their owners became bored with looking at the Moon or M31 for hours on end, did not know what else to do, and abandoned astronomy for some other pursuit. It was thinking about the latter group of observers that the germ of the idea for this book was sown: such amateurs need to be shown the vast range of useful observations that are possible, even with modest equipment, and how they should proceed.

Amateur astronomy

Just as professional astronomers may be divided into two broad groups, observers and theorists, the same applies to amateurs. Some are purely observers, and others are happier in a library than in an observing dome. Apart from these two classes of amateur, however, there are two more categories. There are the 'Telescope Nuts' (as they are often termed in North America), who spend their lives building bigger and better telescopes, but who often have no time to use their latest equipment before they start on something new, and there are the society members, who are the backbone of all the local and national astronomical groups and keep them in existence, but who have no time to look at the sky in between organising meetings, exhibitions, or conferences.

Let us concentrate on the observers. There are three types of observations made by amateurs:

Sight-seeing   This consists of looking at objects in the sky purely for their beauty, and this is where most people begin. This is often the initial reason – and for some people the only reason – for taking up astronomy. Some amateurs never progress beyond this stage, but many others feel an urge, after a certain time, which may be months or years, to take things further, and frequently they are unable to find any guidance either in their circle of acquaintances or in the literature. Even if 'sight-seeing' is, by definition, unproductive, apart from the pleasure that it brings, it should not be neglected, because it enables amateurs to gain experience and become accustomed to observing; both qualities that are of considerable value when they turn to more 'serious' work.

Educational   Here, the astronomer carries out an experiment, which may require special equipment or measurements, or else a greater or lesser degree of sophistication in analysing the results. There are many different examples, but unfortunately the results are foregone conclusions, with the outcome having been known for decades – even for centuries – and to a degree of accuracy that is far beyond amateurs' reach.

True observation   This often forms part of a specific programme of observation, and the results contribute – albeit sometimes in a very modest fashion – to our overall knowledge of the universe. People frequently have little idea of the range of observations within the capabilities of amateurs, who are thus able to make really useful observations.

In fact, amateur and professional astronomy truly complement one another. Although professionals have powerful methods at their disposal, amateurs are able to cover those fields that require little equipment or small instruments (such as binoculars, small, wide-field telescopes, etc.), mobile resources (such as the observation of grazing occultations), or a large number of observations (examples are variable stars, double stars, sky patrols, etc.). The result has been an increase in the collaboration between professionals and amateurs, with the latter having greater access to large telescopes such as the 600-mm reflector at the Pic du Midi Observatory, or participating in meetings such as the one organised by the Société Astronomique de France in Paris in 1987, which was supported by the International Astronomical Union (IAU Colloquium 98).

If we consider the fields for which amateurs are noted, we find that France is among the first rank on an international level when it comes to the study of planetary surfaces, double stars and variable stars. It is also relatively well placed as regards the Sun, comets and meteors.

Occultations appear to be a Belgian speciality, but in many other fields such as astrometry, deep-sky work or the discovery of new objects (comets, novae, supernovae, and minor planets), the United Kingdom, United States, Australia or Japan set an example. It is time that French astronomers woke up to these particular fields, where their forebears were extremely prominent. One gets the impression that nowadays, apart from a handful of top-notch observers, whose names appear from time to time in the International Astronomical Union's telegrams and circulars, or in specialized international journals, that French astronomy has lapsed into somnolence, and has left it to other countries to make all the discoveries. (The balance-sheet when it comes to comets is particularly eloquent.)

The observer's guide

Somehow we need to convey to amateurs who have grown bored with seeing the same few notable objects that they look at every night, that there are fascinating observational programmes, in which they can also achieve extremely useful results.

It became obvious that it would be of great value if descriptions of the useful work that may be carried out could be brought together in one publication. This is what gave birth to the idea of this '*Observer's Guide*'. To avoid overawing beginners, it does include some 'sight-seeing' and 'educational' projects, but because it aims to guide observers into making useful 'scientific' work, it does concentrate on the latter.

Any such project is highly ambitious and could not be realized by a single person. In fact, each subject needs to be described by a specialist, and as no one can hope to be expert in every field, this naturally led to the idea of a collaborative work.

This book examines all the different types of object available to amateur astronomers, and for each one it describes the types of observation that are possible (paying particular attention to their scientific value), the equipment required, the methods to be employed, where appropriate information may be found, and the organisations to which the results should be reported. Each chapter has been written by a French or Belgian expert (or experts) in the subject concerned; the authors include both professional and amateur astronomers.

The first fifteen chapters describe observational programmes, divided into individual classes of astronomical object. The last five chapters discuss techniques that relate to several of the subject-areas previously described. This arrangement avoids unnecessary repetition.

An effort has been made to standardize the text provided by the various authors, and to limit the amount of editing required; some subjects that are considered to be particularly important, however, are deliberately included in more than one place in the book.

Despite its size, the '*Observer's Guide*' cannot include every aspect of amateur astronomy. It has been assumed that the reader is familiar with all the basic techniques, such as setting up a telescope, finding objects, etc. If not, details may be found in any one of a number of books intended for beginners, so it would be pointless to repeat them here.

For similar reasons, there is no chapter devoted to astrophotography as such. The basic techniques that are required are described in a number of books, which are given in the bibliography. It therefore seemed pointless to include information about this general area in this book, and to do so would have required 100 to 200 additional pages. On the other hand specific photographic techniques required by particular types of observation are discussed in considerable detail in some of the individual chapters.

Advice to the reader

The '*Observer's Guide*' is not meant to be read straight through from cover to cover. It has been designed so that the individual chapters are relatively independent, so that the reader may concentrate on the subjects of interest, and in any desirable order. Relevant cross-references are included to other chapters and sections as required.

This book should be regarded as a catalogue, where amateurs may find out what types of observation are possible given the facilities that they have available, and from which they may choose those fields that seem most attractive to them. The text describes all the steps and equipment necessary to take the first step. If this serves to confirm the initial attraction for any particular field of interest, addresses of various organisations are given that may be contacted for further information and guidance.

So now ... good reading, and good observing!

Translator's preface

In a multi-author work, there are always problems in ensuring consistency in presentation, style, and usage. Not surprisingly, there were some disparities in the original French text. In translating this work, I have tried to ensure that the whole text is consistent. In accordance with recommended practice, for example, all wavelengths in the visible and adjacent spectral ranges are now given in nanometres (nm) rather than the older, perhaps more familiar, ångström units – 1 nm (10^{-9}m) = 10 Å (10^{-10}m). At longer wavelengths, the unit used is the micrometre μm (10^{-6}m). Similarly, the dimensions of telescopes, etc. are consistently given in millimetres or metres (not in centimetres or decimetres).

The symbols for physical variables and constants have been altered to those most commonly used in English-language works and, whenever possible, agree with the recommendations of the IAU or the various IAU Commissions. Annotations and captions to the figures have also been changed to forms more readily related to the English-language terms for the terms and items designated.

Dates and times have been expressed consistently in the accepted international scientific format: Year, month, day, hour, minute, and second, with decimal forms where appropriate. In accordance with this standard format, the name of the month is given in full or as a three-letter abbreviation, not in numerical form.

One major alteration from the layout of the original volumes concerns the positioning of notes, references, glossaries and bibliographies. The original chapters were inconsistent, some giving this material within the text, some in footnotes, and some in 'appendices' at the ends of the chapters concerned. The majority of notes, glossaries, and appendices have now been incorporated in the text. Where it has been thought advisable to insert a translator's note, this is enclosed in square brackets. Section numbering and cross-referencing have also been simplified, and do not correspond precisely with the arrangement of the original edition. A few items (including some Figures) have been moved to more appropriate positions. Two general appendices on 'Time-scales' and the interesting 'T60 Association' may be found at the end of Volume 2.

Discussion of the merit (or demerit) of specific charts, atlases, etc. for a particular branch of practical work has been retained in the text, usually within a special sub-section. References to papers, articles, etc. – i.e., items with no accompanying discussion or comments – may be found with a more general bibliography (and a few notes) for each chapter in sections at the back of each volume. Please note that where books are cited as specific references, they are not repeated in the general bibliography.

The original work contained many references to French-language material, in particular to papers and articles in French journals. Obviously, many English-speaking readers will be unable to understand the original language. It was felt, however, that some items were not adequately treated in English, so some references to major French-language publications, particularly books and the journal *l'Astronomie*, have been retained. Most references to publications issued by local groups have been removed, because the majority of readers will find the material impossible to obtain or consult. Various references to easily available English-language books and journals have been added, including some material not published when the original text was written.

There are known to be inconsistencies in the details given for some of the references, many of which were missing volume, issue, or page numbers – and even in some cases date, author or title! Regrettably, it was not possible for me to spend the extremely large amount of time that would be required to check and correct every individual reference. (Many dozens of corrections have been incorporated, and I would particularly like to thank Mr Peter Hingley, the Royal Astronomical Society's Librarian, for all the work that he carried out in this respect.) It is believed that the information given will enable the appropriate papers to be located without too much trouble, even though the information may be incomplete.

All necessary amendments to the French edition have been incorporated, and in addition, I have corrected many minor, previously undetected errors, and included the latest data wherever possible – such as revised totals for numbered minor planets, and the number of supernovae discovered by Bob Evans – without reference to the authors. Many major discrepancies and errors have been referred back to the original editor, Patrick Martinez, and the authors, and appropriately corrected.

In a few cases, the precise methods used by observational groups in Francophone countries differ from those employed by the major, English-speaking organisations. No attempt has been made to 'correct' these points, not only because the various methods may be equally valid, but also because the differences may be instructive. However, attention is sometimes drawn to any disparity in a translator's note.

It is impossible for me to mention individually the many amateur and professional astronomers in all parts of the world who have so unstintingly helped me with details of methods and organisations, or with more general advice. There is not a single chapter that has not benefited from their assistance, which is greatly appreciated. If I may single out one person in particular, whose advice has extended far beyond the field of just minor planets and comets, it is Dr Brian Marsden, of the IAU Central Bureau for Astronomical Telegrams in Cambridge, Massachusetts. His advice, often received almost instantaneously, thanks to the wizardry of modern computer networks, has been invaluable.

Storm Dunlop

11 Aurorae

M. Blanc and J. Mäkinen

11.1 The physical phenomenon

11.1.1 The nature of the aurorae

Anyone who has had the opportunity of travelling to high northern or southern latitudes during the long winter night, and has discovered the polar aurorae for themselves has undoubtedly been left with an indelible impression. Because they are usually too faint to be seen in the presence of city lights, you have to go to a really dark, remote place, far from any urban activity, before you are able to see their true beauty. They then appear in an incredible variety of forms, from a very faint, diffuse, greenish patch to a spectacular show of coloured curtains dancing in the sky, apparently set in motion by an invisible wind (Fig. 11.1). If you are lucky enough to see the aurora (which just takes a little bit of patience and effort), you can expect a wonderful experience: the beauty of the aurora is enhanced by the darkness of the night, the silence of the polar wilderness and, very often, the cold that you have to withstand to watch the show until it comes to a (temporary) halt.

Although the true beauty of the aurorae can perhaps be appreciated only from high latitudes, it is certainly wrong to suppose – as the vast majority of people imagine – that it cannot be seen elsewhere. It is certainly visible (although more rarely) at lower geographical latitudes. The aurora is associated with the Earth's magnetic field, and has even been seen at the geomagnetic equator, so essentially anyone might see an aurora, wherever they are on Earth, although it could be a once-in-a-lifetime experience. Events such as the great auroral storms of 1938 January and 1989 March covered such a vast area and were so spectacular that they generated considerable interest among the public and the news media. In any case, amateur astronomers, who naturally seek out dark skies for their other observing programmes, are ideally placed to observe any aurorae that do occur. But what is the physical nature of the aurora?

The aurora has been known since ancient times, and has attracted the curiosity of some of the great names in the history of science and astronomy, such as Aristotle and Seneca (who both personally witnessed the appearance of an aurora), and later Gassendi, Galileo Galilei and Edmond Halley. The first use of the term 'aurora borealis' (northern dawn) is commonly attributed to the French astronomer Gassendi, although recent research suggests that his famous 17th-century contemporary Galileo may have been the first to use that form of words. It is also possible that Gregory of Tours used the description aurora borealis as early as the 6th century.

The first scientist to have an approximately correct idea of the origin of the aurora was the French scientist Jean-Jacques Dortous de Mairan (1678–1771), whose paper *Traité Physique et Historique de l'Aurore Boréale*, published in 1733 by the French

Fig. 11.1. *A typical auroral display occurring during the polar night. (Photo: Heikki Ketola, Ursa Astronomical Association)*

Académie des Sciences, was the very first monograph on the aurora. De Mairan interpreted the aurora as the result of the interaction between the Earth's upper atmosphere and the extended atmosphere of the Sun (what we now call the corona), along the surface in space where the gravitational forces of the Sun and the Earth are in balance. If a 'bubble' of solar atmosphere were pushed over to the Earth's side of this boundary, it would fall into the Earth's atmosphere, and its reaction with the constituents of the terrestrial atmosphere would produce the fire that was seen as the aurora.

De Mairan was obviously wrong about the physical nature of the interaction between the solar and terrestrial atmospheres, but this was inevitable in his day, because electromagnetism had yet to be discovered. He was right, however, in claiming that this interaction was the primary cause of the aurora.

How do we understand and describe this interaction nowadays? Taken as a local phenomenon, the aurora is rather simple (Fig. 11.2). Energetic electrons originating in the Sun's atmosphere move along the geomagnetic field lines. Their energies range from a few hundred eV to hundreds of keV, with most of the population having energies between approximately 500 eV and 10 keV. At high altitudes, say above about 400 km, because of the magnetic forces acting on charged particles, these electrons move in helical trajectories around the Earth's magnetic lines of force. When these electrons reach lower altitudes, they move through a progressively denser and denser atmosphere, because the density of the Earth's atmosphere, like that of any planet, increases exponentially with decreasing altitude. The approximate density at any altitude is given in Fig. 11.3, which shows, for example, that at 400 km there are between 10^6 and 10^8 atoms or molecules per cubic centimetre. The precise value depends on the exospheric temperature, i.e., the temperature of the upper atmosphere above 200 km, which varies with solar activity. At 100 km, however, this density has increased to over 10^{12} atoms or molecules per cubic centimetre. The probability of a collision between an electron that is moving down a magnetic field line and the ambient atoms and molecules therefore increases exponentially with decreasing altitude. At a certain height, a collision becomes inevitable; lower down, collisions become more and more frequent, until the incident electron finally loses all its kinetic energy.

Where does that energy go? A small fraction is transformed into kinetic energy of the atoms and molecules with which the electron has collided. Most of the energy, however, goes to other electrons, namely the electrons in the electronic shells of those atoms and molecules, because in collisions the maximum amount of energy is exchanged between particles of similar masses. Some of the electrons in the shells receive enough energy to escape from their atom or molecule. They become free electrons, leaving behind newly created positive ions.

The X-ray and ultraviolet components of ordinary sunlight carry enough energy to eject electrons from the outer shells of atmospheric oxygen and nitrogen, giving rise to populations of ions and electrons at altitudes greater that 80 km, and thus forming the ionosphere. Radio communications over long distances exploit the reflective properties of ionospheric layers. During auroral activity, localized clouds of increased ion and electron concentration may disrupt, or sometimes enhance, the propagation of radio signals.

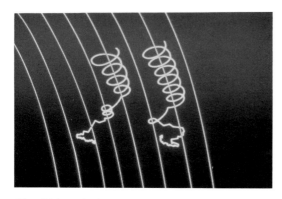

Fig. 11.2. *The basic mechanism generating the aurora. An energetic electron originating in the Sun precipitates into the Earth's upper atmosphere after spiralling round the lines of the Earth's magnetic field, which are nearly vertical over the polar regions. Collisions with atoms and molecules modify its trajectory, slow it down, and transfer energy into excitation of those atoms and molecules, which later release this excitation energy by radiating a photon. This causes the auroral emission.*

Most of the energy from the incoming electrons is transferred to electrons that do not completely escape from their atoms or molecules, but simply move to a different shell with a higher energy level. This produces an excited atom or molecule, a physical system that is always unstable, and which tends to return to a state of lower energy at some later time. When this happens, the excited electron drops to a lower energy level and the excess energy is carried away by an emitted photon. The photon's energy corresponds to one of the characteristic, spectroscopic emission lines of the emitting atom or molecule.

The aurora is thus a set of emission lines of the components of the Earth's upper atmosphere excited by incoming electrons of solar origin. (A somewhat similar process, involving excitation by sunlight, and lower particle energies, produces the much fainter phenomenon known as airglow.) Examination of a typical auroral spectrum (Fig. 11.4) shows that it consists of a few major lines and bands produced by transitions of molecular nitrogen, atomic nitrogen, and atomic oxygen, which are the main components of the Earth's upper atmosphere. The group of blue and green lines and bands below 500 nm is mainly caused by permitted transitions of molecular nitrogen N_2. They are the main component of auroral emission below 100 km, in the lower part of the aurora. But the most intense auroral emission lines are primarily the yellow-green lines at 557.7 nm and then the red line at 630.0 nm, both caused by an excited state of oxygen, O I. There is a special story about these two lines. During early studies of the auroral spectrum, considerable efforts were made to identify the components of the spectrum. If the aurora were just atmospheric airglow, the auroral lines obviously had to lie at the known wavelengths for emission from the main components in the atmosphere, oxygen and nitrogen.

The mystery arose because the lines at 557.7 and 630.0 nm just do not exist in the atmospheric spectrum as seen at the surface of the Earth, or measured in a

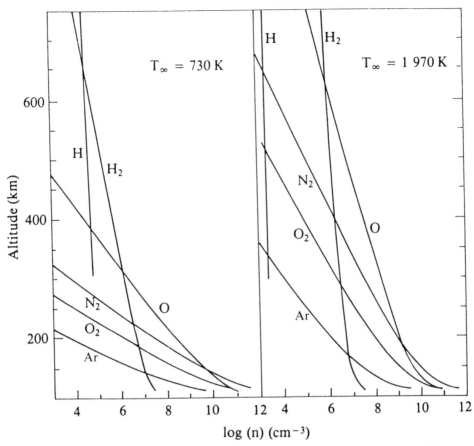

Fig. 11.3. *Typical vertical variation of the main components of the Earth's upper atmosphere. (After Giraud & Petit, 1978)**

laboratory! The reason is that these lines are so-called 'forbidden' lines. They are produced by a special type of excited state of oxygen, known as a 'metastable state', which has a very long lifetime before it is de-excited by the emission of a photon. These long lifetimes – 0.75 s for the green line and 110 s for the red line – may be compared with the typical lifetime of about one millionth of a second for the other emission lines. Because of their long lifetimes, these metastable states have absolutely no chance of radiating a photon at ground level, because the excess energy is lost through collisions with surrounding molecules in the dense atmosphere long before it can be radiated away. It is only in a very rarefied medium, where the time separating successive collisions with neighbouring molecules is longer than the lifetime of these metastable species, that they have a chance of radiating a photon.

At auroral heights, the atmospheric density is sufficiently low for the forbidden emissions to be produced. The 577.7 nm oxygen emission occurs above 95 km

---

* Notes references and bibliography to volume 2 commence on p. 1141

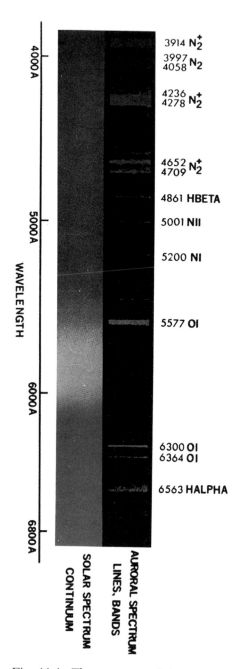

Fig. 11.4. *The spectrum of the aurora in the visible region showing the locations of the main emission lines and bands, and indicating the atom or molecule responsible for each emission. (Photo: A. Vallance Jones, in Eather,* Majestic Lights, *1979.)*

Table 11.1. *Typical altitudes reached by auroral electrons*

Electron energy (eV)	Minimum altitude (approx.) (km)	Emission	
100	300	Red	(630.0 nm)
1000	130	Green	(557.7 nm)
20 000	95	Blue & purple	

altitude. Red oxygen emission at 630.0 nm occurs higher in aurorae, at altitudes in excess of 250 km, but is suppressed (quenched) at lower altitudes. In bright aurorae, the rays may therefore show colour differences along their length. It was only in 1923 that this phenomenon was understood, and therefore that the main emission lines of the aurora were thus definitely identified as produced by atmospheric oxygen.

As may be seen from its spectrum, the aurora may display a fantastic variety of colours, with its emission lines including violet, blue, green, yellow and red. But these different colours are produced at different altitudes as we have just seen. (Although auroral colour is largely controlled by the energy of the incoming electrons, with the more energetic electrons penetrating deeper before they are stopped by collisions, the condition of the atmosphere also plays a part, such as whether the region is actually in sunlight. For example, purple N_2^+ rays occur in sunlight.) A few typical electron energies, and the altitudes at which they produce auroral emissions are given in Table 11.1. Conversely, when we look at an aurora, we can guess from its colours the average energy of the electrons producing it. Because the typical energy of auroral electrons is a few keV, the most common colour is yellow-green. The altitude of maximum emission, which has been studied for several decades by triangulation techniques, averages 107 km for yellow-green aurorae.

11.1.2 The global phenomenon: aurorae and the magnetosphere

Despite centuries of observations from the ground, a global view of the polar aurora was not achieved until the advent of the space age, when satellites were able to carry cameras and take pictures from space. NASA's DYNAMICS EXPLORER-1 satellite, launched in 1981, had an eccentric polar orbit that lay several Earth radii above the poles. This allowed it, for the first time, to observe the global distribution of the aurora over an entire hemisphere. Figure 11.5 shows one of the pictures taken by DYNAMICS EXPLORER's UV camera. The ultraviolet emission seen in the photograph comes entirely from the Earth's upper atmosphere, because the lower atmosphere is not transparent to UV light. The upper left-hand portion of the picture shows the crescent of the sunlit side of the Earth, extending from the limb at the top to the day/night terminator at the bottom. But the most spectacular feature is the oval ring of light visible on the dark side of the Earth in the polar region. This luminous ring simply corresponds to the overall distribution of the aurora, as seen from space. Examining its geometry in detail, it is found that the centre of the oval

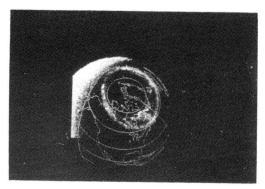

Fig. 11.5. *One of the overall pictures of the auroral oval, taken by the ultraviolet camera on board NASA's* DYNAMICS EXPLORER *satellite. The sunlit side of the Earth is to the left. (Image: Dr L.A. Franck, University of Iowa, Iowa City.)*

is located close to the magnetic pole, but displaced about 500 km towards the night side. Consequently, the aurora lies at lower magnetic latitudes on the night side than on the day side. It is also noticeable that the extent of auroral emission is broader in latitude towards the night side than it is towards the day side.

One of the major findings from DYNAMICS EXPLORER studies is the permanence of the auroral oval. An oval is indeed present at any time around each of our magnetic poles, but the intensity of the emission, and the radius and width of the oval vary with time. We shall see shortly the consequences of this for ground-based observers. But first we have to understand the origin of the auroral oval.

As we have seen, the aurora is produced by the interaction of energetic electrons with the upper atmosphere. It may be likened to the visible image produced by a giant cathode-ray tube, where the screen consists of our atmosphere. Behind – or in our case, above – the two-dimensional reality of the screen there is the three-dimensional electron accelerator that generates and guides the electron beams. In the example of a TV monitor, it consists of electron guns and electrostatic optics. In the case of the aurora, the electron accelerator is the Earth's magnetosphere, and its energy source is to be found in the interplanetary medium and the solar wind. Let us briefly describe how it works.

The magnetosphere is the region of the Earth's environment in which the dynamics of particles are dominated by the Earth's magnetic field. The degree to which the terrestrial magnetic field extends into space is limited by its interaction with the solar wind, which is a continuous outflow of electrons and protons from the Sun, moving at an average speed of 400 km/s, with typical densities of a few particles per cubic centimetre. When the charged, solar-wind particles meet the Earth's magnetic field, they are reflected by it, so the terrestrial magnetic field forms a real obstacle, which compresses and deflects the flow of the solar wind. The geometry of this interaction is represented in Fig. 11.6. This figure shows a cross-section of the magnetosphere in a plane containing the axis of the Earth's magnetic dipole, and the Earth–Sun line. The Sun is to the left, so the solar wind flows from left to right towards the obstacle

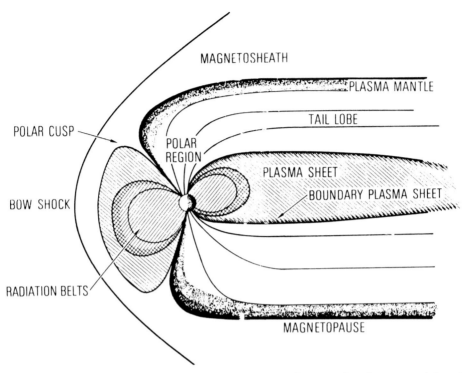

Fig. 11.6. *A cross-section of the Earth's magnetosphere in the plane containing the Sun–Earth line and the axis of the Earth's magnetic dipole. The diagram shows the Earth's bow shock in the solar wind, the magnetic field lines, and the main regions and particle populations of the magnetosphere.*

presented by the Earth's magnetic field. The line farthest to the left represents the 'bow shock'. Because the solar wind is flowing at supersonic speeds with respect to the obstacle, a steady shock forms upstream of the Earth – just like the bow shock in front of supersonic aircraft. Here the solar wind is decelerated to a subsonic value, and at the same time it is compressed. In the region between the bow shock and the magnetopause (which is the external boundary of the magnetosphere), the compressed solar wind flows around the obstacle, before it again accelerates to supersonic speeds. Because of this interaction, the magnetic field of the Earth is confined inside a comet-like cavity, which forms what we call the magnetosphere. A few typical magnetospheric field lines are shown in the figure. In the polar regions, the Earth's field lines emerge from the northern and southern polar caps, before running back in two lobes in the elongated magnetotail. It is well-known that the Earth's magnetotail, like those of all planetary magnetospheres, is very elongated. It extends well beyond the orbit of the Moon, and is probably at least 1 AU long.

Except for the set of magnetic field lines that are connected to the tail lobes, all the other geomagnetic field lines are organised in closed loops which connect approximately symmetrical points on the Earth's surface in the two magnetic hemispheres.

Energetic electrons and ions may become trapped on these closed field lines, and some may be precipitated into the upper atmosphere to produce the aurora. Where do they come from? They come from the solar wind. As may be seen from the figure, there are two special points on the day side of the magnetopause where the magnetic-field intensity may become very small, and thus allow solar-wind particles to penetrate into the Earth's magnetosphere. These regions in the northern and southern hemispheres are called the polar cusps. Solar-wind electrons that drift into the upper atmosphere via the polar cusps have relatively low energies, and so give rise to only diffuse, weak aurorae on the Earth's day side.

Electrons and protons that do enter at the polar cusps subsequently drift towards the night side of the magnetosphere in the lobes, forming a domain in the magnetosphere plasma called the plasma mantle. They then converge towards the central region of the tail, where they accumulate to form a giant reservoir of particles called the plasma sheet. Finally, plasma-sheet particles move back towards the Sun. As they do so they also drift around the Earth, extending the plasma sheet to all longitudes and local times and forming a circular belt that completely encircles the Earth. During this transport, solar-wind particles gain a significant amount of energy. They have an energy of 10–100 eV in the solar wind, and reach several keV in the regions of the plasma sheet that are nearest to the Earth. It is the plasma-sheet particles that form the immediate source of the aurora.

To really understand the connection of these magnetospheric particle domains with the aurora, it is necessary, however, to visualize them in three dimensions, and to see how they project down (along the field lines) onto the upper atmosphere. This is shown in Fig.11.7. The left-hand side shows a slightly different view of the magnetosphere to that in Fig. 11.6, but with the same particle domains. On the right-hand side, each of these particle domains is projected onto a polar map of one terrestrial hemisphere. It will be seen that the projection of the plasma sheet forms an oval, nearly centred on the geomagnetic pole, but slightly displaced towards the night side. It is on this oval that, at any given time, solar-wind electrons may cascade into the atmosphere and excite an aurora. This oval obviously corresponds to the auroral oval shown in the Dynamics Explorer ultraviolet images.

11.1.3 Spatial and temporal variations in the aurora

Now that we know how the aurora is distributed over the Earth, it is easier to understand when and how it may be seen by an observer on the ground. The general pattern shown by the auroral ovals is essentially steady in a reference frame fixed relative to the Sun. They may expand towards lower latitudes, or contract radially towards the pole, depending on the changing conditions in the solar wind, but they do not rotate with the Earth. Under quiet geomagnetic conditions, the ovals may be regarded as fixed in space above the rotating, solid body of the Earth. Terrestrial observers therefore rotate beneath the auroral ovals. This is illustrated in Fig. 11.8, which, by way of example, shows the motion of Scandinavia under the auroral oval in the course of one rotation of the Earth.

This diagram also shows the variation in the type of aurora that may be seen with local time. When Scandinavia is on the day side of the Earth, the auroral

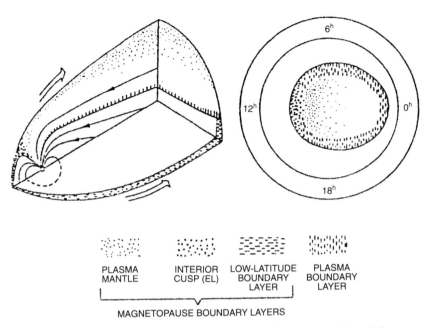

PLASMA MANTLE   INTERIOR CUSP (EL)   LOW-LATITUDE BOUNDARY LAYER   PLASMA BOUNDARY LAYER

MAGNETOPAUSE BOUNDARY LAYERS

Fig. 11.7. Left: *An attempt at a three-dimensional representation of the magnetosphere, showing the main domains and plasma populations as in Fig. 11.6. Right: Map of the projection of the plasma mantle and the plasma sheet along the magnetic field lines onto the polar regions of the Earth. The projection of the plasma sheet essentially corresponds to the auroral circle seen in the photograph in Fig. 11.5.*

oval lies far to the north. The day side 'polar cusp' aurorae may be seen only from very high latitudes, for a short period around the winter solstice, when full darkness prevails at noon. In the northern hemisphere, Spitzbergen Island in the Barents Sea is a suitable location for such observations in December and January. The day side aurora is red, because it is produced by low-energy electrons precipitating into the atmosphere directly from the polar cusps. These electrons come from the solar wind without having been accelerated by passing through the magnetotail.

As time passes, Scandinavia – to continue with our example – moves through the afternoon, then the night, and finally reaches the morning sector. During this process, the auroral oval initially appears to come closer and closer. At some point of the evening, the auroral oval lies overhead as seen from Scandinavia. The region thus remains under the oval for some fraction of the evening, through midnight and into the early morning, until the aurora again appears to recede to the north, and Scandinavia leaves the auroral zone. On a typical night, the northern part of Scandinavia may stay in the auroral zone for 4 or 5 hours around local magnetic midnight.

During this passage through the auroral zone, there is a considerable variation in the shape of the auroral forms as a function of local time. The evening sector is usually dominated by bright, discrete, single or multiple auroral arcs (Fig. 11.9). This

605

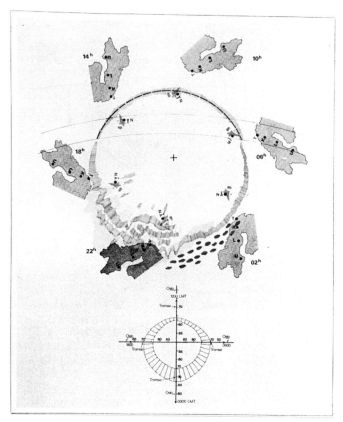

Fig. 11.8. *The motion of Scandinavia beneath the auroral oval, as seen in a frame of reference fixed with respect to the Sun. Different auroral forms are seen in different local-time sectors. (After Egeland and Stoffgren, Norwegian Academy of Sciences and Letters.)*

type of aurora is relatively stable. In the midnight sector, the auroral forms become more complicated. They form curls, which move rapidly towards the west. One of them, known as the 'midnight auroral surge' or the 'westward-travelling surge', is a giant, spectacular curl whose development and very fast westward motion marks the most dynamic phase of what we call an auroral substorm, a typical auroral event that we shall describe shortly. In the early-morning sector, the auroral forms are very turbulent, often appearing in discrete patches of light, and form different types of curls, called 'Omega-bands' because they resemble the shape of the Greek letter, which move eastward towards the daytime sector. Overall, the motions of the auroral forms in the different local-time sectors follow the general motion of the plasma-sheet electrons through the magnetosphere, as they move towards the Sun from the magnetotail of the day-side magnetopause.

Fig. 11.9. *An example of a multiple auroral arc, which usually appears before local midnight. (Photo: Olli Karhi, Ursa Astronomical Association)*

11.1.3.1 Auroral substorms

In the preceding description, the auroral oval was stationary relative to the Sun, while Scandinavia moved beneath it. Thus the apparent temporal variations seen by an observer on the ground were really *spatial* variations. This situation, however, is only true in a very general sense, because large temporal variations are superimposed on this spatial variation. Some of these changes are very slow motions, corresponding for instance to contractions or expansions of the auroral oval, which themselves often correspond to variations in the solar wind. But there is a particular class of temporal variation of the aurora which is very spectacular and deserves special mention because of its importance in the physics of the magnetosphere. This is the auroral substorm. If you are in Scandinavia or Alaska (for instance) around local magnetic midnight, you can quite often watch the development of an auroral substorm. Before it starts, there is a rather weak aurora visible on the northern horizon. Then the aurora moves slowly south towards you. It then appears as a discrete arc. At a specific instant, the arc suddenly brightens. In the minutes that follow, the bright arc starts to recede towards the pole, but at the same time a large portion of the sky brightens, and very complicated and dynamic forms appear. The auroral oval becomes very wide, and is full of a series of very turbulent auroral forms, such as the midnight surge, which moves westward, and the Omega-bands, which move eastward. Figure 11.10 shows the geometry of the auroral oval at two important times during an auroral substorm. The top diagram shows the aspect at the time of the sudden brightening of the auroral arcs at midnight. This is known as the auroral break-up. Immediately after the break-up, there is the expansion phase, which is characterized by the development of turbulent forms. The overall shape of the auroral oval as seen from space in the middle of the expansion phase is shown in the bottom diagram. This shows the discrete arcs to the north of the main region of the oval, and a midnight surge, moving westward, on the left of the midnight sector.

An auroral substorm is the visual result of large-scale and very dramatic changes in the magnetic configuration of the magnetosphere. These changes basically corre-

AT BREAK UP

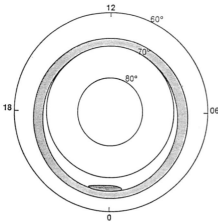

AFTER EXPANSION

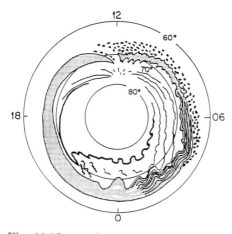

Fig. 11.10. *A schematic representation of the general appearance of the auroral oval at two important times during the development of an auroral substorm: at break-up (top), and at the end of the expansion phase (bottom).*

spond to a disruption of the magnetic configuration of the magnetotail (Fig. 11.11). During a substorm, energy accumulated in the magnetic field of the tail during the preceding hours is suddenly released and converted into kinetic and thermal energy of the plasma-sheet electrons and ions. Some of these are injected into the inner part of the magnetosphere, and some precipitate into the upper atmosphere, where they generate the auroral substorm just described. This disruption of the magnetic configuration in the tail is very similar to the process that occurs in solar flares, which correspond to a sudden change in the magnetic configuration of a region of the solar corona.

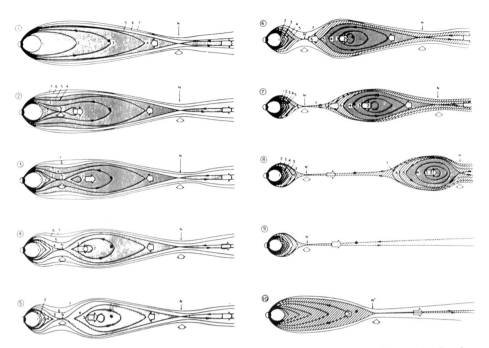

Fig. 11.11. *The magnetic configuration of the tail at different times during the development of an auroral substorm, as described by one of the current (competing) theories of substorms.*

11.1.3.2 Geomagnetic storms

Solar-flare activity in the lower atmosphere of the Sun above sunspot regions is the source of the more extensive 'geomagnetic storms', which cause the night side of the auroral ovals to expand down to lower latitudes. Geomagnetic storms are the result of magnetic 'reconnection' processes that follow the arrival, near the Earth, of energetic particles ejected into the solar wind by flares. Not all flares are necessarily followed by geomagnetic storms. The controlling influence appears to be the magnetic polarity of the ejected material: the interplanetary magnetic field (IMF) must be southwards-pointing for reconnection between solar-wind and terrestrial field lines to be efficient.

Prolonged periods of southerly IMF lead to stress on the plasma sheet in the magnetotail, and a consequent ejection of plasma from a point 'downwind' of the Earth, both back into the solar wind and – more importantly for observers – into the upper atmosphere.

During magnetic storms the increased mobility of the particles and ions in the magnetospheric plasma induces electrical currents at ground level. At high latitudes, these may lead to disruption of power-grid systems. During the Great Aurora of 1989, for instance, large areas of Canada suffered power failures lasting several days.

Geomagnetic storms are most frequent and extensive about a year before sunspot numbers reach a maximum. Typically, they may bring the auroral ovals sufficiently far towards the equator for activity to be visible from central England, the middle of the United States, or New Zealand. More rarely, very major storms may bring the aurora to the skies of observers in Australia, central Europe and the southern United States.

11.1.4 Solar activity and the aurora

The intensity and frequency of auroral substorms is a measure of geomagnetic activity, which reflects short-term fluctuations of the magnetic-field intensity and direction at the Earth's surface that are produced by the electrical currents flowing in the upper atmosphere and magnetosphere.

Auroral and magnetic activity are clearly affected by solar activity, and therefore they vary in response to the two basic cycles arising in the Sun: the 27-day solar rotation, and the 11-year sunspot cycle. Recent measurements, using data obtained by satellites in the solar wind, seem to indicate that the intensity of auroral activity is directly related, in a statistical sense, to the intensity of the solar wind. Because the solar-wind intensity at the Earth is modulated by the two basic cycles of the Sun, these affect the level of auroral activity. First, there is a tendency, at least during certain phases of the 11-year sunspot cycle, for there to be a 27-day variation in auroral activity. Second, as shown in Fig. 11.12, there is a clear 11-year modulation of geomagnetic (and therefore auroral) activity, which follows the sunspot number, at least in the sense that the maxima of geomagnetic activity are always located in the vicinity of sunspot maxima. The period around sunspot maximum is therefore the best time to see spectacular aurorae.

In addition to the variations caused by these two solar cycles, transient phenomena that occur in the corona may trigger short periods of intense auroral activity. For instance, solar flares, which are the most spectacular phenomenon in this category, suddenly release shells of dense, hot solar-wind plasma, which propagate at high speeds into interplanetary space, and usually generate a shockwave ahead of them. If this shockwave hits the Earth (which does not always happen) some 3–4 days later, it compresses the magnetosphere and may trigger intense auroral activity. Some of the most spectacular auroral storms in the past, during which aurorae were seen at middle latitudes, were caused by intense solar flares. The compression of the magnetosphere causes an effect known as Sudden Storm Commencement (SSC), which may be detected at ground level by magnetometers (*see* Sect. 11.2.1.4, p. 620).

11.2 Observing the aurora

Aurorae are undoubtedly some of the most impressive natural phenomena, and are therefore one of the most interesting to observe. Many may consider the polar aurorae as just another beautiful sight, like a sunset over a Mediterranean holiday resort, but it is not only possible, but interesting and scientifically valuable to make observations of the aurorae. These mysterious lights in the sky differ greatly from most other astronomical phenomena because of their size and brightness – sometimes aurorae spread over the whole sky and are so bright that they cast

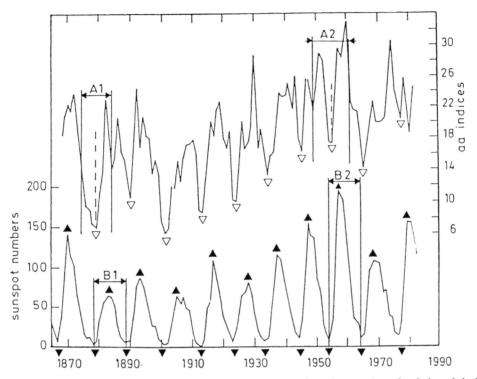

Fig. 11.12. *Long-term variations in the sunspot number* (bottom) *and of the global geomagnetic 'aa' index* (top) *from 1865 to 1982, clearly showing the modulation of geomagnetic (and thus auroral) activity by the 11-year solar cycle. (After Simon & Legrand, 1990) [The aa index is a measure of activity at antipodean points in the Earth's magnetic field. – Trans.]*

shadows. In countries that are regularly under the auroral oval, deep-sky observers are sometimes really furious because auroral glow prevents them from observing; as somebody once said, aurorae are another form of light pollution ...

As in many other fields of astronomy, the observational methods and aims of amateurs and professionals are quite different. In amateur observations, the aurora is regarded as a light-emission phenomenon, which may be classified according to its shape, brightness, colour, and dynamic behaviour. The occurrence of aurorae may be the subject of statistical studies, which are useful, and feasible for amateur observers. In professional auroral research, on the other hand, the aurora is treated as a physical phenomenon, which may be studied not only in its visible form, but also through all the physical mechanisms that are involved. It is therefore observed by a wide variety of ground-based, rocket, and satellite instruments, which will be briefly described later. Our main concern, however, will be with amateur observational techniques.

11.2.1 Amateur auroral observations

11.2.1.1 Visual observations

The best way of observing aurorae is with the naked eye; optical equipment is not only unnecessary, but it may even be a hindrance. Viewed through binoculars, for example, the aurora becomes a diffuse blur of light. All that is required are your eyes and a notebook in which to enter your observations. Some very simple, non-optical equipment (described later) may also be of assistance. The principal aim of amateur observation is to observe and classify every auroral display and storm seen from a particular location. When observations from several sites are combined, it is possible – at least in the ideal cases – to build up a picture of the whole situation over (for example) the northern auroral oval, whether it was disturbed, and if it was, what kind of disturbance took place and how it developed. It is true that a satellite can see all this at a single look, but amateur observations also have their value – even if it is only in giving the observers that sense of contributing to science that is so important for amateurs.

Although at first glance the structure and behaviour of an active auroral display might appear chaotic, most displays in fact consist of a relatively small range of forms, whose brightness and extent vary on time-scales ranging from fractions of a second to several minutes. Auroral activity as seen at a given time and location may be described concisely using a standard, international reporting code that has been in use since the International Geophysical Year in 1957–8. This classification is arranged in categories according to specific features and properties of the aurora: form, structure, type, state (or condition), brightness, and colour. It allows precise and rapid notes to be made of the necessary information and also enables one to describe the evolution of an auroral display without recourse to photography or other types of visual records. (You may see a magnificent display just when you do not have a camera with you.) And because this classification is used for statistical work it is needed to classify the display when you take photographs.

The individual features of aurorae are abbreviated to symbols consisting of a letter (or a letter and a digit). The features begin with the auroral form and this symbol is preceded or followed by additional symbols indicating the precise nature of the display. (The order of the features is given below.) The descriptions that follow of the major forms of aurorae are given in the order in which they are typically seen on an active night. Sometimes, however, activity may go no further than the arc or band phase before declining. On other nights the whole cycle from glow to arc to coronal structures may recur several times. No two aurorae are ever exactly the same!

Form Form is the term applied to the general shape of a particular auroral activity that is present at any time. It is described by a single letter:

Glow (G) The earliest phase of an auroral display (particularly at temperate latitudes) often consists simply of a faint glow, low over the poleward horizon. In appearance, this is usually similar to the brightening sky before dawn (hence the description *aurora borealis*). The glow represents the uppermost parts of a display that will be more extensive at higher latitudes;

sometimes – even frequently – it will be the only form of activity at lower latitudes. Observers should beware of misidentifying local light pollution for such auroral glows.

Arc (A)   At high latitudes, a common form of aurora, especially in the early evening, is the arc, a bow-shaped arch of discrete auroral light, spanning the sky from east to west, and usually in the half of the sky that is towards the pole. Because of the sharp cut-off in auroral emissions in the denser atmosphere below about 100 km altitude, arcs usually have a sharp base and a more diffuse upper limit. During auroral storms at lower latitudes, the arc may develop as the earlier glow rises higher in the sky and assumes a more definite form.

Band (B)   As activity increases, a quiet arc may fold on itself, giving the ribbon-like form of a band. Bands are often active, and move around. This is the active stage when the aurora appears like curtains moving in the wind. Rayed activity frequently develops at this stage.

Rays (R)   Rays are approximately vertical streaks of light that follow the geomagnetic field lines. This form is often seen in association with bands, but may also appear with other forms of aurora. Rays may also appear in isolation, or in small bundles, and are sometimes very long in comparison with other forms.

Patch (P)   Patches are typically diffuse, cloud-like areas of auroral light, which normally appear after a geomagnetic storm, when the auroral display often pulsates. They are also called 'surfaces' in some auroral textbooks.

Veil (V)   In some displays, the whole sky may become suffused with a weak, even background of auroral light, described as a veil.

Not classifiable (N)   Sometimes it is not possible to classify aurorae, because they are (for instance) partially covered by clouds or hidden by trees.

Structure   Once the observer has decided on the basic auroral form, this may be amplified by classifying the basic structure that is visible. Three subdivisions are used:

Homogeneous (H)   No internal structure is visible, and the brightness appears to be nearly the same throughout the auroral form.

Striated (S)   A striated structure may sometimes be clearly seen in wide bands that are nearly overhead (approaching the zenith). These small striations run parallel to the direction of the band itself.

Rayed (R)   When structure appears running perpendicular to the main extent of an auroral form (i.e. following the geomagnetic field lines) it is described as rayed. A subscript 1–3 is used to indicate the length of the rays. R_1 means that the rays are quite short, and R_3 indicates that they are very long – perhaps even stretching from horizon to zenith.

The letter designating structure precedes the auroral form: HA for homogeneous arc; RB for rayed band, etc.

Fig. 11.13. *A spectacular auroral corona, photographed by Peka Parvïainen.*

Type   When relevant, an additional symbol is used to describe specific features that may be present. Again there are three types:

Multiple (m)   Two or more associated and roughly parallel forms are described as multiple. A subscript indicates the number of such features present (e.g. m_2 shows that the form is double).

Fragmentary (f)   This designation is used when a form splits into two or more parts, or only a portion is present.

Coronal (c)   During very active storms the rays may appear to radiate from the magnetic zenith, which is the direction of the Earth's local magnetic field. At a geomagnetic latitude of 60°, the magnetic zenith is at an angular distance of approximately 20° from the geographical zenith. At 70° geomagnetic latitude, it is only 10° from the geographical zenith. An example of a corona is shown in Fig. 11.13. Although most commonly seen with rays, even homogeneous forms may show coronal structure when they move overhead.

The symbol for type precedes those for form and structure: m_2RA indicates a double rayed arc, for example.

State   Motion of the aurora may sometimes be highly complex, and symbols are used to describe the various forms that this may take. Once again, the symbols precede all the symbols previously described.

Quiet (q)   A quiet auroral form is one that is stable or practically stable in position or shape. Homogeneous arcs (HA) are nearly always quiet, and are thus described as qHA.

Active (a)   The auroral display moves rapidly across the sky, or changes its shape dramatically. In practice, the most active auroral displays are usually very bright ones. Activity is estimated in four categories, indicated by a subscript. These subdivisions do not refer to actual speed of motion, but rather to the form that this takes:

Table 11.2. *Auroral brightness estimation*

Code	Description
1	Weak: comparable with the brightness of the Milky Way
2	Brightness similar to moonlit cirrus clouds
3	Brightness similar to moonlit cumulus clouds
4	Very bright (much greater than 3). May cast shadows.

a_1 Motion of folds or irregularities along the lower boundary of a band.

a_2 Rapid changes in the lower border of a form, while the form itself may move rapidly across the sky.

a_3 Rapid motion of rays horizontally along a form. This movement may be in one direction, in both directions simultaneously, or it may alternate.

a_4 This classification applies to the display as a whole, when some forms may fade quickly, and other similar forms appear in another part of the sky. This type of activity is typical of the late stages of major displays.

Pulsing (p) it is quite common for rapid, and often rhythmic fluctuations in brightness to occur. These changes may have periods of a fraction of a second to several minutes. Again this type of activity is subdivided into four categories, indicated by subscripts:

p_1 Slow pulsations in brightness, which are in phase throughout the form.

p_2 A 'flaming' aurora, with waves of light passing upwards from the horizon towards the zenith, with the various forms brightening as the wave passes over them.

p_3 'Flickering', in which large areas of the display undergo very rapid, more or less irregular, changes in brightness. This type of activity is relatively rare.

p_3 'Streaming', a variation in brightness that passes rapidly along homogeneous forms.

Examples of the usage of this description of state are: a_2RB for a rayed band that moves rapidly across the sky, and p_1HA for regular changes in brightness of a homogeneous arc.

Brightness The overall, visual brightness of a display (or any individual feature) is indicated by a number, which *follows* the symbol for form. Estimation of brightness is relatively crude and uses the scale shown in Table 11.2.

Colour The colours in auroral displays may be complex for the reasons mentioned earlier. Various colours may be present at the same time. Rough estimates of colour may be made by eye, however, and are a useful indication of the type and height

Table 11.3. *Colour of auroral forms*

Code	Description
a	Red upper portion of a form, with green below
b	Red lower border, green above
c	White, green or yellow
d	Red only
e	Red and green spread unevenly over the auroral form
f	Blue or violet

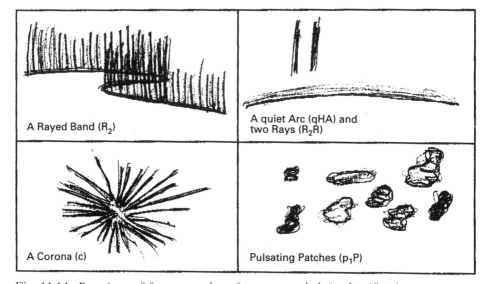

A Rayed Band (R_2)

A quiet Arc (qHA) and two Rays (R_2R)

A Corona (c)

Pulsating Patches (p_1P)

Fig. 11.14. *Drawings of four examples of aurorae and their classification*

of activity. The classification takes the form of a single letter, which follows the particular form and brightness symbols, using the designations shown in Table 11.3.

With this classification scheme practically all kinds of auroral shapes and displays may be identified and reported. For instance $a_3f\ R_2B3c$ indicates a fairly bright white, green or yellow, active fragmented, rayed band. It might later become calmer, and change its form, becoming a quiet arc in the northern sky, when it might be written as qHA1c. Four examples of the basic forms of aurorae are shown in Fig. 11.14 with their classifications.

Other information to be reported Other information should also be recorded when observation of an auroral display is made. The most important, of course, is the exact time of the observation (preferably in UT) to the nearest minute. It is also necessary to report the date, and to prevent confusion this should be shown as the double-date for the night concerned, e.g., 1991 August 24–25.

Another significant piece of information that should be reported is the azimuth of particular features of a display. This may be estimated roughly from the known position of features on the horizon, and reported as simple compass directions (NE, NNW, etc.), or may be obtained with a compass and specified in degrees. Make sure that a direction obtained with a compass is corrected by the magnetic deviation (if any) to give a true (geographical) azimuth, which should be specified in the 0–360° convention, where 0° is north, 90° is east, etc.

Two fundamental height estimates are important. The first, designated by h, is the altitude of the highest point of the base of an auroral feature, i.e., the elevation above the true horizon. (With homogeneous arcs, for example, this is generally in the magnetic north.) When combined with similar measurements of h from elsewhere, the data may allow accurate triangulation of the aurora's height and geographical extent. The second important height is a measure of the vertical extent of an auroral feature, and is designated with the symbol $\nearrow$. This altitude may sometimes be difficult to determine with diffuse displays.

Estimation of azimuth and altitude may involve the only 'equipment' required by auroral observers. A homemade device (known technically as a clinometer) is quite adequate and may consist of a pair of sights fixed to a protractor, which is fitted with a plumb line against which the elevation angle may be read. Some forms of compass are available with a built-in clinometer, so they may be used to give both altitude and azimuth.

Other information to be recorded and passed to any coordinating organisation includes the observer's location, and – essential, but sometimes forgotten! – the observer's name and address.

The simplest observation that may be made is to note the presence of aurora at a given time from a particular location. As the observer becomes more experienced, detailed reports using the reporting code outlined above may be prepared. Observations should be made as frequently as necessary (or possible!) during an active display. Reports should be sent to the appropriate national or international organisation, more details of which are given in Sect. 11.4.

An example of the use of the classification may be found in the yearly reports of the Aurora Section of the Finnish-based URSA astronomical association. The report for 1988, for instance, shows that in that year 66 auroral displays were observed over Finland. Of these, 33 % were very active and 41 % quiet; pulsations were observed in 11 % of the cases. A corona was seen in 32 %, and 70 % were green or white. The most common brightness index was 2 or 3, with 41 % and 38 % respectively. Bands were the most common form, with arcs in second place.

It might be worth mentioning that there is also the highly controversial fact that some people claim that they can hear (!) the aurora. Many stories occur in the mythology of Finnish Lapland (and from peoples elsewhere in the world) of a high-pitched buzzing or whistling sound that may be heard during an active auroral display. These reports are persistent, despite there being no scientific evidence to confirm them, and there have been suggestions that the sounds arise from other causes but are wrongly attributed by the observer to the aurora. It has, however, been proposed that the magnetic fields and electric currents associated with aurorae might induce effects in nearby objects to produce faint sounds. As always, observers should keep an open mind ... and report their findings as objectively as possible!

11.2.1.2 Photographing the aurora

Naked-eye classification of auroral features is undoubtedly a good method of observing, but even more scientific and valuable results may be obtained with photography. Auroral photography, however, is very simple, because all you need is a camera that is able to take exposures with durations of up to 1 minute, a relatively wide-field lens, and a tripod. This is all that is required to begin, but more sophisticated systems may be used once you gain experience.

It is hard to give details of how best to photograph a particular auroral display, because the size and brightness are not stable or predictable. In general, fast films are required so that short exposures may be made of rapidly moving arcs or bands, and normally films slower than ISO 100 are not very useful. Although everyone should really experiment for themselves, only those at high latitudes can depend on reasonably frequent auroral activity with which they can optimize exposures. Observers at lower latitudes lack the opportunity to experiment. The following guidelines are based on past experience by observers in Scotland, and should enable even infrequent observers to obtain reasonable results.

Best results are obtained on fast colour films, preferably ISO 400. The best emulsions at present appear to be Kodak Ektachrome or Fujichrome slide films. The sensitivity and colour response shown by some emulsions is such that auroral displays that appeared pale and relatively featureless visually may reveal subtle coloration on any photographs. However, it is worth remembering that with quiet aurorae, slower films may be used without problems.

Exposures should be carried out with the lens set to an aperture of (say) f/2 or f/2.8. The length of time obviously depends mainly on the brightness and activity of the aurora. Faint, weak horizon glows may require exposure times of between 30 and 60 seconds. Bright rapidly moving aurorae should be given shorter exposures (perhaps as little as 5 seconds) to avoid loss of detail.

Different techniques may be required for photography at high latitudes, such as in Scandinavia, Canada, or Alaska, where the aurora may be very bright and move extremely rapidly (Simmons, 1988). Exposures of less than a second can effectively record some of the brightest displays.

With a 35-mm camera, the standard 50-mm lens, or a 28-mm wide-angle lens will give good results. Shorter focal-length lenses may also be used, but may introduce a degree of image distortion. This may cause problems if it is intended to measure *h* (or the top of a display) from the photographs, although obviously this may be overcome by suitable calibration. The field of fish-eye lenses often covers most of the sky, but such lenses are very expensive, although less than the true fish-eye optics that do include a complete hemisphere. The answer is an all-sky camera, which in principle may be no more than an ordinary camera photographing the image of the sky, reflected in a spherical mirror. (In fact, this type of camera may frequently be arranged to include a field that is slightly more than 180° across.) Figure 11.15

Fig. 11.15. *A simple, home-made, all-sky camera of a type easily constructed for amateur use. (Photo: Peka Parviainen, Ursa Astronomical Association)*

shows a simple home-made model that is quite suitable for amateur or professional use.

Observers should also consider the possiblility of using interference filters, selected to provide maximum transmission of specific auroral lines. This technique is capable of detecting auroral emissions that are quite invisible to the naked eye. (Full details may be found in Simmons, 1985.)

11.2.1.3 Moving pictures of the aurora

Because auroral displays are rarely static, it is interesting to film them with a cine or video camera. The use of these cameras is similar to the use of normal cameras, so wide-angle or fish-eye lenses are needed. But we should first point out that normal cameras may also be used to take good moving pictures. By taking repeated photographs with an ordinary camera at regular intervals, say once every minute or two, it is possible to generate a short 'movie' showing how the aurora evolves with time. Even a 36-exposure film would cover more than an hour if one exposure were made every two minutes. If one possesses one of the special camera backs, such as those used by sports photographers, even longer runs may be made. Such photographs may be recorded onto video tape later for viewing on a television.

Currently available camcorders and home video cameras are not sensitive enough for all kinds of aurorae, unlike ordinary still cameras, although the brightest displays may be recorded by modern equipment. Professionals normally obtain videos by using photomultiplier or CCD cameras, but these are too expensive or too complicated for widespread use among amateurs. But we can thank home videos for providing us with a new opportunity: the prices of cine cameras (especially used ones) have dropped to very low levels, and so a cine camera may be a serious, and sensible alternative to a traditional still camera. But there are some problems associated with the use of cine cameras: the speed of 24 frames per second is too high, and the exposure time of about 1/24 second is not enough. Some makes of camera allow single frames to be taken with a variable exposure time, and a simple timer may be purchased or devised to take time-lapse photographs. With others, it may be possible to slow down the speed of the motor by adding a resistor, for instance. Most cine cameras will need some modification before they may be used successfully for auroral observation, but the results are worth the effort.

11.2.1.4 Magnetic and radio techniques

The considerable disturbances in the geomagnetic field associated with auroral events may be detected with suitable equipment, known as magnetometers. The variations consist of changes in the horizontal and vertical field strengths, and also in the direction of local magnetic north. The alterations in the latter are measured as the angular deviation D from the direction that applies under undisturbed conditions.

The simplest equipment appears quite crude, but is capable, with care, of providing surprisingly useful results. It is commonly known as the 'jam-jar magnetometer' (Livesey, 1982 & 1989). It basically consists of a freely swinging (and preferably fairly powerful) magnet, suspended inside a transparent container (jam jars, coffee jars, plastic tubes, and other containers have all been used). The container serves to protect the magnet from local air currents. A small, but bright spot of light from a suitable lamp is reflected from a small mirror fixed to the magnet. In the simplest arrangement, the reflected spot of light falls on a suitable scale at a distance of 2 m or more. Small fluctuations in the position of the magnet are amplified by the 'optical lever' and may be read from the scale. (The principle is the same as that used in early galvanometers.)

This extremely simple device has proved to be remarkably effective in detecting major auroral events and a number have been constructed by various amateurs around the world. The principal disadvantage is that, if well-made, they are very sensitive to minor, local alterations in the magnetic field that may be caused by the movement of various objects, whether these be passing vehicles, garden equipment, or even the change in position of metal tools! A 'magnetically quiet' site is therefore essential, but anyone who has a relatively isolated house, shed or observatory – or even a spare bedroom – should find that they can control the movement of metal objects sufficiently to produce useful results.

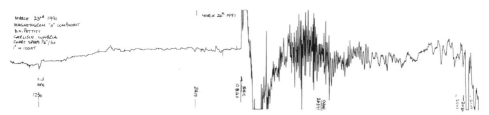

Fig. 11.16. *A flux-gate magnetometer record of changes in the horizontal component of the Earth's field during a magnetic storm, which began on 1991 March 24 and lasted for 3 days. The cause, a Solar Flare Effect (SFE), was noted. The magnitude of the SSC on March 24, 03:40 was +347 nT. Visual aurorae were recorded from the same site (Carlisle, Cumbria, United Kingdom) on March 24–25. (Record obtained by D. O. Pettit.)*

Readings may be made by eye, determining the position of the spot of light on the scale, but it is also possible to convert the deviations into an electrical signal that may be recorded automatically. Although not all changes in *D* result from auroral activity, it is quite possible to detect SSCs (*see* p. 610), and auroral storms and sub-storms. Major storms may cause the geomagnetic field to be in a disturbed state for several days, and such patterns have been fully recorded by simple jam-jar devices.

Changes in *D* serve as an indication of possible auroral activity, and on a number of occasions observation of fluctuations in *D* has enabled warnings of possible or current activity to be issued to amateur auroral observers – sometimes when the magnetic observers themselves were clouded out! It has, in fact, been found that magnetic disturbances are detectable at lower latitudes than the associated auroral activity, so this is one field where observers at lower latitudes can play a definite part in monitoring geomagnetic activity.

A somewhat more sophisticated instrument is a magneto-resistive magnetometer (Smillie, 1992). This again uses a suspended magnet, but this time in a partially oil-filled container. Hall-effect sensors detect the vertical and horizontal motion of the magnet in response to the changing geomagnetic field. Such a device is, obviously, more easily converted to automatic recording than a simple jam-jar magnetometer.

A few amateurs with electronic knowledge have constructed more complicated instruments known as flux-gate magnetometers (Pettitt, 1984). These are able to detect variations in the strength of the local magnetic-field components, rather than just the deviation *D*. This means that they are more sensitive, and are thus able to detect SSC at an earlier stage. A typical record is shown in Fig. 11.16.

Finally, it should be mentioned that amateur radio operators regularly make use of the increased ionisation that occurs with auroral activity. The enhanced ionospheric density acts as a reflector and enables radio hams to establish contact with others in countries that are not normally accessible. The best conditions occur when an arc lies along a line of geomagnetic latitude, and contacts over the greatest distances take place when a strong arc lies to the geomagnetic north or south of the

621

operators' sites. In addition, auroral regions reflect signals from transmitters and beacons that are usually too distant to be detectable. Once again, such information may act as a warning that an auroral storm is in progress.

11.2.2 *Professional observing methods*

11.2.2.1 *Optical techniques*

Before the space age and the development of high-power radar techniques, optical instruments were obviously the only tools available to study the aurorae. Even today they continue to play an important, although less central, role in auroral research.

The primary optical technique is that of the all-sky camera. There is a world-wide network of all-sky cameras, and these are deployed around the world at auroral and polar latitudes. They usually take photographs at 5-minute intervals and their records make it possible to reproduce the development of interesting auroral events.

Photometric measurements through narrow-pass filters that isolate a single line of the auroral spectrum (*see* Fig. 11.4), or spectroscopic equipment have been extensively used. These techniques have enabled the physical nature of the aurora to be described, and determined the composition of the atmospheric gases involved, and the excited states of atoms and molecules that contribute to auroral light. They have also given indications of the nature and energy of the particles that precipitate into the atmosphere and excite the aurora, because as we have seen earlier, the different colours are produced at different altitudes and, in addition, the depth to which the particles penetrate is a function of their energy as well as of their physical nature. The photometers, sets of filters, and spectrographs used in auroral research are mounted on small telescopes, and are very similar to those used for general astronomical observations.

During the last 20 years, high-precision interferometers of the Fabry-Perot or Michelson designs have been developed for detailed studies of the shape and position of auroral emission lines. If an interferometer is used with a telescope to look at a particular emission line in a specific direction, the width of the line (which is broadened by the Doppler effect) gives a measure of the atmospheric temperature in the emission region. The Doppler shift of the centre of the line relative to its normal position gives the speed of the wind of the neutral gases (again in the emission region) along the telescope's line of sight. This technique is actually the only one that gives us access, from the ground, to temperatures and winds in the upper atmosphere above approximately 90 km. Figure 11.17 shows the MICADO Michelson interferometer used in Scandinavia by CNRS, the French scientific research agency. This type of instrument is able to determine how the upper atmosphere is heated by the aurora, and how this auroral heating produces winds that generally blow from the heated regions (the polar upper atmospheric region) to the cooler mid-latitudes during periods of intense auroral and geomagnetic storms.

Complementing work in the visible region of the spectrum, in recent years photometers, spectrometers and imaging systems have been developed for the infrared, UV and X-ray domains. The picture of the complete auroral oval shown in Fig. 11.5, for instance, was taken from space by a UV camera fitted with a filter. X-rays are

Fig. 11.17. *The* Micado *Michelson interferometer used in auroral research to measure winds and temperatures in the upper atmosphere.*

also produced via what is known as the bremsstrahlung mechanism when auroral electrons encounter atomic nuclei of the various atmospheric components. Auroral X-rays are not detectable from the ground, because they are absorbed by the thickness of the atmosphere, but they may be detected from stratospheric balloons or from satellites. Like the visible airglow, they may be used as indicators of the energy of the precipitating electrons.

11.2.2.2 Radar techniques

Since the 1970s, radar has become more and more important as a tool for auroral research. Radio waves emitted by a radar may be partially backscattered by the free electrons present in the upper atmosphere, which, with ions, are the components of what we call the ionosphere. These free electrons are produced by the ionizing power of solar ultraviolet light and auroral electrons, acting on the atoms and molecules in the upper atmosphere. They then act as tiny targets – with each electron being equivalent to a reflecting surface of 6×10^{-25} cm^2 – for radar waves. These targets may nevertheless be seen from the ground if the radar transmits enough power and has a sensitive enough receiver, and if the number of free electrons per unit volume is high enough. In practice, auroral precipitation in the auroral ionosphere may produce populations of as many as one million electrons per cubic centimetre at altitudes above 100 km. Different types of radars are used in the auroral zone to study the ionosphere produced by the aurora.

At present, there are two high-powered radar systems (called 'incoherent scatter'

623

Fig. 11.18. *A view of the* EISCAT *installations at Tromsö in Norway, showing the UHF radar's parabolic dish aerial, the VHF radar's parabolic cylinder aerial, and an auroral form in the sky overhead.*

radars) in operation, one operated by the U.S.A. at Sondre Strömfjörd in Greenland, and one operated by six European countries (Finland, France, Germany, Norway, Sweden, and the United Kingdom) in northern Scandinavia, called EISCAT (for European Incoherent SCATter facility). Figure 11.18 shows the EISCAT installations as Tromsö in Norway, with the operations and transmitter buildings and the two radar aerials (a UHF 30-m parabolic dish, and a VHF parabolic cylinder). These radar facilities usually transmit peak powers of over a megawatt, and use cryogenic receivers to detect the very faint echo from the ionosphere, which is sometimes only a small fraction of the background sky noise. The detection techniques used are very similar to those employed with radio telescopes. An example of the result of a radar scan through an auroral arc is shown in Fig. 11.19, which plots the iso-contours of the concentration of free electrons in the vertical plane scanned by the moving radar beam (the x-axis shows the horizontal distance across the auroral arc, and the y-axis the altitude). The location of the arc corresponds to a spectacular increase in the number density of free electrons, which, as may be seen, is easily detected by the radar.

11.2.2.3 Rockets and satellites

With space vehicles and their on-board instrumentation, it is possible to go *inside* the aurora, or to monitor it from above. Several earlier scientific satellites have been devoted to the study of the auroral and polar ionosphere, and its coupling with the magnetosphere and the solar wind. Among the latest, and most productive ones were NASA's two DYNAMICS EXPLORER satellites (launched in 1981), the Soviet/French AUREOLE-3 satellite (also launched in 1981), and more recently, the Swedish VIKING satellite. Both DYNAMICS EXPLORER-1 and VIKING carried UV cameras for imaging the complete auroral oval, and provided a wealth of data on the dynamic evolution of auroral forms during storms and sub-storms. In addition to the cameras, they performed *in-situ* measurements with a set of instruments that is basically the same

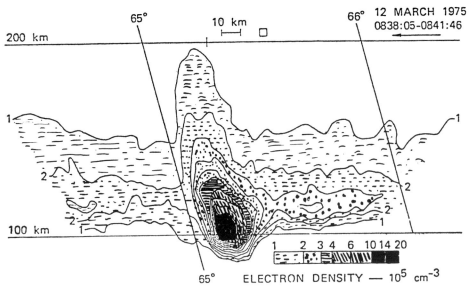

Fig. 11.19. *An auroral arc as detected by radar scans across it in a vertical plane. The location of the arc may be seen as a strong enhancement in the measured concentration of free electrons.*

on all satellites of this type. As an example, Fig. 11.20 shows a diagram of the VIKING satellite, which carried the following instruments:

- particle detectors, which were able to measure all the particles (both electrons and ions) passing through the medium, and which obviously included the precipitating auroral electrons. The detectors were able to measure the fluxes of these particles as functions of their kinetic energies;
- antennae and receivers to analyse the average values and the fluctuations of the electric and magnetic fields in the medium. In the diagram, the electrical antennae are the long booms attached to the body of the spacecraft;
- systems to measure the medium's background free-electron density.

Because they could simultaneously measure the precipitating auroral particles, the background electron density in the medium, and the detailed spectrum of the electromagnetic radiation present, the VIKING experimenters were able to make considerable progress in understanding the acceleration mechanisms for auroral electrons, and of the mechanisms generating what we call the Auroral Kilometric Radiation (AKR). The AKR is a powerful radio emission, in the kilometric-wavelength domain, which is emitted into space by the auroral zones of the Earth. This phenomenon makes the Earth a powerful radio source, which other civilizations in the Galaxy may have already detected! In fact, other magnetospheres in the Solar System are also powerful radio sources, Jupiter being the most important one. In all cases these radio sources are associated with the presence of aurorae. The VIKING investigations into the nature of this terrestrial radio source were particularly interesting, because there

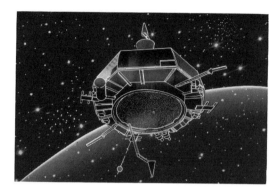

Fig. 11.20. *A schematic view of the* VIKING *satellite used by Sweden for auroral research*

was no doubt that the results would give indications of the nature of aurora-related radio sources on other planets, where *in-situ* measurements will not be available for a long time. VIKING was able to show that our AKR has its source inside so-called 'auroral cavities', regions of anomalously low density of free ionospheric electrons and ions, where downward-flowing auroral electrons, and some upward-flowing accelerated ionospheric ions interact constructively with the various electromagnetic waves that can propagate inside the cavity. Overall, the system behaves as a giant maser, which amplifies the natural low-level electromagnetic noise to produce AKR.

Rockets have also been used extensively for auroral research. They carry the same types of instruments as the satellites just described. Although they have the limitation that they stay in space for a limited time, their advantage in comparison with satellites, apart from the much lower cost, is that they cross the auroral region at a much lower speed, and give a more detailed view of the small-scale (hectometric and kilometric) structures within the auroral forms. A good way to use a rocket is to launch it through an active arc. Anyone who has ever observed an aurora can understand the practical difficulties of this: it takes a fast, and critical, decision to find the appropriate launch time, and you may have to take this decision after days – actually, nights – or weeks of waiting on the part of all the scientific and technical staff for just this right moment to arrive!

Professional auroral research has shown that beyond the first simple principle of precipitating electrons exciting atoms and molecules in the upper atmosphere to produce auroral light, the aurora is a very complex physical phenomenon involving the global interaction between our upper atmosphere, our magnetosphere, and the solar wind, as well as rather complicated mechanisms of plasma physics. There is still a lot to understand about the aurora, and this is the reason why new satellite missions to understand the magnetosphere and its interaction with the interplanetary medium and the solar wind are planned. Starting in 1993, a new generation of magnetospheric satellites built by the European Space Agency (ESA), NASA, and the Japanese Space Agency will be launched to explore critical regions in the near-Earth solar wind and the magnetosphere. ESA's SOHO mission will be positioned in the solar wind ahead of the Earth and monitor the Sun, its corona, and how coronal activity controls the local solar wind at the Earth's orbit. NASA's WIND mission will study the solar wind and its interaction with the magnetosphere in detail. Another NASA mission POLAR will monitor our magnetosphere and auroral activity from an eccentric orbit culminating over the polar regions. GEOTAIL (Japanese Space Agency, ISAS) will explore the Earth's magnetotail. Finally, ESA's four CLUSTER satellites, to be launched at the end of 1995, will cross all the critical regions of the magnetosphere (the bow shock, cusps, magnetopause, magnetotail, and auroral magnetic field lines) in a coordinated way. With their four adequately spaced measurement points, they will resolve these critical regions in three dimensions for the first time, and provide new insight into some of the remaining mysteries of the aurora.

11.3 Noctilucent clouds

Observers who live at high enough latitudes to see aurorae regularly have to contend with the fact that in summer it may never become completely dark, and so observation of aurorae (or any astronomical objects) is impossible. They do, however, have the compensation that during the midsummer months – typically from late May to the middle of August in the northern hemisphere – they may be fortunate enough to observe displays of noctilucent clouds (NLC). These clouds are quite distinct from ordinary clouds, which occur in the troposphere, where the major weather systems are concentrated. The maximum height of normal clouds is about 15 km – actually up to about 20 km extreme height over the equatorial region, but nearer 10 km or less at higher latitudes – but noctilucent clouds occur far above, at about 80 km. (In fact, noctilucent clouds show a remarkably small range of altitudes: the vast majority of measured heights fall between 80 and 85 km, and the range in a single display is usually less than 1 km.)

Noctilucent clouds are visible only when the observer is in shadow (generally in twilight), but the clouds themselves are illuminated by the Sun (Fig. 11.21). This means that they may be observed from latitudes of approximately 50–65°. (For some time it was unknown whether they occurred at lower latitudes, but could never be visible because of the geometry of the situation, but it now appears that they do not actually occur much closer to the equator.) Generally, they begin to be visible when the Sun has reached an altitude of −6° (i.e., at the end of civil twilight), and fade a similar time before sunrise. The brightest portion of the display is usually vertically above the position of the Sun, and thus moves throughout the night.

The appearance of NLC is distinctive and very beautiful, because they are a silvery-blue colour or brilliantly white against the deep blue of the twilight arch. When low on the horizon they may appear golden. The appearance distinguishes them immediately from ordinary, lower, tropospheric clouds, which appear dark against the sky. NLC displays as a whole, or features within them, frequently move under the influence of upper-atmosphere winds. Because of the extreme height, such changes appear slower than those observed in tropospheric clouds, and are certainly far slower than the fluctuations that may occur in aurorae.

NLC are very thin and tenuous and consist of minute ice particles, which appear

Fig. 11.21. *Noctilucent clouds, photographed from Finland by Peka Parviainen, on 1982 July 17–18. It should be noted that NLC do not reproduce well in black-and-white photographs, but some of the characteristic structure may be seen here.*

brilliant in the sunlight. It is a still unresolved problem how water vapour can reach such extreme heights. Transport from lower levels appears to be unlikely, because temperature inversions act to block any upward transfer of water vapour. Even exceptionally violent volcanic eruptions appear to be incapable of injecting water vapour to such extreme altitudes. What is more, for ice crystals to form they require suitable nuclei, and similar objections may be raised to the various mechanisms proposed for transporting the tiny solid particles up to NLC heights. Among explanations proposed (but quite unconfirmed), there are suggestions that the water may originate in cometary ice particles, and that the nuclei may be meteoric dust, or ions from the solar wind.

The normal temperature at an altitude of 80 km is very low, about 143 K ($-130°$C). Although aurorae occur at higher altitudes, the upper atmosphere is, of course, strongly heated during an auroral display, and for many years it was believed that this accounted for the fact that NLC and aurorae were not seen together: the auroral heating vaporized the ice crystals (or prevented them from forming). Recently, however, aurora and NLC have been observed simultaneously in the sky, and confirmatory photographs obtained. This does not, of course, invalidate the general anti-correlation between the two phenomena, but it does suggest that the true explanation is more complicated than originally believed.

11.3.1 Observing NLC

The methods of observing NLC are very similar, although much simpler, than those used for visual auroral observations. The organisations that receive auroral observations generally also have a sub-section that deals with observations of noctilucent clouds. The most extensive reports currently appearing are those prepared by Dr D. Gavine of the British Astronomical Association's Auroral Section. His yearly summaries of activity appear (for historical reasons) in *Meteorological Magazine.*

NLC change very slowly, and observations at 15-minute intervals are normally perfectly adequate. It is very convenient if observations are made at precise times, preferably on the hour, and at 15, 30 and 45 minutes past. If photographs are taken (see later), it is even more important that they should also be taken at these same times. As with auroral observations, it helps to prevent confusion if the double date is always used, and times should be given in UT: e.g., 1990 June 27–28, 00:45 UT.

Structure   Although NLC superficially resemble cirrus clouds, their forms are actually quite distinct and readily recognized. The classification is very simple, with only four different forms, although more than one of these may, of course, be present at any one time.

Veil (I)   A simple, featureless sheet of cloud, which sometimes occurs as a background to other forms.

Bands (II)   Lines or streaks of cloud, sometimes parallel to one another or else crossing at small angles. Unlike the next form, they are not particularly organised.

Waves (III)   These are sometimes called 'billows'. There is a fine, herring-bone-like structure or regular, rippled appearance. This is a very characteristic form of NLC.

Whirls (IV)   These are large-scale loops and swirls of cloud, sometimes with a twisted appearance, but without any particularly regular structure.

Amorphous (V)   Although generally resembling veils, this form is generally brighter and occurs in patches.

These forms are usually fairly easy to recognize, and binoculars will often reveal the characteristic waves in a distant display. Because NLC are so thin, they may be difficult to see when they are overhead, and sometimes observers farther north or south may see a better display than observers situated directly below the clouds.

Other information to be recorded   As with aurorae, the azimuths of the eastern and western limits of the display are useful. However, unlike aurorae, azimuths of individual features within the display are not required. The elevation of the *top* of the display is important and, if it is discernible, that of the bottom should also be noted.

A three-point scale may be used for brightness: 1, faint or visible only with binoculars; 2, readily visible; 3, extremely bright. It will also help to record brief details of the observing conditions, such as the amount of (ordinary) cloud cover, moonlight, haze, etc. If parts of the sky are obscured by lower cloud, it is useful to record its maximum vertical and horizontal extent.

For the reasons described earlier, it is most important to record full information if an aurora happens to be visible at the same time.

11.3.2 Photographing NLC

The photography of NLC is relatively simple, and very similar to the methods used for aurorae. There are less problems in choosing a suitable colour film, because the light is reflected sunlight rather than emission at specific spectral lines that may not suit a particular emulsion. It is usually of advantage to include some of the foreground, because this helps to give a sense of scale. With a lens aperture of f/2.8 and a film speed of ISO 400, exposures of 2–4 seconds are frequently suitable. The relatively static nature of noctilucent clouds, however, does mean that very much slower films may be used. Particularly beautiful renderings have been obtained with slow-speed Kodachrome slide films.

There is still a need for precise triangulation of noctilucent clouds, particularly to investigate upper-atmosphere winds and vertical motions. Such photography may best be achieved by camera stations some tens of kilometres apart. For this work it is advisable to employ fixed camera brackets, which ensure that the camera is always pointing in the same direction. This direction may be established by photographing a star field at a precisely known time during the darker months of the year. With this type of parallactic photography it is particularly important that (as previously described) the fixed times for the exposures are rigorously followed. In this way any images may be directly compared with those obtained by other observers.

11.4 Amateur auroral observing networks

Although only a small percentage of the world's population lives under the auroral ovals, observations are still possible, in the northern hemisphere, for amateur astronomers in Scandinavia, the British Isles, Iceland, Canada, the northern United States (and especially in Alaska), and northern Russia. Southern-hemisphere landmasses are less favourably disposed with respect to the auroral oval, but useful observations are made by amateurs in New Zealand. In addition, observers at lower geomagnetic latitudes may be able to observe aurorae occasionally during major storms. Even though they may lack experience, careful observations, following the guidelines given earlier, are all the more valuable. Reports should be forwarded to one of the national or international organisations that collate auroral data (addresses are given with the chapter notes at the end of this volume.) It is interesting to note that at present European groups predominate, because, unfortunately, auroral observations are not formally coordinated by any North-American organisation.

Finnish observers collaborate through the Aurora Section of the Ursa Astronomical Association, which produces annual reports of activity. In 1988, for example, 13 observers in Finland contributed results. In Sweden, reports are collected by

Svenska Astronomiska Sallskapet in Stockholm. Norwegian observations may be sent to Norsk Astronomisk Selskap, Stavanger.

Observations from the United Kingdom and Ireland are collected by the Aurora Section of the British Astronomical Association. This section was first formed in 1899 and is the oldest auroral group still in existence. (It is also the largest.) Annual reports, including results from the meteorological observers on merchant shipping, are published in the BAA *Journal*.

For North-American observers there is a somewhat less formal Aurora Alert Hotline with around 100 members, run by David A. Huestis, 25 Manley Drive, Pascoag, RI, 02859. Tel: (401) 568-9370. The Royal Astronomical Society of Canada has a number of active observers in various Centres across the country.

The Royal Astronomical Society of New Zealand also has a thriving Aurora Section.

12   Meteors

C. Buil

12.1 What is a meteor?

On a fine, clear night our gaze is drawn by the sky full of stars. It is difficult
to tear one's eyes away, because the sky appears to be infinitely rich in detail on
such a night; subconsciously we begin to pick out the brighter stars and draw lines
between them, marking out the constellations. Suddenly, a bright light appears on
the edge of our field of view. We have scarcely turned our heads to look at it when
it has disappeared. All that remains is the memory of a bright fleeting trail of light
between the stars (Fig. 12.1). We have just seen what is poetically called a 'shooting
star'. To anyone unfamiliar with the sky, it is just the pretext for the old custom of
making a wish, but to an informed astronomer ...

12.1.1 Grains of dust

The trail observed in the sky has been caused by the ablation of a particle of
cosmic dust, entering the atmosphere at high velocity. Such particles are known as
meteoroids. This term should not be confused with 'meteorite', which is used to
describe any body that reaches the Earth's surface.

The 'shooting star' is the result of the Earth encountering the path of the
meteoroid, which was peacefully orbiting the Sun. The meteoroid heats up rapidly
when it encounters the dense layers of the Earth's atmosphere, partly because it has
a very considerable velocity, typically 40 km/s, and partly because of atmospheric
resistance. The kinetic energy is transformed into heat, which usually vaporizes the
celestial visitor in less than a second. The temperature reaches around 3000°C, and
a bright light is emitted, causing the meteoroid's death-throes to be visible from the
ground. All this happens at a height of around 100 km.

The expression 'shooting star' is completely incorrect, and it is preferable to use
the term meteor to describe the luminous event itself created as the meteoroid passes
through the atmosphere.

A meteor of magnitude −5 or more – in other words brighter than Venus –
is known as a fireball or bolide. Fireballs are sometimes so bright that they cast
shadows at night and may be seen easily during the daytime. A famous case is
the fireball that was seen over the western United States on a fine afternoon on
1972 August 10. The object was particularly massive (its mass is estimated to be at
least 1000 tonnes, with a diameter of some tens of metres), but it did not reach the
ground because its trajectory was practically tangential to the surface of the Earth.
It skimmed the upper atmosphere and sped back into space. The path of this object
in the atmosphere extended for some 1500 km, which meant that many Americans
saw it, photographed it, and even filmed it. Another example of an exceptional

Fig. 12.1. *A meteor, captured on 1980 August 12 by a 35-mm f/2.8 lens (Photo: G. Varenne).*

fireball was the one that was seen over Vladivostok on 1947 February 12, the light from which cast shadows even in broad daylight!

Sometimes fireballs are multiple, either because they fragment as they encounter the atmosphere, or because they consisted of separate components that were travelling together in space. It is, in fact, quite common to see meteors appear in groups, which is a sign of repeated fragmentation.

The passage of very large fireballs is sometimes accompanied by sound effects corresponding to explosions of the bolide. Because of the distance of the object, several minutes may elapse between the time of a fireball's passage and when one hears the detonations.

The mass of the meteoroids that produce most of the meteors visible to the naked eye is extremely low. It is estimated that a meteoroid entering the atmosphere with a velocity of 30 km/s and an initial mass of 0.4 g will produce a bright meteor of magnitude 0. A meteor at the limit of naked-eye visibility has a mass of about 0.002 g, and one as bright as Venus a mass of 50 g. The density of meteoroids is quite difficult to determine and varies according to different authors, but everyone agrees that it is frequently less than 1 g/cm^3. Most of the meteors that strike the atmosphere throughout the year have the consistency of cigarette ash.

Naturally there are meteoroids that are much larger in size, and which may even reach the ground. Luckily, they are fairly rare. Indeed, meteoroids become rarer the more massive they are.

12.1.2 Conditions required for a meteorite fall

When the mass of the meteoroid is more than 10 kg, and the path is suitable, the body may reach the surface of the Earth. Because of the violent friction with the atmosphere, at the end of its track the meteorite will have lost 80 % of its initial mass.

It is estimated that there are five meteorite falls per year over an area equal to that of France. Such meteorites arriving at the ground have a mass of slightly more than 1 kg. But we should emphasize that this is only an estimate, because practically none are recovered. Over the same area, one meteorite weighing 50 tonnes falls every 8000 years (with a diameter outside the atmosphere of 2 m). A particularly destructive object of 50 000 tonnes probably arrives every 100 000 years. So there is no need to get worried; there is not much chance of being killed by a meteorite. The only known case of anyone being injured by a meteorite fall is that of Mrs E. H. Hodges, who lived in Sylacauga, Alabama. On 1954 November 30, while Mrs Hodges was resting on her sofa, a 4-kg meteorite crashed through the roof of her house, bounced off the radio and hit her arm, resulting in a minor injury.

Any meteorite with a mass less than 1 tonne that arrives at the ground is very strongly decelerated. At the time of impact, its velocity will be approximately 0.1–0.2 km/s, whereas its velocity when it encountered the atmosphere was several tens of kilometres per second. In such a case, the meteorite buries itself between several centimetres and one or two metres into the soil. The diameter of the hole caused by the meteorite is similar to the diameter of the meteorite itself. Contrary to what one might think, a few seconds after its fall, the meteorite is not particularly hot. In fact, although the surface has been heated to incandescence as it passed through the atmosphere, the centre of the rock did not have time to heat up, and after being violently decelerated, the surface cools as the body falls through the lower layers of the atmosphere. On the other hand, when meteorites are larger than some 10 tonnes, they retain most of their original, cosmic velocity, which causes a considerable crater in the surface. (A 10-tonne meteorite arrives at the ground at a velocity of at least 4 km/s.) The size of the crater that is produced is directly linked to the mass of the meteorite and the impact velocity. When the body is a meteorite of more than about 100 tonnes, the amount of energy produced is such that the object is completely vaporized and none of it remains to be found. The largest meteorite known is a single body that fell at Hoba, in South Africa. It is a rectangular body, 3×2.8 m across and 0.8 m thick, and has a mass of 60 tonnes. The archetypal meteorite crater is 'Meteor Crater' in Arizona, which has a diameter of 1300 m and depth of 175 m. This crater is the result of the impact of a meteorite that had a mass of approximately 50 000 tonnes.

Quite often a large meteorite fragments before reaching the ground. A veritable shower of stones may result, which may extend over several square kilometres. This is what happened on 1803 April 25 close to the small village of Laigle in the district of Orne in Normandy. This fall was the first meteorite fall to be studied scientifically (by Jean-Baptiste Biot). On that occasion about 3000 meteoritic fragments were collected.

Given the mass of meteorites that reach the ground, they probably have a minor-planet origin. Studying meteors therefore enables us to examine cometary material, and studying fireballs and meteorites allows us to touch material from minor planets. [A few meteorites have now been shown to have originated on the Moon and Mars. – Trans.]

12.1.3 The mineralogical composition of meteorites

Chemical and mineralogical studies have enabled two major classes of meteorites to be distinguished:

- aerolites, which consist of stony meteorites (containing minerals like olivine, pyroxene, etc.). Among the aerolites, two sub-classes are recognized: the chondrites, which contain within their crystalline structure spherical bodies of about 1-mm in diameter, which are known as chondrules (57 % of meteorites); and the achondrites, which do not contain chondrules (4 % of meteorites);
- siderites, which consist of metallic meteorites (usually containing 91 % iron and 9 % nickel). This class represents 35 % of the meteorites that are found.

It was found necessary to add a minor class to allow for meteorites that have some characteristics of both stony and iron meteorites: these are the aerosiderites (4 % of all meteorites).

12.1.4 Micrometeorites

We have described objects that are of considerable size, but at the other end of the scale, the Earth encounters large numbers of tiny particles: these are the micrometeorites. A micrometeorite is typically about 10 μm across. A micrometeorite has a large surface area relative to its mass, so the heat produced by kinetic energy when it enters the atmosphere is radiated away and the particle is not vaporized. So the arrival of a micrometeorite does not give rise to any luminous phenomenon. The particle is decelerated at the very top of the atmosphere and then floats gently down to the ground, being carried along by the winds.

The study of micrometeorites is carried out by aircraft flying at heights of 20 km, or by spacecraft. They may be collected by exposing a suitable trap coated with an adhesive resin, or by registering their impact by sensors somewhat similar to microphones. The particles trapped in this way are of two distinct types:

- dark aggregates, which resemble a particular type of meteorite known as a carbonaceous chondrite;
- collections of silicate crystals and minute particles of iron sulphide. Such a microscopic structure is never found in meteorites collected from the surface of the Earth.

A micrometeorite takes several weeks to reach the ground from the time it enters the upper atmosphere. Industrial pollution completely ruins any attempt to study these tiny particles at the end of their wanderings. But some 20 tonnes of cosmic material falls over the whole Earth every day, mainly in the form of micrometeorites.

At sites that are still unpolluted, it is possible, using a magnet, to collect minute, iron particles that are the remnants of meteorites that have been completely destroyed in the atmosphere. To do this, all that is required is to dry out a container filled with rain-water and to pass a small magnet, held behind a microscope slide, over the residue. It is thought that meteoritic dust may be one source of the condensation nuclei found in raindrops. Seen under a microscope the particles appear as tiny spheres that sometimes cluster in a circlet. One might try to see if there is a correlation between the frequency of these spherules and the activity of the major meteor showers.

12.1.5 The orbits of meteors

It was only at the beginning of the 19th century that the celestial origin of meteors was finally established. Until then, various more or less far-fetched theories had been advanced. One was that meteors were falling stones that had been violently ejected from volcanoes, and were illuminated by the Sun in the upper atmosphere.

In 1798, the Germans Brandes and Benzenberg carried out the first triangulations, which proved that meteors appeared at heights of around 100 km. This enabled them to deduce that the velocities of the objects that were the cause of the phenomenon were very considerable and made them realize that meteors were of cosmic origin.

During the night of 1833 November 12–13, a fantastic 'shower of shooting stars' occurred. A single observer would have seen around 50 000 meteors per hour. Spectators compared the sight to a snowstorm. One witness of the event was the astronomer Schiaparelli, who noticed a resemblance to a very similar shower of meteors that occurred in 1799. The details observed were sufficient for Schiaparelli to decide that the shower was following an elliptical orbit around the Sun, and subsequently it was established that this orbit was close to that of Comet Tempel. High activity occurs every 33 years, and the shower may be described as a periodic shower. The 1833 event so impressed witnesses that the study of meteors became an integral part of astronomy overnight, and systematic studies began thenceforth.

A meteor 'shower' therefore occurs when the Earth crosses a trail of cometary debris. This debris is regularly shed from the comet at its successive returns to the Sun, and the particles follow essentially the same orbit as their parent comet. The immense belt of particles created in this way is known as a meteor stream. The Earth crosses each stream on the same date each year.

Meteors belonging to a single stream appear to come from a single point on the sky, known as the radiant (Fig. 12.2). This is the result of perspective, because, to all intents and purposes, the meteoroids in an individual stream are travelling along parallel paths. The effect is exactly the same as that observed with railway lines, which although parallel and remaining the same distance apart, appear to converge to a single point at the horizon. If the paths of meteors belonging to a single shower

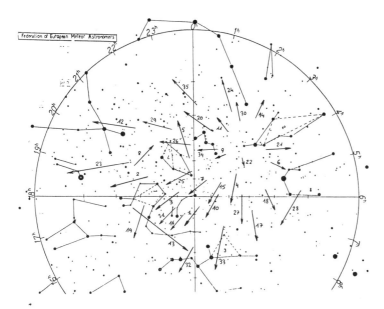

Fig. 12.2. *The meteors observed over the course of a single night have been plotted as arrows on this chart. Many meteors appear to originate at a single point on the sky, the radiant. This is seen to be in Perseus, which is why these meteors are called the Perseids. Note that some meteors are not associated with the shower and are thus sporadic.*

are extended backwards, they appear to intersect at a single point, which is the radiant. Because of the extraterrestrial origin of meteors, the celestial coordinates of this point on the celestial sphere are essentially fixed. [There is a slight change from day to day, caused by the Earth's motion around the Sun. – Trans.] Because, by definition, the radiant is the point on the celestial sphere from which a meteor appears to travel, every meteor, even an isolated one, has its own radiant.

Over the course of the year, the Earth encounters several showers. It is customary to describe a shower in terms of the constellation that contains the radiant. For example, the meteors seen on 1833 November 13 came from the constellation of Leo, so they are known as the Leonids. Another very well-known shower is that of the Perseids, where the radiant is in Perseus, etc. Sometimes there are several radiants in one constellation, and then they are described in terms of the brightest star that is close to the radiant (such as the δ Aquarids, for example).

In the densest part of the Leonid stream, the average distance between the meteoroids is only 20 km. For the Perseids, which are nevertheless still a spectacular shower, this distance is as much as 1000 km. The Perseid stream is about 7 million km wide, however, and the Earth takes several weeks to pass through it (*see* Fig. 12.18, p. 681). Its activity is quite constant from year to year – it is known as a permanent shower. This enables us to say that the Perseids are a very ancient shower, because the particles forming it are distributed uniformly around the orbit, and the activity experienced has essentially no connection with the position of the parent comet.

With the Andromedids, on the other hand, there is a very strong concentration around the parent body, Comet Giacobini–Zinner. (The meteors in this shower are sometimes called the Giacobinids.) If the Earth is close to the comet when the latter crosses the plane of the ecliptic, then we may experience a spectacular display of meteors. This occurred on 1933 October 9, when the Earth crossed the orbit of the comet just 7 days after Comet Giacobini–Zinner had passed the same point. People witnessed one of the finest meteor showers seen in the 20th century, with an hourly rate of about 5000 at maximum.

Although the link between major meteor showers and comets was relatively easily established, for a long time the precise orbit of a certain type of meteor remained a mystery. These were the sporadic meteors, which appear throughout the year at a more or less constant rate, and which seem to be completely independent of one another. Until the 1930s, it was thought that the orbits of the objects that gave rise to sporadic meteors were hyperbolic – in other words that they were not members of the Solar System. It was even believed that evidence had been found for interstellar meteor streams. The principal observational problem is to determine the linear velocity of meteors. Until the 1930s, such determinations were made visually, with all the subjectivity that this entails. (This is despite the fact that the first successful meteor photographs were taken at Harvard in 1896.) In 1934, in the United States, Öpik tried to improve the visual observational technique by placing oscillating mirrors in front of the observers, so that persistence of vision caused the linear tracks of the meteors to appear as spirals. By estimating the number of loops visible along a meteor's path, the observer could deduce the apparent velocity, and thence the geocentric velocity, making certain assumptions about the distance of the meteor. But nothing came of this, because velocities were consistently over-estimated, which irrevocably led to hyperbolic orbits.

It was Whipple, also in the United States, who, in an observing campaign in 1938 that has remained famous, proved that the meteoroids that produced sporadic meteors were members of the Solar System, and were therefore travelling in elliptical orbits. Whipple photographed meteors with specially designed Schmidt cameras. The velocities were obtained by rotating sectors that regularly interrupted the path of the light entering the cameras. The actual velocity was known by photographing the same meteor from two points, 38 km apart, and calculating the distance of the meteor by triangulation.

Meteorites that reach the ground are of course distinguished by their mass from the minute particles that produce the meteors with which we are all familiar. One of the rare cases where the orbit of a meteorite that has reached the ground is known is that of the Pribam chondrite, which fell during the night of 1959 April 7. The fall was photographed from two different sites. It was found to have an elliptical orbit with perihelion between the orbits of the Earth and Venus, and aphelion around the minor-planet belt.

It is now thought that most of the meteorites recovered on the ground have the same origin as the minor planets. In fact many minor planets are known to have very eccentric orbits, with perihelion within the Earth's orbit. Some are classified as 'Earth-Grazing Asteroids' (EGA), which are discussed in Vol. 1, pp. 292 & 310. These are good candidates for turning up in our gardens one of these days!

12.1.6 The path of a meteor in the atmosphere

A meteor's path in the atmosphere is a straight line, except in the very rare instance of grazing incidence or of a violent explosion that disperses several fragments in different directions.

The altitudes of the points where a meteor appears and disappears are mainly a function of the meteoroid's velocity, its angle of incidence with respect to the ground, its mass, and its consistency. On average, a meteor becomes luminous at an altitude of about 110 km, and the ablation of the meteor is complete by the time it has arrived at a height of about 70 km. A meteor's brief life sometimes ends in one or more explosions, corresponding to fragmentation of the meteoroid.

According to a study based on a large number of meteors (Sarma and Jones, 1986), it is possible to calculate the altitude (H) of the appearance and disappearance by more or less empirical formulae of the form:

$$H = a + b \log(\text{mass}) + c \log(V) + d \log(\sin h)$$

where a, b, c, and d are constants and:

mass: the extraterrestrial mass of the meteoroid;
V: the meteor's observed velocity;
h: angle of incidence of track with respect to the horizontal.

If the mass is expressed in grammes and the velocity in km/s, the altitude at which the meteor appears (in km) is obtained for the following values of the constants:

$a = 45.25$; $b = 3.06$; $c = 43.32$; $d = 1.13$.

For the altitude of disappearance we have:

$a = 45.73$; $b = 1.26$; $c = 25.38$; $d = -13.99$.

These constants apply to a typical meteor, disregarding the mechanical properties of the material of which it consists. It will be noted that the higher their kinetic energy, the greater the altitudes at which meteors begin and end.

Note also that it is possible to estimate the average magnitude of a meteor by identical formulae. To do this, H is replaced by M, the magnitude. The constants then become:

$a = 9.88$; $b = -2.02$; $c = -7.17$; $d = 0.10$.

Conversely, from observed properties such as the velocity and magnitude, it is possible to calculate the mass.

As we have already said, the velocity is very high. Most meteoroids are following elliptical orbits with semi-major axes that are greater than 1 AU. It may be shown that their velocity at the orbit of the Earth is then between 30 km/s and 42 km/s. If the velocity is 30 km/s, the orbit is practically circular. If the velocity is 42 km/s, the orbit is parabolic: beyond that the orbit would be hyperbolic.

In fact the observed velocity in the atmosphere is a combination of the meteoroid's own velocity (the heliocentric velocity) and the Earth's orbital velocity. The latter is more or less constant at around 30 km/s. If we assume that the meteoroid is

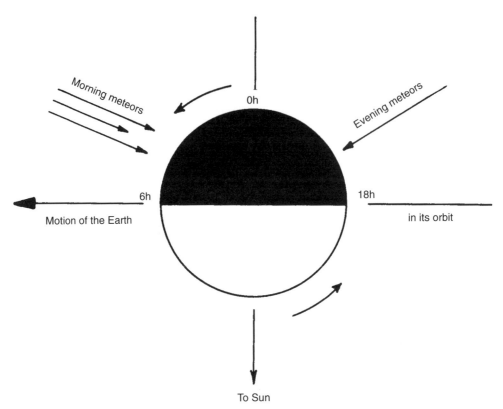

Fig. 12.3. *Why the number of meteors encountering the Earth varies with local time.*

moving along a parabolic orbit, we can see that depending on the orientation of its orbit to that of the Earth, the relative velocity of the two bodies may lie anywhere between 12 km/s and 72 km/s. The combination of the velocities of the Earth and of the meteoroids also produces a diurnal variation in the frequency of meteors. Figure 12.3 shows that at 18:00 hours (local time), the meteors seen are those that are 'chasing' the Earth in its orbit around the Sun. They therefore encounter the Earth at low velocities, and the hourly rate is low. On the other hand, at 06:00 hours (i.e., in the morning), meteors are encountered head-on, and the hourly rate is a maximum. The ratio between the numbers of meteors at these two extremes is approximately 2.

The diurnal variation in the hourly rate is a characteristic feature of sporadic meteors. As far as meteors belonging to a particular shower are concerned, the height of the radiant above the horizon is the determining factor. It is obvious that if the radiant is below the horizon, we cannot expect to see any meteors at all. On the other hand, the higher the radiant is in the sky, the greater the number of meteors. If h is the height of the radiant above the horizon, the relative number N of meteors that one may expect to see approximately follows the relationship:

$$N = \sin h.$$

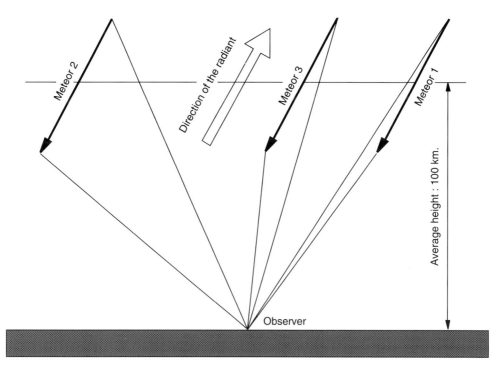

Fig. 12.4. *Variations in the visibility of meteors from a single shower.*

Figure 12.4 shows some important aspects of the geometry of meteors in the Earth's atmosphere:

- The angle of incidence of a meteor with respect to the ground is equal to the angular height of the radiant above the observer's horizon (the Earth is assumed to be flat, which is legitimate in this case).
- In observing meteors near the direction of the radiant, it will be seen that the trails subtend small angles on the celestial sphere. At the limit, it is possible to see meteors coming straight towards the observer. One then sees a brief flash from a stellar object. These are known as point meteors (case 1 in Fig. 12.4).
- The trails are longest when observing at 90° to the radiant. Under these circumstances the angular velocity is also greatest (meteor 2 in Fig. 12.4).
- Meteor 3, observed near the zenith, is closer to the observer than meteor 1, which is observed near the horizon. (It is assumed that the altitudes of both meteors are the same, which will be approximately true.)
- The observer will see an ascending trail for meteor 3, which is obviously only a geometrical effect, but which may confuse the beginner.

12.2 The most important showers

The number of recorded showers is very high. Some catalogues have listed nearly 1000 radiants, but most of these are minor and must therefore be regarded as of doubtful validity. Any radiant giving only one meteor every 2 or 3 hours could well be the result of chance.

The catalogues also contain showers that were once important, but which have ceased to be active; others that have only been observed once; and finally others that have been discovered recently. To understand these changes, it must be realized that, like every other body orbiting the Sun, meteoroids are subject to gravitational perturbations caused by a giant planet such as Jupiter, and their orbits may be altered. The perturbation may affect a major portion of the stream, especially if the orbital period is short. A meteor stream that previously intersected the Earth's orbit may thus be deflected so that it passes at such a distance that activity ceases. For the same reason, a previously unknown shower may suddenly appear. We have already seen that the density of meteoroids within a stream is not necessarily homogeneous. Activity will only be detectable if the Earth crosses the dense part of the stream. Such a crossing may occur at regular intervals, as with the Leonids which cause a spectacular shower every 33 years.

Table 12.1 is a list of the major showers (Cook, 1973). The first column gives the name of the shower, and subsequent columns show:

- the period of activity;
- the date of maximum;
- the zenithal hourly rate (*ZHR*). This is the activity that the shower would exhibit if the radiant were at the observer's zenith. The sky is considered to have a standard limiting magnitude of 6.5. Such a rate is never actually observed under ordinary conditions. In addition, for a given shower, the *ZHR* is variable from year to year;
- the solar longitude of maximum (epoch 1950.0). The solar longitude $\lambda_\odot$ is used as a measure of time in meteor work. The calendar date is not reliable, because of the effect of leap years. To determine the time of maximum, the solar longitude should be compared with the calendar date (use an ephemeris or the method of calculation given on p. 710);
- the right ascension α of the radiant (in degrees);
- the declination δ of the radiant;
- a letter D in the last column indicates that the shower is a diurnal one and that it may be observed with radio or radar methods only.

Table 12.1. *The principal meteor showers*

Name	Dates	Max.	ZHR (max.)	$\lambda_\odot$	α	δ	
Quadrantids	Jan.01–04	Jan.03	140	282.7	230.1	+48.5	
δ Cancrids	Jan.13–21	Jan.16		296	126	+20	
Virginids	Feb.03–Apr.15				186	0	
δ Leonids	Feb.05–Mar.19	Feb.26		338	159	+19	
Camelopardids	Mar.14–Apr.07				118.7	+68.3	
σ Leonids	Mar.21–May 13	Apr.17		27	195	− 5	
δ Draconids	Mar.28–Apr.17			35	281	+68	
κ Serpentids	Apr.01–07				230	+18	
μ Virginids	Apr.01–May 12	Apr.25		35	221	− 5	
α Scorpiids	Apr.11–May 12	May 03		42	240	−22	
α Boötids	Apr.14–May 12	Apr.28		36	218	+19	
ϕ Boötids	Apr.16–May 12	May 01		40	240	+51	
April Lyrids	Apr.20–23	Apr.22	12	31.7	271.4	+33.6	
η Aquarids	Apr.21–May 12	May 03	30	42.4	335.6	− 1.9	
τ Herculids	May 19–Jun.14	Jun.03		72	228	+39	
χ Scorpiids	May 27–Jun.20	Jun.05		74	247	−13	
Daytime Arietids	Jun.01–17	Jun.07		76	44	+23	D
Daytime ζ Perseids	Jun.01–17	Jun.07		76	62	+23	D
Librids	Jun.08–09, 1937	Jun.08	10 (1937)	78.2	227.2	−28.3	
Sagittariids	Jun.08–16, 1957-8	Jun.11		80	304	−35	
θ Ophiuchids	Jun.08–16	Jun.13	9	82	267	−28	
June Lyrids	Jun.11–21, 1969	Jun.16		84.5	278	+35	
Daytime β Taurids	Jun.24–Jly 06	Jun.29		96	86	+19	D
Corvids	Jun.25–30, 1937	Jun.26	13 (1937)	95.2	191.9	−19.1	
June Boötids	Jun.28, 1916	Jun.28	100 (1916)	97.6	219	+49	
July Phoenicids	Jly 03–18	Jly 14		113	31.1	−47.9	
o Draconids	Jly 07–24	Jly 16			271	+59	
Northern δ Aquarids	Jly 14–Aug.25	Aug.12	20	139	339	− 5	
Southern δ Aquarids	Jly 21–Aug.29	Jly 29	30	125	333.1	−16.5	
α Capricornids	Jly 15–Aug.10	Jly 30	30	126	307	−10	
Southern ι Aquarids	Jly 15–Aug.25	Aug.05	15	131	333.3	−14.7	
Northern ι Aquarids	Jly 15–Sep.20	Aug.20	15	147	327	− 6	

Table 12.1. – Continued. *The principal meteor showers*

Name	Dates	Max.	ZHR (max.)	$\lambda_\odot$	α	δ	
Perseids	Jly 23–Aug.23	Aug.12	70	139	46.2	+57.4	
κ Cygnids	Aug.09–Oct.06	Aug.18	5	145	286	+59	
Southern Piscids	Aug.31–Nov.02	Sep.20		177	6	0	
Northern Piscids	Sep.25–Oct.19	Oct.12		199	26	+14	
Aurigids	Sep.01, 1935	Sep.01	30	157.9	84.6	+42.0	
κ Aquarids	Sep.11–28	Sep.21		178	338	− 5	
Southern Taurids	Sep.15–Nov.26	Nov.03	7	220	50.5	+13.6	
Northern Taurids	Sep.19–Dec.01	Nov.13	< 7	230	58.3	+22.3	
Daytime Sextantids	Sep.24–Oct.05	Sep.29		184	152	0	D
Annual Andromedids	Sep.25–Nov.12	Oct.03		190	{ 5 / 20	+ 8 / +34	
Andromedids	Nov.27, 1885	Nov.27	13 000 (1885)	246.7	25	+44	
Orionids	Oct.02–Nov.07	Oct.21	30	207.7	94.5	+15.8	
October Draconids	Oct.09	Oct.09	30 000 (1933)	196.3	262.1	+54.1	
ϵ Geminids	Oct.14–27	Oct.19		206	104	+27	
Leo Minorids	Oct.22-24	Oct.24		211	162	+37	
Pegasids	Oct.29–Nov.12	Nov.12		230	335	+21	
Leonids	Nov.14–20	Nov.17	20 [155 000 (1966)]	234.462	152.3	+22.2	
Monocerotids	Nov.27–Dec.17	Dec.10		258	99.8	+14.0	
σ Hydrids	Dec.03–15	Dec.11		259	126.6	+ 1.6	
Northern χ Orionids	Dec.04–15	Dec.10		258	84	+26	
Southern χ Orionids	Dec.07–14	Dec.11		259	85	+16	
Geminids	Dec.04–16	Dec.14	70	261.7	112.3	+32.5	
December Phoenicids	Dec.05, 1956	Dec.05	100	253.55	{ 15 / 15	−55 / −45	
δ Arietids	Dec.08–14				52	+22	
Coma Berenicids	Dec.12–Jan.23				175	+25	
Ursids	Dec.17–24	Dec.22	20	270	217.06	+75.85	

Four daytime streams (D) are included in this list. All dates are for 1950, unless otherwise specified. [Adapted from (Cook, 1973)]

Additional details of some of the showers listed in Table 12.1 follow. The orbital elements are all given with respect to Epoch 1950.0.

Quadrantids One of the major showers. The name comes from the constellation Quadrans Muralis, which no longer exists, and which was located between Draco and Boötes. Sometimes this shower is called the Boötids. The period of activity is remarkably short, at most about 10 hours centred around a very sharp maximum. Most of the shower occurs in just 4 or 5 hours. The Quadrantids may, of course, go completely unseen if by bad luck the maximum occurs during daylight at the observing site. The velocity of the meteors in the atmosphere is 41 km/s. This means that they appear moderately fast. The Quadrantids are fairly faint. Orbital elements: $\Omega = 282.7$; $\omega = 170.0$; $i = 72.5$; $e = 0.683$; $q = 0.977$; $a = 3.08$.

Camelopardids Few meteors, perhaps 1 or 2 per hour at maximum. The most remarkable feature is that these meteors are extremely slow; their velocity is only about 10 km/s.

April Lyrids This shower was noted in Chinese records of 687 BC. It was previously very active ($ZHR = 700$ in 1803 and 100 in 1922). Unfortunately, because of planetary perturbations, the number of meteors visible has declined greatly. Velocity: 48 km/s. Orbital elements: $\Omega = 31.7$; $\omega = 214.3$; $i = 79.0$; $e = 0.968$; $q = 0.919$; $a = 28$. This stream is linked with Comet Thatcher, the last apparition of which was in 1861.

Eta Aquarids Associated with Comet P/Halley. The ZHR reached 100 in 1980. Velocity: 64 km/s. Orbital elements: $\Omega = 42.4$; $\omega = 95.2$; $i = 163.5$; $e = 0.958$; $q = 0.560$; $a = 13$.

June Lyrids Not to be confused with the April Lyrids. The radiant was discovered in 1966 and since that time the shower has been regular. Orbital elements: $\Omega = 84.5$; $\omega = 237/231$; $i = 44/50$; $e = 0.67/0.92$; $q = 0.83/0.84$; $a = 2.5/10$.

June Boötids This meteor stream is associated with Comet P/Pons–Winnecke. The ZHR was spectacular in 1916 with 100 meteors per hour. Since then activity has been very weak, but a sudden increase may occur at any time.

Delta Aquarids Broad shower (with a double radiant), visible at the same time as the Alpha Capricornid shower, which is very similar. Both showers enliven the warm nights at the end of July. Orbital elements for the northern Delta Aquarids: $\Omega = 139$; $\omega = 332$; $i = 20$; $e = 0.97$; $q = 0.07$; $a = 2.62$. Orbital elements for the southern Delta Aquarids: $\Omega = 305.0$; $\omega = 152.8$; $i = 27.2$; $e = 0.976$; $q = 0.069$; $a = 2.86$.

Alpha Capricornids Linked with Comet P/Honda–Mrkos–Pajdusakova. [This association is now regarded as doubtful because of the deviation in the directions of

645

perihelion. – Trans.] Very slow meteors. Orbital elements: $\Omega = 127$; $\omega = 269$; $i = 7$; $e = 0.77$; $q = 0.59$; $a = 2.53$.

Perseids The best-known shower, both because it is very active and also because it is visible on fine nights in August. Rapid meteors, with persistent trains. Often very bright, with explosions. The first observations of the Perseids were reported by the Chinese in 36 BC, when the maximum occurred on July 17, rather than August 12 currently. The shower is associated with Comet Swift–Tuttle which should have a period of 120 years. The last apparition of the comet was in 1862. The return in 1982 was not observed. Velocity: 61 km/s. Orbital elements: $\Omega = 139.0$; $\omega = 151.5$; $i = 113.8$; $e = 0.965$; $q = 0.953$; orbital period $= 91.20$ years. For comparison, the orbital elements of the parent comet are: $\Omega = 138.7$; $\omega = 152.8$; $i = 113.6$; $e = 0.960$; $q = 0.963$; orbital period $= 120$ years.

Kappa Cygnids Very bright meteors (often fireballs) with a slow velocity (26 km/s) and frequent persistent trains. Orbital elements: $\Omega = 145$; $\omega = 194$; $i = 38$; $e = 0.68$; $q = 0.99$; $a = 3.09$.

October Draconids Shower associated with Comet P/Giacobini–Zinner (the meteors are sometimes known as Giacobinids). Has given rise to some spectacular storms: $ZHR = 5400$ in 1933, 10 000 in 1946. These high rates correspond to occasions when Comet Giacobini–Zinner has passed close to the Earth. The shower is probably young, and has not had time to disperse around its orbit. It still forms a compact cloud near the comet. The period of the peak displays is 13 years. Very slow meteors with a velocity of 23 km/s. Orbital elements: $\Omega = 196.3$; $\omega = 171.8$; $i = 30.7$; $e = 0.717$; $q = 0.996$; $a = 3.51$.

Orionids One of the most active showers. Probably associated with Comet P/Halley. Velocity 66 km/s. The Orionids are fast, relatively faint meteors with persistent trains. The maximum is flat in comparison with that of the Geminids or the Perseids. Orbital elements: $\Omega = 28$; $\omega = 82.5$; $i = 163.9$; $e = 0.962$; $q = 0.571$; $a = 15.1$.

Taurids A shower whose activity extends over a very long period. [It actually consists of two streams that cannot be distinguished from one another visually. The data are for the northern Taurids. – Trans.] The meteors are often bright (fireballs) and very slow. The time of maximum is poorly defined. Associated with Comet P/Encke. Orbital elements: $\Omega = 230.0$; $\omega = 292.3$; $i = 2.4$; $e = 0.861$; $q = 0.359$; $a = 2.59$.

Leonids An extremely active shower, but unfortunately only intermittently. The periodicity of the 'meteor storms' is 33 years. The Leonids were spectacular in 1799, 1833 and 1866. But few meteors were observed in 1899 and 1933, which gave rise to fears that the stream had been lost. In 1966 there came a surprise, the ZHR reached the amazing value of 155 000 (a record!). In the few minutes around maximum, 140 meteors could be seen every second. The next time the Earth encounters the dense part of the stream should be in November 1999. Until then we must rest content with moderate activity and a *ZHR* of 20 at most. Extremely fast meteors with a velocity of 71 km/s. The Leonid stream follows the orbit of Comet P/Tempel–Tuttle. Orbital elements: $\Omega = 234.5$; $\omega = 172.5$; $i = 162.6$; $e = 0.915$; $q = 0.985$; $a = 11.5$.

Andromedids Associated with Comet P/Biela, famous for having broken up when it passed close to the Sun in 1845. Gave fine meteor storms in 1872 ($ZHR = 6000$), in 1885 ($ZHR = 13\,000$) and 1892 ($ZHR = 300$). Today the activity of this shower is low, probably because of perturbations caused by Jupiter.

Geminids An active and regular shower, comparable with the Quadrantids and the Perseids. Velocity 35 km/s. The mass-distribution in the stream is such that the number of faint meteors visible before maximum is greater than it is during maximum. It has therefore been possible to link the average magnitude (m) of the Geminids and the date of maximum expressed as solar longitude ($\lambda_\odot$): $\lambda_\odot = 261.29 - 0.118 \times m$. The shower seems to have been observed only since the middle of the 19th century. Recently it has been established that the stream is apparently associated with minor planet 3200 Phaethon, which may actually be an extinct or quiescent cometary nucleus. The orbit of the Geminids is unique: it has a very small perihelion distance, extreme eccentricity and short orbital period (1.6 years). Orbital elements: $\Omega = 261.0$; $\omega = 324.3$; $i = 23.6$; $e = 0.896$; $q = 0.142$; $a = 1.36$.

Ursids Shower associated with Comet P/Tuttle. Velocity 34 km/s. Orbital elements: $\Omega = 270.66$; $\omega = 205.85$; $i = 53.6$; $e = 0.85$; $q = 0.9389$; $a = 5.70$.

The coordinates of the radiants given in Table 12.1 are only valid for the time of maximum. In fact the motion of the Earth around the Sun produces a slow movement towards the east (about 1° per day). Some showers do not show this motion and they are said to be 'stationary'. They violate the laws of celestial mechanics, and to explain this the theory has been advanced that we are dealing with a series of showers that follow one another with radiants at the same position.

Table 12.2 gives the daily motion of the radiant for a number of showers (right ascension and declination for epoch 1950.0), as a function of solar longitude ($\lambda_\odot$).

A radiant is not a precise point, because of the spread of meteoroids within the stream. As we have seen, this diffusion is caused by planetary gravitational perturbations, and also by collisions between the meteoroids, and by the effects of the solar wind. Another effect, known as the Poynting–Robertson effect, tends to diffuse the meteor stream. It causes the particles to lose kinetic energy as they re-emit solar radiation, and tends to make their orbits evolve towards a circle. Because of these effects, the paths are not identical for all the particles within a stream, which accounts for the radiant's apparent diameter. In the case of the Andromedids, a young stream that has yet to be affected by diffusion, it is estimated that the radiant is 0.1° in diameter at most. For the Perseids, the diameter is approximately 3°. Even more 'diffuse' radiants are those such as the Taurids or the Alpha Capricornids, which are 3° in diameter.

Table 12.2. *Motion of radiant for a number of major showers*

Shower	Right ascension, α	Declination δ
April Lyrids	$271.9 + 1.18 \, (\lambda_\odot - 32.0)$	$+33.3 + 0.17 \, (\lambda_\odot - 32.0)$
η Aquarids	$336.5 + 0.65 \, (\lambda_\odot - 42.4)$	$-1.0 + 0.24 \, (\lambda_\odot - 42.4)$
Northern		
δ Aquarids	$334.8 + 0.75 \, (\lambda_\odot - 126.0)$	$-2.4 + 0.21 \, (\lambda_\odot - 126.0)$
Southern		
δ Aquarids	$338.9 + 0.75 \, (\lambda_\odot - 126.0)$	$-16.4 + 0.21 \, (\lambda_\odot - 126.0)$
Perseids	$46.8 + 1.38 \, (\lambda_\odot - 139.3)$	$+57.7 + 0.18 \, (\lambda_\odot - 139.3)$
Orionids	$94.7 + 0.65 \, (\lambda_\odot - 208.0)$	$+15.9 + 0.11 \, (\lambda_\odot - 208.0)$
Northern Taurids	$58.3 + 0.82 \, (\lambda_\odot - 230.0)$	$+22.3 + 0.20 \, (\lambda_\odot - 230.0)$
Southern Taurids	$50.5 + 0.82 \, (\lambda_\odot - 220.0)$	$+13.6 + 0.20 \, (\lambda_\odot - 220.0)$
Geminids	$112.3 + 0.97 \, (\lambda_\odot - 261.0)$	$+32.4 - 0.08 \, (\lambda_\odot - 261.0)$

12.3 Visual observation of meteors

It is not difficult to observe meteors visually: you only need to look up and watch the sky. When a meteor is seen, all its characteristics are carefully noted. When professional astronomers observe meteors they do just the same. The study of meteors is one of the rare fields of astronomy where both professionals and amateurs are on an equal footing.

Except in a very few countries, systematic observation of meteors by professional astronomers has ceased. Amateurs are therefore in a position to carry out original work. All the more so, because the lack of professional interest is not justifiable from a scientific point of view. The subject is far from being exhausted. This lack of interest is probably the result of one of the fashions that affect professional astronomy from time to time. Yet very considerable resources are employed in studying comets and minor planets, which are only some of the minor bodies in the Solar System. Their study is being pursued because they hold the key to the history of the Solar System and even to that of the Universe itself. Meteors definitely belong to this class of objects that is evoking so much curiosity and their study may be considered very topical. And, because little equipment is required to observe these cosmic visitors, we are able to study the interstellar medium at little cost.

Visual observation forms the basis of meteor astronomy. Thanks to the eye's sensitivity and wide field, the number of meteors that may be detected is such that it is possible to carry out statistical studies that avoid the problems normally inherent in visual work: inconsistency, subjectivity, and scatter. Naturally to carry out any scientific work, rigorous methods must be used. We shall now consider these.

12.3.1 Preparations

In order for an observation to be of use statistically, it is necessary to observe for at least an hour at a time. So a certain degree of observer comfort is by no means a luxury. Ideally, one should use an adjustable garden chair, which enables one to change the direction in which one is looking as required. It is best to watch at an altitude of between 45° and 60°, rather than at the zenith, to see most meteors. Naturally, in winter it is essential to wear warm clothing. Observers who are cramped, because they are not comfortably seated, or shivering, because they only put on three pull-overs, will find their efficiency decreases markedly. It should be unnecessary to stress that artificial lights are to be avoided, because the darker the sky, the more meteors one can observe, especially as the faintest meteors are the most numerous.

When a meteor is seen, all its characteristics have to be memorized in a fraction of a second. The position of the trail should be noted with respect to the stars and drawn on a special chart. Other details such as the time, magnitude, apparent velocity, etc. should also be noted. Not too much time should be lost in recording these details with one's head down, because other meteors may occur. Everything required for writing down the details should be to hand. Generally one or more charts are fixed to a suitable board, with, on the other side, the sheet for recording meteor details. In the dark, a pencil is always to hand if it is attached to the board by a piece of string.

Details of the meteor need not be written down if a tape-recorder is used. This technique is often used by meteor observers. It is, however, essential to ensure that the tape-recorder is functioning: when the temperature falls, surprises sometimes occur!

For illumination, use a hand-lamp, the brightness of which has been reduced to avoid being dazzled each time one writes down the details of a meteor, so that one makes full use of dark-adapted vision. For preference, the lamp should be red.

Among the indispensable points that should not be forgotten are having a hot drink available. Music can also help to while away the time between two meteors.

12.3.2 Charts

The paths of meteors across the sky are arcs of great circles. A great circle is one where its plane passes through the centre of the celestial sphere, centred on the observer. The celestial equator and the meridians of right ascension are arcs of great circles. On the other hand, the declination circles (except the equator) are small circles. It is convenient if, on a chart, arcs of great circles are represented by straight lines; meteor tracks may therefore be drawn with a ruler. This makes it easy to extend the paths backward to determine the position of the radiant. A chart-projection that does have this feature is one called gnomonic. Such a projection is obtained by placing the centre of the plane of the chart tangent to the celestial sphere (*see* Fig. 12.5). A straight line from the centre of the celestial sphere is then used to project the position of any star onto the plane of the chart. The farther one moves away from the centre of the chart the greater the distortion. The

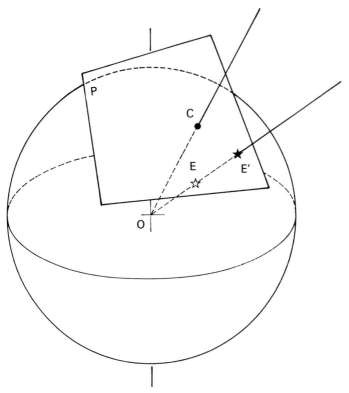

Fig. 12.5. *Gnomonic projection. The plane of projection (P) is tangent at point C to the celestial sphere with centre O. A star at E on the celestial sphere is projected to point E' on the plane of the chart.*

resulting deformation of the constellations sometimes confuses beginners, but like most things, one soon becomes accustomed to this with experience.

Most charts of the sky that are available commercially do not use the gnomonic projection. Gnomonic charts may be obtained from various national meteor-observing groups and are supplied to members of the American Meteor Society (AMS). Typically four to six charts are required to cover the area of the sky observed on any one night. If a strong shower is active, it is important to have several charts of the same area of sky to hand to avoid overcrowding, and thus rending it more difficult to carry out the reduction later. These additional charts need to be obtained before the observing session begins, because it is rather difficult to find a photocopier at three o'clock in the morning!

If gnomonic charts are not available, one can, if necessary, use traditional charts. It is then usually necessary to draw the path of a meteor as an arc of a circle. In practice, the radius of curvature is very large and any attempt to draw it is fraught with potential error. In order not to introduce additional errors, one should be content with just drawing a short section as a straight line.

The track may be drawn freehand, if you have a steady hand with a pencil –

and this is quicker. But otherwise, use a ruler. Conventionally, the direction of the meteor's flight is shown by an arrow, whose length is identical to the angular distance of the track. A number alongside the path indicates the order of appearance. Figure 12.6 shows a typical chart.

When observing a major shower, it is absolutely essential to avoid drawing the meteors and aligning them with the radiant, the position of which one knows in advance. On the contrary, one should try to ignore its position, in order not to be influenced by it, thus avoiding mixing sporadic meteors with those belonging to the shower, which would falsify the statistics.

Drawing the path of a meteor is not particularly easy. The observation of several hundred meteors is necessary to produce a reliable report. Note in passing that it is not possible to keep up one's expertise if one observes only for 3 or 4 nights a year! Frequently beginners have a tendency to draw paths as too short, or all of the same length. The most difficult thing, when beginning, is to draw meteors independently of one another. But this is essential to avoid inventing a radiant, just because one has the impression that three or four meteors all come from the same part of the sky.

If the hourly rate is very high, it becomes impossible to draw the paths on the charts. Above about three meteors a minute, observers generally tend to panic, because they do not have enough time to record all that they see. If one is lucky enough to witness such meteor activity, one should content oneself by just making a simple count over periods of approximately 15 minutes, taking care to distinguish sporadic meteors from those belonging to the shower. If the shower is extremely active, one may resort to just counting the meteors in a well-defined region of the sky – the Square of Pegasus, for example.

12.3.3 Details to be recorded

When a meteor is seen, the first thing to do is to note down the track, because this is the feature that is most difficult to retain in one's mind. One then enters the following details on a sheet prepared in advance:

- the sequential number for that night;
- the time (in UT). An accuracy of one minute is sufficient when the activity is weak. On the other hand, when the activity is high, and the results of several observers are to be compared, then the time may be noted to the nearest second. Sometimes two or more meteors may appear separated by just a few seconds. Then the time in seconds between each event may be recorded instead;
- the apparent velocity of the meteor. This information is very dependent on the individual observer's judgement. However, after having observed a number of meteors, one begins to be able to decide which are fast, which are moderate in speed, and which are slow. To take notes rapidly, abbreviations may be used: VS for very slow, S for slow, M for moderate speed, F for fast, and VF for very fast. This method of determining the speed is purely subjective, but it is difficult, when observing visually, to go any further. It is

651

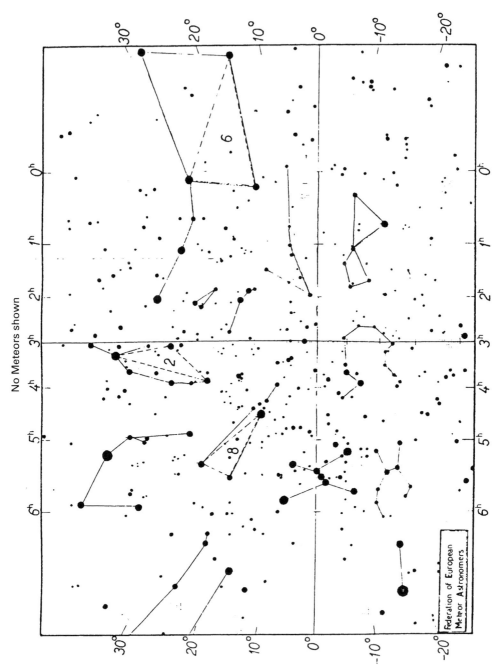

No Meteors shown

Federation of European
Meteor Astronomers

Fig. 12.6. *Example of a gnomonic chart. The distortion of the Square of Pegasus at top right is particularly noticeable.*

Table 12.3. *Magnitudes suitable for meteor estimates*

Magnitude	Object
−4	Venus
−2	Jupiter
−1.5	Sirius
0	α Lyr (Vega), β Ori
+1	α Aql (Altair), α Tau (Aldebaran)
+2	α UMi (Polaris), γ Gem
+3	δ Per, ι Ori, γ Sgr
+4	ε UMi, μ UMi, β Del
+5	ζ Cas, ν Leo, π Cap

quite impractical to equip oneself with a stop-watch to estimate velocities. Take care, because with identical actual velocities, a meteor with a long path often appears slower than one that only has a short trail across the sky;

- the duration of the train. Sometimes, after a meteor's passage, a luminous train persists. If this train is hardly perceptible, it is described as being of short duration. If the train is visible for 0.5–1 second, it is considered persistent. If the duration is 1–3 seconds, it is considered very persistent and should be noted accordingly. Beyond that the train is described as being extremely persistent and the time in seconds should be recorded. Some fireballs do give rise to trains that persist for as long as an hour. When a train disappears to the naked eye, it may still be followed with a pair of binoculars. So it is not a waste of time having binoculars to hand. In certain cases one can follow changes in the shape of the train caused by high-altitude winds. It is estimated that just after the passage of a meteor of magnitude 0, the train of ionized gas is about 1 m in diameter. The train expands rapidly and its width is sometimes perceptible to the naked eye;

- the average magnitude of the meteor. This is generally rounded to the nearest whole magnitude. The brightness may be compared with that of nearby stars. This is not easy, given the fleeting nature of the event. But, with a bit of experience, on comparing observations with those made by others, one may be surprised to find that the magnitudes are in fair agreement. (A typical accuracy is ±0.4 magnitude.) The most difficult to estimate are the brightest meteors, because suitable comparison stars are lacking. One has to resort to extrapolating magnitudes. Do not forget that the magnitude scale is logarithmic: a meteor that appears six times as bright as Jupiter (magnitude −2) is not magnitude −12, but magnitude −4. As a guide, Table 12.3 gives typical magnitudes of some planets and stars, which may be used to provide a comparison sequence.

In general, it is rarely possible to see meteors fainter than magnitude +5, because of their motion. The limiting magnitude strongly depends on the field of view. A bright fireball of magnitude −7 may be seen wherever it may be in the sky, because of the brilliant light that it emits. A meteor of magnitude 0 must be less than 55° from the observer's line of sight for it to be noticed. This angle drops to 38° for a meteor of magnitude 3, and to just 7° for a meteor of magnitude 5.

- the weighting given to the path of the meteor shown on the chart, on a scale of 1 (good) to 3 (poor). Weight 3 is often given to meteors seen at the edge of the field of view. In extreme cases, one only sees a fleeting light amongst the stars, and any details that can be recorded for the object are obviously very incomplete;
- the constellation where the meteor appeared. This helps reduction by enabling the meteor to be found easily on the charts;
- in the remarks column, note down any other information such as explosions, colour, etc.

It is obvious that it is essential to draw the meteor's path on the chart and to note down the details as quickly as possible so that a minimum time is lost before being able to resume watching. A good knowledge of the sky and using abbreviated notes reduces the time required to less than 10 seconds. Remember that a tape recorder may be used.

Figure 12.7 shows a typical observing form. Apart from the columns corresponding to the details already mentioned, the top of the form shows:

- details of the observer: to enable easy identification when the report is submitted to a coordinating organization;
- the date, and the times of the beginning and end of the watch. Note that the beginning and the end of a watch do not necessarily coincide with the observation of a meteor;
- it is, obviously, necessary to rest from time to time, but it is essential to record the length of such interruptions so that the overall time may be accurately known;
- the limiting magnitude allows the quality of the night's seeing to be assessed. For this a technique developed by the Federation of Meteor Associations (FEMA) should be used. It consists of determining the number of stars (N) visible in a small triangular zone of sky defined by three bright stars (these three stars are also laid down). From this number the limiting magnitude (L_m) is determined. Because meteorological conditions may change during the night, the limiting magnitude is determined at regular intervals;
- the existence of any natural obstacle, or the arrival of clouds, that limits the area of sky observable. In such a case one notes down the fraction of the sky that is unobservable.

Name:

Address

Night of: to:
Starting time of observations (UT) :
Finishing time of observations (UT) :

Breaks:
              from   to
              from   to
              from   to

Limiting magnitude   Fraction of sky not visible

Time	zone	N	L_m

Time	F

Time for notes: sec:

Number	Time	Magnitude	Velocity	Train	Const.	Weight	Notes

Fig. 12.7. *Visual observing form*

12.3.4 *Reduction of visual observations*

12.3.4.1 *Determining the activity of a shower*

If a shower is active, it is important to distinguish between shower members and sporadic meteors by examining the charts. Naturally, more than one shower may be active at the same time, so it is also necessary to know which meteor belongs to which shower.

In extending the paths of meteors backwards, one ought to find an approximate convergence at the radiant point. Sometimes, when studying sporadic meteors, it is possible to detect a new radiant. To be able to assume such a discovery, however, it is essential to check with the results of other observers working at the same time. Finding three paths that converge on a single point is just not sufficient proof, given the possibilities of coincidences and errors in drawing the tracks. Naturally, the existence of similar characteristics (e.g. identical velocities) may lead one to suspect the existence of a radiant, but in this field it pays to be prudent. If, however, the evidence appears strong, then it is necessary to check if the activity extends over several days. It is worth recalling that several hundred minor showers have been catalogued and one may well have rediscovered one of these streams, which is itself not bad, and· is of a certain scientific interest. But to conclude that a new shower is permanent, it needs to be observed over several years.

Sometimes meteors are seen far from the radiant, which may lie on a different chart. It is then necessary:

- either to extend the tracks to the edge of the chart that contains the meteor, and then to try to carry the tracks over to the chart containing the radiant. To do this one can make use of stars common to both charts that are adjacent to the meteor's track;
- or, calculate the equation of the meteor's track expressed in Cartesian coordinates that are linked to the chart, and then determine the intersection of the straight lines thus defined. It is not very difficult to calculate mathematically the intersection of two straight lines. On the other hand, if one is trying to determine the position of the radiant from more than two tracks, it is essential to use least-squares techniques to obtain the position that has the highest probability of being the radiant. To transform Cartesian coordinates into equatorial coordinates and vice versa the equations given in the section on gnomonic charts (*see* p. 702) may be used.

12.3.4.2 Calculating hourly rates

After having differentiated between sporadic meteors and those belonging to a shower, we need to calculate the activity of the two populations. The first thing that comes to mind is simply to determine the number of meteors visible in unit time. Such a procedure would be incorrect, because the activity of a shower changes as a function of the zenith distance of the radiant. This dependence of the observed activity and the zenith distance is one that is relatively simply explained in geometrical terms. If F is the meteor flux measured in a plane at right angles to the incident flux; F' the corresponding flux in a horizontal plane; and Z the zenith distance of the radiant, we have:

$$F' = F \cos Z.$$

This equation may be understood as follows: when the radiant is below the horizon, no meteors are visible – meteors are arriving beneath our feet and the Earth prevents our seeing them in the sky. Then, as the shower rises above the horizon, the activity increases. This activity is at a maximum when the radiant culminates. Naturally, only rarely does a radiant pass through the zenith, so the value of the rates measured never reaches F.

An important operation in reducing observations consists of calculating the flux F at a specific time as a function of the observed flux F' and the zenith distance of the radiant at that time. In practice, it may be shown that the equation just given is simplistic. Effects such as the curvature of the Earth and the change in the density gradient along the meteors' paths as a function of the zenith distance also have to be taken into account. The last parameter modifies the intrinsic luminosity of the meteors and therefore their visibility from the ground. Numerous authors have tried to determine the best reduction formula, with greater or lesser degrees of success. The standard work nowadays is Zvolankova (1983). It is based on a study of 17 000 visually observed meteors belonging to the Perseid shower. The reduction formula determined experimentally is:

$$F' = F \cos^{1.47} Z.$$

In practice, the zenithal hourly rate (*ZHR*) is calculated. The *ZHR* is the number of meteors observed in an hour when the radiant is at the zenith and the limiting magnitude is 6.5. The full equation for calculating the *ZHR* is:

$$ZHR = N[K C_{L_m}/T_{\text{eff}}] \times [1/(1-C)],$$

where N is the number of meteors belonging to the shower that are observed (i.e., excluding sporadic meteors); C is the average fraction of the sky that is unavailable during the observation period (because of natural obstacles, clouds, etc.); K is a correction factor that takes account of the effect of the zenith distance (Z) of the radiant. It is given by: $K = 1/\cos^{1.47} Z$.

The zenith distance is calculated from:

$$\cos Z = \cos \phi \cos \delta + \sin \phi \sin \delta \cos(\theta - \alpha),$$

where

$\phi =$ geographical latitude,
$\delta =$ declination of the radiant,
$\theta =$ local sidereal time in degrees,
$\alpha =$ right ascension of the radiant in degrees.

C_{L_m} is a correction factor which depends on the limiting visual magnitude L_m:

$$C_{L_m} = r^{(6.5-L_m)}.$$

The parameter r is a function of the average magnitude of the meteors belonging to the shower. It will be explained later (p. 660).

T_{eff} is the effective time of observation in hours:

$$T_{\text{eff}} = \text{Time of end of observation} - \text{Time of beginning} - tN_t,$$

where t is the time taken to draw the path on the chart and note down the details of the meteor, and N_t is the total number of meteors observed (sporadics and shower meteors). The effective duration of an observing session should exceed an hour whenever possible.

The error in the *ZHR* is calculated from:

$$\text{Error} = ZHR/\sqrt{N},$$

and the value will be given as $ZHR \pm \text{Error}$.

The activity of sporadic meteors is given by the hourly rate (*HR*), defined by:

$$HR = N_s[C_{L_m}/T_{\text{eff}}] \times [1/(1-C)],$$

where N_s is the number of sporadic meteors observed.

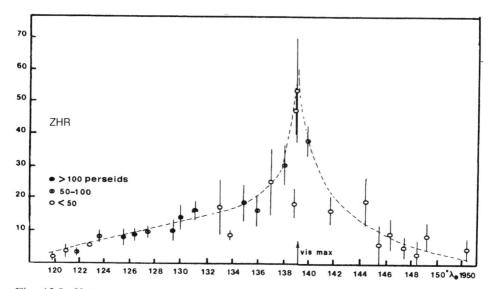

Fig. 12.8. *Variation in activity of the Perseid shower in 1984. Note that the curve is asymmetrical: this is normal for many showers. The time scale is given in solar longitude $\lambda_\odot$ (Dutch Meteor Society).*

The evolution of a shower's activity may be followed by calculating the *ZHR* every hour and then plotting the results on a graph. Figure 12.8 shows the typical changes in the Perseid shower, whose activity extends over several days.

The *ZHR* and the sporadic meteors' *HR* are closely linked with the limiting magnitude. It is therefore extremely important to record these details at the time of observation. Working under an average sky, for example, with a limiting visual magnitude of 6, one can expect to see between 6 and 8 sporadic meteors per hour. If the sky is very clear (limiting magnitude 6.5), this rate may reach 10 meteors an hour. Under an exceptionally fine sky, some particularly experienced observers may see as many as 14 meteors per hour.

An abnormally high number of sporadic meteors for given conditions is a sign that one or more showers are active. Careful study of the tracks recorded on the charts usually enables the radiants to be 'discovered'. This is why it is of value to observe meteors outside the times when major showers are active.

An example of the calculation of *ZHR* and *HR*:

- During observation of a shower, over a period of 1^h30^m, 18 meteors belonging to the shower and 9 sporadics were recorded. The time taken for recording was 10 seconds. The average zenith distance of the radiant during the session was 52°. During the first hour of observation, a natural obstacle blocked observation of 10 % of the sky, then during the last half-hour, some clouds appeared, and the percentage of the sky hidden increased to 30 %. Finally, the limiting magnitude was 5.8. (For the shower under consideration, r is taken to be 2.6, and for the sporadics $r = 3.42$.)

- Total number of meteors observed: $N_t = 27$
- Effective duration: $T_{eff} = 90 - 27 \times 0.6 = 73.8$ min $= 1.23^h$
- Average portion of the sky covered: $C = (60 \times 0.1 + 30 \times 0.3)/90 = 0.17$
- Limiting magnitudes: for the shower, $C_{L_m} = 2.6^{(6.5-5.8)} = 1.95$ for the sporadics, $C_{L_m} = 3.42^{(6.5-5.8)} = 2.36$
- Zenith distance: $K = 1/cos^{1.47} 52° = 2.04$.

The *ZHR* is then:

$$ZHR = 18[(2.04 \times 1.95)/1.23] \times [1/(1 - 0.17)] = 70,$$

and the hourly rate for sporadics:

$$HR = 9(2.36/1.23) \times [1/(1 - 0.17)] = 21.$$

12.3.4.3 The magnitude distribution
The visual magnitude m of a meteor mainly depends on three factors:

- the intrinsic magnitude;
- the distance of the observer, which is a function of the meteor's zenith distance (Z), as shown in Fig. 12.9. If we assume that a meteor becomes visible at around a height of 100 km, the distance between the meteor and the observer is:

$$d(\text{in km}) = 100/\cos Z$$

- the variation in atmospheric transparency, itself a function of the radiant's zenith distance. The decrease in magnitude relative to a meteor observed at the zenith may be written as:

$$dm = k[(1/\cos Z) - 1]$$

where k is a factor that depends on the horizontal visibility (Vi) expressed in kilometres:

$$k = 0.236 + 2.9/Vi.$$

The horizontal visibility rarely exceeds 100 km. A good average value is 25 km.

To compare the magnitude of one meteor with that of another, the magnitudes are converted to the value that they would have had if the meteors had appeared at the zenith, that is, at a distance of 100 km from the observer. For this we calculate the zenith magnitude or absolute magnitude (m_a) of the meteor. If m is the observed magnitude, and Z the average zenith distance of the meteor, we find:

$$m_a = m + 5\log(\cos Z) - dm.$$

To take an example: calculating the zenithal magnitude of a meteor of apparent magnitude -1 observed at a zenith distance of $30°$ when the visibility is 30 km, we have:

$$m_z(= m_a) = -1 + 5\log(\cos 30°) - 0.05 = -1.4.$$

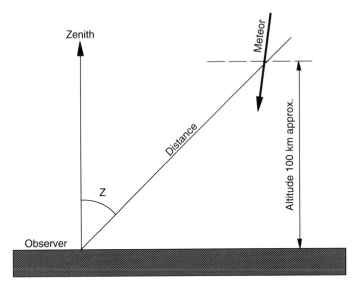

Fig. 12.9. *Distance of a meteor as a function of its zenith distance Z.*

It is not always very easy to calculate systematically the absolute magnitude of all the meteors observed. In addition, the reduction method just described does not take account of the observer's personal equation, which has a considerable effect on the apparent magnitude. For reducing a large amount of data obtained by several observers, another method should be adopted.

Let $m_{\rm sp}$ be the average magnitude of observed sporadic meteors, and $m_{\rm sh}$ the average magnitude of the meteors belonging to a particular shower – note that a different average magnitude is calculated for each shower – that are reported by a particular observer under given atmospheric transparency. We want to calculate the magnitude difference $\Delta m = m_{\rm sh} - m_{\rm sp}$.

We may consider this value as being independent of both the observer's personal equation and atmospheric conditions, because $m_{\rm sh}$ and $m_{\rm sp}$ are affected similarly. The magnitude difference is systematically negative for most showers. This indicates that a greater proportion of the population of meteor streams are bright meteors than for sporadics.

The average magnitude of sporadic meteors has a remarkably stable, and well-established constant value: $m_{\rm sp} = 3.25 \pm 0.03$. So we can deduce that the average magnitude of the meteors in the showers is:

$$m_{\rm sh} = \Delta m + 3.25.$$

In 1966, M. Kresakova established a relationship between the average magnitude of a population of meteors and the ratio (r) between the number of meteors in two adjacent magnitude classes of meteors (Kresakova, 1966). If N_m is the number of meteors of magnitude class m, and N_{m+1} the number of magnitude class $m+1$, then we have:

$$r = N_{m+1}/N_m.$$

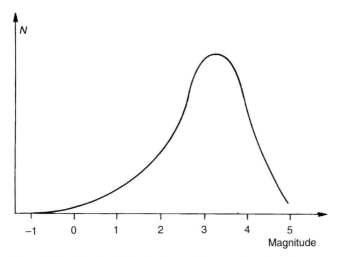

Fig. 12.10. *Typical distribution of the number N of meteors observed versus magnitude.*

The quantity r is constant for a given population – at least for all magnitudes visible to the naked eye. The distribution of meteors as a function of magnitude is a geometrical progression with argument r. It is important to realize that N_m represents the real number of meteors of magnitude class m. The number of meteors increases as their magnitude decreases. For example, if one observes one meteor (r^0) of magnitude 0 then one should, in theory, observe r^5 meteors of magnitude 5. In practice, visual sensitivity decreases as meteors become fainter. In addition, the distribution of meteors observed has a bell-shaped distribution with a maximum around magnitude 3, and with a relatively low number of meteors of magnitude 5 (Fig. 12.10). It will be seen that theory does not fully agree with reality. Individual observers' personal equations also have an effect (Fig. 12.11).

Table 12.4 shows the relationship between the ratio r and the average magnitude. For sporadic meteors, a value of $r = 3.42$ is adopted (average magnitude 3.25).

If we know the probability (P_m) of seeing a meteor of a given magnitude, it is possible to determine the parameter r in a different manner. If dN'_m is the number of meteors of magnitude class m, the real number of meteors (dN_m) is:

$$dN_m = dN'_m / P_m.$$

The distribution is then given by:

$$dN_m = dN_0 r^m,$$

where dN_0 is the real number of meteors of magnitude 0. The equation may be written in logarithmic form as:

$$\log dN_m = (\log r) \times m + \log dN_0.$$

This is the equation of a straight line of the form $y = ax + b$ with $y = \log dN_m$ and

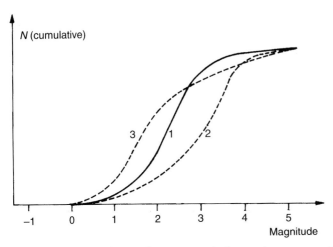

Fig. 12.11. *Cumulative frequency of observed meteors. If the curve of observer 1 is taken as the standard, it will be seen that observer 2 always under-estimates the brightness of the meteors. Observer 3, on the other hand, over-estimates bright meteors and under-estimates faint ones. If this type of graph is plotted for the results from a group of observers, it enables discrepancies to be seen, and appropriate corrections to be made.*

Table 12.4. *Variation in r as a function of mean magnitude*

r	Mean magnitude
2.50	2.66
2.80	2.91
3.10	3.09
3.40	3.24
3.70	3.37
4.00	3.47

$x = m$. This may be drawn from experimental results and the coefficients (a, b) may then be determined by linear regression. Finally:

$r = 10^a$ and $dN_0 = 10^b$.

The probability of seeing a meteor of magnitude m was established by Kresakova in the paper already mentioned, and is given in Table 12.5. It will be seen that a meteor of magnitude -6 will not escape an observer's notice, whatever its position in the sky, because of its brilliance. On the other hand, the probability of seeing a meteor of magnitude 6 is very low. Naturally the probability varies according

Table 12.5. *Probability (P_m) of observing a meteor of magnitude m*

m	P_m
-7	1.00
-6	1.00
-5	0.98
-4	0.95
-3	0.87
-2	0.73
-1	0.57
0	0.48
1	0.420
2	0.343
3	0.232
4	0.064
5	0.008
6	0.00007

to the observer's experience and degree of concentration. The values given may be considered as normal.

This method of calculating r is not valid unless a large number of magnitude estimates are available. Frequently it is necessary to compare the results from several observers to obtain the necessary accuracy.

Assuming that the brightness of the meteor is proportional to the mass of the meteoroid, a relationship may be deduced between the magnitude m and a quantity that is known as the differential mass index s:

$$\log_{10} r = 0.4(s - 1).$$

If dN is the number of meteors with a mass between μ and $d\mu$, then the differential mass index is defined by:

$$dN = C_1 \mu^{-s} d\mu,$$

where C_1 is a constant.

For any one population of meteors it is possible to calculate a cumulative number (N) of meteors up to mass μ by integrating the above equation:

$$N = C_2 \mu^{(s-1)}.$$

The quantity $(s - 1)$ is known as the integrated mass index.

The mass index may be interpreted as follows: the relative number of small particles decreases as the index decreases. In such a case most of the mass in a particular meteoroid population is to be found in massive particles.

The ratio r, and thus the mass indices, are different for meteors belonging to a shower and sporadic meteors. Similarly, it has been established that:

- the value of r is not identical for all showers;
- the value of r may change, depending on the position of the Earth in a stream. (This effect is, for example, clearly shown for the Geminids.);
- there may be variations from year to year for any given shower (enables the mass distribution around the stream to be determined);
- in some showers, r varies as a function of magnitude class (r decreases slightly towards fainter magnitudes).

It will be seen that determination of just a single average magnitude is a particularly powerful way of studying the mass-distribution of meteoroids within a stream.

12.4 Group observing

One individual observer can carry out useful work, but as we shall see, by observing as a group it is possible to determine additional information. The field of vision of a single observer only covers a small part of the celestial sphere and the number of meteors seen is frequently very low (typically 8 per hour when no major shower is active). On the other hand, if several observers are watching under the same conditions, the overall hourly rate increases in proportion to the number of observers. That increases the statistical accuracy, because one then has a larger sample available. In addition, the fact of working in a group acts as a stimulus, which an individual observer does not necessarily experience, and which often results in greater efficiency. When a colleague sees a magnificent fireball behind your back and tells you about it, experience shows that you tend to pay even more attention to the sky! Psychology does play a certain part in the observation of meteors. On the other hand, it should be taken seriously – misplaced chatter can definitely affect concentration. The group leader should ensure that everyone understands that scientific work is being carried out.

The members of the group may spread themselves over several observing sites, with the aim of carrying out triangulations. The aim is to calculate the altitude of the meteors. The triangulation technique will be described later when photography is discussed (*see* p. 680). Here we may mention that determining altitudes from visual observations is problematic because of the relatively low accuracy with which tracks may be drawn on charts. For visual observations, triangulation should not be attempted unless the observers are at least 50 km apart.

When working as a group at a single site, various observers are assigned different areas of the sky. For example, four observers looking towards the four cardinal points of the compass can achieve almost complete coverage of the sky.

The observers should note down the meteors that they see, irrespective of the observations made by the other members of the group. Sometimes meteors are seen by more than one of the observers, at the edges of the fields of vision. Such meteors are also noted down by each observer, because the report sheets should be analyzed separately. During the reduction, examination of all the tracks recorded by the observers usually allows the activity of a shower to be determined or even permits the existence of a previously unknown radiant to be established.

When observers cover the same area of sky, it is possible to use the double-counting technique first tried by Öpik in 1922. We have already described the relationship that exists between the observable limiting magnitude for a meteor and the angle of view. This relationship is, in fact, variable between one observer and another, and this frequently explains why some observers see more meteors than others. The main reason for the difference is fatigue, which decreases one's attention, but also the physiology of the eye. Öpik introduced a coefficient of perception (P) unique to each observer. Let us assume that two observers are looking in exactly the same direction for a sufficiently long period of time, say 1 hour. Let N_a and N_b be the number of meteors seen by each observer, and N_{ab} the number seen by both. Öpik defined the perception coefficient as:

$$N_{ab} = P_a N_b = P_b N_a.$$

Using the perception coefficient, we can deduce the true probable number (N_p) of meteors that would be seen by an observer with a perception coefficient of unity:

$$N_p = N_a/P_a = N_b/P_b = N_a N_b/N_{ab}.$$

If more than two observers are involved in the double-counting method, the perception coefficients are calculated by an equation of the type:

$$P_a = (N_{ab} + N_{ac} + N_{ad} + \ldots)/(N_b + N_c + N_d + \ldots).$$

The probable number of meteors is then:

$$N_p = S/[1 - (1 - P_a)(1 - P_b)(1 - P_c)\ldots],$$

where S is the number of different meteors seen by the whole group of observers.

It is always very instructive to carry out double counting, because then one realizes that no one notes all the meteors that pass through their field of view. If, at the same time, the tracks are drawn on the charts and the details are noted down, later comparison of the results enables the errors to be detected. It is a good way of improving one's techniques.

12.5 Observation of telescopic meteors

Telescopic meteors are meteors that are invisible to the naked eye, and therefore require the use of optical equipment to be observed. They are thus meteors fainter than about magnitude 6. These naturally include sporadic meteors, but there are also shower meteors among their number. There are even some meteor streams that consist of telescopic meteors only.

Observation requires wide-field instruments, with magnifications that give the maximum exit-pupil size. This means binoculars or very wide-aperture telescopes with wide-field eyepieces (Erfle-type eyepieces, for example). For a given instrument of aperture D (in cm) and magnification G, the magnitude of the faintest meteors that may be seen is given by:

$$M_{lim} = 5 \log D - 0.036G + 5.52.$$

This indicates that the lowest possible magnification should be used, provided the exit pupil does not exceed that of the eye. This consideration also led to the conclusion that even with the largest instruments, it would not be possible to see meteors fainter than magnitude 11.5. [In fact, this is not borne out by experience, because image-contrast is a factor. Fainter meteors are often visible with slightly higher magnifications (and thus smaller exit pupils). The limiting magnitude is, in any case, strongly observer-dependent. In addition, other considerations, especially plotting accuracy, suggest that the apparent field should be between 50° and 60° – Trans.]

The instrument needs to have a rigid mounting, the most important point being that the observer must be comfortable. If possible, the instrument should be binocular, so that both eyes are working under the same conditions. A pair of 80 × 11 binoculars is a good choice.

After 30 minutes observation, fatigue is generally significant and it is advisable to rest for at least 10 minutes before resuming the watch. Telescopic observation of meteors is not easy and requires more attention than naked-eye observations. It is certainly not the sort of work for beginners who want to take up observation of meteors. On the contrary, it needs an experienced observer who has seen more than 1000 meteors to be able to make useful telescopic observations.

In undertaking a routine watch for sporadic meteor activity, any azimuth may be chosen, but as for naked-eye work, the altitude should be between 30° and 50°. On the other hand, if one is studying a stream with the aim of establishing the radiant position, watches should be carried out for periods of 30 minutes in two different directions at 90° to the presumed radiant, but still keeping to an altitude of around 45°. If two observers are working simultaneously, each should take one of these two areas.

The principal stars in the telescopic field should be drawn on the report sheet. This field should be identified in a celestial atlas to determine the magnitude of some of the stars, so that the magnitude of the meteors may be estimated. The field covered should be the same throughout the watch; so an equatorial mounting is therefore strongly recommended. If one observes for several consecutive nights (which is practically essential in any serious work), the same field should be chosen to avoid the process of identifying new stars. When a meteor is seen, the track should be drawn on the chart that one has prepared, as well as the same details as those recorded for naked-eye meteors (time in UT, magnitude, speed, duration of any train, colours, etc.). [Note that the main advantage of telescopic work is the plotting accuracy that may be attained, rather than the detection of low-luminosity meteors, the determination of rates, etc., important though such factors are. Accurate, telescopic determination of radiants should enable many minor showers (which cannot be adequately studied by naked-eye methods) to be investigated. – Trans.]

The number of meteors visible telescopically is very close to the figure found using the naked eye, provided one uses the appropriate magnification to give the maximum exit pupil.

The equipment used for some types of observation, such as variable-star work, is also ideal for the observation of telescopic meteors. So it is a good idea for variable-star observers to note down all the meteors that they see passing across the field. [A further advantage is the availability of comparison stars with known magnitudes. – Trans.] Comparison with observations made by others may provide interesting information about meteor activity.

12.6 Meteor photography

Compared with visual work, photography provides additional accuracy that allows accurate photometry of meteors, determination of their velocities and, as a result, calculation of their orbits. On the other hand, a camera can only record the brightest meteors, so statistical studies, and work aimed at establishing radiants largely remain within the domain of visual observation.

12.6.1 Choice of photographic equipment

The basic method of photographing meteors consists of pointing a camera towards the sky, leaving the shutter open for several minutes, and hoping that during that period of time a sufficiently bright meteor will leave a record on the emulsion. The main features of a camera suitable for photographing meteors are:

- a lens with the widest-possible aperture that will capture faint meteors, which are also the most numerous;
- a fairly short focal length, so that a wide area of the sky may be covered, thus increasing the chance of capturing a meteor;
- a B setting, enabling the shutter to be kept open for as long as required.

A sophisticated camera is not necessary for meteor photography; in fact, a degree of simplicity is preferable. It is essential to avoid cameras that have electrical shutter operation, because such shutters require a continuous current to keep them open, which may cause batteries to be drained in a very short time (perhaps just 3–5 minutes). The price of batteries precludes such extravagance. It is, of course, possible to tinker with the camera and arrange for it to be provided with a fully independent, stabilized DC supply.

Cameras with non-interchangeable lenses may be used. Unfortunately, the lenses are not normally very fast, rarely exceeding f/2.8, which, as we shall see, is a handicap. Ideally, a good, old-fashioned, manual reflex camera should be used. Unfortunately, such bodies are rare nowadays.

To choose the most suitable lens, we need to look at some of the criteria. The efficiency (E) of a lens is:

- proportional to the collecting area of the lens, i.e., to the square of the diameter D of the entrance pupil;
- inversely proportional to the focal length F, because the greater the latter, the greater the linear velocity of the meteor's image at the film-plane, and the less the emulsion is affected;
- proportional to the angular area of the sky that is covered, which (to a first approximation and for a given, constant, film format), is inversely proportional to the square of the focal length.

Table 12.6. *Theoretical efficiency of lenses*

Focal length (mm)	Efficiency (E)				
	$F/D = 1.2$	1.4	1.8	2.0	2.8
17					1.21
24		3.42	2.08	1.68	0.85
28		2.93	1.77	1.43	0.72
35		2.35	1.42	1.14	0.58
50	2.24	1.64	1.00	0.81	0.41
55	2.03	1.48	0.90	0.73	0.37
85	1.32	0.97	0.58	0.47	0.24

Efficiencies are given relative to a 50-mm, f/1.8 lens as standard

This gives us:

$$E = D^2/F \times 1/F^2 = D^2/F^3 = F/\text{numerical aperture}.$$

In this equation the D^2/F term reflects the speed of the lens, and the $1/F^2$ term the field covered. Table 12.6 shows the result for several current lenses, taking a 50-mm, f/1.8 lens as a standard. This table shows that telephoto lenses are to be shunned because their small fields are a great handicap. Medium focal lengths enable significant results to be obtained because fast lenses are available. Short focal lengths (fish-eye-type lenses) compensate for their slower speeds with wide fields.

The efficiency criterion that we have just described does not take account of the distribution of meteors as a function of magnitude. A lens with a low D^2/F ratio will only allow the brightest, and therefore rare, meteors to be captured. If we want to refine our model, we need to take that distribution into account. We know that for sporadic meteors we may write:

$$R = 3.42^{(m-m')},$$

where R is the ratio between the magnitude classes m and m'.

If we compare the speed of a 50-mm, f/1.8 lens and a 35-mm, f/2.8 for meteor photography, we have:

for the 50-mm: $D^2/F = 15.43$,
for the 35-mm: $D^2/F = 4.46$.

So the 50-mm lens will record meteors $15.43/4.46 = 3.46$ times fainter than the 35-mm. This ratio may be turned into a magnitude difference (dm) by using Pogson's equation, giving:

$$\mathrm{d}m = 2.5 \log 3.46 = 1.35.$$

For identical photographic fields, the ratio of meteors recorded, in the 50-mm lens' favour will be:

$$R = 3.42^{1.35} = 5.26.$$

Table 12.7. *Theoretical efficiency of lenses for meteor work*

Focal length (mm)	Efficiency (E)				
	F/D = 1.2	1.4	1.8	2.0	2.8
17					0.63
24		3.19	1.63	1.23	0.50
28		2.88	1.47	1.11	0.45
35		2.48	1.27	0.96	0.39
50	2.95	1.96	1.00	0.75	0.31
55	2.77	1.84	0.94	0.71	0.29
85	2.07	1.37	0.70	0.53	0.21

Efficiencies are given relative to a 50-mm, f/1.8 lens as standard, and taking the distribution of meteor magnitudes into account

Naturally, the fact that the 35-mm has a larger field than the 50-mm has to be taken into account. The final efficiency of the 50-mm lens relative to the 35-mm lens is:

$$E = 5.26[F^2(35\text{-mm})/F^2(50\text{-mm})] = 5.26(1225/2500) = 2.6.$$

More generally, if a 50-mm, f/1.8 lens is taken as standard, the relative efficiency E_R of a lens with a focal length F and a diameter of entry pupil D will be:

$$E_R = 1/[(F^2/2500)3.42^{2.5\log(15.43F/D^2)}].$$

Table 12.7 enables lenses to be compared, because it takes account of the distribution of meteor magnitudes.

Extreme wide-angle lenses lose some efficiency. In practice, as indicated in Table 12.7, a 55-mm, f/1.2 lens will detect a fairly considerable number of meteors. Such a lens suffers from three defects, however:

- Because of the wide aperture, the aberrations at the edge of the field remain significant unless the optics are aspheric, and thus quite expensive. Ultimately, this means that there is a loss of efficiency because the images are enlarged.
- The average length of a meteor trail is about ten degrees. Because the field of the lens is only about 20°, many meteors will not be photographed in full.
- The speed of the lens means that only short exposures may be used if one wants to avoid the film being fogged through light-pollution. The consumption of film is therefore quite considerable.

Nevertheless, if a 55-mm f/1.2 lens is lying around unused, don't hesitate to start photographing meteors with it.

Extreme wide-angle lenses pose problems of a different sort: if the focal length is too short, the meteor trails will also be short, which means that measuring the photographs is not particularly easy. Experience shows that a good compromise between spatial resolution and light-grasp is obtained with a focal length of 35 mm. There are lenses of this focal length that have apertures of f/2, which are highly effective, but the price of such lenses is high.

'Fish-eye' lenses should not be discounted – on the contrary. Naturally, only the brightest meteors may be photographed (magnitudes exceeding −2), but such a lens covers practically the whole sky. The low light-grasp means that exposures of about 30 minutes may be made, which reduces the effort required. Carrying out a fireball survey on a national level is a very important programme, requiring such lenses, and we will discuss it later (Sect. 12.10). We will just mention here that there are supplementary lenses (anamorphic lenses) that may be mounted in front of ordinary lenses, such as 28-mm lenses, and which convert these into a form of fish-eye. Although this solution is not very good optically, because aberrations are significant, the cost is low, and such lenses may sometimes be of considerable use.

12.6.2 Mounting the camera

There are three cases:

- a driven equatorial mount is available;
- a photographic tripod may be used;
- no suitable mounting is available.

12.6.2.1 An equatorial mounting

This is the ideal solution, because then stars appear as points on the photograph, which enables the equatorial coordinates of meteors to be determined accurately when reducing the observations.

Adjustment of the equatorial mounting does not need to be of extreme accuracy in view of the focal lengths that are used for meteor photography. Nevertheless, the adjustment should not be skimped to such an extent that the stars no longer appear as points, because the advantages of an equatorial mounting are then lost. With a little practice, it is possible to set up the mounting with sufficient accuracy in less than 10 minutes. It is easier to set up mountings that have a polar alignment telescope. For others, you need to use a small auxiliary telescope that may then be removed so that several cameras, pointing in different directions, may be fixed to the mounting.

The mounting should obviously be fitted with an electrical drive, because it is out of the question to use a manual drive as the observing period may extend over several hours.

12.6.2.2 A photographic tripod

The camera is fixed, so the stars will cause trailed images on the film (Fig. 12.12). It is obviously impossible to determine the position of the meteor with respect to the stars without adopting a special technique. As soon as a meteor bright enough to

Fig. 12.12. *A Perseid photographed on 1978 August 10. Note the terminal explosions that are characteristic of members of this shower. Lens: 35-mm, f/2.8; film: Tri-X 400 ASA. (Photo C. Buil)*

be recorded has passed across the field of view, the exposure should be ended. The position of the stars at the time the meteor occurred with therefore correspond to the end of the trails on the photograph.

A problem arises when the observer failed to see the meteor. This typically happens when a single observer is using several cameras. Keeping watch on several camera fields at once is a difficult task. There are two remedies: either call on friends, persuading them of the fascination of observing meteors, or reduce the exposure times so that the star trails are kept very short. This second method will reduce the amount of error in determining the positions of meteors with respect to the stars.

12.6.2.3 No photographic mounting
Rush off to your favourite photographic shop and buy a tripod before it gets dark. Don't just rest the camera on the ground. If it is covered in grass, the camera is not necessarily going to be steady, and in any case there is a serious risk of the optics becoming covered in dew. Again, in the dark, it is very easy to tread on the camera ...

If after all this, you do not necessarily want to buy a tripod, then some form of support should be built. This need not be very elaborate, for example a strong post set in the earth with a platform on top will serve quite well.

12.6.3 Choice of film

For preference 400-ISO, black-and-white film should be used (such as Kodak Tri-X or Ilford HP5). Colour does not add much from a scientific point of view, and development is usually more difficult to master. Films faster than 400 ISO have

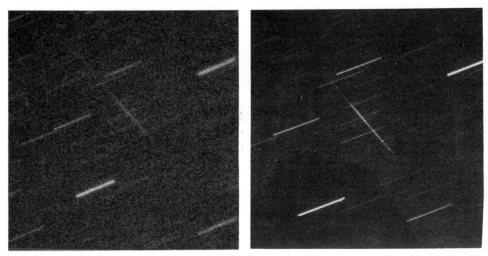

Fig. 12.13. *A Quadrantid meteor photographed simultaneously on Tri-X (left) and High Speed Infrared film, no filter (right), 1987 January 3. The infrared image is distinctly clearer than that obtained with a traditional film (photos P. Martinez).*

been found to be counter-productive, because they show a lack of contrast as a result of rapid fogging, and have a low resolution caused by their coarse grain. Theoretically, the new Kodak TMax-400 would seem to be better than Tri-X and HP5, but sufficient experience has not yet been gained with using it on meteors for us to comment here.

Certain trials with infrared film have shown interesting results: background fogging is diminished because the sky is very dark at infrared wavelengths and meteors appear at least as bright as with normal film, probably because of the presence of emission lines in that part of the spectrum (Fig. 12.13). Further experiments need to be carried out to investigate this possibility.

It is absolutely useless to try hypersensitizing the film to reduce reciprocity failure, because meteors are moving objects and the exposure time is approximately 1/100th second. It is even undesirable, because of the increase in background fogging.

The best efficiency/price ratio is obtained using 24×36 mm format. Cassettes are generally sold with 20, 24 or 36 exposures. Assuming that we begin a new exposure every 15 minutes, and that the observing session lasts 5 hours, a 20-exposure film would suffice, but if meteor activity is very high, it will be advisable to curtail many exposures, and a 36-exposure cassette would be appropriate.

Changing the film in a camera during the night is to be avoided. This operation may take a few minutes and, during that time, a fine fireball could quite well appear in the field of view. This tends to put the observer in a very bad temper! Moral: it is better to lose 4 or 5 frames on an unfinished film and change films during the day, than it is to miss the meteor of the century, by changing films during the night.

Larger-format $6 \, \text{cm} \times 6 \, \text{cm}$ cameras can be (and have been) used successfully, but given the cost of material, it is often difficult to cover the sky using a battery of cameras. [U.K. observers have had considerable success using batteries of the cheap, Russian 'Lubitel' cameras. – Trans.] However, 6 × 6 cameras are capable of giving highly detailed images of meteors, enabling complete light-curves to be obtained (showing, for example, small flares caused by repeated fracturing of a meteor or by its rotation during flight).

Whenever a film is changed, the date of the exposures and the camera used should be carefully and immediately noted down on the end of the film, particularly if several camera bodies are being used. A lot of care is required to ensure that films do not become inextricably confused; this is spoken from bitter experience!

12.6.4 Exposure times

The length of exposure that may be used is a function of the optics, the meteorological conditions, the presence of the Moon, etc. There are several factors involved and it is difficult to give any general rule. Typically, with a lens of f/2.8 and a clear dark sky, 15-minute exposures are possible. With a lens of f/1.8 exposures cannot exceed 10 minutes. Longer than that, fogging becomes bad enough to affect the identification of faint meteors.

We have already said that the exposure should be terminated when a bright meteor appears in the photographic field. This is essential when the camera is mounted on a simple photographic tripod, but is also advisable, for safety's sake, when an equatorial mounting is being used – in case a colleague carrying a handlamp wanders in front of the camera, and thus ruins the photograph of a lifetime!

But there is one important point: the meteor may appear only a few moments after the shutter has been opened. Under these circumstances, the exposure should not be halted immediately, because the stars would not have had enough time to affect the emulsion. Calmness is required: if using a tripod, continue the exposure for about 30 seconds; with an equatorial, about 2 minutes are required to obtain a good star field.

The minimum time should be taken to reset the camera, because a fine meteor may well appear at just that moment, which is most annoying. Generally, it is best not to look at the sky whilst resetting the camera – that way one has nothing to regret!

A special sheet should be used to record when the pictures are wound on. If using three cameras, for example, the record might appear as in Table 12.8. Here photography began at 22:36, the exposures were wound on at 22:50 and 23:02 for all three cameras, then at 23:12 a bright meteor appeared in the field of camera 2 and just that exposure was stopped. The number of the meteor in the visual record has been written in the remarks column, as well as its magnitude. The cameras were wound on at 23:20, 23:32, etc.

It is a good idea to make a very short exposure in the middle of the observing session on all the cameras, carefully noting down the time on the record. This short exposure is easy to locate among the other pictures and allows the photographic sequences to be recalibrated if an error has been made in noting down the details.

Table 12.8. *A typical record for three cameras operated simultaneously*

Time	Camera 1	Camera 2	Camera 3	Remarks
22:36	–	–	–	Start
22:50	–	–	–	
23:02	–	–	–	
		23:12		No.39, $M = -1$
23:20	–	–	–	
23:32	–	–	–	

Table 12.9. *Limiting magnitude photographically obtainable*

Optics	Limiting magnitude
55-mm, f/1.2	2 to 1
50-mm, f/1.8	1 to 0
35-mm, f/2.8	0 to −1
28-mm, f/2.8	−1 to −2

12.6.5 Photographic limiting magnitudes

It is very important to estimate the magnitude of a meteor that passes through a photographic field to judge whether it is likely to affect the emulsion, and whether the exposure should thus be terminated. Table 12.9 gives the limiting magnitudes obtained with various lenses and 400-ISO film.

The values given in Table 12.9 are approximate, because the limiting magnitude largely depends on the meteor's apparent velocity. But there is a ratio of 30 between the fastest and slowest velocities. So it is by no means impossible to photograph a very slow meteor of magnitude 1 with a 35-mm lens. One should also beware of meteors that show brief flares in brightness, which are difficult to see visually but are recorded by the film.

For the record, it may be noted that shortly after World War II, Harvard Observatory used cameras that were specially built for meteor work. They were called 'Super-Schmidt' cameras because their optical design was similar to that of Schmidt telescopes. They had a relative effective aperture of 0.85, a focal length of 203 mm, and a circular field 55° across. They were capable of recording meteors as faint as magnitude 4! There are no such cameras in use today. Could a skilled amateur optician perhaps take up the challenge and produce equivalent cameras?

12.6.6 The likely chance of photographing a meteor

Experience shows that a 35-mm, f/2.8 lens is able to photograph one meteor every 5 hours when no major stream is active. Naturally, if two cameras are used, the probability is doubled, and so on. When people start photographing meteors in a serious way, they typically have a battery of five or six cameras set up at the same time.

When a stream is active, the chances of success increase considerably. Around Perseid maximum, for example, a meteor may be photographed every quarter of an hour with just a single camera, and it is not uncommon for several meteors to be present on a single frame.

12.6.7 Where should the cameras be pointed?

Although it is quite impossible to know where a meteor may appear, you cannot just point the camera in any direction you choose. We have already discussed the effects of distance when observing near the horizon: meteors are then farther away and they appear fainter. But, as Fig. 12.14 shows, when the camera is pointed fairly close to the horizon, the area of sky covered at a height of 100 km is considerably greater than it is when the camera is pointed towards the zenith, and the number of meteors visible will be greater in the former case than in the latter. Two opposing effects therefore influence the number of meteors that may be photographed: the distance effect, which causes meteors to appear fainter when looking near the horizon; and the area effect which increases the lower one points the camera. Once again it is a question of a compromise, so the camera should be pointed at an altitude of 45°. Experience shows that the finest meteors appear at that altitude.

If a shower is active, do not point a camera directly towards the radiant, because any trails recorded will be too short to be measured accurately. If several cameras are being used then one can make an exception to this rule, because one may be lucky enough to capture a cluster of meteors coming from the radiant, any such picture being both of aesthetic value and instructive.

Contrary to a widely accepted idea, the camera should not be pointed towards the zenith, except when a fish-eye lens is being used, which must be directed towards the zenith if half of the field is not to be occupied by the ground.

12.6.8 Some practical advice

Photographing meteors is not difficult, but there are a few elementary precautions that should be taken to ensure that a session is successful. It is a good idea to check at the beginning of the session that the cameras are correctly set; in other words:

- that the focus ring is set to infinity;
- that the diaphragm is set to the maximum aperture;
- that the shutter is set to B.

These may appear trivial points, but experience shows that such checks are by no means superfluous, especially when one has to cope with a whole battery of cameras.

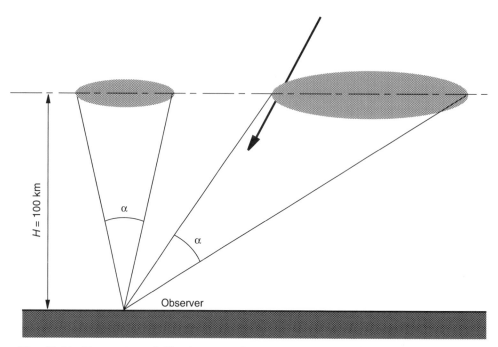

Fig. 12.14. *For a given field of view, a larger area is covered closer to the horizon than towards the zenith.*

It is also essential to check that the film is passing through the camera. This may be done when loading the film by checking that the feed spool rotates as the film is wound on.

Dew and frost on the optics are among the worst enemies of astronomers, and particularly of meteor observers. It is important to check regularly that neither of these afflictions has affected the optics. If trouble has arisen, the lens must be wiped or heated gently with a hair-dryer, but the situation is usually critical and operation generally has to be restarted from scratch. A good method of defeating condensation and frost is to surround the end of the lens with a small resistance heater. This technique is regularly used by meteor observers who have encountered this problem a few times.

The observational site should be laid out sensibly. In particular, it is absolutely essential to avoid putting cameras alongside visual observers, because the latter have the habit of illuminating their surroundings with lights, even if they are dim ones. Finally, it is, of course, essential to have material to hand on which to note down the exposures; one or two spare cable releases; and . . . a few biscuits for the middle of the night.

Fig. 12.15. *Two meteors photographed using a 6 × 6 camera, with an 80-mm, f/2.8 lens, Tri-X film. (Photo C. Buil)*

12.7 Determining photographic magnitudes

Photography allows remarkably accurate study of the variations in luminosity of meteors. Even without using a microdensitometer, it is possible to recognize certain specific characteristics. For example, the Perseids often end in one or two terminal bursts, whereas other meteors such as the Geminids show just a smooth variation along their tracks. The range of light-curves is very wide, which is an indication of significant structural differences in the particles encountering the Earth. A meteor that shows several flares, for example, is a sign that the meteoroid is fragile, fragmenting easily when it hits the atmosphere.

It is possible to relate the brightness of recorded tracks to stellar magnitudes by means of an exposure made without a sidereal drive. To do this one has to find a star trail that has the same width as the part of the meteor track that is being investigated. Let M_s be the magnitude of the star. It may be shown that the magnitude of the meteor (M_m) is given by:

$$M_m = M_s - 2.5p \log(V_m/V_s),$$

where V_m and V_s are the angular velocities of the meteor and of the star respectively, and where p is a coefficient that takes the film's reciprocity failure into account.

V_m is calculated as follows: let t be the time taken for the meteor to go from one coordinate point (A_1, D_1) to another point (A_2, D_2), these points being determined from the track. The angular distance covered during time t is ω, such that:

$$\cos\omega = \sin D_1 \sin D_2 + \cos D_1 \cos D_2 \cos(A_1 - A_2).$$

From this we can deduce the meteor's angular velocity:

$$V_m = \omega/t.$$

The angular velocity of the star depends on its declination:

$$V_s(\text{deg/s}) = \cos D/240.$$

Typically, $V_m/V_s = 2000$. The reciprocity failure of the film may be quite considerable; typically a value of $p = 1.06$ is used.

12.8 Measurement of velocities

The velocity of a meteor is an essential factor, which, together with the position of the radiant, enables its orbit in the Solar System to be calculated. Visual estimates are far too subjective to give suitable results. On the other hand, as we shall see, it is very simple to measure velocities using photographic techniques. The astrophysical importance of this piece of information is such that an attempt should be made to measure the velocity of any meteor that is captured on film.

The basic principle behind measuring velocities is the use of a rotating sector, which occults the camera lens at regular intervals (Fig. 12.16). The sector is driven by an electric motor, with a precisely known, constant speed. Any meteor photographed with such an arrangement will show a broken trail.

To measure a velocity accurately, the meteor's photographic trail should show more than five interruptions. A meteor's typical duration is half a second, so the lens should be occulted at least ten times a second. But it is important not to go to the other extreme: if the frequency of the interruptions is too high, the trail will consist of closely spaced dots, which will not be resolved by the lens, and will therefore make it impossible to determine any velocity. This is particularly true if short-focal length lenses are being used. As a general rule, one should not go beyond 50 interruptions per second, the ideal being between 25 and 30 per second for most modern lenses. With a very wide-angle lens, the frequency may be approximately 10 interruptions per second (Fig. 12.17).

This all serves to determine the speed of the motor. Naturally, the number of sectors in the rotating shutter may be changed, but there is one limitation. Care must be taken to ensure that the lens is completely unobstructed when one of the cut-outs is in front of it (thus avoiding vignetting); and also that it is completely covered by the blades, giving a sharp, distinct break in the trail. The problems are reduced by placing the shutter as close to the lens as possible, so that the latter is almost brushed by the blades. Four sectors is a good choice. The opaque portions may differ in size from the open sectors. If so, it is easier to study the light-curve when the visible portions of the trail are larger than the breaks.

The shutter is made of thin, sheet metal, painted matt black to prevent stray

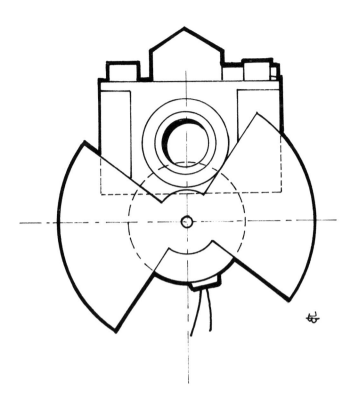

Fig. 12.16. *How a rotating sector is arranged in front of a camera.*

reflections. Cardboard should be avoided as it warps with dampness at night. Because of the speed of the shutter, it is essential that it should be completely balanced to avoid any vibration. To cover a single lens a shutter might be about 15 cm in diameter. It is possible to use one shutter in front of several lenses. The diameter should be adjusted to suit, and the power of the motor should be sufficient to prevent it stalling.

The motor must be of the synchronous type, because then the speed of rotation is directly proportional to the frequency of the mains power supply (50 or 60 Hz), which is very stable. Direct-current motors should not be used, because they require sophisticated electronics to ensure a constant, known speed that is accurate to less than one per cent. Similarly, the power should be taken from the mains, and not through a variable-frequency controller, unless the latter has an accurate quartz oscillator. (It would be possible to check the rotation speed with a stroboscope.)

Because the lens is regularly obscured by the rotating shutter, it is possible to increase exposure times by a factor of two without having to worry about an increase in sky background fogging.

Fig. 12.17. *An example of a meteor trail broken by a rotating shutter. The breaks here occur 48 times a second. This is a sporadic meteor, photographed on 1980 August 12 by C. Buil, using a 35-mm, f/2.8 lens.*

12.9 Triangulation

The velocities measured with a rotating shutter are expressed in degrees per second. But the important information is the linear speed of the meteor in the atmosphere in km/s. This velocity cannot be determined from the apparent motion without a knowledge of the distance covered between two points on the trail in a precise interval. This may be derived by employing intermediate steps: determining the distance between the observer and the meteor, and the latter's altitude.

A single observer can only make certain assumptions about these values. For example, we have seen that factors such as magnitude, the angle of incidence, and the atmospheric velocity (none of which are known initially, and must therefore be derived by iteration), may give an idea of the altitude at the beginning and end of the trail. If we assume that a specific point on the trail is at altitude h and zenith distance Z, then the distance d to the observer is:

$$d = h/\cos Z.$$

Let h_1 and h_2 be the altitudes of two points that the meteor passed in interval t. If Z_r is the zenith distance of the meteor's radiant, Fig. 12.9 shows that the distance (l) between these two points is:

$$l = (h_1 - h_2)/\cos Z_r.$$

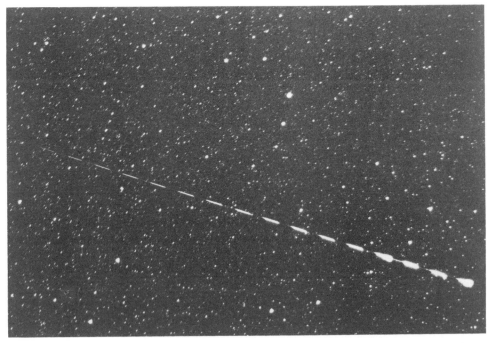

Fig. 12.18. *A bright Perseid observed on 1980 August 6; the maximum was on August 12. This shows how observations away from the maximum may be well worthwhile. (Photo C. Buil, 35-mm, f/2.8 lens, interruption every 0.08 second.)*

To determine the atmospheric velocity accurately, we have to calculate the height of at least two points on the track. The only means of doing this is by triangulation. The principles of triangulation are:

- the same area of sky is photographed from two different sites (or stations), some tens of kilometres apart;
- when a specific meteor is photographed from both stations, a comparison of the results will show that the tracks are not in the same position relative to the stars. This is an effect of parallax;
- measuring the parallax of a point on the track and knowing the precise geographical positions of the stations enables the altitude of the point to be determined.

12.9.1 Some geometry

Let S_1 and S_2 be the two observing stations (Fig. 12.19). The line S_1–S_2 is the baseline for the triangulation. Let M be a point on the meteor's trail. The parallax of the point M is the angle π. This is the angle subtended by the baseline at M. Determining the elements of the triangle S_1–M–S_2 enables the height of the point M to be determined (whence the name triangulation). Drop a perpendicular from M

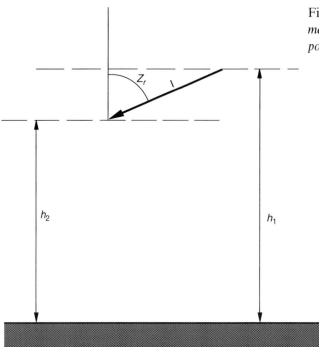

Fig. 12.19. *The distance a meteor covers between two points.*

onto the plane that includes both stations and is tangent to the surface of the Earth. The distance MN is approximately the height of the point M. (To be completely rigorous, the curvature of the Earth should be taken into account, and the altitude should be corrected to sea level.)

A_1 and A_2 are the azimuths of M as seen from the stations and measured with respect to the baseline (the angles are measured in the triangle S_1–N–S_2). The angles H_1 and H_2 are the angular heights of M from S_1 and S_2. These elements are calculated by converting the equatorial coordinates of point M into horizontal coordinates for each of the pictures (*see* p. 700). Using a map, the orientation of the baseline and the distance (*b*) between the stations are determined.

We then have (see Fig. 12.20):

$$\gamma = 180° - (A_1 + A_2)$$

and

$$a_1 = b \sin A_2 / \sin \gamma$$

$$a_2 = b \sin A_1 / \sin \gamma.$$

The altitude of M is independently derived from the following expressions (taking the mean):

$$MN = a_1 \tan H_1, \text{ and } MN = a_2 \tan H_2.$$

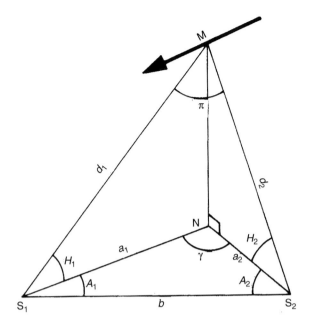

Fig. 12.20. *The principle of triangulation*

We now only have to express the position of point M in a set of Cartesian coordinates with its origin at one station, the *x*-axis pointing south, the *y*-axis pointing west and the *z*-axis pointing towards the zenith. If (A, H) are the horizontal coordinates of the point M as seen from one station, and d is the distance between that point and the station, we have:

$x = d \cos H \cos A$

$y = d \cos H \sin A$

$z = d \sin H$.

Note that z is the height of the meteor.

If (x_1, y_1, z_1) are the Cartesian coordinates of the meteor at time t_1 and (x_2, y_2, z_2) the coordinates at time t_2, the distance (L) covered by the meteor in time $t_2 - t_1$ is given by:

$L^2 = (x_1 - x_2)^2 + (y_1 - y_2)^2 + (z_1 - z_2)^2$

and the atmospheric velocity (V) by:

$V = L/(t_2 - t_1)$.

Triangulations may be carried out from more than just two stations and indeed these are preferable, because cross-checking improves the accuracy of the measurements (Fig. 12.21).

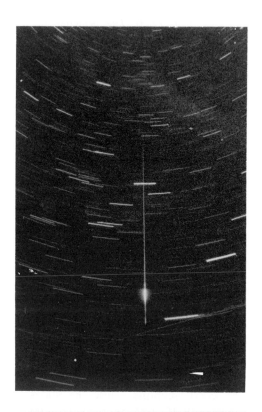

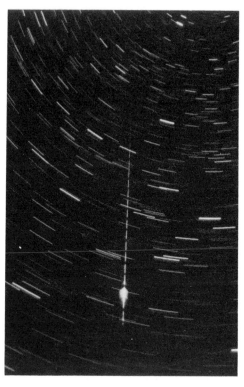

Fig. 12.21. *An example of a Perseid seen from three different points. The stations were about 25 km apart. This type of record enables a very accurate measurement of the altitude and velocity of the meteor to be obtained. (Photos by G.L. Carrié, C. Buil and S. Chevrel, 1986 August 12.)*

The point M for which the altitude is derived corresponds to a very precise point along the meteor's real trajectory. Obviously this point must be identified unambiguously on the tracks obtained from each station. This problem of the relationship between the tracks is tricky, and requires fairly complicated mathematical treatment. Determining the relationship between the tracks may be assisted when some specific feature, such as a flare, is visible somewhere along the paths. The beginning and end of a track may be used, but the positions of these points on an exposure are always poorly defined, because the conditions under which the pictures are obtained are not necessarily the same for both stations.

12.9.2 How to obtain good triangulations

The greater the distance between the stations, the better the accuracy obtained by the calculation, because the parallax increases. A baseline of between 25 and 100 km may be considered satisfactory. Below 15 km, one cannot hope to obtain a sufficiently accurate measurement, given the short focal length of the lenses. If the baseline is very long, the parallax increases and the regions to be covered differ very considerably, which may lead to other problems.

Let us assume that the height and the baseline are 100 km and that a meteor's path has the same alignment as the baseline. We then have the situation shown in Fig. 12.22. If the meteor is observed at an angular height of 40° from station 1, the distance S_1N would be:

$$S_1N = 100/\tan H_1 = 100/\tan 40° = 119 \text{ km}.$$

From this we obtain $S_2N = 119 - 100 = 19$ km, so that from station 2 the meteor would be observed at a height H_2 such that:

$$\tan H_2 = 100/19,$$

whence $H_2 = 80°$. The value of the parallax π will be: $\pi = H_2 - H_1 = 40°$.

It is obvious from this example that cameras should not be pointed towards the same area of sky to detect meteors that are common to both stations. On the contrary, if the camera at S_1 is directed at an altitude of 40°, the one at station S_2 should be pointed at an altitude of 80°.

If the cameras were directed at right-angles to the baseline, the parallax angle would obviously lie in a horizontal plane. For other camera directions, the parallax angle assumes intermediate orientations. It is easy to determine the directions in which cameras should be pointed by plotting the positions of the stations on a map:

- choose a point (C) representing the centre of gravity of the various stations;
- from this central point, draw a straight line representing 100 km at some specific azimuth. Let the end of this line mark the foot of the perpendicular from a meteor seen at an altitude of 45° from point C;
- then draw straight lines joining each station (S_1, S_2 ...) to point C. The orientation of these lines gives the azimuths (A_1, A_2 ...) in which the cameras should be directed. By measuring the length of the lines (L_1, L_2

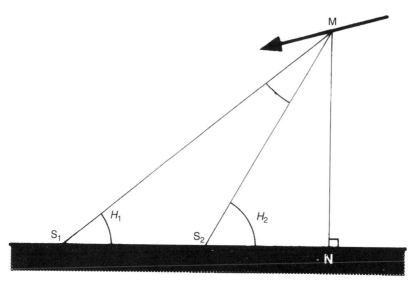

Fig. 12.22. *A triangulation example*

...), one can derive the heights (H_1, H_2 ...), at which the cameras should be pointed, from $\tan H = 100/L$.

Such a procedure is pointless if the baseline is less than 20 km, because then the parallax is sufficiently low for all the observers to be able to monitor the same area of sky.

The azimuth as seen from site C is determined following the general recommendations given for photographic observations: for preference the direction should be 45° from the radiant at an altitude of 45°. If several cameras are available at each station, obviously the same number of triangulation zones may be defined. It is then advisable for the photographic fields to overlap slightly to ensure complete coverage of an area of the sky, thus eliminating any slight errors in alignment.

[A somewhat simpler method is employed by the Meteor Section of the British Astronomical Association. A number of fixed points on the ground are chosen (in practice, these are railway stations) and triangulation is carried out on the volumes of sky vertically above these points. One slight advantage is that prior consultation between observers in not essential to obtaining proper triangulation. – Trans.]

It is obviously advisable to visit all the observing sites at night to ensure that there are no obstacles or stray light that prevent observations from being made in a certain direction. If this should be the case, then the parallax should be recalculated for all the stations to determine more favourable orientations.

Given that it is difficult to orient oneself using horizontal coordinates when actually at the observing sites, there is a danger that such problems may introduce discrepancies in the area covered. An effective solution, and one that is highly recommended, is to agree to set the cameras by pre-determined bright stars at a specific time, and not to shift them during the observation. The stars that were used for setting will eventually move out of the field, but that is of no significance, the main thing being that the horizontal coordinates should be constant and exact. The azimuth and altitude of these stars are calculated from the parallaxes that were determined. If each camera is on an equatorial mounting, this may be realigned to its original position roughly every half-an-hour. Another method is to point towards the only fixed point in the sky: the region around the celestial pole. The area around Polaris is, for example, ideal when studying the Perseid shower in August.

When using a camera with a 50-mm lens, an error of 10° in alignment at each station may mean that the triangulation fails. Such an error is easy enough to make unless one is particularly careful: hence the advantage of using wide-angle lenses.

It is a good idea to have one or more preliminary meetings before the actual observations take place, to clarify which areas will be covered – prepare a sketch map of the sky with the regions to be covered from each station – the times of observation, how the photographic equipment should be shared out, etc.

12.10 An all-sky monitoring programme

Fish-eye lenses offer the possibility of carrying out routine monitoring of the whole sky. Indeed, just a single camera-body is required if the lens has a 180° field of view. The slow speed of the lens means that exposures of at least half-an-hour may be taken before winding on the film. Only very bright meteors will be photographed, but such fireballs often cause spectacular trails (Fig. 12.23).

If several observers cooperate in an observing programme, it becomes possible to obtain important information about these comparatively massive objects and, in particular, to study their distribution in space by calculating their orbits. In a few rare cases it has even been possible, by analyzing the trajectories of exceptionally bright fireballs, to determine the site of the fall. Then both the orbit and the mineralogical composition are known. The observation of the Pribam fireball over Czechoslovakia on 1959 April 7 was the first to result in a meteorite being recovered.

Because the field covered by a single lens is so large, the stations may be as much as 100–150 km apart. It is therefore not too difficult to set up an observing network covering the whole (or a large part) of a country. The large parallax produced by such long baselines compensates for the low spatial resolution of the lenses used.

At the Ondrejov Observatory in Czechoslovakia, Ceplecha heads one of the few professional teams monitoring fireballs. The cameras are automatic, and are sometimes supervised by non-professionals such as meteorologists or even farmers. The work involved is minimal, because whenever it is fine, all that has to be done is to switch on the automatic equipment and then switch it off at dawn. The wide-field cameras have a 9 × 12 cm film format, and are fitted with 30 mm, f/3.5 lenses. A rotating shutter is located between the lens and the film. [A magnificent fireball was captured by this network in 1991 May (Witze, 1992), and should be enough to inspire anyone. – Trans.]

There are similar programmes in Belgium, the Netherlands and the United Kingdom, run by amateurs, who use more conventional wide-field lenses (24 × 36 mm format). Nothing of the sort exists in France, although only seven or eight points

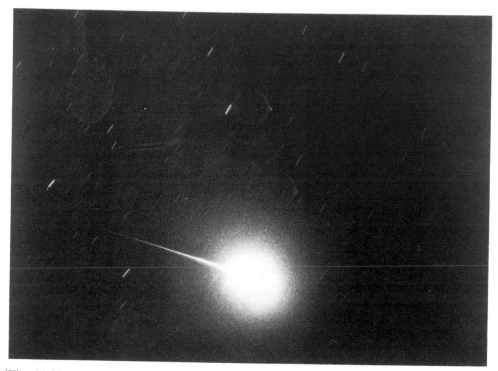

Fig. 12.23. *A spectacular, magnitude* −10 *fireball, photographed on 1983 August 12 at 22:17:14 UT by I. Cognard. This fireball was also observed from southern England and Belgium.*

would suffice to cover the whole of the country [and proposals for a new network in North America have met with little response – Trans.].

Effective work means that photographs should be taken every clear night that is free from moonlight. To prevent operation from being too demanding, the camera should be equipped to take the photographs automatically. Motor drives are available for certain cameras and solve the problem of advancing the film. The shutter-release may be controlled by a set of cams driven by a motor, or by an electromechanical arrangement. Overall control may be by some form of mechanical arrangement (for example, the shutter may be controlled by a synchronous motor and released once every complete rotation) or, even better, by a system controlled by a microprocessor, which allows far greater flexibility in that it may be reprogrammed (the exposure time may be altered in accordance with sky clarity, for example). [A basic, automatic, all-sky camera system is described by Betlem and Mostert (1982). – Trans.]

With a fish-eye lens, the rotating shutter needs to be very large to ensure that the whole of the field is covered. It may be constructed of some light material such as balsa wood.

It is not possible to leave an equatorially mounted fish-eye lens to its own devices throughout the night, because it will not be long before it is photographing the ground. In any case, the time of a meteor's passage would be poorly determined (the error being equal to the duration of the exposure). If several observers work together, it is a good idea for them to begin exposures at different times, rather than together. This enables the uncertainty to be reduced when the photographs are compared.

We should also mention that it is possible to monitor the sky with photomultipliers linked to appropriate electronics that will detect any sudden increase in brightness. In general, such detectors are used without any optics in front of them, except for a diaphragm to limit the field to a few tens of degrees. Comparison of signals from detectors pointing in different directions enables a meteor to be distinguished from a sudden flash of light arising from other sources (vehicle lights for example). Such an arrangement was first used successfully in 1951 (by McKinley, in Canada) and is still used by some professionals and amateurs for determining rates, and to obtain the timing of events in association with wide-field surveillance cameras.

By digitizing the signal from the detector and recording it with a microcomputer, it is possible to sample it at short intervals of time, and thus obtain an accurate light-curve. (It is possible to make hundreds of measurements of the brightness per second, for example.)

12.11 Meteor spectroscopy

Meteors show a typical plasma emission spectrum (particles are ionized by some of the meteoroid's kinetic energy). The continuum spectrum is very faint or non-existent. The first spectra were recorded visually. A. S. Herschel appears to have been the first to observe a meteor spectrum (accidentally) in 1864, when he was studying Capella. A few, very patient observers, such as J. Browning and N. Von Konkoly, tried to detect visual spectra in the late 19th century. Only one line in the green and one in the yellow were detected, which were attributed to magnesium and sodium respectively. Spectroscopes in those days contained only prisms.

After the advent of photography, the light emitted by the plasma could be studied far more easily. The first photograph of a meteor spectrum was obtained at Harvard Observatory in 1897. One of the first photographic observation programmes was organized in Moscow in 1904. As late as 1930, however, only eight photographic spectra had been obtained. Advances in photographic emulsions then caused the number of spectra to increase rapidly. The Canadian astronomer Peter Millman compiled a list of spectra obtained prior to 1958, and this catalogue contained nearly 400 spectra. Millman used the catalogue to establish the first spectral classification.

After 1950, gratings became the most commonly used form of dispersive element. In addition, the use of advanced optics such as Schmidt telescopes enabled the efficiency of the spectrographs to be increased by a factor of 100. In just 4 years' observation (1968–72), an intensive programme carried out in the U.S.A. obtained 764 spectra. Since then, no truly systematic investigation has been carried out by professional astronomers. This is a chance for amateurs to make their mark, especially because, although there is a spectral classification, the range of spectra recorded suggests that considerable investigation is still required.

12.11.1 Spectral classification

Millman's classification is still used. It is based on the appearance of the spectra, and not necessarily on the meteors' chemical composition. In fact, the same material may show different spectra according to the plasma's degree of excitation. Spectral studies confirm the great variety of components found in meteors. Here are the spectral types as defined by Millman:

Type X: the most intense lines are those of magnesium and sodium. This class has a low degree of excitation, and covers 20 % of spectra.

Type Y: intense lines of ionized calcium in the blue-violet. Such spectra are rare, and are only found in 2 % of those recorded. This type indicates a high degree of excitation.

Type Z: numerous lines of iron or of chromium. These are the principal components of siderites. The class includes 66 % of all recorded meteor spectra.

Type W: spectra that do not belong to any of the above classes.

The principal spectral lines are as follows:

- H and K lines of ionized calcium (Ca II) at 393.4 nm and 396.9 nm;
- neutral calcium, which gives a line in the blue at 422.7 nm and in the red at 616.2 nm;
- iron lines between 404.6 and 414.4 nm, between 426.8 and 442.7 nm, between 488.6 and 498.8 nm and between 537.1 and 545.6 nm
- ionized magnesium (Mg I) which gives three principal lines at 516.7, 517.3 and 518.4 nm (another line of neutral magnesium is visible at 385.5 nm);
- the forbidden, neutral oxygen line at 557.7 nm. This line is of atmospheric origin (it is the one seen in aurorae);
- the sodium (Na I) doublet at 589.0 and 589.6 nm;
- neutral oxygen (O I) at 615.7 nm;
- the ionized silicon (Si II) doublet at 634.7 and 637.1 nm.

12.11.2 The dispersive element

To obtain the spectrum of a meteor, a suitable dispersive element that will separate the various wavelengths must be placed in front of the objective. Either a glass prism or a diffraction grating may be used.

12.11.2.1 Prisms

The deviation of rays of light by a prism depends on the refractive index of the glass. The value of this index varies as a function of the wavelength of the light, hence the rainbow colours that emerge from the prism. The latter is placed in front of the photographic lens, and thus functions as an objective prism.

Figure 12.24 shows the path of light in a prism. The angle D between the incident ray and the refracted ray is known as the deviation angle. The angle A is called the apex angle. The shorter the wavelength of the light the greater the amount of deviation.

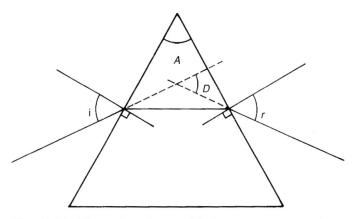

Fig. 12.24. *The paths of rays of light in a prism. i and r are respectively the angle of incidence on entering the prism and the angle of refraction on exit. For minimum deviation at a given wavelength, i = r.*

To obtain the best quality image, the inclination of the prism should be adjusted to give minimum deviation. Minimum deviation may be determined by tilting the prism in front of the lens while looking through the finder: the point at which the image begins to move backwards indicates minimum deviation. Note that such an adjustment is really only valid for a single wavelength. At minimum deviation, the angle of the incident ray with respect to the normal at the entry face of the prism is given by:

$$\sin i = n \sin(A/2),$$

where n is the refractive index of the glass for a specific wavelength. This angle is also equal to the refractive angle at the exit from the prism.

The deviation (D) is expressed as:

$$D = 2i - A.$$

The change in deviation (dD) as a function of a change in index dn is given by:

$$dD/dn = 2\sin(A/2)/\cos i.$$

This equation enables one to calculate the angular distance between two rays in different spectral regions.

If D_r and D_b are the calculated deviations for a red ray and a blue ray (the angles being expressed in radians), and if F is the focal length of the lens, the size of the spectrum on the film is given by:

$$L = F(D_b - D_r).$$

Let us assume that we have a prism made of BK7 (the most common optical glass), with an apex angle of 45° and that it is placed in front of a lens with a focal length of 50 mm. We may calculate the length of the spectrum on the film, given

Table 12.10. *Refractive index of BK 7 as a function of wavelength*

λ(nm)	n
400	1.530
550	1.519
650	1.514

that the latter can only record visible wavelengths. The refractive index of BK7 as a function of wavelength is given in Table 12.10.

We first calculate the angle of incidence for minimum deviation of the green line (the centre of the spectrum).

$$\sin i = n \sin(A/2) = 1.519 \sin(45°/2) = 0.5813$$

whence $i = 35.54°$.

The deviation for a line of wavelength 550 nm is then:

$$D = 2 \times i - A = 2 \times 35.54 - 45° = 26.08°.$$

The dispersion of the overall spectrum relative to the central line is given by:

$$\begin{aligned} \mathrm{d}D &= [2\sin(A/2)/\cos i]\mathrm{d}n \\ &= [2\sin(45°/2)/\cos 35.54°](1.530 - 1.514) \\ &= 0.015\,05 \text{ radian} = 0.86°. \end{aligned}$$

The linear size of the spectrum on the film is:

$$L = F(D_b - D_r) = 50 \times 0.015\,05 = 0.75\,\text{mm}.$$

This is very small, and only the principal lines could be examined. The dispersion is enough, however, for the spectral type of a meteor to be determined. Note that the spectral dispersion is not linear with wavelength and that it is greatest in the blue.

It may, therefore, be of advantage to use 6×6 format cameras with lenses of 80-mm focal length, thus increasing the spectral resolution. Similarly, one could seek to use glasses with a higher refractive index than BK7 (e.g., flint glass).

The size of the prism should be sufficient to cover the aperture of the objective without vignetting, and the prism should be mounted as close as possible to the lens. Finally, the incident rays are deviated by the angle D: this should be taken into account in pointing the camera.

12.11.2.2 Gratings

A diffraction grating consists of a glass substrate on which a series of closely spaced grooves have been ruled (several thousand per millimetre). Such a structure diffracts light in such a way as to disperse an incident beam into various spectra known as orders. For a beam incident normal to the plane of the substrate (Fig. 12.25), the angular position of the different orders is given by:

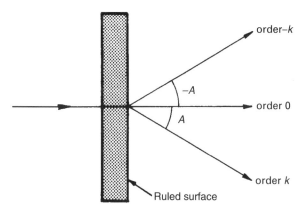

Fig. 12.25. *The path of rays of light through a transmission grating*

$\sin A = k\lambda N$

where k = the diffraction order, which is an integer $(\ldots, -2, -1, 0, 1, 2, \ldots)$, λ is the wavelength, and N is the number of lines per millimetre in the grating.

For $k = 0$, there is a white-light, undeviated image. As the absolute value of k increases, the dispersion and the resolution both increase. In practice, for meteor work one can only use the first order or, occasionally, the second order.

It is obvious that the luminous flux is divided into several spectra, which is a disadvantage. If the grooves are given a specific profile (in a process known as 'blazing'), it is possible to concentrate the flux in a given order. The grating is then said to be blazed for order k. Typically, it is possible to concentrate about 70 % of the incident energy into a specific order (usually -1 or 1) by this technique. A blazed grating is practically essential for obtaining meteor spectra, because otherwise the luminosity is far too low.

There are both transmission and reflection gratings. The first type is preferable, because of the greater field covered. The ruled surface may completely cover the field of the lens. It is usually necessary to use a transmission grating that is 50 mm square. A good quality grating of this sort will cost around £600 ($1050).

Let us calculate the size of the first-order spectrum $(k = 1)$ obtained with a 35-mm lens with a grating having 300 lines per millimetre. The deviation in the red $(\lambda = 650\,\text{nm})$ is:

$A_{\text{r}} = \arcsin(0.65 \times 10^{-3} \times 300) = 11.24°.$

In the blue $(\lambda = 400\,\text{nm})$, we have:

$A_{\text{b}} = \arcsin(0.40 \times 10^{-3} \times 300) = 6.89°.$

The dispersion is therefore: $11.24° - 6.89° = 4.35° = 0.0759$ radian. The linear dimension of the spectrum on the film will be:

$L = 35\,\text{mm} \times 0.0759 = 2.66\,\text{mm}.$

This dispersion is sufficient to carry out detailed study of spectra, and the field covered by the 35-mm lens means that such a combination is highly efficient.

Unlike prisms, gratings produce dispersions that are almost linear with wavelength, which help considerably in identifying spectral lines.

12.11.3 Making the exposure

Everything that has been said about normal meteor photography applies to spectroscopic work. A few other precautions need to be taken:

- The deviation of the light as it passes through the dispersing element means that aiming the camera has to be carried out with particular care. Despite the presence of the grating or the prism, it is usually possible to see the spectra of bright stars, which makes things easier. With a grating, obviously the camera needs to be aimed in the direction appropriate to the blazing. (Take care not to confuse the positive and negative orders.)
- The existence of additional optical elements in front of the lens may give rise to spurious images. For example, light from a distant lamp behind the camera may produce a ghost image on the film after having been reflected from one face of the prism or from the grating. A cardboard baffle surrounding the dispersion element and the lens is sometimes required.
- If shower meteors are being photographed, it is very important to orient the apex of the prism, or the lines on the grating, parallel to the direction in which the meteors are travelling. This gives spectra that are easier to interpret (Fig. 12.26). When the observation lasts for a long time, the camera should be moved to compensate for the motion of the radiant. With sporadic meteors, of course, which may arrive from any direction, it is not possible to take special precautions about the alignment of the dispersing element (Fig. 12.27).
- It is advisable to operate an ordinary camera in parallel with the camera having the objective prism or grating, so that the position of the meteors may be determined, because this is very difficult to do with an image in the form of a spectrum. If this second camera is fitted with a rotating shutter, it is possible to study all aspects of the meteor simultaneously.

12.11.4 Reduction of the observations

The negatives need to be examined with a magnifying loupe. The presence of stellar spectra does not simplify the investigation of the meteor spectra. However, use of another camera not fitted with a dispersing element enables meteors likely to give a spectrum to be easily identified. With normal sorts of photographic lenses, a fine, negative-magnitude meteor is required for a spectrum to be recorded.

The calibration of wavelengths of spectra obtained with a prism is not easy. The lines in meteors' spectra may be identified by comparison with the spectrum of a ordinary mercury-vapour street light obtained with the same optical combination.

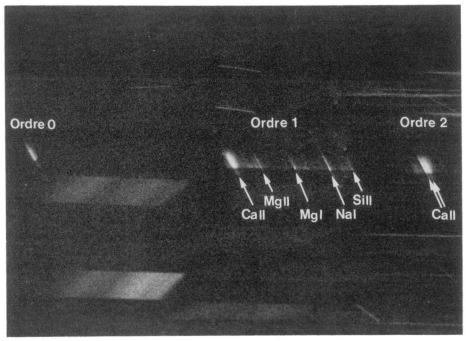

Fig. 12.26. *Spectrum of the meteor shown in Fig. 12.9.1, obtained by placing a grating with 300 lines/mm in front of a 35-mm, f/2.8 lens. The photograph shows the zero and first orders, and part of the second order. Only the terminal flare gave a detectable spectrum. (Original French photo, by C. Buil).*

The principal lines in a mercury lamp occur at wavelengths of 366.3 nm, 404.7 nm, 435.8 nm, 546.1 nm, and 579.1 nm.

With a grating spectrum, the identification of the lines is simple if the zero-order spectrum is also present on the frame. First the image-scale should be calculated by measuring the distance between the zero orders of several stars. The linear distance between the zero order and the various lines is then obtained. Because the scale is known, the linear distance may then be converted into an angle (A); and finally the wavelength is then given by:

$$\lambda = \sin A/kN.$$

12.12 Developing and examining photographs

It is strongly advisable to develop one's own films. For 400-ISO films, a developer such as HC 110 – or even better, DIAFINE (a two-bath developer) – may be used. Generally, it is advisable to overdevelop to increase the sensitivity of the film. The latter should be handled with great care when wet and also when dry, to avoid any scratches that an inexperienced eye might mistake for meteors. Films should be carefully stored in the usual envelopes designed to take strips of six pictures.

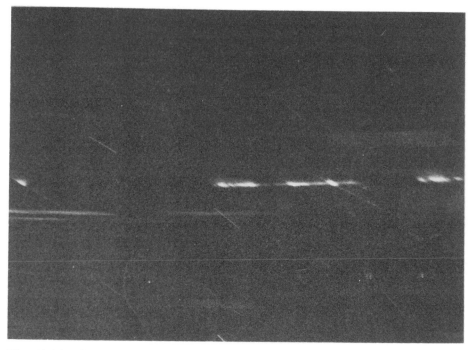

Fig. 12.27. *Photographed with the same set-up as for Fig. 12.26, a sporadic meteor shows a very different spectrum.*

Examination of the negatives for meteors is done with the aid of a loupe against a bright, diffuse, even background. This is often the moment of truth, when one finds out whether one has a good 'haul' of meteors. Careful scrutiny of the negatives is required to detect the very weak trails left by faint meteors. It is important not to be deceived by scratches; these show reflections under certain conditions of illumination which enable them to be identified unambiguously. The most pernicious problem is caused by artificial satellites. Some satellites cross the sky slowly, vary in brightness, and are visible only over a short arc. On a photograph they may therefore be easily mistaken for meteors. (Luckily, most satellites show a very long, even track that may be recognised without difficulty.) There are three ways of identifying the intruders:

- the satellite has been observed visually, and this fact has been carefully noted down in the observing log;
- if a rotating shutter has been used and the track shows no breaks, this is a sure sign that the object moved too slowly to be a meteor;
- the object has been identified by triangulation, and found to be at an altitude of 400 km, which excludes a meteor. (Yes, you can make use of the opportunity by measuring the altitude of artificial satellites as well!)

If none of these three methods of checking are possible, it may be reassuring to know that such events are very rare.

12.13 Measuring the photographs

The aim is to measure the equatorial coordinates of certain specific points along the meteor's track: the centre of the breaks to determine the velocity of the meteor (in the case of a meteor captured with a rotating shutter); the position of any flare, so that the altitude may be calculated at that point; the positions of two well-separated points on the track so that the radiant may be determined by extrapolation, etc.

The traditional method of astrometric reduction may be used. This takes place in three stages:

- transformation of the equatorial coordinates of certain stars on the photograph into gnomonic coordinates (*see* p. 702);
- establishing the relationship between the calculated gnomonic coordinates and the Cartesian coordinates of each of the stars, as measured on the image (i.e., calculating the image constants, taking the orientation, scale, etc., into account);
- measuring the Cartesian coordinates of a point on the meteor's track, and finally calculating the equatorial coordinates, using the coordinate-conversion factors given by the image constants.

The optics used in meteor photography often produce distorted images, so the constants for a series of non-linear equations have to be determined. In addition, because of the large field of such optics, it may be shown that it is essential to determine the centre of the photograph accurately. This makes the reduction method more onerous and complicated (because it is essential to measure a whole series of reference stars).

A less accurate, but much faster, method may be used. This consists of choosing two stars A and B straddling the track (*see* Fig. 12.28). A line is drawn connecting these two stars and intersecting the track at point M. If (α_1, δ_1) and (α_2, δ_2) are the equatorial coordinates of the two stars A and B, then the coordinates (α, δ) of the point M are:

$$\alpha = \alpha_1 + [m/(m+n)](\alpha_2 - \alpha_1) = \alpha_2 - [n/(m+n)](\alpha_2 - \alpha_1)$$
$$\delta = \delta_1 + [m/(m+n)](\delta_2 - \delta_1) = \delta_2 - [n/(m+n)](\delta_2 - \delta_1).$$

To obtain the greatest accuracy, the line A–B should intersect the track as close to a right-angle as possible. The chosen stars should be immediately adjacent to the track. Wil Tirion's *Sky Atlas 2000.0* shows stars down to magnitude 8, which is generally sufficient. If not, the *SAO Atlas* (which is very large and contains the positions of most stars down to magnitude 9), or *Uranometria 2000.0* could be used.

By measuring several points on the track, and transforming the measurements into gnomonic coordinates, it is possible to obtain the best fit for the straight line passing through all the points. The orientation of the track is thus determined very accurately.

The method does not allow the position of a point to be determined directly, because the point of intersection obviously lies on the line joining the two stars.

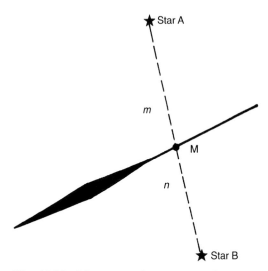

Fig. 12.28. *Measuring the position of a point on the meteor trail using two stars*

However, the interpolation method that we have just described may be used along the track once the coordinates of two points on it are known.

To make accurate measurements, don't hesitate to make giant enlargements from the negatives. The only equipment required to make these measurements is a good-quality rule, a calculator, and an acute eye.

12.14 Radio observation of meteors

The free electrons in the plasma created by a meteor are capable of reflecting radio waves. This fact has been profitably used since 1940 to detect meteors using radar. A powerful, short-wave transmitter illuminates the sky and an associated receiver picks up the signal scattered by the ionized trail. The same aerial may be used for both transmitter and receiver.

The duration of such a radar echo lies between a fraction of a second and several minutes. Various techniques may be used to determine the length of the ionized column (by measuring the time a pulse takes to make the double journey); the velocity of the meteor (by measuring the Doppler–Fizeau effect from the frequency shift); and even sometimes the deceleration. The error in determining the distance of meteors by radar methods is approximately 1 km. By using three receiving stations, the orientation of the meteor trail in space may also be determined, as well as the position of the radiant (to a precision of 2–3°).

Obviously radar methods are not available to amateurs, but it is still possible to study meteors by radio. The method consists of listening for a distant transmitting station, the signal from which is normally inaudible. When a meteor appears, it reflects the radio waves from the transmitter, and the signal strength increases. The meteor serves as a sort of relay, as shown in Fig. 12.29. This technique, known as 'meteor scatter', is well-known to radio hams, who are able to communicate

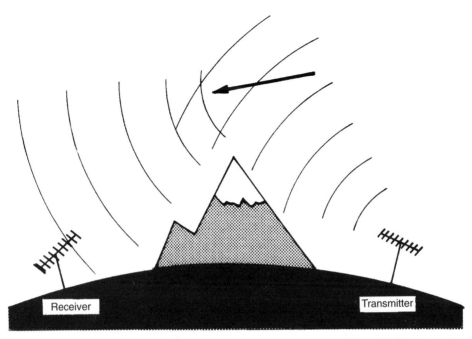

Fig. 12.29. *Although the receiver is not in a direct line of sight with the transmitter, the meteor trail acts as a reflector, enabling the radio signals to be detected.*

intermittently over long distances at wavelengths that are not normally reflected by the layers in the ionosphere. In practice, the frequency bands between 30 and 100 MHz have to be employed, and these are partly covered by the FM band used by ordinary domestic radio receivers (which is generally between 88 MHz and 108 MHz), or by TV stations. To avoid receiving the ground wave, it is important to use a Yagi-type aerial that is pointed at the zenith. If possible, automatic gain control (AGC) should be disconnected. It is useful to have a PLL (phase-locked loop) receiver, which allows the wavelength to be set very accurately. One may then tune in to an FM transmitter that is some 200–500 km away. The most inconvenient part of having to work in the FM band is the fact that many transmitters go off the air during the middle of the night.

When listening, sudden bursts are heard when the transmitter's modulation may be detected. These bursts rarely last more than a few seconds. A recorder connected to the detector enables a permanent record to be obtained. It is also possible to use a counter that only records echoes above a certain amplitude (i.e., that discriminates against noise). It is essential, of course, to avoid working from the centre of a large town, where strong ghost signals may be present, and where local radios may often prevent one from finding any suitable, free, frequency bands.

It is very important – especially when setting up the equipment – to observe visually at the same time as the radio watch is being carried out. This enables one to be sure that meteors are being detected. The majority of meteors visible to the naked eye produce an echo.

Radio observation of meteors, once the techniques have been mastered, offers a powerful method of carrying out statistical investigations of meteors, even in daylight or when it is cloudy. This is how the diurnal radiants were discovered.

12.15 Ensuring the best use of the observations

Many studies of meteors are based on statistical analysis of observations. This means that a considerable amount of observational material is required, often far more than a single observer can provide. It is therefore very important to form part of a larger, meteor-observing group. Meteor observation is an ideal activity for amateur societies. Even if the members are beginners, observing meteors quickly introduces them to scientific methods; familiarizes them with the sky, etc. The competitive aspect often motivates observers – such as seeing who can observe the most meteors or take the best photograph. Then people realize the wide range of types of meteors, and their previously unsuspected beauty. This is often the trigger that turns them into meteor observers. The process of observing, reducing one's observations oneself, and arriving at a scientific result, is fairly rare in amateur astronomy, and this is one reason why people become converts to meteor astronomy.

When instigating a meteor-observing programme in a society, begin with observation of the major showers. If this is well-received, then one can move on to study less-active radiants and to sophisticated methods of observation. Note that meteor observation lends itself to thorough investigation of the observations. This is not only a suitable task for rainy nights, but may be a chance to generate interest in the way in which computing methods may be applied to astronomy.

Isolated amateurs, who may be some tens, or even hundreds of kilometres apart, can still collaborate in triangulation programmes. To set up such a group, it is sufficient to have a coordinator who will undertake to keep centralized records. As far as other details go, the telephone may be used to arrange observing sessions, the areas to be covered, etc.

The major astronomical organisations in most countries have meteor-observing programmes. In English-speaking countries, particularly active are the American Meteor Society, (Dept. of Physics & Astronomy, State University College, Geneseo, N.Y. 14454); and the British Astronomical Association's Meteor Section (Burlington House, Piccadilly, London W1V 9AG). Other groups making major studies of meteors are the Dutch and Belgian Meteor Associations, and the Japanese Meteor Association. Many of these collaborate with groups and individuals outside their own immediate countries.

12.16 Mathematical techniques

12.16.1 Conversion of equatorial to horizontal coordinates

Coordinate conversion is frequently required in meteor astronomy, for example, in calculating the *ZHR* or the altitude of a meteor. Let α, β be the equatorial coordinates of the point whose horizontal coordinates (altitude a; azimuth A) are required. The observation is made from a point, with geographical coordinates (λ,

ϕ). First the hour angle (h) is calculated from the Greenwich sidereal time (GST) at which the observation is made, the longitude of the site (λ), and the right ascension (α), using the same unit for all these terms:

$$h = GST - \lambda - \alpha.$$

The altitude above the horizon is given by:

$$\sin a = \sin \delta \sin \phi + \cos \delta \cos h \cos \phi.$$

The zenith distance is given by:

$$Z = 90° - a.$$

Finally, the azimuth A is obtained from:

$$\tan A = \sin h / (\cos h \sin \phi - \tan \delta \cos \phi).$$

The reverse operations may be carried out by the following equations:

$$\sin \delta = \sin \phi \sin a - \cos \phi \cos a \cos A,$$

$$\tan h = \sin A / (\cos A \sin \phi + \tan a \cos \phi).$$

12.16.2 Radiant determination from two tracks

The position of the radiant is given by the intersection of the tracks of two meteors' from the same shower. The position may therefore be calculated from two images of the same meteor obtained from two different geographical locations. This second method is the only one that is truly rigorous, because with two independent meteors there is no guarantee that they come from exactly the same area of the radiant.

Figure 12.30 indicates the two sets of coordinates measured from the tracks of the two meteors. The pairs of coordinates define arcs of two great circles, one intersection of which (A_r, D_r) is to be calculated. The auxiliary points (Ak_1, Dk_1) and (Ak_2, Dk_2) are first determined:

$$\cot Ak_1 = -(\cot D_{11} \sin A_{11} - \cot D_{12} \sin A_{12})/(\cot D_{11} \cos A_{11} - \cot D_{12} \cos A_{12})$$
$$\tan Dk_1 = -\cot D_{11} \cos(Ak_1 - A_{11}) = -\cot D_{12} \cos(Ak_1 - A_{12})$$
$$\cot Ak_2 = -(\cot D_{21} \sin A_{21} - \cot D_{22} \sin A_{22})/(\cot D_{21} \cos A_{21} - \cot D_{22} \cos A_{22})$$
$$\tan Dk_2 = -\cot D_{21} \cos(Ak_2 - A_{21}) = -\cot D_{22} \cos(Ak_2 - A_{22}).$$

Then we may derive:

$$\cot A_r = -(\cot Dk_1 \sin Ak_1 - \cot Dk_2 \sin Ak_2)/(\cot Dk_1 \cos Ak_1 - \cot Dk_2 \cos Ak_2)$$
$$\tan D_r = -\cot Dk_1 \cos(A_r - Ak_1) = -\cot Dk_2 \cos(A_r - Ak_2).$$

The angle β between the two tracks is given by:

$$\cos \beta = \sin Dk_1 \sin Dk_2 + \cos Dk_1 \cos Dk_2 \cos(Ak_2 - Ak_1).$$

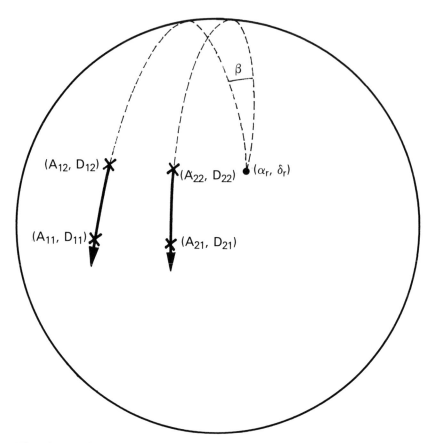

Fig. 12.30. *The intersection of two meteor trails enables the position of the radiant to be determined.*

12.16.3 Rapid method of calculating the height of several points on a track

We want to determine the altitude of several points along the meteor's track from the altitude of one of these points and the horizontal coordinates (A_r, Z_r) of the radiant.

Let alt_0 be the altitude of the known point with coordinates (A_0, Z_0), and (A, Z) the coordinates of a point for which the altitude (alt) is required. We then have:

$$alt = alt_0(\tan Z_0 \cos A_0 - \tan Z_r \cos A_r)/(\tan Z \cos A - \tan Z_r \cos A_r).$$

The distance between the observer and the meteor is: $d = alt/\cos Z$ and the distance between the two points: $l = (alt - alt_0)/\cos Z_r$.

12.16.4 Gnomonic projection

Let (A_0, D_0) be the equatorial coordinates of the centre of the projection. The Cartesian (x, y) coordinates of a point with the coordinates (A, D) are calculated from:

$a = \sin D \sin D_0 + \cos D \cos D_0 \cos(A - A_0)$

$x = R \cos D \sin(A - A_0)/a$

$y = R[\sin D \cos D_0 - \cos D \sin D_0 \cos(A - A_0)]/a,$

where R is a coefficient that determines the scale of the chart.

Equatorial coordinates may be derived from the (x, y) Cartesian coordinates, whose origin is the centre of the chart, from the following equations:

$L = R \cos D_0 - y \sin D_0$

$M = R \sin D_0 - y \cos D_0$

$N = x$

$\sin D = M/\sqrt{L^2 + M^2 + N^2}$

$\tan(A - A_0) = N/L.$

12.16.5 Mathematical method of optimizing the line of sight when triangulating

The problem to be solved is as follows: From one station with the geographical coordinates (L_1, ϕ_1) a camera is pointed at azimuth A_1 and altitude a_1. What is the direction in which another camera should be pointed from a station at L_2, ϕ_2 to give successful triangulation?

Step 1   Calculate the longitude (L) and the latitude (ϕ) of the point where the observed meteor – with horizontal coordinates (A_1, a_1) from station 1 – is at the zenith:

$\sin \phi = \sin \phi_1 \cos \omega_1 - \cos \phi_1 \sin \omega_1 \cos A_1$

$\tan(L - L_1) = \sin A_1/(\cot \omega_1 \cos \phi_1 + \sin \omega_1 \cos A_1)$

where $\omega_1 = 90° - \zeta - a_1$ and $\zeta = 0.984\,544 \cos a_1$.

Step 2   Calculate the horizontal coordinates (A_2, a_2) as seen from station 2 of the point (L, ϕ) directly beneath the meteor.

$\tan A_2 = -\sin(L - L_2)/[\cos \phi_2 \tan \phi - \sin \phi_2 \cos(L - L_2)]$

$\sin a_2 = (82\,427\,800 \cos \omega_2 - 81\,153\,800)/(12\,740 \times \sqrt{82\,437\,800 - 82\,427\,800 \cos \omega_2})$

where

$\cos \omega_2 = \sin \phi \sin \phi_2 + \cos \phi \cos \phi_2 \cos(L - L_2).$

An example   From a site whose geographical coordinates are $L_1 = -2.53°$, $\phi_1 = 43.33°$, a camera is pointed due North $(A_1 = 180°)$ at an altitude a of $45°$. At what azimuth and altitude should a camera be pointed from a second station at $L_2 = -3.60°$ and $\phi_2 = 43.69°$?

We find:

$\zeta = 44.12°$

$\omega_1 = 90° - 44.12° - 45° = 0.879°$

$\phi = 44.21°$

$L - L_1 = -2.53°$
$A_2 = 124.4°$
$a_2 = 43.4°.$

12.16.6 Calculating the orbits of meteors

The orbit of an object around the Sun is entirely defined as soon as one knows its position relative to the Sun at a given instant, and also its heliocentric velocity. For meteors, the position relative to the Sun is obviously that of the Earth. Calculating the orbit then comes down to determining the velocity vector of the meteoroid just before it encountered the Earth. A number of corrections have to be applied to the value of the atmospheric velocity of the meteor to obtain the heliocentric velocity (its velocity in the solar reference frame).

12.16.6.1 Atmospheric deceleration

As it passes through the atmosphere, a meteor is subject to braking. The deceleration is several km/s^2. This is often very difficult to measure because the meteor is visible over such a short track. We will not include it in our calculations here. The observed atmospheric velocity is denoted V_1.

12.16.6.2 Correction for the Earth's rotation

Figure 12.31 represents the Earth seen from one of its poles (P). Let U be the Earth's tangential velocity at the point vertically below the meteor (point A). The velocity, corrected for the Earth's diurnal rotation is V_2, the vector sum of U and V_1.

If S is the local sidereal time at A, the vector U intersects the celestial sphere at a point whose equatorial coordinates are:

$\alpha_U = S + 90°$

$\delta_U = 0.$

It may be shown that the angle ϕ between the velocity vector of point A and V_1 may be expressed as:

$\cos \phi = \cos \delta_r \cos(\alpha_r - \alpha_U),$

where α_r and δ_r are the equatorial coordinates of the radiant.
Finally, we obtain the velocity V_2 from:

$$V_2 = \sqrt{V_1^2 + U^2 - 2V_1 U \cos \phi}.$$

The tangential velocity at point A is calculated from the Earth period of sidereal rotation and the latitude (ϕ) for the point of observation, i.e.:

$U = 0.4651 \cos \phi$ (in km/s).

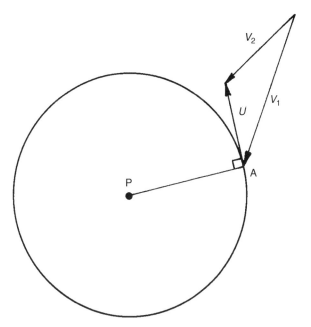

Fig. 12.31. *The velocity resulting from the combination of the Earth's diurnal rotation and the meteor's own motion.*

12.16.6.3 Correction for diurnal aberration

In this case, the aberration is a shift in the apparent direction of arrival of the meteor, caused by the relative motion of the meteor and the observer, the latter being associated with the rotation of the Earth on its axis. If α_0 and δ_0 are the observed coordinates of the radiant, and (α, δ) the coordinates of the same radiant corrected for aberration $(d\alpha, d\delta)$, we have:

$$\alpha = \alpha_0 + d\alpha$$
$$\delta = \delta_0 + d\delta,$$

and

$$d\alpha = -(26.65/V_2)(\cos\phi\cos h_0/\cos\delta_0)$$
$$d\delta = -(26.65/V_2)\cos\phi\sin h_0\sin\delta_0$$

where h_0 is the hour angle of the apparent radiant, and where the velocities are expressed in km/s.

12.16.6.4 Correction for terrestrial gravitation

Close to the Earth, the path is curved because of the effect of the Earth's gravitational attraction. The path tends to become a part of a hyperbola. This effect is known as zenithal attraction and has two consequences:

- The meteor accelerates. The velocity before it encounters the Earth may be calculated from:

$$V_g^2 = V_2^2 - 797\,200/p$$

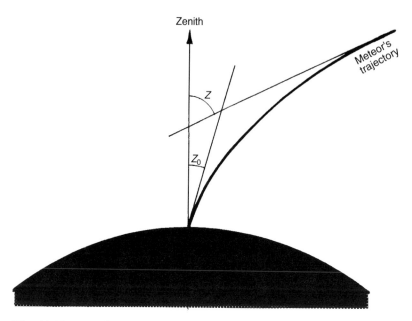

Fig. 12.32. *Zenithal attraction*

where p is the geocentric radius of the Earth at the observing site, plus the height of the meteor (p may be taken as 6467 km). V_g is known as the meteor's geocentric velocity.

- The zenith distance of the radiant is altered, as shown in Fig. 12.32. If the path were unaffected by the Earth's gravitational attraction, the radiant would have a zenith distance Z. In fact, the zenith distance observed Z_O is given by:

$$Z = Z_0 + dZ.$$

The zenithal attraction dZ is calculated from:

$$\tan(dZ/2) = [(V_2 - V_g)/(V_2 + V_g)]\tan(Z_0/2).$$

The equatorial coordinates of the radiant are then determined, using this corrected zenith distance.

12.16.6.5 Correction for the effect of the Earth's orbital motion
This is the most important correction, because the Earth's orbital velocity is approximately equal to that of the meteor (about 29 km/s). Figure 12.33 shows the intersection of the Earth's orbit with that of the meteoroid. The two objects meet at A. V_h is the heliocentric velocity of the meteoroid and V_E is the Earth's heliocentric velocity.

The vector V_g points in the direction of the apparent radiant, and the angle E is the elongation of the apparent radiant (the radiant corrected for diurnal aberration and zenithal attraction). The true motion of the meteoroid in the Solar System, and

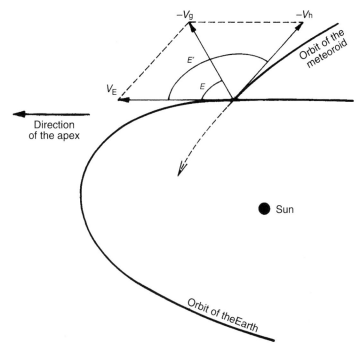

Fig. 12.33. *Determination of the heliocentric velocity from the geocentric velocity of the meteoroid and the Earth's orbital velocity.*

close to the Earth is shown by the vector V_h. The elongation of the true radiant is called E'. The direction of the vector V_E is known as the apex (the point to which the Earth is moving as it orbits the Sun).

To calculate V_h, it helps to use ecliptic coordinates. Let (λ, β) be the ecliptic coordinates of the apparent radiant, and $\lambda_\odot$ the longitude of the Sun at the time of observation. The longitude of the apex (A) is then:

$$A = \lambda_\odot - 90° + \eta,$$

where the angle η is the angle between a line perpendicular to the Earth radius vector and the Earth's velocity vector. This angle, which is always less than $\pm 1°$, arises because of the Earth's orbital eccentricity. To determine its value, we first calculate the longitude of perihelion for the Earth (ω_0), which varies slowly as a function of the epoch of observation (t):

$$\omega_0 = 102.075 + 0.0172(t - 1950)$$

then:

$$\eta = 0.9579° \sin(\lambda_\odot - \omega_0).$$

Figure 12.34 shows the positions (projected onto the celestial sphere) of the directions of the apparent radiant TR, of the apex TA, and of the Sun TS, the Earth

707

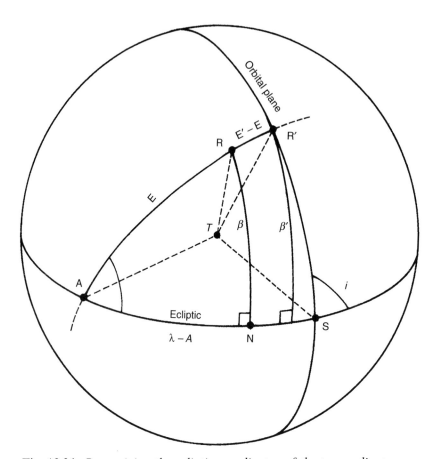

Fig. 12.34. *Determining the ecliptic coordinates of the true radiant*

(at T) occupying the centre of the sphere. It may be shown that we have:

$$\cos E = \cos \beta \cos(\lambda - A)$$
$$\sin E \cos \gamma = \cos \beta \sin(\lambda - A)$$
$$\sin E \sin \gamma = \sin \beta,$$

where the intermediate angle γ lies between the plane of the ecliptic and the plane containing the direction of the apex and the direction of the radiant.

Thanks to this set of formulae, the elongation E may be calculated with no ambiguity as to sign. The heliocentric velocity of the meteoroid is then given by:

$$V_{h}^2 = V_{g}^2 + V_{E}^2 - 2V_{g}V_{E} \cos E$$

and the elongation of the radiant may be calculated from:

$$\sin(E' - E) = (V_{E}/V_{h}) \sin E.$$

The ecliptic coordinates of the true radiant (λ', β') are then:

$\cos E' = \cos \beta' \cos(\lambda' - A)$
$\sin E' \cos \gamma = \cos \beta' \sin(\lambda' - A)$
$\sin E' \sin \gamma = \sin \beta'$,

whence we calculate the equatorial coordinates of the true radiant.

12.16.6.6 Calculation of the orbital elements

The velocity (in km/s) of an object moving in an elliptical orbit in the Solar System is given by:

$$V^2 = 886.1(2/r - 1/a),$$

where r is the radius vector (the distance between the object and the Sun), and a is the semi-major axis of the orbit. These two quantities are expressed in astronomical units.

By setting $V = V_h$ in the above equation, we can easily calculate the semi-major axis of the meteoroid's orbit. At the time of observation, the meteoroid's radius vector r is obviously that of the Earth. The latter value may be found in astronomical ephemerides (and is always close to 1). (Note that it is easy to calculate the orbital velocity of the Earth (V_E) by setting $a = 1$. If $r = 1$, $V_E = 29.767$ km/s.)

The inclination of an orbit is the angle between the plane of the orbit and the plane of the ecliptic. This angle is shown in Fig. 12.34. By solving the spherical triangle R'SN, we find:

$$\tan i = \tan \beta' / \sin(\lambda' - \lambda_\odot),$$

where $\lambda_\odot$ is the solar longitude.

We then calculate the cosine directors of the true radiant (L, M, and N) in a set of coordinates aligned with the ecliptic, where the x-axis is directed towards the apex:

$L = \cos \beta' \cos(\lambda' - A) = \cos E'$
$M = \cos \beta' \sin(\lambda' - A) = \sin E' \cos \gamma$
$N = \sin \beta' = \sin E' \sin \gamma$,

and then the following intermediate elements:

$A = M \sin \eta - L$
$B = M + L \sin \eta$
$C = 1 - B^2$.

The orbital parameter (p), the eccentricity (e), and the eccentric anomaly (v) are then given by:

$$\sqrt{p} = V_h r C$$

$$e \sin v = V_h B \sqrt{p}$$

$$e \cos v = p/r - 1.$$

The value for the eccentricity may be checked from the equation:

$e^2 = 1 - p/a.$

The perihelion distance is given by:

$q = a(1 - e) = p/(1 + e).$

The argument of perihelion in latitude (ω) is identical (to within 2π) to the eccentric anomaly, because the meteoroid is at one of its nodes when it encounters the Earth. If

$\beta' > 0$, then $\omega = 180° - v$,

or else

$\beta' < 0$, then $\omega = 360° - \eta.$

All that remains is to determine the longitude of the ascending node (Ω). Once again, because the meteoroid is crossing the ecliptic when it encounters the Earth, it is easy to see that Ω is equal to the longitude of the Sun $\pm\pi$. If

$\beta' > 0$, then $\Omega = \lambda_\odot$,

or else

$\beta' < 0$, then $\Omega = \lambda_\odot \pm 180°.$

12.16.6.7 Longitude of the Sun for Epoch 1950.0
In meteor astronomy, the period of a shower's activity is related to the position of the Earth in its orbit. It is therefore logical to describe the date of the activity in terms of the solar longitude, and thus avoid the vagaries of the civil calendar (leap days, etc.).

To calculate the longitude of the Sun ($\lambda_\odot$) for the date YEAR, MONTH, DAY, the value of DAY includes decimal fractions of a day if the longitude is to be determined for any time other than 00:00 UT.

First, we calculate the Julian Day and Julian Century for the given date.

If MONTH> 2, then y =YEAR and m =MONTH. If MONTH< 1 or MONTH = 2, then y = YEAR $- 1$ and m = MONTH $+ 12$.

Then:

$A = \text{INT}(y/100)$

$B = 2 - A + \text{INT}(A/4).$

The Julian Day (JD) is given by:

$\text{JD} = \text{INT}(365.25 \times y) + \text{INT}[30.6001 \times (m + 1)] + \text{DAY} + 1\,720\,994.5.$

The Julian Century (T) may then be derived:

$T = (\text{JD} - 2\,415\,020)/365\,25.$

The following values are then calculated:

Long-term inequality (IN):
$$IN = 0.001\,777\,8 \times \sin(231.0 + 20.2 \times T) + 0.000\,522\,78 \times \sin(57.2 + 150.3 \times T).$$

Mean longitude of the Sun:
$$\lambda = 279.475\,83 + 36\,000.768\,92 \times T + 0.000\,302\,5 \times T^2 + IN.$$

Mean solar anomaly:
$$M = 358.475\,83 + 35\,999.049\,75 \times T + 0.000\,003\,3 \times T^2 + IN.$$

Equation of the centre:
$$\begin{aligned} C \quad = \quad & (1.919\,460\,3 - 0.004\,788\,9T)\sin(M) \\ & + (0.020\,093\,9 - 0.000\,100\,3T)\sin(2M) + 0.000\,293\,8\sin(3M). \end{aligned}$$

Finally, we obtain the longitude of the Sun for Epoch 1950.0:

$$\lambda_\odot = \lambda + C - 0.013\,96(\text{YEAR} - 1950).$$

In the last equation, the variable YEAR is given with decimal places.

13 Double and multiple stars

P. Durand

13.1 Historical

13.1.1 The first discoveries

When people first started looking at the sky, they noticed that certain stars were very close together: v Sgr was expressly mentioned as being double by Ptolemy in the 2nd century (this was the first time that the term was specifically used); and the pair Mizar and Alcor were referred to by Al Sufi in the 10th century. Both pairs have separations of about a dozen minutes of arc. After the invention of the telescope, Riccioli examined Mizar in 1651 and found that it was itself double (separation 10.45 arc-seconds). Then came θ Ori (Huygens), ψ Ari (Hooke), α Cen (R. P. Richaud), and γ Vir, which were all discovered, more or less by accident, in the following decade.

These early observers undoubtedly considered such close pairs as exceptional, and caused by perspective. We now class these objects as optical doubles, which may indeed be expected to occur on a purely statistical basis. We now know some hundreds (representing about 1 % of all catalogued doubles). If it had been possible to carry out sophisticated statistical calculations in those days, the accepted explanation would doubtless have been that proposed by two men – who were actually ahead of their time: J. H. Lambert (1761) and John Mitchell (1767). They were certain that many stars that were close together must be gravitationally bound. Some time later, in 1777, the German, Christian Mayer began a systematic search with his quadrant – equivalent to a 60-mm refractor today – very rightly assuming that such a class of specific objects did exist.

In fact the true, modern concept of a binary star is to be credited to Sir William Herschel, even though this came about by chance, as do many major discoveries. Trying to determine the parallax of stars, he measured the position of bright stars (which were assumed to be nearby), against fainter neighbouring stars (which were assumed to be distant, and thus essentially fixed). These measurements consisted of determinations of position angle – using the diurnal motion of the sky as a means of reference – and also of separation. Observing Castor, γ Leo, ϵ Boo, ζ Her, and γ Vir with his famous 20-foot and 40-foot reflectors, he arrived at the conclusion, in 1803, that these stars were bound by their mutual gravitational attraction. From being simple curiosities, these objects contributed to the new view of the universe that was arising at that time. Newton's contributions were vital, however, in enabling Herschel to describe the elliptical orbits of the five binaries that he had examined. For Castor, in particular, he determined the period (the time for one star to complete an orbit around the other) as 342 years. (Modern studies give a value of about 500 years.) This was the beginning of a new branch of astronomy. The only thing lacking was an instrument for making the delicate measurements that were required. The device known as the filar micrometer had originally been envisaged by the Frenchman Azout around 1666, and thanks to advances in instrument-making techniques, it was to be available to Herschel's successors.

13.1.2 The pioneer period

Using his father's telescopes, John Herschel, William's son, and Sir James South began a survey of the sky in 1816, concentrating on the southern sky, where they catalogued 2195 doubles. At the same time, F. G. W. Struve undertook the same task for the northern sky and, thanks to having the first of the accurate modern refractors – the Dorpat telescope, a 24-cm doublet, on an equatorial mounting fitted with an accurate drive – published the fundamental work on binary stars in 1837. Between 1824 and 1836, in his *Stellarum duplicium et multiplicium mensurae micrometricae*, Struve described his methods: systematic searching, use of a filar micrometer, and detailed measurements. These methods subsequently became standard. He catalogued 3134 objects between declinations +90 and −15°, laying the basis for later investigations.

Numerous observers began to add to the foundations thus laid, and most of these were amateurs: Mädler, Smyth, the eagle-eyed Dawes, and Dembowski. The last, working from Naples with 13–18-cm refractors, recorded 21 000 observations. Struve's own son, Otto, should not be forgotten – as well as being the amateur's province, double stars also tend to 'run in the family'. Using what was, for the time, a very powerful refractor, 38 cm in aperture, installed at Pulkova, he discovered 514 doubles, most of which were quite close, between 1843 and 1850.

In 1827, at about the same time that these discoveries were being made, J. Savary, who worked with ζ UMa, and, a little later, John Herschel, published the mathematical methods of determining the orbits of observed binaries. These orbital calculations were required to determine the orbital characteristics: the period, the shape of the ellipse and its inclination to the plane of the sky.

A new era began when American observers entered the scene. Another amateur, S. W. Burnham, discovered his first double in 1870, using a 13-cm refractor. His reputation grew in subsequent years, and he was eventually able to use a powerful telescope (the 47-cm at Dearborn), in collaboration with Hough. He continued his searches, with various large instruments and eventually discovered 1336 doubles. Most importantly, he recognized the need for a comprehensive review of the field and gathered together all the discoveries and measurements that had been obtained, publishing them in 1906 as the *General Catalogue*.

13.1.3 Modern times

Burnham's work began the reign of the large refractors, and this was continued by W. J. Hussey and particularly by R. G. Aitken, who used the marvellous 36-inch (91-cm) at Lick. In 1899 these two observers undertook a methodical survey of the sky, using the AGK listing, beginning at declination −22° and moving towards the North Pole. By 1915 they had discovered 4434 new doubles. Many were very close pairs, with separations of just 0.01 arc-seconds. A new stage in binary-star studies had been opened.

In Great Britain, the Rev. R. P. Espin also discovered 4000 doubles, although, in general, these were less interesting, because he had less powerful instruments. His compatriot Milburn made a similar contribution, using comparable methods. In France, another amateur, R. Jonckheere, earned a deserved reputation, discovering 3350 doubles by 1945, using his 33-cm refractor (which is now at Lille), and also working at the observatories of Greenwich and Marseille. Most of these doubles were faint, specially selected objects.

In the Southern Hemisphere, following John Herschel's work, Innes continued searching from the Cape and Johannesburg. He discovered 1600 doubles. The torch was kept alight by Van den Bos and Finsen, who by 1935 had catalogued 3200 objects. At Bloemfontein, Rossiter achieved a total of over 7000 by 1946.

Once again it became essential to compile a comprehensive catalogue of all these binaries. In 1932, Aitken published the *New Catalogue of Double Stars Within* 120° *of the North Pole* (known as the ADS), containing 17180 double and multiple systems, giving the measurements of each as annual means. Innes had published a catalogue covering the Southern Hemisphere in 1899.

Other observers also played a notable part: Van Biesbroeck at Yerkes; Rabe in Germany; Arend in Belgium; and Voûte in Java represented the Netherlands. In France, interest was sustained by Giacobini, Perrotin and Flammarion (who published the first catalogue of orbits), and then closer to our own time, by exceptional amateurs such as V. Duruy and P. Baize. The latter, who has made more than 40000 measurements, most at the Paris Observatory, has acquired an international reputation. He is a member of the IAU (the International Astronomical Union) and continues to publish orbits of binaries in *Astronomy and Astrophysics*. (He has calculated about 150.) He is honorary president of the Double Star Section of the Société Astronomique de France, and is an example every amateur should seek to emulate.

13.1.4 Other methods of observation

The successive development of new forms of detector has created various sub-classes of binaries. In 1857, Bond applied photography to double-star work by making observations of Mizar (another first), and Rutherfurd developed the method of making photographic measurements. Thanks to the size of the field covered, the positions of the components could be related to those of other stars, enabling absolute positions to be calculated. This was one method by which the wavy path of certain stars could be detected, proving that they had invisible companions, the most famous being Sirius B. These objects became known as astrometric binaries (whether discovered from transit observations or from photographs, as at Sproul Observatory). Photography has been responsible for the detection of numerous faint, wide pairs. It has also increased the accuracy of measurements of binaries that are of moderate separation: between 1.5 and a few tens of arc-seconds.

An even richer field is that represented by spectroscopic binaries, opened up by Pickering in 1889 with Algol, and with Mizar (yet again) in 1890. When the spectrum of a star is examined, a series of bright (emission) lines or dark (absorption) lines is visible. The true positions of these lines are known from laboratory observations. If the star is moving relative to the observer, the lines will be shifted (because of the Doppler–Fizeau effect) by an amount $\Delta\lambda/\lambda = V_r/c$ where λ is the wavelength, V_r the radial velocity, and c the speed of light. If the star is a binary, the lines appear doubled when the orbital motion causes one star to move towards us and the other away from us. This method enables binaries with separations way below a telescope's resolving power to be discovered and, by measuring the spectra, for partial calculations of the orbits to be carried out. Unfortunately it requires large telescopes and expensive equipment, so it is not really suitable for amateurs. Nevertheless, spectroscopy has enabled information about some one thousand close binaries to be gathered.

At about the same time, Fizeau and Michelson suggested the use of interferometry to measure close binaries, and eventually, in 1919, Pease and Anderson at Mount Wilson obtained interference fringes of Capella A and B, which were already known to form a spectroscopic binary. This method is also capable of exceeding the resolution of a telescope, and it has led to the discovery of new, close doubles.

Finally, ways of studying eclipsing binaries have been developed. The variations in the brightness of β Per (Algol) have been known since 1670, and in 1783 Goodricke suggested that the phenomenon could be explained as the effect of a body passing in front of a brighter star. Spectroscopy confirmed his theory. We now know of about 5000 eclipsing binaries: they are of particular interest to astrophysicists. They also give double-star and variable-star observers a common ground of investigation.

13.1.5 *The new wave of observers*

The period 1945–60 was one of consolidation. It was felt that the large American surveys had essentially catalogued all the objects that existed, and that there was nothing left to discover. A team of observers at Nice realized, however, that the searches were incomplete. In the 1960s, P. Muller and P. Couteau began searching again, using a fine 500-mm refractor. Their harvest currently amounts to some 3000 doubles, most of which are very close pairs. Other observers joined in such as Heintz, Worley, and workers at the Belgrade Observatory, proving that after the flurry of activity by Aitken and Hussey, there were still binaries to be discovered. (Which is why someone said to Muller that he had undertaken a labour of Sysiphus.)

Still in France, a new method of measuring binaries, based on interferometry, was introduced around 1970 by A. Labeyrie. This enables large reflectors to be used to the very limits of their resolution. Workers at CERGA [Centre d'Etudes et de Recherches Géodynamiques et Astronomiques – Centre for Geodynamic and Astronomical Studies and Research], (particularly Bonneau and Foy) have specialized in this method, which is known as speckle interferometry. It has also been used by MacAlister at Kitt Peak, and Belega and Tokovinin at Zelenchukskaya. For about a decade it has provided highly accurate measurements of close binaries.

We may also mention P. Lacroute, the main proponent of the Hipparcos mission, which, despite the problems caused by it failing to attain the desired orbit, should still measure many thousands of binary stars, thanks to the work of J. Dommanget (at

Uccle in Belgium). The latter and his team have compiled a modern catalogue (the *Catalogue of Components of Double and Multiple Systems* or CCDM), in machine-readable form, which contains confirmed objects with accurate positions. It replaces the enormous compilation known as the *Index Catalogue of Double Stars* by Jeffers *et al.*, published by Lick in 1963 and containing some 64 000 entries. A catalogue of 847 orbits of visual binaries, by Worley and Heintz appeared in 1983. [The various catalogues are discussed in more detail in Sect. 13.3. – Trans.]

Overall, the number of binaries catalogued amounts to some 100 000, of all classes, which suggests that three stars out of every four are members of binary or multiple systems.

13.2 Why study double stars?

13.2.1 The current situation

Ever since the middle of the last century, when the first catalogues appeared, astronomers have tried to analyze the observations recorded. Statistical information about the frequency distribution versus magnitude, separation, or spectral class has indicated new lines of investigation. One factor that gradually became evident from results was that of observational selection. Because of the limitations of the observational methods there were 'holes' in the coverage. The missing material indicated new fields that needed to be investigated.

Specialists agree that multiplicity occurs in a constant percentage of stars, regardless of magnitude and separation. Currently, however, only 27 % of known binaries are fainter than magnitude 11. A majority of these are distant objects, with components whose true separations are large, and thus have very long orbital periods. But if an object has a large proper motion we are obviously dealing with a nearby, faint star, generally a dwarf: a companion ought to be detectable. Similarly, a pair with a large magnitude difference – known as 'large delta-m' (Δm) in the jargon – must be nearby. This is the case with Sirius and with η Cas. More distant pairs must have magnitude ratios that are closer to unity for them to be detectable.

Table 13.1 shows the distribution versus spectral class and separation, drawn from a total of 70 000 pairs. It illustrates that binary stars have the same distribution as single stars. The deficit at faint magnitudes is rapidly being removed, thanks to recent surveys (by workers such as Couteau, Muller, and Heintz).

It is worth noting that photometric measurements (by P. Muller) show that every spectral class is found among the components of multiple stars.

As far as visual orbits are concerned, we now know about 1000, but this number is likely to rise rapidly with recent discoveries of close pairs, which obviously contain a large number of short-period objects.

One might wonder if there is a connection between multiple stars and open clusters: whether they form a continuous sequence. Figure 13.1 shows that there is a fundamental difference between these two types of object, and this difference is illustrated by Fig. 13.2. Multiple stars are, by their very nature, hierarchical systems, like the Solar System, where we have: Sun–planet–satellite. Thanks to the variety

Table 13.1. *Statistical distributions of 70 000 pairs, after Couteau (1975)*

Spectral class	O	B	A	F	G	K	M
% of double stars	a few	8	25	26	22	16	3

Separation	≤ 0.25″	< 0.5″	< 1″	< 2″	< 5″
% of double stars	2 ↗	7 ↗	14	23	47

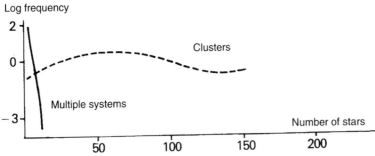

Fig. 13.1. *Multiple stars and clusters: two completely different types of objects (after J. Dommanget).*

and complementary nature of methods of observation, we are able to observe as many as four levels (Fig. 13.3). And by no means all such cases are known ...

These facts give some indication of what still remains to be discovered and to be done, and also what any observational programme should aim to cover. There are numerous blank spots, where amateurs can undertake useful work with the (frequently modest) means at their disposal. We shall discuss these possibilities in following sections.

[Whether observing is being carried out visually or photographically (Fig. 13.4), the basic observational techniques for double-star work are fairly simple, and these are shown in Fig. 13.5. The orientation of the field must be known unambiguously and this may be determined by using the diurnal motion to indicate the East–West direction. It is customary to describe the approximate positions of components in terms of the quadrant in which they are found. This convention is also shown in the Figure.

Scientific work consists of determining the separation, ρ, between the components (normally in arc-seconds), and the position angle (PA or θ), measured 0–360° from North through East, South and West. — Trans.]

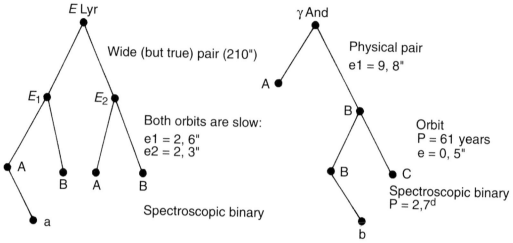

Fig. 13.2. *Two hierarchical multiple systems*

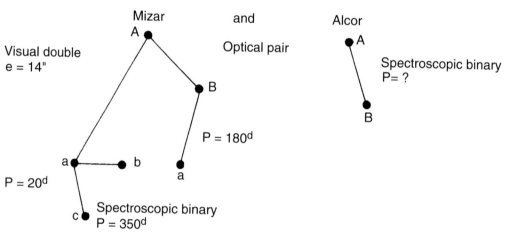

Fig. 13.3. *A complex family, with eight known members.*

13.2.2 Observing for one's own curiosity

The initial interest in observing double stars comes from the satisfaction gained by observing anything unusual in the sky. They are the prime target for any town-dwelling amateur astronomer who wants to move beyond the confines of the Solar System. It is not, indeed, essential to have brilliantly clear skies to start searching for 'those celestial diamonds' (as Flammarion called them). Provided turbulence and image motion are not too great, moderate seeing is adequate. Every form of observation can be satisfying, provided one is realistic in one's expectations of what is possible with the equipment being used. Doubles like ϵ^1-ϵ^2 Lyr, which have separations of up to 30 arc-seconds, are generally accessible with binoculars. Under

Fig. 13.4. *A visual double, photographed with a 200-mm reflector, F = 12.5 m, enlarged 15 times during printing. This is γ Del (magnitudes 4.5 and 5.4). Measurement gives θ = 266.7°, ρ = 9.45″. Date: 1984 August 31. Fujichrome 400, exposure 2 s. (Photo: Durand-Simier)*

good conditions, an optically correct, 400-mm telescope may show a binary with a separation of just 0.25 arc-second. Between these two extremes there are tens of thousands of objects accessible with all sorts of equipment (*see* the brief listings given in Tables 13.4–13.11).

No one can fail to be struck by the gem-like, contrasting colours shown by Albireo or ε Boo, nor by the sight of two adjacent Airy disks, surrounded by distinct rings, as seen on an exceptionally calm night. Neither should we forget the psychological satisfaction of detecting a binary at the limit of the telescope's resolution. Initially we may be unable to tell if the star is round or slightly elongated, and then a sudden moment of calm enables us to detect the shape that is indicative of two components. This is usually enough to cause any observer to shout in triumph.

13.2.3 A test of equipment and observer

Double stars are a good criterion of the optical quality of an instrument. The image of a star, we may recall, consists of a false disk, or Airy disk, surrounded by a series of rings of decreasing brightness. Generally, only the first of these is visible. The radius of the Airy disk mainly depends on the diffraction caused by the circular aperture of the telescope. There is a simple formula for this radius: R (in arc-seconds) = $140/D$ (in mm). This value applies to bright stars, because in fact the brightness of the disk decreases from the centre to the edge. A series of stars of decreasing brightness therefore shows smaller and smaller disks. Beyond a certain threshold, the disk ceases to be visible: all that can be seen is a faint glimmer of light. This is why it is difficult to separate a close binary when the disks are no longer distinctly visible. Experience shows that the disks disappear when stars are less than 4.5 magnitudes above the instrument's theoretical limit. But a lot depends on the observer's eyesight, so Table 13.2 should be taken merely as a guide. The other assumption is that adequate magnification is used. This should be approximately twice the diameter of the objective measured in millimetres.

Table 13.2. *Empirically determined conditions for observing the diffraction disk with a given telescope*

Limiting magnitude	10	9.5	8.5	8	7.3	6
Objective diameter (mm)	400	300	210	150	110	60
Magnification	800	600	420	300	220	120

Under such conditions, it will be found that the diffraction disk forms a natural micrometer. It might be objected that because the size of the perceived disk decreases with fainter stars, it ought to be possible to see closer pairs. In practice it is found, however, that the decline in brightness causes the eye to lose resolution and that the ability to separate two close stars is practically constant whatever the brightness. Table 13.3 indicates how the appearance of a close double may be converted into separation (in arc-seconds). It takes into account the fact that, because of an effect that makes the photocentres appear closer together – *see* the description in Couteau (1981), p. 37 – the geometrical image of the two unresolved components no longer corresponds to the theoretical one.

Something must be said about the limiting resolution of $120/D$ (where D is measured in mm) suggested by Dawes. It corresponds to an arbitrary optimum where the centres of the two objects are separated by 85 % of the radius of the Airy disk. The visible image consists of two, superimposed patches of light, the central area remaining fairly faint. This enables the eye to decide that the object is double. A good objective, good conditions, and an experienced observer can do better. All that is required is to memorize Table 13.3 and compare one's own observations with the values given in the catalogues. The ideal situation occurs when the components are of equal magnitude; some modification is justified in asymmetrical cases. The most sensitive case is when one component is close to, or on, the first ring: i.e., when the separation in arc-seconds is $185/D$ (in mm). (For the second ring, the value is $300/D$.) This is yet another case when the separation may be detected with great accuracy.

For faint pairs, it is essential to remember that resolving power decreases with decreasing brightness. Peterson has published a diagram relating separation and magnitude (Peterson, 1954) derived from examination of about one hundred diverse pairs, using a 75-mm refractor with a magnification of 45×. Such a survey would be of more value if it were to be repeated with an instrument of 100–150 mm in diameter, and a magnification of $2D$.

13.2.4 *Learning to judge separations*

Examining double and multiple stars simply to satisfy one's curiosity can serve a useful purpose in educating the eye to recognize fairly large separations, such as 5, 10, and 20 arc-seconds, where round numbers are of particular interest. Nothing subsequently prevents you from trying to estimate any separations that are encountered, and from noting down the values and checking them in the

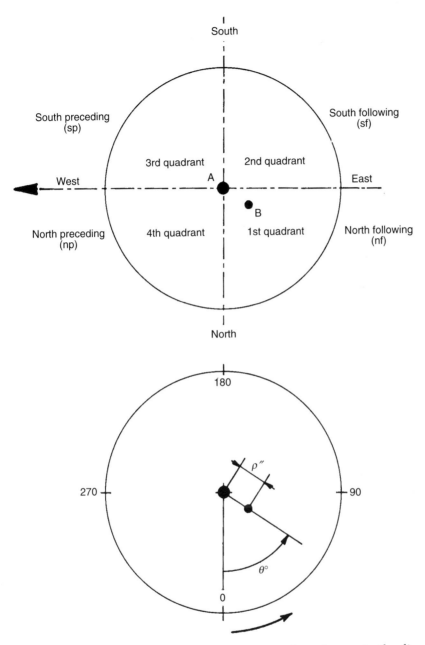

Fig. 13.5. *Aligning a double star in the eyepiece by reference to the diurnal motion; and the principle of making micrometer measurements in polar coordinates. Here the companion (or component B) is in the first quadrant (North following). The position angle θ is 58°, the separation ρ will be read by the micrometer.*

Table 13.3. *Separation and image appearance* [*]

Appearance	Description	Separation relative to spurious disk	Separation for $D = 200$ mm (arc-seconds)
	separated	1	0.7
	tangent	0.95	0.66
	figure of eight	0.90	0.63
	flattened eight $(12/D)$	0.85	0.60
	narrow rod	0.80	0.56
	rod	0.75	0.52

[*] As seen in a 200-mm telescope. For diameters other than 200 mm, the values given should be multiplied by $200/D$ (where D is measured in mm).

catalogues, provided, of course, that you always use the same telescope and the same magnification. This is an excellent way of educating the eye, and may prove of considerable help in making useful observations in fields other than double-star work. Knowing what 10 arc-seconds looks like in the field of an eyepiece is very valuable in making drawings, for example. (*See* Tables 13.9, 13.10 and 13.11 for suitable doubles.)

13.2.5 Continuing work on doubles

People have been measuring double stars for a couple of centuries. This is a long time, and it is a good example of the continuity of scientific observations. But it is very short when it comes to determining orbits that have periods of several centuries, or even a thousand years. Generations of double-star observers have contributed their measurements to ensure continuity. It is up to us to accept the challenge, and ensure that there is no break in the series. There is all the more work, because we also have to cover the binaries that have been discovered recently.

Until the 1950s, many observers, both amateur and professional, were still making measurements. Since then, the number has gradually declined, until now there are hardly any left. Traditionally, amateurs have always occupied a prominent place in this field. We only have to mention the names of Dembowski, Espin, Jonckheere, Baize, and Duruy to realize this. Some of the national amateur organisations have specific double-star sections, but few have many observers. At least two or three times as many are needed to ensure that the required number of re-observations are carried out.

13.2.6 *Educating the eye*

Amateurs starting to observe binaries are frequently disappointed in not being able to resolve a pair of stars that they are shown. But they should not lose heart. The eye is an exceptionally sensitive instrument, capable of adapting to major changes in illumination with factor of about one million. It works rapidly, and can follow rapid, tiny movements seen in the eyepiece. Even better, because it is connected to the finest computer, which analyses the information received in real time, it can adapt and learn, provided it is kept in good condition. But it also needs regular practice to maintain its abilities.

When everything is in its favour, learning may take place in just a few weeks. One friend, using a 115-mm reflector, spent his holidays examining the doubles in the small constellation of Equuleus, and managed to resolve the components of ϵ Equ (with a separation of $1''$) after just a few observations, at intervals of several days. Others were unable to resolve it initially, even with a 200-mm reflector. But the price that has to be paid for this ability is that the eye may be influenced by the brain, and be mistaken in what it sees. Some ghost images are caused by subconscious suggestion, and this effect may even go so far as to falsify measurements. So be wary! It is best to repeat an observation over several nights if there is any doubt.

13.3 Catalogues

The choice of observational programmes, whether they are chosen simply to satisfy one's curiosity, or with a view to making actual measurements, are always based on catalogues of double stars. There are many of these. Even though the catalogues used by professional astronomers may be hard to obtain, it is easy enough to acquire others that give specific selections from the larger catalogues, perhaps containing orbital pairs, binaries particularly suitable for amateurs, spectacular pairs, etc.

At present, the oldest catalogue that one may sometimes be lucky enough to obtain is the ADS, *New General Catalogue of Double Stars Within 120° of the North Pole* by Aitken and Doolittle, which was published in 1932 in two volumes (Aitken, 1932). It covers 17 180 objects down to declination $-30°$, giving the details, and often the mean, of the measurements of a pair made since its discovery. It is useful for anyone calculating orbits who wants to reconstruct the motion of the companion.

Although very difficult to obtain (but microfiche copies are held by certain astronomical libraries), the next catalogue to consider is the IDS, *Index Catalogue of Visual Double Stars, 1961.0*: two volumes published by Lick Observatory (Jeffers *et al.*, 1963). This covers 64 250 objects over the whole sky, but only gives single measurements, or two if motion is noticeable. Unfortunately, it suffers from certain errors, and includes a large number of poorly determined photographic pairs. (The magnetic version of this catalogue is constantly revised at both Washington – which holds the primary index – and Nice.)

There is also the *Cinquième Catalogue d'Ephémérides d'Etoiles Doubles*, published by the Paris Observatory in 1986 (Couteau *et al.*, 1986). It contains data on 886 objects (975 orbits) and ephemerides for the years 1988–98. The three catalogues just mentioned may be held by some of the sections that coordinate double-star observations within some of the national amateur societies.

Other catalogues are drawn from those just mentioned. The most important are: the catalogue of 2500 double stars compiled by J. Minois (1978), arranged according to constellation and with orbital ephemerides; and *Sky Catalogue 2000.0*, Vol. II (Hirshfeld & Sinnott, 1985), which contains – as well as other information – several thousand binaries, generally drawn from the ADS.

There are various books and handbooks dealing with practical astronomy in general, and with this subject in particular, that give fairly extensive catalogues of double stars. Mention must be made of Paul Couteau's book (Couteau, 1981), which contains a list of 744 objects classified by right ascension, including close pairs down to 0.2 arc-seconds, which will test the largest amateur instruments.

Finally, we should mention the new, accurate catalogue prepared by Jean Dommanget at Uccle, which is in machine-readable form. The IDS was taken as a basis, but was checked, purged of erroneous data, and enhanced. The catalogue does not list systems, but gives a line for each component, whence its name *Catalogue of Components of Double and Multiple Systems* (CCDM). Unfortunately, there is little hope that it will ever be published in paper form.

The catalogues mentioned, and the details given in Tables 13.4–13.11, generally contain the following columns of information. First, the designation of the star: its name if it is bright and the abbreviation indicating the discoverer, followed by his identification number – H for Herschel, STF or Σ for F. W. Struve, STT or OΣ for his son Otto, A for Aitken, Cou for Couteau, Mul for Muller, etc. Sometimes the ADS number is given. Next comes the position of the object in equatorial coordinates α and δ (together with the epoch: 1900, 1950, or 2000), to a tenth of a minute in α and one minute of arc in δ. This is sufficient to identify the object visually through the eyepiece. Then come the visual magnitudes m_1 and m_2 of the components, and finally the companion's position relative to the primary, in polar coordinates, where N is the origin of the angle θ, and ρ is the separation in arc-seconds. Additional notes are sometimes given.

13.4 Techniques: Visual methods

13.4.1 Choosing a suitable instrument

Measuring double stars is astrometric work of a very particular kind. A long focal length is required to obtain the necessary accuracy. For the image to be measurable, it needs to be sufficiently large for the method of measurement that has been selected. The linear size at the focal plane that corresponds to an angle of 1 arc-second is

Table 13.4. *Ten wide, binocular doubles* *

Name	ADS	2000.0 coordinates	m_1, m_2	θ (for 1987)	ρ	Notes **
α Cas H 18	561 AD	$00^h40.5, +56°32'$	2.5, 8.5	280°	64.4″	optical, widening
ψ Psc Σ 88	899 AB	$01^h05.6, +21°28'$	5.6, 5.8	159°	30.0″	
λ Ari H 12	1563 AB	$01^h58.1, +23°35'$	4.9, 7.7	46°	37.4″	difficult
θ_2 Ori Σ 16 Ap.1	4188 AB	$05^h35.4, +05°25'$	5.2, 6.5	92°	52.5″	triple, difficult, in Nebula
ι Cnc Σ 1268	6988	$08^h46.7, +28°46'$	4.2, 6.6	307°	30.5″	probably optical
ν Dra Σ 35 Ap.1	10628	$17^h32.2, +55°11'$	5.0, 5.0	312°	61.9″	optical
ζ Lyr	11639 AD	$18^h44.7, +37°36'$	4.3, 5.9	150°	43.7″	
β Cyg Σ 38 Ap.1	12540	$19^h30.7, +27°58'$	3.2, 5.4	54°	34.4″	contrasting colours
61 Cyg Σ 2758	14636 AB	$21^h06.3, +38°39'$	5.6, 6.3	147° 148° 150°	29.5″ 29.7″ 30.3″	binary: (1990); (2000)
δ Cep Σ 58 Ap.1	15987 AC	$22^h29.2, +58°25'$	3.8–4.6, & 7.5	191°	41.0″	A is var.

* Assumes mounted binoculars, and comfortable observing conditions.
** Dates in parentheses indicate predicted values of θ and ρ.

F/206 265. Figure 13.6 shows that a focal length of more than 2 m is required for the errors in the measurements (which are about 0.01 mm) to be approximately 1 arc-second. However, this is inadequate for our purposes. The solution is therefore either to use an instrument with a long focal ratio (F/D) and a focal length of at least 3 m, or, in the case of a Newtonian, to increase the primary focal length by 2.5–3 times by using a Barlow lens. An effective focal ratio of f/15 to f/20 is a suitable value.

The finder should be relatively powerful, with an aperture of roughly one-quarter of that of the main telescope: for example 60 mm for a 210-mm aperture, or 76 mm for a 310-mm objective. It should enable one to see stars whose Airy disks are visible in the main telescope and, in particular, by use of an appropriate magnification, should allow objects to be found easily even when using the most powerful magnification on the main telescope.

Although any instrument may be used for initial sight-seeing of double stars, the criteria described and the necessity for good definition means that 100 mm is the

Table 13.5. *Six doubles for 60–70-mm refractors*

Name	ADS	2000.0 coordinates	m_1, m_2	θ (for 1987)	ρ	Notes[*]
α Pis	1615	$02^h02.0, +02°46'$	4.2, 5.2	275°	1.6″	binary
Σ 202				268°	1.5″	(1995)
τ Oph	11005 AB	$18^h03.0, -08°11'$	5.2, 5.9	279°	1.8″	binary
Σ 2262				281°	1.7″	(1995)
ϵ^1 Lyr	11635 AB	$18^h44.3, +39°40'$	5.1, 6.1	353°	2.6″	binary
Σ 3282				352°	2.6″	(1995)
ϵ^2 Lyr	11635 CD	$18^h44.3, +39°40'$	5.1, 5.4	88°	2.3″	binary
Σ 2383				78°	2.4″	(1995)
2 Equ	14556	$21^h02.2, +07°11'$	7.7, 7.4	218°	2.8″	rather faint
Σ 2742						in a 60-mm
ζ Aqr	15971 AB	$22^h28.8, -00°02'$	4.3, 4.5	208°	1.8″	binary
Σ 2909				199°	2.0″	(1995)

[*] Dates in parentheses indicate predicted values of θ and ρ.

Table 13.6. *Five doubles for a 100-mm refractor or 110-mm reflector*

Name	ADS	2000.0 coordinates	m_1, m_2	θ (for 1987)	ρ	Notes[*]
–	1709	$02^h14.1, +47°29'$	6.6, 7.1	272°	1.1″	binary
Σ 228				275°	1.1″	(1995)
52 Ori	4390	$05^h48.0, +06°27'$	6.0, 6.1	211°	1.4″	closing slowly
Σ 795						
57 Cnc	7071	$08^h54.2, +30°34'$	6.1, 6.6	315°	1.4″	C, 9^m at 55″
Σ 1291						
η CrB	9617 AB	$15^h23.2, +30°18'$	5.6, 5.9	18°	0.9″	binary: $P = 41$ yr
Σ 1937				42°	1″	(1995);
				64°	0.8″	(2000)
ϵ Equ	14499	$20^h59.1, +04°18'$	5.8, 6.3	285°	1.0″	binary
Σ 2737				248°	0.9″	(1995);
						C, 7^m at 10.5″

[*] Dates in parentheses indicate predicted values of θ and ρ.

minimum size for any serious measurement work. Paul Baize, as a young beginner in the 1930s, may have found it easy enough to use an altazimuth mount, but it is much faster with an equatorial, which is more common nowadays. The latter type enables any object to be followed easily, and also means that it may be located using equatorial coordinates. If the star cannot be seen by the naked eye, this form

Table 13.7. *Five doubles for a 200-mm reflector*

Name	ADS	2000.0 coordinates	m_1, m_2	θ (for 1987)	ρ	Notes[*]
66 Psc OΣ 20	746 AB	$00^h 54.6, +19°12'$	6.2, 6.9	215° 205°	0.49″ 0.51″	$P = 360$ yr, (1995)
γ And OΣ 38	1630 BC	$02^h 03.5, +42°23'$	5.5, 6.3	106° 105°	0.56″ 0.52″	A-BC = 9.8″ (1995)
– OΣ 235	8197	$11^h 32.4, +61°05'$	5.5, 6.3	270° 315° 4°	0.57″ 0.61″ 0.7″	$P = 71$ yr (1995); (2005)
– Σ 2315	11334	$18^h 25.0, +27°23'$	6.5, 7.5	127° 125°	0.72″ 0.76″	$P = 775$ yr (1995)
52 Peg OΣ 483	16428	$22^h 59.2, +11°44'$	6.1, 7.4	311° 323°	0.67″ 0.67″	$P = 286$ yr (1995)

[*] Dates in parentheses indicate predicted values of θ and ρ.

Table 13.8. *Five doubles for a 300-mm reflector*

Name	ADS	2000.0 coordinates	m_1, m_2	θ (for 1987)	ρ	Notes[*]
– BU 4	1097 AB	$01^h 21.3, +11°32'$	7.4, 8.0	109° 105°	0.43″ 0.45″	$P = 180$ yr (1995)
– Σ 511	3098	$04^h 17.9, +58°47'$	7.5, 7.9	95° 83°	0.39″ 0.45″	$P = 254$ yr (1995)
– STT 298	9716AB	$15^h 36.1, +39°48'$	7.4, 7.6	257° 272° 28° 171°	0.32″ 0.29″ 0.26″ 0.8″	$P = 55$ yr (1988); (1995); (2005)
73 Oph Σ 2281	11111AB	$18^h 09.5, +04°01'$	6.1, 7.0	309° 285°	0.40″ 0.45″	$P = 270$ yr (1995)
72 Peg BU 720	16836	$23^h 34.0, +31°20'$	6.0, 6.0	87° 93°	0.53″ 0.51″	$P = 240$ yr (1995)

[*] Dates in parentheses indicate predicted values of θ and ρ.

of finding is very rapid, whether one uses the offset method (locating an object by setting the telescope on an easily visible star of known position, and then using the difference in right ascension and declination), or by employing sidereal time (ST), where HA = ST − α, where HA is the hour angle to be set on the circle and the meridian is obviously 0^h.

Perfect tracking is not required, because it is often necessary to shift the position

Table 13.9. *Five doubles with separations of about 5 arc-seconds*

Name	ADS	2000.0 coordinates	m_1, m_2	θ (for 1987)	ρ	Notes*
– Σ 48	582	$00^h42.6, +71°21'$	8.0, 8.2	334°	5.4″	C, 7^m at 55″
– Σ 958	5436	$06^h48.3, +55°43'$	6.3, 6.3	257°	4.8″	
ζ Cnc Σ 1196	6650 BC	$08^h12.2, +17°40'$	5.0, 6.3	78°	5.9″	primary double: 0.6″
π Boo Σ 1864	9338	$14^h40.7, +16°26'$	4.9, 5.8	110°	5.5″	
α Her Σ 2140	10418 AB	$17^h14.6, +14°23'$	3.5, 5.4	106°	4.6″	primary variable

* Dates in parentheses indicate predicted values of θ and ρ.

Table 13.10. *Five doubles with separations of about 10 arc-seconds*

Name	ADS	2000.0 coordinates	m_1, m_2	θ (for 1987)	ρ	Notes
γ And Σ 205	1630 A-BC	$02^h03.5, +42°23'$	2.2, 5.1	63°	9.8″	given for BC, Table 13.7
1 Cam Σ 550	3274	$04^h32.0, +53°55'$	5.7, 6.8	308°	10.3″	
– Σ 1245	6886 AB	$08^h35.8, +06°37'$	6.0, 7.0	26°	10.2″	
– Σ 1904	9493	$15^h04.2, +05°30'$	7.0, 7.0	347°	10.1″	
γ Del Σ 2727	14279	$20^h46.7, +16°08'$	4.3, 5.2	267°	9.6″	

of the object in the field of view, but the slow motions should be accurate. It is particularly important to keep a check on the polar alignment, otherwise errors in position-angle measurements may accumulate.

A question often asked is: refractor or reflector? Traditionally, double-star observers have used refractors in Germany, France, the U.S.A., etc. – and reflectors in Great Britain (in particular). With instruments above 200-mm aperture, there can be no doubt that a refractor gives images of better quality. This arises from the closed tube, and the lack of secondary obstruction. The Airy disk is also sharper, because of the effects of the secondary spectrum. On the other hand, a 300-mm reflector has a greater light-grasp and is less expensive than a 200-mm refractor.

Table 13.11. *Five doubles with separations of about 20 arc-seconds*

Name	ADS	2000.0 coordinates	m_1, m_2	θ (for 1987)	ρ	Notes
χ Tau Σ 528	3161	$04^h22.6, +25°38'$	5.5, 7.6	24°	19.4″	
24 Com Σ 1657	8600	$12^h35.1, +18°23'$	5.2, 6.8	271°	20.2″	
α CVn Σ 1962	8706	$12^h56.1, +38°18'$	2.9, 5.5	229°	19.5″	Well-known double
61 Oph Σ 2202	10750 AB	$17^h44.6, +02°35'$	6.2, 6.6	93°	20.6″	
8 Lac Σ 2922	16095 AB	$22^h35.8, +39°38'$	5.8, 6.6	186°	22.3″	5 components in system

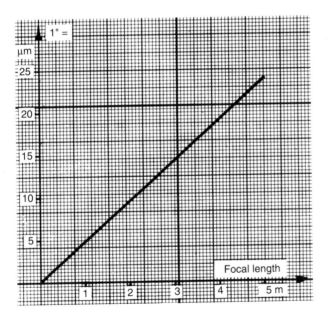

Fig. 13.6. *Focal-plane scale of an instrument. The ordinate gives the size of the image, expressed in μm, represented by an angle of 1 arc-second.*

There are undoubted advantages in having a closed tube, but few amateurs have tried this form, and then made comparative tests. The possibility of fixing the secondary mirror to an optical window, and thus eliminating the diffraction spikes caused by the legs of the spider should offer considerable advantages. With bright stars, the diffraction reinforces the rings and may prevent a faint companion from being observed.

13.4.2 *Simple equipment for starting to make measurements*

Quite apart from their value to the scientific community when they are published, measurements made over several years – and in a few cases even over several months – enable one to detect changes in the appearance of binaries that are in motion. One way of living in suspense – albeit suspense that is easily withstood – is to wait and see if the motion is in accordance with predictions.

All that is required is a protractor and a graduated eyepiece. The eyepiece should be 6 or 8 mm in focal length, giving a magnification of between 2 and $2.5 \times D$. A ring, which carries a nylon thread stretched across a diameter, should be cemented to the eyepiece's diaphragm, so that it is in sharp focus. A good way of obtaining a fine thread is to extract one from a piece of nylon or Terylene fibre. By first fixing it to a bowed piece of cardboard to keep it under tension, it may then (under a magnifying glass) be carefully cemented to the ring with two drops of glue. After being carefully centred with the eyepiece itself, the protractor is fixed to a ring that may be rotated around the focussing mount and locked in any position. The protractor should be located sufficiently far back so that it does not touch the observer's nose. An index, parallel with the cross-hair, is marked on the outside of the eyepiece. (In fact, two indexes at 180° that are read simultaneously will counteract any error in centring.) The field or the cross-hair may be illuminated by an LED supplied by a 4.5 V battery in series with a variable potentiometer rated 10–100 Ω.

Measurement of the position angle θ is obtained by taking the difference between the angle indicated by the diurnal motion (adding or subtracting 90° to obtain North), and the direction indicated by the line between the two components (see Fig. 13.7). The image is allowed to trail by halting the drive, and the eyepiece is turned until a chosen star close to the equator exactly follows the cross-hair from one side of the field to the other. The value given by the index against the protractor is noted. The latter may also be adjusted to indicate 270° (West). If this procedure is carried out three or four times and the average calculated, the direction of North will be known accurately. The protractor should now be left undisturbed, the telescope pointed at the double and the images bisected several times; in other words the cross-hair is aligned so that it bisects the centres of the images of the two stars (the quadrant in which the companion lies being noted previously). A mean is taken, from which the value, previously obtained (i.e., for North), is subtracted, giving the angle θ, the binary's position angle.

Measurements of the separation are made by the transit method. This consists of halting the drive, aligning the cross-hair North–South, and in timing the interval between the transits of the two stars across the cross-hair. This duration corresponds to the separation between the components and is also dependent on declination. The separation is calculated from $\rho(\text{arc-seconds}) = t \times 15 \cos \delta / \sin \theta$, where t is the interval in seconds (to an accuracy of at least 0.1 s), δ is the declination, and θ the value just derived. Four or five timings will give a fairly sensible average value for ρ if the star has a moderately high declination, and if the observer's reaction time is the same as each star transits the cross-hair. This method can give separations to about 0.5 arc-second, or even better.

This method becomes imprecise close to 0° or 180°. The cross-hair may then be

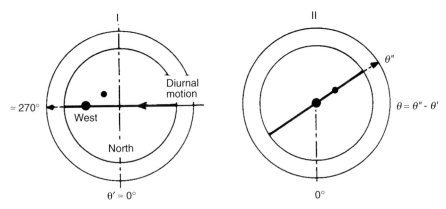

Fig. 13.7. *Measuring the angle θ with a cross-hair reticle*

inclined at 45° relative to the meridian, as shown in Fig. 13.8. The procedure is as follows:

A   the interval of time, $t = t_2 - t_1$, between the two bisections is measured; the image has then moved through an angle x, such that $x = 15t \cos \delta$, where x is in arc-seconds, and t in seconds of time.

B   illustrates the relationship between the measured value of x and the separation ρ between the two stars. The diagram shows how, when the cross-hair is oriented as the first bisector (i.e., approximately SE–NW):

$$x = \rho \mid \cos \theta + \sin \theta \mid$$

When the cross-hair is oriented as the second bisector (NE–SW), this becomes:

$$x = \rho \mid \cos \theta - \sin \theta \mid$$

The orientation of the cross-hair should be chosen to give the largest value of t.

This transit method only applies to pairs that are separated by at least 15 arc-seconds. It does have the advantage of introducing the observer to quantitative measurements with the minimum equipment. For doubles in contact, the diffraction figure serves as a scale (as already shown in Table 13.3).

13.4.3 The rotatable micrometer

The system just described may be improved easily by cementing a second cross-hair near, and parallel to, the first. The two threads should be completely parallel, and this may be achieved by first cementing them against one another, and then separating them with a small piece of hair that is pushed up to each end and cemented to the ring. This delicate task may be carried out using a fine pair of tweezers, working under a loupe or linen-counter.

731

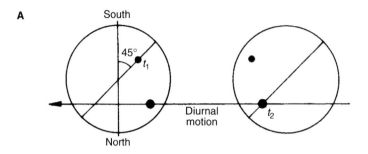

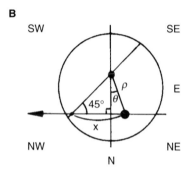

Fig. 13.8. *Method of measuring transits with the cross-hair at 45° to the North–South line, when θ is close to 0° or 180°. (For description, see text)*

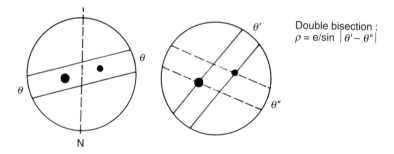

Fig. 13.9. *Using the rotatable micrometer*

When measuring θ, this arrangement avoids having to bisect the images, which is difficult when a thread hides the stars. The two stars are located between the cross-hairs and aligned parallel with them. For ρ, the micrometer is next turned until each star is bisected by a thread (Fig. 13.9), when the angle θ' or θ'' is read. The difference between these two angles i, enables one to calculate ρ from a knowledge of the separation e between the two cross-hairs (obtained by calibration using pairs with well-established separations, such as those given in Table 13.5): $\rho = e/\sin i$.

Four or five readings are still required to obtain a valid mean. This method is more sensitive than the preceding one for separations of about 2–7 arc-seconds, if the projected separation of the cross-hairs against the sky is 2 arc-seconds. Beyond that, as will be readily understood, a small uncertainty in the angle soon leads to an unacceptable repercussion on ρ. Used in conjunction with the first method described, however, it can cover all possible separations.

13.4.4 The 'V' micrometer

In this design the threads are not stretched across the eyepiece, but consist of lines on a transparent support (Fig. 13.10). The most difficult task consists of photographing the test image on high-definition film – similar to Recordak 5786, and developed in fine-grain developer – in such a way that the support remains as clear as possible. The image consists of a V on a background of four sheets of black, high-quality drawing paper, glued together, and fixed to a panel. The V consists of very white thread, 0.5 mm in diameter, stretched between pins. The branches of the V form an angle of 1.6°. On the panel it is 480 mm high and 13.4 mm wide. It is graduated from 1 to 10 and crossed by a thread at right angles that serves as the 5th graduation. To increase its range of application, we can add a thread that runs parallel to one side, 13.4 mm away. The same procedure may be used on the other side. [The dimensions and angles are, of course, arbitrary, and may be varied as the observer sees fit. The graduations must be calibrated by observing doubles with known separations. – Trans.] This arrangement is photographed – for example, at a distance of 6.6 m, using a 55-mm lens, set at f/5 to f/8 to obtain the best definition. (For more details, *see* Durand, 1979.)

Indirect lighting, located either in the eyepiece tube or close to the objective, and linked to a potentiometer, is fitted as described previously (Fig. 13.11). The V is mounted close to the focal plane of the eyepiece, and very carefully focussed. Take the precaution of cutting out a free sector, before mounting the reticle in place, so that any necessary checking may be carried out.

The position angle is measured using the transverse line, after having determined the direction of diurnal motion – which establishes the position of the North point – at the beginning of the session. Separations are measured by finding the point at which the two branches of the V bisect the stars, one beginning from 0, and again beginning from 10. This alternation is repeated, thus obtaining four values, from which one derives the mean. With a focal length of 3 m, the scale is about 0.7 arc-second per division, which allows a range of measurement from 1–14 arc-seconds; the fourth line may be used to give measurements up to 21 arc-seconds, which enables most of the interesting pairs to be included.

The only problem is that the illuminated negative, even when it is very transparent, absorbs light, and may cause as much as one magnitude to be lost. The accuracy that can be attained, about one-tenth of a graduation, means that the error is about 0.1–0.2 arc-second, depending on the care with which the system is calibrated (*see* Tables 13.12 to 13.14, pp. 758–759).

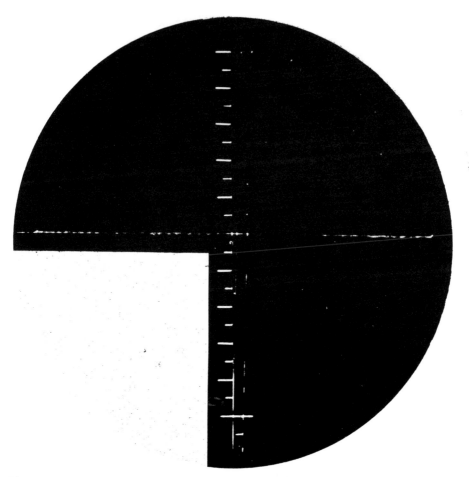

Fig. 13.10. *A 'V' micrometer. A negative print of the actual graduations. The scale runs from 0 to 10 arc-seconds or from 10 to 20 arc-seconds, depending on which of the two parallel lines is used.*

13.4.5 The diffraction micrometer

Diffraction phenomena may be used as a standard by inserting a grating consisting of a regular series of opaque bars and intervening spaces in the light-path (Fig. 13.12). Such a grating placed in front of the objective causes a star to appear as a series of successively fainter satellite images, aligned perpendicular to the axis of the grating. (In fact, the images are short spectra.) The angular distance between the main (zero-order) image and the first-order images is α (in arc-seconds) $= 206\,265\lambda/P$ where λ is the wavelength of peak visual sensitivity (560 nm) and P the spacing of the grating (one bar and one space). For example, for $P = 10$ mm (5-mm strips), $\alpha = 11.6$ arc-seconds (which is of the same order as the values to be measured).

One rapid way of making such a grating is to cement strips of magnetic tape, either from a cassette (3.8 mm wide) or of the reel-to-reel type (6 mm wide), next

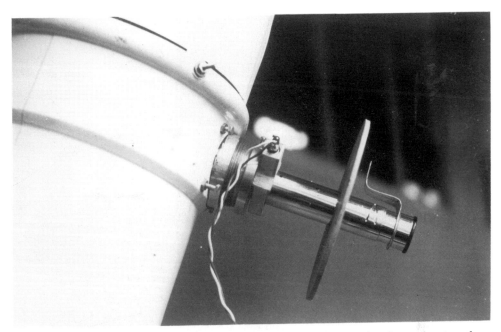

Fig. 13.11. *A 'V' micrometer mounted on a telescope with a Barlow lens, showing the indirect lighting arrangement on the focussing mount, the protractor that is also fixed to it, and the index mounted on the eyepiece.*

to one another on a suitable ring, and then remove alternate strips. The grating is mounted on a support with 360° graduations that may be rotated, and the whole assembly is mounted in front of the objective. (The grating obviously needs to be slightly greater in diameter than the objective itself.)

A fairly strong eyepiece is required, graduated as described on p. 730, with an index reading against a suitable protractor mounted on the eyepiece tube. By turning the grating, some striking image patterns may be created (triangles and rectangles), which have to be solved for the angles. There is, however, a faster method.

Once the direction of diurnal motion, and hence of North, has been established by using the reticle, all that has to be done is to turn the grating until a suitable triangle is obtained, with the double, AB, and a secondary image A′ as the corners (Fig. 13.13). The reticle and its index enable the angles θ, θ_1, and θ_2 to be established. Then:

$$AB = AA' \sin | \theta_1 - \theta | / \sin | \theta_2 - \theta | .$$

AA′ corresponds to the scale of the grating in seconds of arc. Several measurements are used to obtain means of θ and ρ.

The two main disadvantages with this method are: two or three gratings are required to cover all the separations between 1 and 30 arc-seconds; and it requires plenty of light (there is a loss of 1.5 magnitude in the primary image). It is therefore

735

Fig. 13.12. *A diffraction grating mounted on a reflector*

only suitable for bright pairs with low Δm, unless one is using telescopes that are 350–400 mm in aperture.

One possible improvement is to make the bars with copper wire or nylon thread, 1–2 mm in diameter. However, the first of the secondary images then becomes fainter. It is also possible to remove some of the strips from the centre and edges to obtain the same result (Pither, 1980).

The accuracy of this device with regard to separations depends on the difficulty in making the observations. Again, it is approximately 0.1 arc-second for an experienced observer. (For more details, *see* Minois, 1984.)

13.4.6 The filar micrometer

This is the device that has been generally used since the time of W. Struve. It has the advantage of making no alterations to the beam of light, thus preserving the full

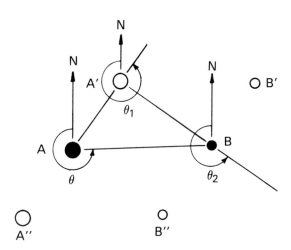

Fig. 13.13. *Measuring the angles (with a cross-hair eyepiece) using a diffraction micrometer (A', A'', B' and B'' are the secondary images).*

quality of the image given by the objective. Forms of this device are manufactured in Great Britain and the U.S.A., but they are expensive and may not satisfy the observer. Similar instruments are made in France.

A micrometer may be built with some help from anyone with some skill in precision engineering, as has been done by a number of amateurs in various countries, including France (see Fig. 13.14) and the U.S.A. (*see* Robertson, 1985, or Ingalls, 1946). The main reference work remains (Danjon & Couder, 1935), but we repeat the basic details here.

It must be light and not bulky. A range of motion of several millimetres for the frame that holds the movable cross-hair is sufficient. This slide must move in a plane that is exactly perpendicular to the optical axis, which means that the guides must be perfectly flat and dovetail in shape. To be sufficiently sensitive, the movement should be governed by a screw of 0.5-mm pitch. If this screw also provides the measuring element, using a micrometer-type arrangement, contact with the slide should be via a small ball-bearing. Tension springs keep the slide in contact with the ways and against the screw. A simpler arrangement is to mount a high-quality dial gauge, reading to at least 0.01 mm, opposite the screw.

The cross-hairs may be reduced to two: one fixed to the outer frame, and the other movable, cemented to the slide. These hairs should be in nearly the same plane, just brushing past one another. They may be cemented with beeswax: fine adjustments may then be made under a linen-counter and with a heated metal point. The wax may be carefully pushed into place to position the hairs exactly. The cross-hairs are traditionally made from spider's web, but artificial fibres of 5–7 μm in diameter may be obtained from textile firms. Side-lighting may be provided by using a yellow LED, controlled by a potentiometer.

The mountings of eyepieces need to be adapted appropriately. With Plössl or orthoscopic designs, they must be shortened so that the optics may be brought sufficiently close to the cross-hairs for these to be sharply seen in the eyepiece. This

Fig. 13.14. *A filar micrometer built by J. L. Agati. A: screw providing the longitudinal motion of the frame carrying the cross-hairs; B: sliding frame with return springs; C: eyepiece; D: dial gauge (measures the movement of the movable cross-hair); E: 360° graduated circle; G: plate, which allows the whole assembly to be rotated.*

means that the diaphragm must be eliminated, and also that the exit pupil needs to be reduced: if the eye is not fully centred parallax errors may arise.

The first adjustment consists of checking that the eyepiece is sharply focussed on the cross-hairs, altering the distance appropriately. The whole assembly is then moved to bring the image of the star into sharp focus, when it is in exactly the same plane as the cross-hairs. This may be checked by moving the head slightly: there should be no shift of the image relative to the cross-hairs.

Like weighing objects with a balance, a method of double measurement is used with a filar micrometer, and this is shown in Fig. 13.15. This procedure has the advantage of only involving the play on the screw once for the two measurements: one series of bisections is made rotating the micrometer screw in one direction, and another in the opposite direction. Four checks (or six if there is any uncertainty) may be averaged to obtain a measure of the separation, accurate to 0.01 mm. This average must be multiplied by a scaling factor, expressing the screw's pitch in arc-seconds. This scaling factor should have been obtained previously (*see* p. 754).

The position angle is measured as described earlier, using the two cross-hairs, which are brought up to the double, more or less from opposite sides. The angle is read on the protractor, which is fixed to a plate, against which the body of the micrometer, carrying the index, is free to rotate. The average of four or five readings

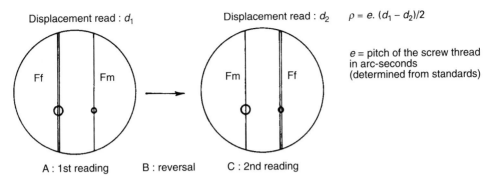

Fig. 13.15. *The double-measurement technique using a filar micrometer. Ff: fixed cross-hair; Fm: movable cross-hair.*

may be calculated to an accuracy of $0.1°$. The procedure is repeated on three or four nights to obtain an overall mean.

Filar measurements should always be made with the line joining the two eyes parallel to that formed by the two stars. The eye is then at its most sensitive, but the positions required are sometimes difficult to hold. This is why making a measurement should not last too long, otherwise fatigue affects the accuracy of the results, which is approximately 0.1 arc-seconds (for separations of 1–20 arc-seconds), but sometimes considerably better if the observer is experienced and has first-class equipment.

13.4.7 The comparison-star micrometer

Observing with both eyes open, a fairly bright double star is projected into the field of view. One can picture this as providing a fixed pair of graduations, alongside the actual double, that may act as a scale for measurements. In this case the scale depends only on the magnification being used. The operation is somewhat like the camera lucida arrangements that enable one to make drawings at a microscope. Using this principle, V. Duruy (who also devised the diffraction micrometer) conceived a binocular arrangement in 1930. In this design, the scale is given by a pair of points of light with variable separation: two slits in a 'V'-shaped arrangement slide behind another slit set at right-angles to the axis of the 'V', which is lit from behind (Fig. 13.16).

The problem here is that the physiological effects caused by differences in accommodation between the eyes is likely to introduce a variable for which allowance cannot be made, thus affecting the measurements. On the other hand, the light-path is completely unobstructed, and there is no need to illuminate the field. It needs to be tried.

In 1931, Davidson and Symms constructed a somewhat more sophisticated arrangement, which had previously been suggested by Hargreaves. A Wollaston prism is used to form two images, whose separation is a function of the distance between the source and the prism. Two Nicol prisms allow the brightness of each bundle of

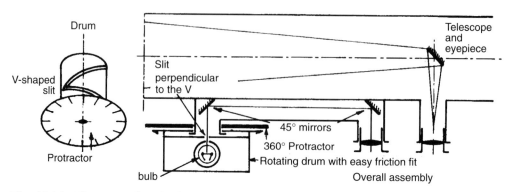

Fig. 13.16. *The principle of a binocular comparison micrometer*

rays to be altered. A 45-degree mirror in the centre of the tube reflects the image towards the eyepiece, so an observer has two, adjustable, comparison images in the same field as the double being measured.

13.4.8 *Double-image micrometers*

In 1935, an amateur, Silva, suggested a system similar to that used in the heliometer: a Barlow lens split across a diameter, one half being fixed and the other being held in a mounting that may be moved by means of a screw (*see* Ingalls, 1946). The device gives two images of a star and, as in the filar micrometer, the separation is proportional to the displacement. With double stars, the images are arranged to fall in specific patterns to which the eye is particularly sensitive: a linear arrangement with regular spacings, a square, or a rhombus. This is possible because the images may be moved by turning the mounting of the device.

The most satisfactory arrangement was discovered by Paul Muller in 1937, and has been widely used since then. It uses two (birefringent) quartz prisms. Their hypotenuse faces are cemented together, with perpendicular cystallographic axes. The whole may be traversed at right-angles to the optical axis, which causes a greater or lesser separation of the images. The prism is mounted on the same sort of movable frame as that in a filar micrometer. The separation of the images is sensitive to the slightest displacement, so the system acts to amplify any translation. For this reason, the device is more sensitive than a filar micrometer. However, the image separation varies across the field, in what is known as the *differential micrometric effect*. This may be largely corrected by using a smaller, similar, compensating prism set close to the eyepiece. This eyepiece needs to be in the form of a small microscope to gain access to the image plane, which lies within the prism. (For a full description of this micrometer *see* Muller, 1939 and 1972.)

Figure 13.17 shows the various ways in which the separation of the pair may be measured. At any one time, the observer sees four points of light: the stars A and B of the binary, and their secondary images, marked A′ and B′ in the figure. In the 4*d* method, the images AB and A′B′ are placed so that they are aligned

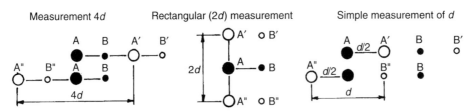

Fig. 13.17. *Specific, easily determined positions used when measuring separations by the double-image technique.* Top: *first position;* bottom: *second position.*

Fig. 13.18. *Measuring θ using the double-image technique.* Left: *faulty alignment;* right: *correct alignment. The closer A and B, the more sensitive this technique becomes.*

and equidistant from one another (first setting); the position of the carriage is then noted. The carriage is then moved so that the positions of the images AB and A'B' are switched, but keeping the images equidistant (second setting). The displacement of the second setting with respect to the first is then four times the separation of the binary (4d), and it may be established by simply reading off the value of the amount that the carriage has been displaced, it having been calibrated in advance. If the images are arranged so that the four points of light form the corners of a rectangle, the displacement of the carriage when the relative positions are reversed corresponds to twice the separation (2d). The images may also be aligned so that A' falls between A and B; A, A', B, and B' are then equidistant. Reversal of the arrangement (A now falling between A and B') corresponds to a displacement equal to the separation of the binary.

Position angles may be measured with a high degree of accuracy by aligning the images (Fig. 13.18), once the direction of doubling has been determined from the position-angle protractor. This method is very sensitive with close pairs.

The 4d method of measurement is used for pairs that are just separated, up to separations of a few seconds of arc. For greater separations, the 2d method (the rectangle) is chosen; beyond that the 1d (simple) alignment. It will be appreciated that such a system is particularly sensitive and accurate for short separations. With the 4d method the error is reduced by a factor of 4: it is only a few hundredths of a second of arc.

A variant was devised by Bernard Lyot and Henri Camichel in 1949, using a plate of spar, cut and polished parallel to the crystallographic axis. Birefringence causes doubled images, and the separation may be varied by varying the inclination

Fig. 13.19. *An amateur-built, double-image micrometer, using a fluorspar birefringent plate, mounted on a reflector. A: graduated semicircle for measuring the inclination of the plate; B: box containing the pivoting plate; C: eyepiece holder with 2.5× Barlow lens.*

of the plate. For a plate thickness e, inclination i, and extraordinary and ordinary refractive index n_e and n_o respectively, the separation of the images di is given by:

$$di = e\sqrt{2}/2 \times \sin 2i \, (1/\sqrt{2n_e{}^2 - 1 + \cos 2i} - 1/\sqrt{2n_o{}^2 - 1 + \cos 2i}).$$

In this form of micrometer the mechanical arrangement is particularly simple (Fig. 13.19). All that is required is that the plate should pivot without any play, that the axis of rotation should be truly perpendicular to the telescope's optical axis, and that it is possible to read the inclination to an accuracy of $0.2°$. These readings may then be converted into separation (in seconds of arc) using an appropriate table, provided the focal length of the telescope is accurately known.

This device is extremely sensitive, even though its range is limited, being 6–7 arc-seconds for a focal length of approximately 3.5 m. The double image means that there is a loss of approximately 1 magnitude (0.75 magnitude from division by 2, plus reflection losses at the two surfaces), which imposes some limits when faint variables are studied.

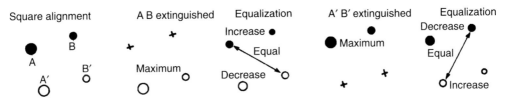

Fig. 13.20. *Measuring Δm using a double-image micrometer and a rotating polarizing filter.*

13.4.9 The double-image photometer

One of the advantages of the last two micrometers described is the use of birefringence, which means that their uses may be expanded to include photometry. The two emergent beams are polarized at 90°. A polarizing filter behind the birefringent crystal enables the difference in brightness of the two images to be varied. The corresponding rotation is converted into the differential magnitude, Δm.

The filter must fully darken the brightest image. It is fitted with an index reading against a protractor divided into 400 grades, which is more convenient in this instance than conventional degrees. The procedure is as follows, with the images being arranged at the corners of a square (Fig. 13.20):

(i)   extinguish AB by turning the polarizing filter – read the angle;

(ii)   continuing to turn the filter in the same direction, try to equalize A and B′ – read the angle;

(iii) now extinguish A′ and B′ – read the angle;

(iv) still turning, equalize A′ and B – read this angle.

The pairs of images are switched – or rather the extraordinary ones are shifted to the opposite side of the ordinary ones – in an identical square arrangement, and the observations are repeated. We then have two sets of comparable measurements, the differences being approximate multiples of 100 grades (this is where the advantage lies). After deriving means, the difference between the angles of extinction and equalization gives the angle α (in grades). Then $\Delta m = 5\log(\tan \alpha)$.

This method allows Δm to be determined to better than 0.1 magnitude Going beyond this, colorimetric measurements may be made by adding coloured filters, with known peak transmissions. P. Muller did this in 1949 and 1951, making measurements, which have never been repeated, on doubles as close as 0.45″.

13.4.10 Interferometric measurements

If a screen with two slits is placed in front of an objective, a star will give a system of fringes parallel to the direction of the slits. The spacing of the slits should be variable symmetrically with respect to the centre of the objective, and it should be possible to rotate the screen around the optical axis, enabling the orientation of the fringes to be altered as required. If the slits are moved apart, the spacing of the fringes becomes smaller; conversely it increases with a smaller separation.

Fig. 13.21. *Measuring double stars by interferometry.*
a: the system of fringes obtained from the A–B pair;
b: superimposing the fringes produced by A and B, when the fringes are sharpest;
c: alternating the fringes from A and B, giving a continuous band of light.

(Reproducing Young's original experiment on interference.) This provides a variable scale that may be used to measure the separation of binaries (Fig. 13.21a).

With a pair of close components of equal (and sufficient) brightness, we obtain two sets of parallel fringes that may be superimposed by rotating the screen. This enables the position angle to be determined to $\pm 180°$, using a graduated scale, once the direction of daily motion has been established.

The separation may then be measured by varying the separation between the slits. If this is increased, the fringes contract until, at a point corresponding to the separation between A and B, the fringes from star A are superimposed on the set belonging to B. The fringes are then sharpest. If one continues to separate the slits, the two sets of fringes alternate, and all that is visible is a continuous band of light (Fig. 13.21b and 13.21c).

In practice, the slits are first set to their maximum separation, producing fringes that are very close together. By rotating the screen first one way and then the other, the two sets of fringes are superimposed, giving a single continuous band of light, noting the angle each time, and thus obtaining θ. The separation between the slits is then slowly reduced. When perfect alternation of the fringes occurs (giving a continuous band of light and the weakest fringes), the separation D between the interferometer's slits is determined. Then:

$\rho(\text{in arc-seconds}) = 206\,265\lambda/2D,$

where λ is 560 nm, and where D is also expressed in nm.

This technique is not very spectacular; the fringes are delicate and require a magnification of at least $2D$ to be observed easily. It also wastes a lot of light, which means that it can be used only with large-diameter telescopes. An interferometer consisting of two slits, each 20×200 mm, on a 310-mm telescope, passes as much light as a small instrument only 90 mm in diameter. In addition, this light is split into the various fringes, which causes a loss of 2–3 magnitudes, so such a device allows only stars down to about magnitude 7 to be examined. What is more, the pair needs to be nearly equal, because even a Δm of 1 makes measurement difficult. On the other hand, the interferometric method is insensitive to turbulence, and may exceed the theoretical resolution: 0.20 arc-seconds as against 0.32 at best for a 310-mm instrument. The brightness of the fringes may be increased by widening

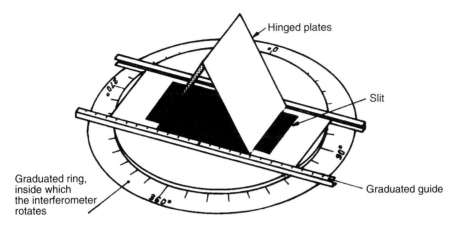

Fig. 13.22. *Interferometer for measuring close binaries*

the slits slightly: this reduces the number of fringes and increases the brightness of each one.

An interferometer consists of four light, rectangular plates (their length equal to the diameter of the instrument, and width equal to the radius), hinged to one another so that they fold up like a concertina (Fig. 13.22). The two outer plates, in each of which there is a slit next to the hinge line, slide in two graduated guides. The plates in the centre only serve to prevent any other light from entering the instrument, and open and close according to the separation of the slits. The whole assembly is mounted on a ring that is a gentle friction fit inside another ring, graduated in degrees, which is fitted to the top of the tube. It is useful if the separation and rotation of the slits may be controlled without removing the eye from the eyepiece.

To make it easier to make the adjustments when using large telescopes, Finsen, in 1933, devised an eyepiece interferometer, but he then had to take the reduction ratio into account. Amateurs are not likely to find this method of interest, because it requires an extremely high degree of accuracy in manufacture.

13.5 Measurements using imaging techniques

13.5.1 Photography

Few amateurs have tried this type of work on double stars, particularly with the aim of obtaining quantitative observations, whereas professionals, particularly in the U.S.A., Italy, and France, have developed programmes of measurement. However, the appearance of new emulsions should be an incentive and an additional help.

Once again, the primary focal length needs to be extended, by a factor of about 10, to obtain a scale that is large enough to ensure that later measurements are easy to make, and also that the Airy disk is about ten times as large as the grain of the emulsion. This amplification may be obtained by using a clean, good-quality eyepiece, extending the draw-tube as much as possible so that the emerging pencil of light is as close to parallel as possible. Naturally, there will be some aberration and

745

distortion, except at the centre of the photograph. The magnification γ is calculated from $\gamma = (t/f) - 1$, where t is the extension, and f the focal length of the eyepiece. The ideal solution would be to replace the eyepiece by a 16-mm photomicrographic objective, which is designed for use under such conditions. But such an objective is very expensive.

Another somewhat cumbersome, but optically correct, possibility is to combine two 3× Barlow lenses to obtain a γ of 9. A Cassegrain telescope is advantageous here, because a single Barlow is enough to produce an overall focal length of 15–20 m. However, heed must be given to the stability of the focal length if the optical elements are not made from material with a low coefficient of expansion. Variations in the focal length greater than one-thousandth part cause similar changes in the scale, leading to a risk of inconsistency in measurements of a specific separation. With all these methods of increasing the focal length, it is, of course, essential that the mechanical arrangements should be such that the optical elements may be replaced in exactly the same place each time.

The camera should be light, not to upset the balance of the instrument and cause it to shake during exposures. A reflex finder is useful for centring and focussing, provided it is bright and that the optical paths between the mirror and focussing screen, and between the mirror and film plane, are equal. Focussing is particularly accurate if a magnifying system is used.

Guiding is risky with such extreme focal lengths, so it is better to rely upon a truly accurate drive in RA over a period of about a minute, and upon accurate alignment (Bigourdan's method).

All this naturally affects the quality of the images that are obtained. Ideally the images should be Airy disks. In practice, turbulence, and diffusion in the emulsion layers mean that the actual image is a diffuse spot. For an F/D ratio of 62.5, the theoretical diameter of 0.08 mm is only rarely achieved. In the most favorable cases the image will be about 0.1 mm across, and will have a whitish peak surrounded by a reddish/orange halo. For this reason, it is interesting to use a colour emulsion, with an exposure time that gives the maximum quantum efficiency (i.e., the film should not be overexposed). Fujichrome 400 is very suitable. An empirical curve that takes reciprocity failure into account is given in Fig. 13.23, the exposure being based on the magnitude of the fainter component.

The following procedure is used to make the exposures. First: locate the object with the finder – it is very useful to have a powerful finder, because the photographic field is so small – then centre it, check and carefully focus the pair using the reflex finder. Second: cock the camera shutter; cover the telescope's aperture with a suitable hand-held occulting 'bat'; release the camera shutter; uncover the tube when any vibrations have had time to die away; count the number of seconds exposure; cover the tube (do not close the camera shutter). Third: repeat the second step after having interrupted the drive for a few seconds to shift the images on the film. Four exposures are made on the same frame, the first two being shorter than the theoretical time, and the second two longer. The drive is stopped for 2 s between the first and second exposure, 3 s between the second and third, and 4 s between the third and fourth. This enables the frame to be oriented later: the longer exposures and the longer intervals are towards the west (Fig. 13.24).

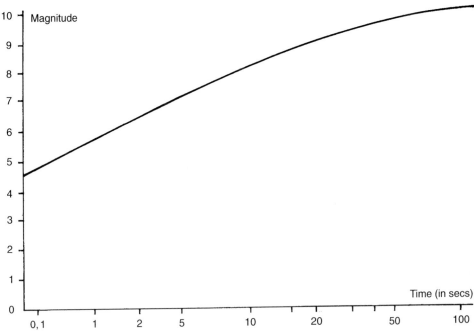

Fig. 13.23. *Limiting magnitude as a function of time of exposure for D = 200 mm, F/D = 62.5, using ISO 400 film. Reciprocity failure has been taken into account.*

Next, wind on the film (recock the shutter) then refocus and repeat the sequence of operations twice more. This gives twelve images, with differing focus and exposures during the same night. These details must be carefully noted down in full in a notebook. The same pair should be photographed on two further, and if possible consecutive, nights.

Measurement of the photographs is preferably undertaken using a micrometer measuring machine, either one obtained from salvage, or constructed like a thread micrometer, on the lines of the one shown in Fig. 13.25. Measurements may be made either centre to centre, or as rectangular coordinates that are then converted into polar coordinates. In both methods four or five measurements should be taken to obtain a satisfactory mean.

It is also possible to use an enlarger set to its maximum, and carefully focussed (check that the plateholder is truly perpendicular to the axis of projection). A stretched black thread, held by adhesive tape, indicates the E–W direction, and another bisects the two disks on the best of the four images, showing the position angle. With the light switched on, this angle may then be determined. The value of θ may be calculated from the mean of four or five measures, having determined the direction of North, which serves as the origin. To determine separations it is best to use a calliper gauge, with a needle-point cemented to each jaw to serve as an extension. These points should be curved as shown in Fig. 13.26, and able to cross. One reading from the centre of the disks should be taken with the points

Fig. 13.24. *Albireo (β Cygni): four successive exposures of 1 to 4 seconds, indicated by increasing intervals, West at left, North at bottom. F = 12.5 m, enlargement about 15 times, Ektachrome 200 film. (Photo Durand-Simier)*

separated, and another with them crossed, the actual separation then being obtained by subtraction. Again, four or five separate measurements should be made to obtain a mean for the best image on each of the three frames, and then an overall average for that particular night.

If the conditions are kept constant, then the overall system (telescope + measuring engine) may be calibrated using the pairs listed in Tables 13.12–13.16. In this particular case, 0.60 mm equals 1 arc-second.

Using this sort of method, professionals obtain errors that are approximately 0.01 arc-second, but an amateur can hardly expect to get results better than 0.1 arc-second, the accuracy to which readings may be taken being just slightly better than one-tenth of a millimetre. It is possible to do somewhat better with equipment that can measure the photographs directly. These values are comparable with those obtained by direct methods such as the rotatable or diffraction micrometers. The advantage is that one can obtain numerous images in a single session, and reduce them when observation is impossible.

Improvements may be gained by increasing the scale or reducing the size of the diffraction disks, requirements that are completely contradictory in practice. Increasing F/D causes a large disk, both through the laws of geometrical optics, and also because the instrument is more sensitive to vibration and turbulence (an increased exposure being obviously essential). One could try to determine the

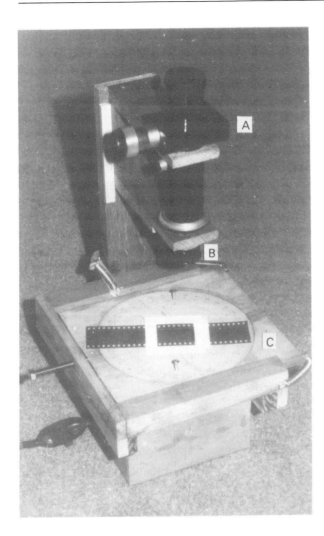

Fig. 13.25. *A device for measuring negatives, consisting of a microscope micrometer (A) fitted to an f = 50 mm, enlarging lens (B), which enlarges the image about twelve times. The film is mounted on a movable table, with two perpendicular motions, and one rotation. (Photo: P. Durand)*

optimum compromise. With a 200-mm telescope, it is around 15 m focal length. Proper interpretation of the images is one way of making progress, especially determining the exact photocentre. Short exposures have a tendency to give spurious peaks that should not be confused with the true centre.

This technique is suitable for fairly wide pairs, with separations greater than 2–3 arc-seconds, because when the spots start to coalesce, the measured separation tends to be less than it actually is.

13.5.2 Other methods of measurement

Speckle interferometry has been considerably developed in recent years (*see* Baize, 1982c, Labeyrie, 1982, and Kitchin, 1984). The resolving power of large telescopes may be attained by capturing the image as distorted by turbulence. Using a laser,

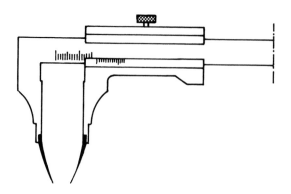

Fig. 13.26. *Calliper gauge modified to measure the distance between two star images.*

the short-exposure images of a binary, obtained through a narrow-band filter, are summed as an optical Fourier transform. From the separation and orientation of the fringes thus obtained ρ and θ may be derived. Nowadays, however, results may be obtained even more quickly by using a photon-counting device linked to a computer, which analyzes the image in real-time by determining the autocorrelation function required to reconstitute the image. Measurements obtained in this way have enabled refined orbits to be obtained for many visual binaries that are at the limit of classical methods. Speckle interferometry has also made inroads into the domain of spectroscopic binaries.

At Uccle, J. Dommanget has tried to use the images produced by a sensitive video camera. Recent progress in this area means that an amateur having such a camera and a video recorder can use it as if it were an ordinary photographic camera. These video cameras are generally fairly sensitive and can record the brightest stars. When the image has been recorded on tape, the best images may be chosen and measured, either from the screen or, better, determined by computer techniques.

The increasingly wide-spread use of CCDs means that they are the ideal choice for making real-time measurements, with an accuracy comparable with photographic techniques. This method is, however, restricted to fairly wide pairs. For any amateur who is well versed in electronics and computation, this is an extremely promising field, which is both fast and capable of producing highly accurate results.

The observation of spectroscopic binaries is another challenge to anyone with access to a 400-mm or larger Cassegrain telescope, and who is looking for a suitable programme. At the Observatory of Toulouse, Bouigue has become a specialist in this field. He has published numerous catalogues of this type of object. An amateur may consider the construction of a spectrograph (*see* Chap. 18), and try observing spectroscopic binaries. In addition, it will be necessary to have access to a measuring engine to determine the separation when the lines are doubled (Fig. 13.27). Naturally, any such equipment will be limited in resolution and will only be able to observe the largest radial velocities (above a few tens of km/s). Nevertheless, there is still a wide enough field of study among the brightest stars (down to magnitude 7 or 8).

Finally, eclipsing binaries that do not form part of multiple systems should also

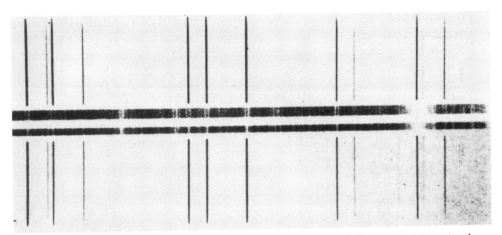

Fig. 13.27. *A spectroscopic binary (negative image). Two different positions in the orbit are shown.* Top: *the motions along the line of sight are in opposite directions, the velocities are at a maximum (here the separation between the lines indicates a velocity of +48 km/s and −48 km/s), P = 26 days.* Bottom: *the displacement is zero relative to the observer and only one line is seen (the two lines are superimposed).* (*Photo: C.N.R.S.-O.H.P.*)

be followed. These, however, are the realm of variable-star observers, and will be discussed in the next chapter.

13.6 Problems affecting measurements

13.6.1 Effects of atmospheric turbulence

Turbulence frequently interferes with tests of the optical quality of an instrument. When observing a double star with a magnification equivalent to at least twice the diameter of the telescope (measured in mm), its deleterious effects may vary. With a faint object, it may give rise to a completely fluid or imprecise image, making it quite impossible to obtain a proper focus. This is the most severe case.

Generally, the structure of the image constantly fluctuates, more or less rapidly, becoming double or patchy, such that the Airy disk is never seen. In fact, it is still present, but in the form of several, superimposed, individual images. (The image frequently doubles in front of one's very eyes.) It needs a highly experienced eye to make sense of this sort of image, but the stakes are high: it is possible to see that each element of the image is double, as Couteau found in 1986 October when using the 2-m reflector at the Pic du Midi.

Overall motion of the image is much less detrimental, because the image sharpness is retained. This situation often occurs with a small refractor. The rings are often broken, but the disk remains more or less circular, despite moving around rapidly and in a random fashion, by about 1 second of arc. When trying to make a measurement under these conditions, it is difficult to bisect the centre of images that are constantly on the move. The eye is able to follow the motion easily, but a photograph would show just an enormously extended disk of light. These rapid movements often arise in the immediate vicinity of the telescope, and result from radiation by the ground, heating effects inside a dome, or eddies occurring behind a hill or nearby building.

Danjon analysed turbulence as a function of the deviation of rays of light from a star. When the angular amount of the deviation t is small, only amounting to $\rho/2$, where ρ (in arc-seconds) $= 140/D$ (in mm), the disk is practically circular, and doubles can be detected. When $t = \rho/4$, the ring is easily seen. Difficult objects – with close components or unequal magnitudes – may be studied under these conditions, which do not occur every night. However, one must learn to be patient and how to cope with the factors that are not under one's control.

A small instrument, especially if it has a closed tube, is less sensitive to turbulence than a larger one. The cells causing the perturbations are approximately one-tenth of a metre across, so it will be appreciated that there is nothing to be gained by switching from a 300-mm telescope to one of 400 mm aperture, except on very rare nights when the conditions are truly superb. On the other hand, a gain in light improves the conditions and enables the eye to separate faint stars.

13.6.2 The choice of site

To make measurements of double stars with small separations, the false disk must be visible, which means that only rarely can faint objects be studied. A perfectly pure sky that allows the theoretical limiting magnitude to be reached is therefore not essential. In fact, the opposite is almost true. Above all, turbulence must be as low as possible, which often applies when the sky is slightly misty, minimizing cooling during the night.

High magnifications are used (2–2.5 × the diameter, measured in mm), so the sky background seen in the eyepiece is very dark. It is therefore possible to work from a town, which is precisely where the permanently slightly hazy conditions tend to lessen turbulence. It is only necessary to be sheltered from direct illumination by artificial lighting.

If tube currents have been eliminated by ensuring good thermal equilibrium and closing any apertures in the tube, the next point to consider is avoiding local turbulence by raising the aperture as much as possible to escape eddies arising from the ground. Three metres is a minimum, and 5–10 m would be better. (Visions of a disused water tower come to mind ...)

An ideal type of installation would be a 3.5-m dome, mounted on a tower 4–5 metres high, sheltering an F/7 Newtonian or an F/15 Cassegrain reflector, 300 mm in aperture – an equivalent 250-mm refractor would be far too unwieldy. A light-weight dewcap extending well outside the slit would ensure that mixing of the interior and exterior air would not degrade the image. Outside and inside temperatures should, in any case, be equalized. If the telescope is mounted at ground level, in the open air, a 1-m pillar surrounded by a wooden floor would be a good compromise, with the top of the tube being at about 3 m from the ground.

The geographical situation also needs to be considered. Valleys should be avoided because they are frequently covered by inversions. It is best to be on the windward side of a hill rather than in its lee. Avoid the neighbourhood of heated buildings, especially if the wind is likely to carry pockets of warm air over the telescope, or if the chimneys of heated buildings occur to the south. The area of sky towards the south is all-important for observation. One should always try to observe doubles when they are close to culmination, because the higher a star is above the horizon the less its image is affected by turbulence. For most people in Europe and in the northern parts of the U.S.A., measurement of close binaries at negative declinations is only possible under good conditions and with some degree of luck.

13.6.3 Choosing a method of measurement

As we have seen, there are numerous, and very varied, methods of observation, relying on different physical principles: image scale at the focus, diffraction, bire-fringence, interferometry, etc. Similarly, we have seen that no technique is capable of doing everything well: i.e., that has high resolution, and is able to deal with faint pairs and wide pairs simultaneously. The programme therefore needs to be chosen with respect to the equipment: or else the techniques selected according to the pro-gramme to be followed. Figure 13.28 summarizes the various possibilities available to an amateur depending on the aims in mind. For example, a programme of orbital determination does not require the equipment to cope with large separations, but does requires accurate measurements of close pairs instead. If one's interests lie in young stars, interferometry or spectrography are preferable. We may note, finally, that all this explains why the filar micrometer has always been highly regarded (see Fig. 13.29).

13.6.4 Identifying the star to be observed

Although doubles that are visible to the naked eye are easily recognized and located, the same does not apply to fainter ones. In the finder, one 9th-magnitude star looks just like another 9th-magnitude star. Before trying to make a measurement, it is essential to check that the object in question is really the one given in the catalogue, and that its magnitude and separation are accessible with the given telescope/micrometer combination.

The positions of doubles are given in the catalogues with their coordinates in RA (α) and Dec (δ) for a given epoch (which may require some calculation of precession). It is sensible to use a reasonably accurate atlas drawn for the same epoch: *AAVSO Atlas, Sky Atlas 2000.0*, or Becvár's *Atlas Borealis* and *Atlas Eclipticalis*. If the star is not actually shown, its position should be determined on the appropriate chart. Rather than risking damage to the sheets of the atlas, it is best to make a tracing and use this for checking at the eyepiece. All this preparatory work may be carried out well in advance, always bearing in mind that a fine night might occur unexpectedly and without giving one enough time to make last-minute preparations. Always be ready to observe.

Once identified from its surrounding field, the double is centred and examined with

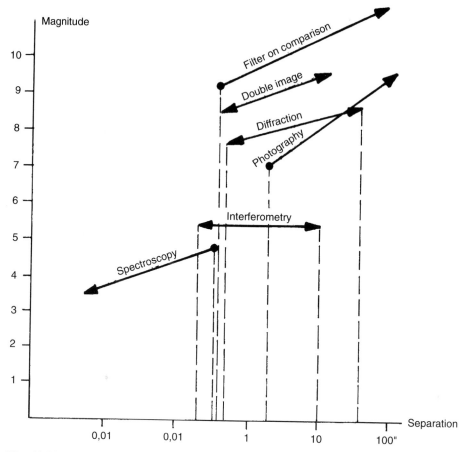

Fig. 13.28. *Measuring techniques that may be used with a given instrument (400-mm aperture).*

the working eyepiece, because it is essential to check its characteristics: magnitude, position angle, and separation. The last two are, naturally, subject to change. If, in addition, the magnitudes differ, an examination of neighbouring stars is called for, and, as a last resort, a further check on the catalogue. Simply by neglecting these common-sense precautions, observers have been fooled into mistaking one double for another, leading to aberrant measurements being recorded.

13.6.5 Establishing the image scale

When measuring the relative positions of the components of a binary or multiple star, determining the position angle θ does not pose any particular problems, provided the mounting is well-made, properly oriented, and that one has taken the trouble to use a star close to the celestial equator to establish which is north from the daily motion.

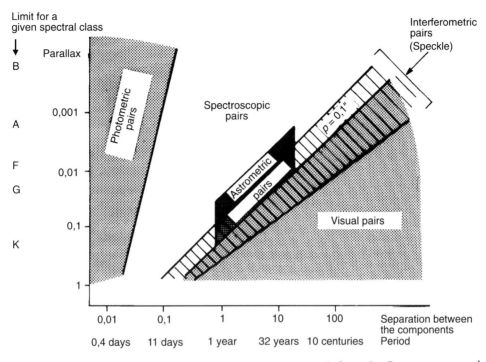

Fig. 13.29. *Performance of given instrumentation (after J. Dommanget and P. Couteau). Period may be related to the separation between the components for the purpose of calculating a period from an observed separation. Similarly, a spectral type may be related to distance, the latter serving as a 'filter', because spectral types tend to have a given luminosity.*

Measuring the separation, on the other hand, requires a knowledge of the image scale at the instrument's focus, and the use of a sensitive, accurate method of measurement at that focus. With reflectors, the use of some form of amplification system means that the effective focal length, and thus the scale, is not known directly. It may also happen that the accuracy of the measuring system is uncertain. The most serious fault is a lack of repeatability, which will affect the overall accuracy. Calibration aims at determining the unknowns.

Common sense suggests that accurate calibration may be obtained by actually using the overall system. All that is required is to measure known separations, and relate these to the values indicated by the measuring instrument (the micrometer). The ratio between them gives the calibration coefficient e. Using a series of determinations plotted on a graph, the function of e may be derived, and we can check to see if it passes through the origin. The value should be known to an accuracy significantly greater than the micrometer's sensitivity, to ensure that it does not increase the errors in any values that are measured (when the errors happen to be in the same sense).

This operation should, therefore, be carried out with the greatest possible care.

This assumes the use of the most accurate standard objects available, which should be as varied as possible, both as to type and separation. This is a long process, but one that naturally does not preclude one from undertaking a regular programme in parallel, and making the reductions only when one considers that calibration is complete. (It took us two years, and 46 measurements, to determine the calibration of our birefringent-spar micrometer.)

Some methods are as follows:

Transits In general a star with a high declination should be chosen (preferably between the zenith and the pole), at a time when it is crossing the meridian. The time that the star takes to cross the two cross-hairs in the micrometer is measured. The stopwatch is started just as the star touches the first cross-hair, and stopped as it touches the second. (With a double-image micrometer, the twin images of the star are timed as they transit a single cross-hair.) The cross-hair(s) must be exactly at right-angles to the direction of diurnal motion. The average of ten successive transits, repeated over three nights, for increasing separations (10, 20, 30, and 60 arc-seconds) will enable the micrometer's consistency to be determined. Using the average of the thirty measurements, the angular separation of the cross-hairs is calculated from d(in arc-seconds) $= 15t \cos \delta$, where t is the interval expressed in seconds of time. It should be obvious that the relative error is better for transits over 10 arc-seconds than it is over 50 arc-seconds.

Wide pairs It is possible to reduce the errors in calibration by using widely separated pairs. This is why, in his book *Lunettes et Telescopes*, Danjon suggests measuring the difference in declination between various stars in the Pleiades: 24–Alcyone, 659.78 arc-seconds in 1984; Alcyone–22, 686.25 arc-seconds (1984); and Caelaeno–Electra, 632.74 arc-seconds (1984). A close pair is Asterope I–II, 152.12 arc-seconds. In Ursa Major, the well-known pair of Mizar and Alcor are 708 arc-seconds apart (Couteau, 1981).

Diffraction grating Using cassette magnetic tape (*see* p. 734), 3.81 mm wide, the pitch of the grating is 7.62 mm, and it gives two images symmetrical about the zero-order image, separated by 30.32 arc-seconds (for $\lambda = 560$ nm). With a bright star, it might even be possible to measure the second-order images. The inaccuracies in this method arise from differences in the eye's sensitivity to colour. This may be avoided by using a filter with a known peak transmission.

Using test objects It is also possible to measure artificial stars: small holes pierced in ordinary kitchen foil with a needle, lit from behind with a suitable bulb. This test object should be placed as far away as possible, so that it is essentially at infinity: several hundred metres in the case of a telescope of 3-m focal length.

Measuring the distance between the holes poses no problems. Two holes, separated by 50 mm, subtend 34.38 arc-seconds when seen at a distance of 300 m. Using a vernier calliper gauge, the separation between the holes may be measured to 0.1 mm, giving a very low degree of uncertainty. On the other hand, measuring a distance of 300 m to an accuracy of a few tenths of a metre is not so easy, but it is possible.

Because the test object cannot be considered to be truly at infinity, its image is not formed precisely in the focal plane, but slightly behind it. As a result, calibration with a test object gives a value for the focal length F that is larger than its real value by ϵ where $\epsilon/F = F/d$, d being the distance of the test object. The image scale is increased by the same ratio, so it is advisable to calculate the correction to be applied. This might, for example, be about 1 % for a focal length of 3 m and a distance of 300 m.

Muller's standard pairs – orbital binaries   In 1949 November, Paul Muller published – in the *Journal des Observateurs* (a publication no longer in existence) – a list of 15 doubles that appeared to have been stationary since Struve's time. Subsequently, the majority of these have been confirmed as unchanging. A few do show some motion. J. Minois has used a computer program to predict the positions of these from all the available measurements. We have done the same ourselves graphically, using just photographic measurements. We have added two pairs with similar characteristics and with RA greater than 20^h, so that the whole sky is covered (Table 13.12).

In addition, some binaries have already completed an orbit since their discovery, and others have nearly done so. The calculation of their orbits has integrated all the available measurements, so their separations are known to an accuracy of several hundredths of a second of arc. Unfortunately, there are not many such cases (Table 13.13).

All these pairs are useful for establishing and checking a micrometer's calibration, provided at least a dozen are measured. Uncertainties in the positions are then more likely to cancel out statistically. Any evidence of a systematic error that is too large or too small indicates that the calibration should be corrected. These objects are extremely varied in nature, so they also have the advantage of enabling one to check the overall accuracy of the telescope–micrometer–observer system. In the same vein, we may mention regular observations of ϵ^1 and ϵ^2 Lyr. P. Muller carried out an investigation of these, which showed the reliability of the double-image micrometer relative to the filar type.

The procedure to be followed in calibrating the system may be summarized as follows: once some material is available, and the scale at the focus is roughly established, a programme of measurements may be started, provided one ensures that a calibration pair, or one with a well-established orbit, is observed during each session. The calibration will become progressively more accurate, and, if there is any doubt about the data gained during an observation, there will always be one element with a known value. For binaries suitable for photography, see Table 13.14.

Finally, we will repeat the advice given by Paul Couteau:

'Start by publishing good measurements, so that you will be taken seriously. If not, you risk being written off immediately.'

Table 13.12. *Some very stable binary systems*

ADS	Σ	Position 1950 (2000)		1840 mean	1960 mean	1985 projected deduced	min.
683	61 = 65 Psc	00^{h}47.2^m (49.8)	+27°26′ (43)	θ 298.5°	296.1°	295.8°	
		vis. mag. 6.3, 6.3		ρ 4.42″	4.53″	4.30″	
3297	559	04^{h}30.6^m (34.6)	+17°55′ (57)	278.1°	276.9°	277.0°	
				3.22″	2.96″	3.07″	
3353	572	04^{h}35.2^m (38.5)	+26°51′ (56)	210.3°	195.0°	192.5°	190.2°
				3.17″	3.96″	4.12″	4.15″
3734	644	05^{h}09.6^m (10.4)	+37°14′ (18)	220.4°	221.0°		
				1.60″	1.59″		
5436	958	06^{h}44.0^m (48.3)	+54°46′ (43)	256.7°	257.0°	256.8°	
				5.07″	4.76″	4.68″	
7034	1282	08^{h}47.6^m (50.9)	+35°15′ (04)	277.4°	278.2°	279.0°	
				3.40″	3.58″	3.60″	
7286	1333	09^{h}15.4^m (18.5)	+35°35′ (22)	40.9°	47.5°	50.0°	48.5°
				1.52″	1.78″	1.80″	1.80″
9338	1864 = ε Boo	14^{h}38.4^m (40.7)	+16°38′ (26)	99.4°	107.5°	110.0°	105.7°
				4.9, 5.8	5.84″	5.62″	5.62″	5.74″
9737	1965 = ζ CrB	15^{h}37.5^m (39.4)	+36°48′ (38)	300.5°	304.1°	304.8°	
				5.1, 6.0	5.97″	6.27″	6.29″	
9969	2021	16^{h}11.0^m (13.3)	+13°40′ (33)	317.9°	345.8°	351.0°	351.3°
				7.5, 7.6	3.32″	4.11″	4.12″	4.31″
10526	2161 = ρ Her	17^{h}22.0^m (23.7)	+37°11′ (08)	308.8°	315.9°	317.0°	317.7°
				4.5, 5.5	3.66″	4.02″	4.03″	4.07″
10597	2180	17^{h}27.8^m (29.0)	+50°27′ (53)	265.3°	260.5°	260.5°	
				7.7, 7.9	3.17″	3.07″	3.07″	
10905	2245	17^{h}54.2^m (56.4)	+18°20′ (20)	294.2°	292.7°	293.0°	
				7.4, 7.4	2.49″	2.49″	2.68″	
11640	2375	18^{h}43.0^m (45.5)	+05°27′ (30)	112.5°	116.7°	118.5°	
				6.2, 6.6	2.34″	2.41″	2.49″	
12962	2583 = π Aql	19^{h}46.4^m (48.7)	+11°41′ (48)	121.6°	110.0°	107.6°	
				6.1, 6.9	1.45″	1.41″	1.43″	
15007	2799	21^{h}26.4^m (28.9)	+10°52′ (65)	331.3°	275.6°	267.0°	265.9°
				7.7, 7.5	1.42″	1.61″	1.74″	1.68″
16270	2944 AB	22^{h}45.3^m (47.9)	−04°29′ (13)	247.6°	275.8°	284.0°	283.8°
				7.3, 7.8	4.03″	2.58″	2.30″	2.31″

Table 13.13. *Four well-known binaries*

ADS	Σ	Position 1900 (2000)		1980	1985	1990
8119	1523 = ζ UMa	11^{h}12.8^m (18.2)	+31°66′ (33)	θ 104.7°	91.1°	60.1°
		vis. mag. 4.4, 4.9		ρ 2.92″	2.33″	1.27″
8630	1670 = γ Vir	12^{h}36.6^m (41.7)	−00°54′ (87)	297.0°	292.9°	287.6°
				3.6, 3.7	3.92″	3.52″	2.06″
9413	1888 = ξ Boo	14^{h}46.8^m (51.4)	+19°31′ (07)	332.9°	329.3°	325.5°
				4.8, 6.7	7.22″	7.16″	7.04″
10157	2084 = ζ Her	16^{h}37.5^m (41.4)	+31°47′ (36)	141.6°	109.6°	84.4°
				3.0, 6.5	1.26″	1.44″	1.57″

(Ephemerides calculated by J. Minois)

Table 13.14. *Eleven, almost fixed, wide pairs**

ADS	Σ	Mags.	Position 1950		θ (for 1987)	ρ
639	1 Ap.1	7.4, 7.5	00^{h}43.7^m	+30°40′	47.7°	47.1″
4000	698	6.5, 8.0	05^{h}21.9^m	+34°49′	347.6°	31.2″
6988	1268 = ι Cnc	4.2, 6.8	08^{h}43.7^m	+28°57′	307.5°	30.41″
8600	1657	5.2, 6.8	12^{h}32.6^m	+18°39′	270.5°	20.24″
8706	1692 = α CVn	2.9, 5.5	12^{h}53.7^m	+55°11′	151.8°	14.42″
8891	1744 = Mizar	2.1, 4.2	13^{h}21.9^m	+55°11′	151.8°	14.42″
9739	1965	5.1, 6.0	15^{h}37.2^m	+36°48′	305.2°	6.29″
9933	2010	5.3, 6.5	16^{h}05.8^m	+17°11′	12.4°	27.50″
10759	2241 = φ Dra	4.9, 6.1	17^{h}42.8^m	+72°11′	16.2°	30.21″
11745	39 Ap.1 = β Lyr	3.4, 7.8	18^{h}48.2^m	+33°18′	148.5°	45.72″
12540	43 Ap.1 = β Cyg	3.0, 5.3	19^{h}28.7^m	+27°51′	54.6°	34.34″

* Because they are relatively static, these pairs may serve as photographic standards.

13.6.6 *Choosing a programme*

When one begins to become involved in double-star astronomy, it is difficult to decide what forms of research are feasible, and one wonders what to do. Here are a few suggestions that may serve as starting points. Later on, everyone tends to specialize.

Verifying identifications Some faint doubles have never been re-observed since their discovery. Their positions are poorly known, and they risk being confused with neighbouring stars. Such a programme was started by a team working on the CCDM. It is not essential to have a micrometer. It is sufficient to have a good atlas and to be able to draw the field accurately. In France this work is coordinated by the Double Star Section of the Société Astronomique de France.

Measurement of neglected pairs Wide pairs with a large value of Δm have frequently been little studied since their discovery. It is possible that some may have undergone movement since that time: new observations are urgently required. The filar and comparison micrometers, and photography are very suitable for such work. A list of 106 objects has been prepared by the SAF's Double Star Section. There can be little doubt that assiduous observers will be rewarded for their work in that re-observations will show relative rectilinear motion, thus establishing that the pairs are optical doubles. It would be possible to extend this work and undertake a more specific programme. The Observatory of Uccle in Belgium has published a catalogue, prepared by Dommanget and Nys, which shows that 20 % of close pairs are optical doubles. Not all have been catalogued, therefore. This would be a specific, and very interesting contribution, because, using transit observations, it would enable proper motions to be established and thus provide more information about the motion of stars within the Galaxy (*see* Table 13.15).

Measurement of known orbital binaries   Another type of specific programme consists of following pairs with orbital motion, using the Fourth Catalogue of Ephemerides prepared by Muller and Couteau as a basis. This assumes that a careful selection of accessible pairs is made before any work is begun. Many of the objects are, in fact, close to the limit for amateur instruments, but nevertheless there remains plenty of work to be done. Accurate observations will provide new data that may be used to improve the preliminary orbits that have been calculated, and many of which are rather uncertain. If the doubles are not too faint, the best micrometer is a double-image type: any one of the three versions described earlier. A selection of objects is given in Table 13.16.

Investigation of new pairs   With visual doubles, this type of work is not really possible for amateurs, unless they possess adequate instrumentation, but discoveries may occur by chance when examining field stars near a known binary. With the same sort of idea in mind, Paul Baize has described several enigmatic binaries in two articles (Baize, 1982a & 1982b). On the other hand, an original research programme might be undertaken on Algol-type variables in known pairs, by making frequent estimates of the Δm of the selected pairs. We saw earlier that there is a whole hierarchy among multiple systems. It is quite possible that in some cases one of the components is actually a photometric double, showing periodic variations in Δm. Frequent observations over several days, or several months, using the Argelander method, or, even better, a double-image photometer (which is more accurate), should detect such changes in quite close doubles (between one and a few arc-seconds in separation). If the overall magnitude m of such a pair is known, together with its Δm, the magnitude of each component may be calculated from Fig. 13.30. Paul Baize published a catalogue of binaries with one variable component in 1962; it remains of interest. [A later catalogue (Proust, *et al.*, 1981), containing details of 300 stars, and which largely supersedes that by Baize, is particularly recommended. – Trans.] Vierbinski (at Wroclaw) has also published a catalogue of Δm.

In all these investigations, it is not a good idea for observers to work in isolation – unlike the popular idea of a lone individual perched on a tower away from everyone. Working as a team provides both the necessary encouragement and an appropriate stimulus.

13.6.7 Publication of Results

Professionals publish their measurements in the *Supplements* to the European journal *Astronomy and Astrophysics*, which is to be found in all observatories. Unfortunately, this journal is not open to amateurs, except in a few exceptional cases (such as Paul Baize). In France the *Journal des Observateurs* was not so exclusive. This is why the Société Astronomique de France created a special journal, the *Observations et Travaux*, which publishes this type of measurements. Because it is sent to observatories that are interested in this field of work, amateurs can expect their

Table 13.15. *Ten neglected pairs requiring measurement*

Σ	ADS	Coordinates 2000		m_1, m_2		First observation			Last observation			Notes
						Year	θ	ρ	Year	θ	ρ	
132 AB	1202	$01^h32.1^m$	$+16°57'$	6.9,	9.9	1783	$28°$	$15.8''$	1961	$344°$	$58.0''$	multiple
510	3054	$04^h12.1^m$	$+00°44'$	6.9,	9.9	1831	$300°$	$10.8''$	1964	$301°$	$11.0''$	
560	3284	$04^h31.5^m$	$-13°39'$	6.2,	9.1	1901	$44°$	$29.8''$	1919	$45°$	$29.8''$	
889	4930	$06^h19.9^m$	$+25°01'$	7.6,	9.9	1830	$222°$	$22.0''$	1957	$237°$	$21.3''$	
1034	6047	$07^h24.0^m$	$-03°59'$	7.1,	9.6	1830	$285°$	$13.3''$	1960	$286°$	$14.6''$	
1803	9111	$14^h06.4^m$	$+38°25'$	8.1,	9.9	1831	$43°$	$17.8''$	1967	$43°$	$18.0''$	
2431	11910	$18^h58.8^m$	$+40°42'$	6.2,	8.5	1829	$236°$	$18.7''$	1969	$236°$	$19.0''$	
2844	15408	$21^h51.8^m$	$+64°53'$	7.2,	9.2	1903	$258°$	$11.2''$	1916	$261°$	$12.0''$	
2982	16550	$23^h09.5^m$	$+08°40'$	5.4,	10.0	1831	$188°$	$32.6''$	1923	$198°$	$32.9''$	
3028	16550	$23^h09.5^m$	$+08°40'$	7.1,	9.6	1829	$205°$	$19.5''$	1961	$201°$	$19.2''$	

Table 13.16. *Ten binary systems that require measurement*

Name	ADS	Coordinates 2000		m_1, m_2	Ephemeris 1988		Notes
					θ	ρ	
Σ 547	48	$00^h05.4^m$	$+45°49'$	8.9, 9.0	177°	5.9″	multiple
Σ 13	207	$00^h16.3^m$	$+76°57'$	6.8, 7.1	55°	0.9″	
OΣ 18	588	$00^h42.4^m$	$+04°10'$	7.8, 9.4	206°	1.5″	multiple
BU 1100	999	$01^h14.7^m$	$+60°57'$	8.3, 8.3	33°	0.5″	
Σ 186	1538	$01^h55.8^m$	$+01°51'$	6.8, 6.8	56°	1.3″	
Σ 305	2122	$02^h47.5^m$	$+19°22'$	7.4, 8.2	308°	3.7″	multiple
Σ 400	2612	$03^h34.9^m$	$+60°02'$	6.8, 7.6	266°	1.4″	multiple
Σ 1110	6175	$07^h34.6^m$	$+31°53'$	2.0, 2.9	78°	2.9″	Castor
Σ 2383	11635 CD	$18^h44.3^m$	$+39°40'$	5.2, 5.5	88°	2.3″	ϵ_2 Lyr, ϵ_1 is in same field
Σ 2737	14499	$20^h59.1^m$	$+04°18'$	6.0, 6.3	285°	1.0″	multiple

observations to be used. This implies that the way in which the data are presented should agree with certain rules.

The items given should be: the star's identification, the discoverer's identification number and the catalogue (ADS or IDS); the position for a given epoch (1950 or 2000); dates of the measurements (years and decimals of years); the values obtained (θ to 0.1 degree, ρ to 0.01 arc-second); estimated magnitudes (optional); comments (observing conditions, etc.); finally, an average of the values obtained over the three or four nights of observation. Each entry is given in order of increasing right ascension. An example is given in Fig. 13.31.

In this field, the author is responsible for whatever is published, which assumes that great care and honesty will be taken in making the observations and in preparing the actual material for publication.

13.7 Reduction and analysis

13.7.1 Orbits

Some observers may be satisfied with simply making measurements. This is by no means unknown, and such people are happy to amass a large number of doubles. If anyone wants to go farther, then the determination of orbits is an obvious step.

This is difficult work, which requires considerable knowledge of the methods involved, because frequently several hundred observations, of differing degrees of accuracy, have to be analysed. It is essential to know how these should be interpreted to obtain a good orbit that will not merely waste space in any catalogue that may be published, and which would be similar to the one by Worley and Heintz in 1983.

The apparent orbit, an ellipse, which is directly obtained from the measurements (Fig.13.32), is the projection of the true orbit, another ellipse, onto the plane of the

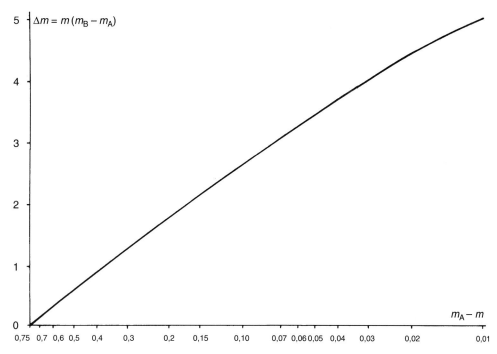

Fig. 13.30. *Combined magnitudes (after P. Muller). Examples of how this diagram may be used:*

If $m_A = 8.2$ and $m_B = 8.7$, than $\Delta m = 0.5$; whence from the curve $m_A - m = 0.52$; therefore $m = 8.2 - 0.52$, or about 7.7.

If $\Delta m = 1.3$ and $m = 4.5$; using Δm, $m_A - m = 0.3$ from the curve, whence $m_A = 4.5 + 0.3$, and $m_B = 4.8 + 1.3 = 6.1$.

sky, perpendicular to the line of sight. From this, geometrical considerations enable one to arrive at the true orbit, which is defined by the following parameters:

- Elements giving the motion and the shape of the ellipse: P, period in years; T, the date of periastron; a, semi-major axis (in arc-seconds); e, eccentricity.
- Elements fixing the position of the orbit with respect to the plane of the sky (known as Campbell elements): i, inclination of the plane of the true orbit to the plane of projection, reckoned from $0°$ to $90°$ if the motion is direct (θ is increasing), and from $90°$ to $180°$ if it is retrograde (θ is decreasing); Ω, the position angle of the ascending node counting from $0°$ to $180°$ (this is the angle between North and the line of intersection of the true orbit with the plane of projection); finally, ω, the angle between the line of nodes and the major axis, reckoned from $0°$ to $360°$ in the direction of motion (see Fig. 13.33).

Calculation of the elements of the true orbit is the classical problem of determining the motion of two bodies that are not orbiting one another but are both orbiting their common centre of gravity, the barycentre. In general, there are no

763

```
            MESURES D'ETOILES DOUBLES A L'OBSERVATOIRE DE NICE
            EN 1979 ET 1982.                    (1ère SERIE  )

                par Pierre DURAND, Président de la Commission
                des Etoiles Doubles de la S. A. F.

.ADS 48AB     0h00,2 N45°16              .COU 150      2h07,2 N21°46
          1979,54  172,6°  5,99" 1n              1982,61   86,5   0,21
Guntzel-Lingner 1954 -0,5  +0,09                 - ,62    85,4   0,17 m=10-10
Hopmann 1961        -1,1  +0,12                   - ,63    94,5   0,18
                                                 1982,62   88,8   0,18 3n
.ADS 48AB                                couple difficile même par bonnes images.
          1982,60  174,5   5,86
          - ,62   174,6   5,91           .ADS 1950     2h28,0 N39°51
          1982,61 174,6   5,89 2n                 1982,62  142,9   6,36 m=9,3-10
Guntzel-Lingner 1954 +0,3  -0,03                  - ,63   142,0   6,17
Hopmann 1961        -0,5   0                      1982,63  142,5 ? 6,27 2n

.ADS 61       0h01,0 N57°53              .ADS 1959     2h29,0 N39°51
          1982,61  293,4   1,33                  1982,62  non vue
          - ,62   292,9   1,41                    - ,63   347,7   0,17 difficile!
          1982,62  293,3   1,37 2n               1982,63  347,7   0,17 2n
Baize 1957          +0,3  -0,06
                                         .BZ           2h30,0 N30°46 dénommée par
.ADS 433AB    0h26,3 N66°42                                             Couteau.
          1982,61  159,0   4,18 m=10-12,5         1982,63  non vue        2n
Hershey             -1,5  +0,17 1n
(4)                                      .ADS 2051     2h35,4 N49°00
.ADS 671      0h43,0 N57°17  êta Cas             1982,63   26,2   0,22 1n
          1982,61  307,6  11,96 mauvaises   un compagnon de 13ème à environ 250° et 30"
          - ,62   308,6  11,99 images
          - ,62   307,4  11,97           .COU 865      2h44,8 N34°44
          1982,62  307,9  11,98 3n               1982,63  124,1   0,17 1n
Strand 1969         -0,9  -0,14
                                         .ADS 9413AB   14h46,8 N19°31
.COU 21       0h13,4 N20°55                      1979,53  331,1   -
          1982,63  179,3   1,29 m=8-12           - ,55   331,1   7,27
          - ,63   180,2   1,01                    - ,55   332,9   7,31
          1982,63  179,8   1,15 2n               1979,54  333,0   7,29 3n
mesure difficile: faible compagnon bleu. Wielen 3 1962       -0,6  +0,08

.ADS 755      0h49,6 N23°05              .ADS 9578AB   15h14,0 N27°12
          1982,63  265,1   0,69                  1979,53  251,5   -
          - ,63   264,2   0,74 dm=0,2            - ,54   251,7   1,36
          1982,63  264,7   0,71 2n                - ,54   250,0   1,38
Muller 1955         -4,5  +0,08                  1979,54  251,0   1,37 3n
mesures par bissection sans doute un peu Heintz 1964         +0,9  -0,01
fortes comparées à l'examen de la tache
de diffraction donnant plutôt 0,6".      .ADS 9617AB   15h19,1 N30°39
                                                 1979,53  307,5   0,35
.COU 1214     1h31,5 N39°45                      - ,54   307,6   0,49
          1982,63  180,8   0,23 dm=0,2 1n         - ,54   109,4   0,13
                                                 1979,54  308,2   0,40 3n
                                         Danjon 1938         -0,6  +0,08
```

Fig. 13.31. *Example of an international form listing double-star measurements. The details given are: the stellar designation(s); a line or lines containing the date, θ, ρ, and (sometimes) Δm. The line with the means follows, with the number of nights, and finally, the $O - C$ if an orbit has been calculated, then remarks.*
The line giving the mean values is the one to be listed in a datafile, together with the observer's name. Any interested amateur should start to compile such a datafile as soon as possible. It should consist of an entry (or computer screen) with all relevant measurements and other useful information (spectra, parallax, etc.) that may be required in subsequent calculations.

absolute measurements of position (although some programmes of observation with astrolabes are being established). The values of θ and ρ merely relate the position of one component relative to the other. Instead of determining the orbits of A and B around the barycentre G (A, G and B always fall on a straight line), we can determine an equivalent orbit, that of B with respect to A. This is a Keplerian orbit,

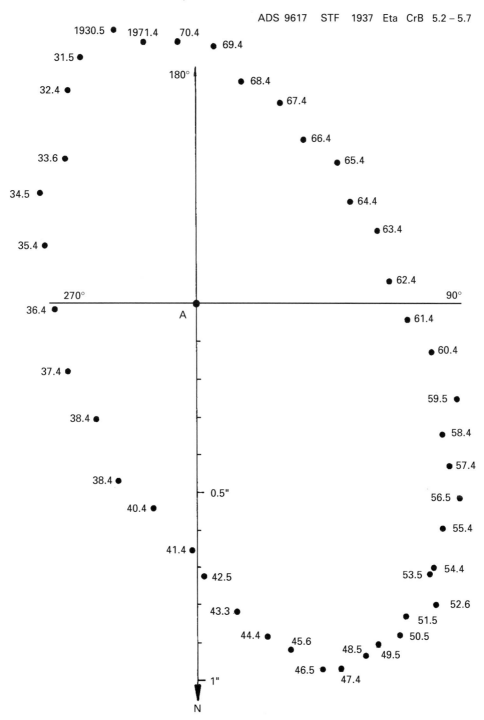

Fig. 13.32. *Apparent orbit of a visual binary between 1930 and 1971. The measurements, derived from the literature, have been reduced to yearly means and plotted, using a protractor and scale rule.*

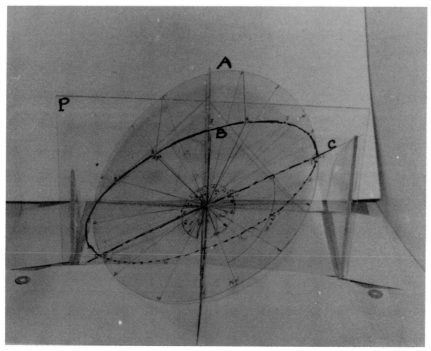

Fig. 13.33. *A model, made by E. Nègre, showing the real orbit A, projected against the plane of the sky (perpendicular to the line of sight), producing the apparent orbit B. The line of nodes (the intersection of the orbits), is marked (C).*

which means that it is an ellipse with A at one of the foci. The motion of B obeys the law of equal areas, the centres of the ellipses coincide, and there are several conjugate diameters.

13.7.2 Why calculate orbits?

It is intellectually satisfying to be able to describe apparently complex motions in terms of simple laws, especially when the equations allow one to determine future positions of the bodies involved. Even more interesting is the use that astrophysics can make of the geometrical and dynamical data a and P. We know that $a^3/P^2 = \mathcal{M}_A + \mathcal{M}_B$ where $\mathcal{M}_A$ and $\mathcal{M}_B$ are the masses of the components expressed in solar masses, a is given in astronomical units, and P in years. The value of a is obtained in arc-seconds: to express it in astronomical units, we need to know the system's parallax. This is the only way of directly determining stellar masses, because if the position of G is known, or the magnitude and spectrum of each component, both masses may be determined. Unfortunately, the small number of definitive orbits and accurate parallaxes means that slightly less than one hundred masses are known from this method.

Because of this fact (inaccurate knowledge of small parallaxes), Baize and Romani developed a method, known as dynamical parallax, of determining the distances of binaries not accessible by the trigonometric method. Based on the mass–luminosity relationship, we can calculate the distances once we know P, a, the magnitudes, and the spectra.

At present, these calculations are based on a few hundred definite orbits: the companion must have nearly completed its orbit, and the measurements need to be numerous and sufficiently accurate. These are definitive orbits. They are often preceded by preliminary orbits. A pair may, therefore, be described in terms of four or five successive orbits. These may be used once a and P no longer vary. Thanks to recent discoveries of binaries, many of which have short periods (of approximately a decade), the total of a thousand orbits currently catalogued will probably increase rapidly.

13.7.3 Calculating ephemerides

Beginning with the true orbital elements, we have to determine the polar coordinates θ and ρ of component B with respect to A. This calculation is useful if no catalogue of ephemerides is available, and if one has information about the orbital parameters. The standard equations for an ellipse are used:

$$\tan(\theta - \Omega) = \tan(v + \omega)\cos i, \tag{13.1}$$

$$\rho = r\cos(v + \Omega)/\cos(\theta - \Omega), \tag{13.2}$$

$$M = n\,(t - T), \tag{13.3}$$

where t is the chosen epoch, v is the true anomaly, M the mean anomaly, r the radius vector, with Ω the position angle of the line of the nodes and ω the angle of perihelion, as before. Kepler's equation gives us:

$$M = u - (180°/\pi)e\sin u \text{ (in degrees)}, \tag{13.4}$$

where u is known as the eccentric anomaly, and

$$r = a\,(1 - e)^2/(1 + e\cos v). \tag{13.5}$$

If we have a programmable calculator (or a computer program) we can proceed as follows:

(i)   calculate $M = 360° \times (t - T)/P$;
(ii) iterate to calculate u;
(iii) calculate v from $\tan v/2 = \sqrt{(1 + e)/(1 - e)} \times \tan u/2$;
(iv) calculate r from (13.5);
(v)   calculate θ and ρ from (13.1) and (13.3).

If preprogrammed conversions to polar coordinates are available, this helps considerably. The reader will find details in the previously mentioned catalogue of 2500 stars (Minois, 1984) and in *Observations et Travaux*, **2** and **3**, together with programming examples.

The calculations may also be carried out using tables (Danjon, 1980), and again thanks to the Tiele–Innes elements (A, B, F, G, C, H), which are frequently given, and which greatly facilitate the work (*see* Couteau, 1981a)

13.7.4 Determining an orbit

There are numerous methods of doing this: see the books by Danjon and Couteau already mentioned. The reader may also be interested in the following simple method described by Baize (1984).

First, all the measurements of the pair to be studied are collected (here it is useful to have access to a complete datafile). The measured values of θ may need to be corrected for precession (subtracting the quantity $0.0056(t - t_0)\sin\alpha\cos\delta$), if they were obtained over a long period of time, and if the binary has a high declination. If the observations are numerous, yearly means are then taken.

One then proceeds as follows:

(i)   Make a large-scale plot of θ versus t, using a large sheet of millimetre graph paper. The waviness is a sign of variations in the velocity of the companion and also the shape of the true orbit: one gets a feeling of the eccentricity.

(ii) Make a large-scale drawing of the apparent orbit, plotting the yearly values of θ and ρ using a rule and protractor. Draw the ellipse with thread and two pins, taking heed of the law of areas: divide the ellipse into sectors of equal time, using appropriate, good-quality observations. These sectors should have the same area and this may be verified by using a planimeter, or sectors of thick, homogeneous cardboard that are accurately weighed. The success of this operation is crucial for the final result.

   The foci of this apparent orbit are marked F' and F''; its centre C coincides with that of the actual orbit (Fig. 13.34);

(iii) Determine the true orbit: the primary, A, lies at one of the foci and the major axis passes through A and C. This gives the position of periastron P; the line of the nodes $\Omega\Omega_1$ is the external bisector of the angle F'AF''.

(iv) Measure the angles NAP (which is the projection of ω) and Ω with the protractor; and the eccentricity $e = CA/CP$ with the rule. Note that because, in general, it is not known which is the ascending node, it is customary to take the value of Ω as being either angle NAΩ or NAΩ_1, whichever is less than $180°$.

(v)   If the companion has not completed one orbit, the period P may be found from the apparent orbit by measuring areas: $P = S/c$ (where S is the total area and c the area constant: $c = dS/dt$, where dS is the area swept out during an interval dt). If an orbit has been completed, the period may be read off from the curve of $\theta(t)$, as may T, the date of periastron passage.

(vi) Having P, T, e, and Ω, all that remains is to calculate ω, using (13.1) – when $\theta - \Omega$ equals $0°$, $90°$, $180°$, or $270°$, $v + \omega = \theta - \Omega$; then i, because at periastron $\cos i = \tan(\theta_P - \Omega)/\tan\omega$; and finally, a, using (13.5).

Example orbits are listed in Table 13.17. Using the values determined for the

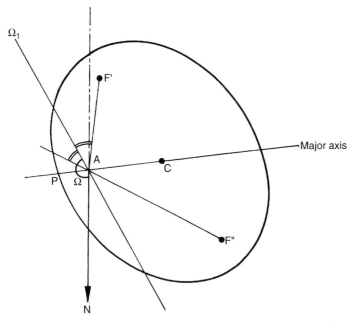

Fig. 13.34. *Method of determining the actual values of e and Ω from an apparent orbit.*

orbital parameters, the $O - C$ residuals are calculated for the data used in drawing the orbit. The residuals should show a random distribution and have no overall trend. If they do, the orbital parameters should be adjusted to obtain smaller residuals. E. Soulié has published a method of carrying out this optimisation on a computer (Soulié, 1986).

With spectroscopic binaries, we only have variations in radial velocities, and these should be plotted to give a curve as a function of time. From this it is possible to calculate elements that describe motion in an ellipse, in a manner somewhat similar to that used for visual binaries. However only $a \sin i$ may be obtained. (The inclination of the orbit cannot be known unless visual or interferometric observations provide a position angle and a separation.) Conversely, if we know the radial velocities of a visual binary, we can determine the inclination of the true orbit, because we are able to tell whether the companion is receding or approaching.

13.7.5 Determining linear trajectories

This field of double-star work is simply concerned with establishing that pairs are purely optical. In fact, we are returning to the original interest in binary stars (in particular that shown by William Herschel), which may be used to determine stellar parallax. We shall see shortly that this is still our aim.

This type of work is seldom undertaken, yet it is simpler than determining an orbit. In fact, essentially all we have to do is to fit a straight line to a set of points, and to check that the velocity is constant, once we are certain that we are dealing

Table 13.17. *Example orbits*

a: Two orbits seen side-on

Σ 1728, ADS 8804
$13^h10.0^m$, $+17°31'$ (2000), 5.2, 5.2

1987: $\theta = 192°$, $\rho = 0.42''$
1990: — $\rho = 0.04''$!
1992: $\theta = 12°$, $\rho = 0.5''$

Orbit: Harting 1950
P: 25 yr   T: 1911.7   e: 0.57   a: 0.68″
i: 89.7°   ω: 277.7°   Ω: 11.9°

Σ 1909, ADS 9494, 44 Boo
$15^h03.9^m$, $+47°39'$ (2000), 5.3, 6.2

1987: $\theta = 44.5°$, $\rho = 1.4''$
1990: $\theta = 47°$, $\rho = 1.7''$

Orbit: Heintz 1978
P: 225 yr   T: 2021   e: 0.43   a: 3.77″
i: 83.9°   ω: 38.8°   Ω: 57.8°

b: Two orbits seen face-on

Σ 1196, ADS 6650 AB, ζ Cnc
$08^h12.2^m$, $+17°40'$ (2000), 5.6, 6.0
Triple system

1987: $\theta = 21.8°$, $\rho = 0.6''$
1990: $\theta = 182°$ $\rho = 0.6''$

Orbit: Gasteyer 1951
P: 59.7 yr   T: 1930   e: 0.32   a: 0.88″
i: 172°   ω: 233°   Ω: 58°

Σ 1338, ADS 7307 AB
$09^h21.2^m$, $+38°12'$ (2000), 6.6, 6.8

1987: $\theta = 267°$, $\rho = 1.1''$
1990: $\theta = 219°$, $\rho = 1.0''$

Orbit: Arend 1953
P: 389.05 yr   T: 1998.59   e: 0.29   a: 1.52″
i: 15.8°   ω: 258.9°   Ω: 28.1°

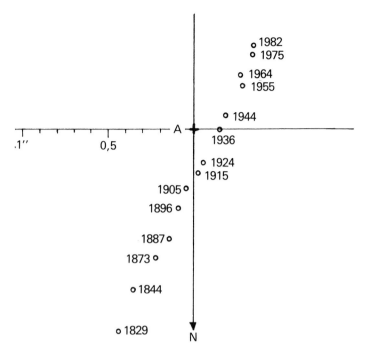

Fig. 13.35. *The linear, uniform motion of the secondary component in ADS 8231 (Σ 1555), an optical double.*

with an optical binary (Fig. 13.35). Care must be taken, however, before we can be certain that we are not viewing a true binary more or less in its orbital plane, and over part of the orbit between periastron and apastron.

When one makes an observation, and measures the relative values of θ and ρ, the area of sky concerned is very small, so there is no need to make any corrections for refraction, nor for annual aberration, because generally the observations are spread throughout the year. On the other hand, with observations made over a long interval of time, and close to the pole, it is wise to correct θ for precession by subtracting $0.0056(t - t_0) \sin \alpha \cos \delta$.

At present there is only one catalogue of linear trajectories, that by J. Dommanget (1964), on which we base most of our remarks. It contains 325 entries (which may be compared with the 536 calculated orbits at the same date). By now, some 500 objects have been studied (as against approximately 1000 calculated orbits). Fairly recently (1977), P. Muller added 17 objects. Either this type of investigation is of less interest to researchers or, statistically, orbits are twice as numerous as optical doubles.

Let us consider the geometrical aspects of this question. Although natural elements, expressed in polar coordinates, are better than the observations themselves in showing the θ and ρ of closest approach and the annual motion of the components, vector elements that are expressed in rectangular coordinates are more suitable for

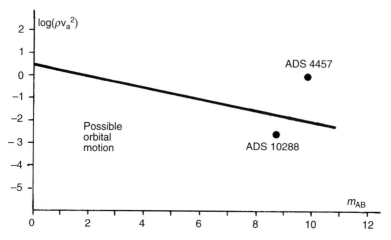

Fig. 13.36. *Diagram for determining whether a binary is an optical double.*

use in the calculation of proper motion μ. It is always possible to revert to equatorial coordinates by using:

$$\Delta\alpha = \mu x/\cos\delta, \text{ and } \Delta\delta = \mu y.$$

What method should be adopted? Rather than keying all the data into a program, where they will no longer be immediately available for consideration, it is preferable to log all the measurements on a large sheet, giving the date and the observer. The linear trajectory may then be drawn (or calculated using a linear-regression method). The advantage of drawing the path is that one can consider all the points immediately, and give each an appropriate weighting as necessary (placing, for example, greater weight on photographic measurements obtained at Washington). Finding $\Delta\alpha$ and $\Delta\delta$ is then simple.

Is it an orbital or optical double? A constant velocity will indicate the latter. J. Dommanget has devised criteria for defining an optical double more precisely. The first is a dynamical criterion. It involves determining a limit π_1 for the parallax of the pair, using the mass–luminosity relationship determined by Paul Baize, at which the observed measurements could still represent elliptical motion. The formula (Dommanget and Nys, 1964) is not very easy to apply if one does not have access to specialized catalogues (magnitudes, spectral classes, and parallaxes). The other criterion, in contrast, is statistical, and was established using the dynamical criterion. It is the relationship: $\log(\rho v_a^2) = +0.39 - 0.207 m_{AB}$, where v_a is the annual velocity, ρ the separation, and m_{AB} the combined magnitude of the pair. Plotted, this relationship gives a straight line separating two regions, the lower being where orbital motion is possible, and the upper where it is not possible and the pair is therefore an optical binary (Figs. 13.36 and 13.37). In addition, a large Δm is a good indication of whether a binary is optical, if there is no white or red dwarf in the system. A high annual velocity and large separations are also good indicators.

In discussing such work, it is essential to give a full identification of the double,

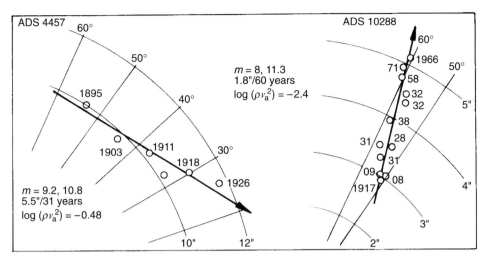

Fig. 13.37. *Two examples of linear, relative trajectories, one of which (ADS 10288) is possibly a true binary.*

the magnitudes, and the calculated $\Delta\alpha$ and $\Delta\delta$ with the epoch t_0 (generally 1950.0). Then one can derive the $O - C$ residuals for the measurements that have been used. The proper motion calculated for one or both components may be given, using the catalogue values (from the SAOC, IDS, ADS, AGK3, etc.), together with any remarks: accuracy, or confirmation of μ (*see* Muller, 1978 and de Froment, 1987 - the latter on ADS 8231).

It will be seen that this type of observation is not currently fashionable. There is room for work by amateurs, provided they have access to a regularly revised file of measurements. In addition, thanks to programmes covering neglected variables, observers will certainly have their own data to add to this type of study.

13.7.6 Amateurs' part in double-star work

Being free from worries about their careers, and the inevitable constraints imposed by astronomical research institutions, and also able to undertake whatever research they like, amateurs are able to invest a large part of their leisure time, and their experience – or are able to develop the latter – in whatever field is of interest. This is why they are able to make a major contribution to double-star astronomy, and carry out scientifically valuable work.

It is not possible to do everything. All major research requires a considerable amount of study, and the compilation of a lot of different information, which is not always feasible for amateurs. The borderline comes somewhere around the calculation of orbits, where vast quantities of data need to be acquired, and which are only available to professionals. There is one exception, however, Paul Baize, whom we have already mentioned. Various statistical studies of stellar populations are also beyond amateurs' reach, because they require the analysis of extremely large samples of data.

On the other hand, at a time when the number of professional observers of double stars is declining, the field is being increasingly left to amateurs. Some professionals fear that after they drop out, amateurs will be the only ones carrying the load of ensuring continuity of observation. Permanence can only be assured by working within those organisations that are currently involved in this field. Rather than working with professionals, it is best for amateurs to collaborate, to work as a team, and perhaps devise ideas for programmes suggested by their own observations. There is also the chance of using instruments at observatories (because, in general, the instruments at amateurs' disposal are naturally rather limited). Such collaboration already exists in various fields, and in particular in double-star work, where there have been opportunities to verify data for the HIPPARCOS mission, and to use large telescopes.

14 Variable stars

M. Dumont & J. Gunther

In 1596, Fabricius observed a star in Cetus that was not shown on any atlas. Some months later the star had disappeared; it was rediscovered in 1603. This is the star Mira, o (omicron) Ceti, which was the first variable star known. (A variable star is any star that changes its brightness.) In 1669, Montanari discovered the variation of Algol (β Per). Discoveries became more and more frequent subsequently: in 1844, 18 variables were known, 2054 in 1920, and more than 30 000 nowadays.

14.1 The classification of variable stars

Three broad types of variable stars are recognized:

- eclipsing variables
- eruptive variables
- pulsating variables

14.1.1 Eclipsing variables

Eclipsing variables are binary stars, the two components of which are in orbit around one another or, strictly speaking, around their common centre of gravity. If the Earth lies in the plane of the orbit, regular eclipses of one star by the other are seen. The two stars are too close for them to be distinguished from one another, so the eclipses appear as a decline in the combined light from the pair. The behaviour of the eclipses is generally a function of the distance between the two components.

14.1.1.1 Variables of class EA
EA-class eclipsing binaries are pairs that are relatively widely separated (Fig. 14.1A). The periods range from 1 day to more than 10 years. The brightness of the variable is (approximately) constant outside the eclipse (Fig. 14.2). If the two stars are similar, the eclipses are also similar in appearance. But generally one component is brighter than the other; primary and secondary minima then alternate (Fig. 14.2). The most famous of this class of stars is Algol, and objects of this type are often called 'Algol-type variables'; some well-known examples are given in Table 14.1.

14.1.1.2 Variables of classes EB and EW
When the components of a binary are close, they are distorted by tidal effects and become elongated (Fig. 14.1B). An equipotential surface is a surface of hydrostatic equilibrium (i.e., the gravitational and pressure forces are in equilibrium); the surface of a lake is an equipotential, for example. Around an isolated star, the equipotentials are spheres centred on the star. Near a binary, the equipotential

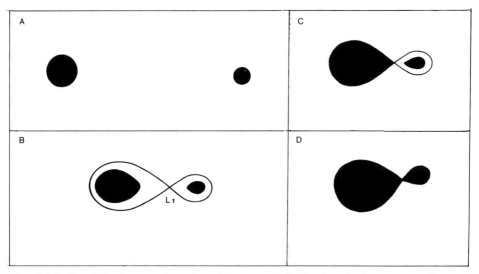

Fig. 14.1. *Possible forms of a close binary. A & B: detached systems; C: semi-detached; D: contact*

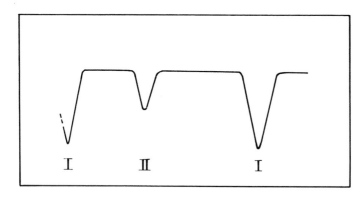

Fig. 14.2. *Schematic light-curve of an EA (Algol-type) eclipsing binary. The primary minimum (I) and secondary minimum (II) alternate.*

surfaces are distorted, and one particular surface forms an envelope around the two stars, the two lobes making contact at a single point, L1: the first Lagrangian point. This surface defines what are known as the Roche lobes, which meet at L1. If neither star fills its Roche lobe, as in Fig. 14.1A or 14.1B, the system is described as being 'detached'. If one of the two stars fills its Roche lobe (Fig. 14.1C), the system is 'semi-detached', and the star filling its lobe may lose material to the other component. If both components fill their Roche lobes, then the stars form a 'contact binary' (Fig. 14.1D). The binary then frequently takes on the form of a dumbbell-shaped object with two cores and a common atmospheric envelope.

The tidal distortion gives rise to continuous variation as the binary rotates; the star is then known as a 'β-Lyrae star', belonging to class EB, with most periods

Table 14.1. *Some well-known EA variables**

Star	Period (days)	Magnitude			Radius		Sep. AB	Total mass
		M	m I	m II	A	B		
Algol	2.867 304 3	2.12	3.40	2.19	2.74	3.60	14.03	4.51
RZ Cas	1.195 247	6.18	7.72	6.43			2.50	
U Oph	1.677 346 2	5.88	6.58	6.48	3.2	3.2	11.9	10.1
AR Aur	4.134 695	6.15	6.82	6.37			4.85	
RW Tau	2.768 835 6	7.98	11.47		2.25	3.00	10.9	3.10
ϵ Aur	27.08 yrs	2.85	3.90	?	2500			

* M = maximum; m I = primary minimum; m II = secondary minimum. The radii and separation of the two components are expressed in solar radii, and the mass in solar masses.

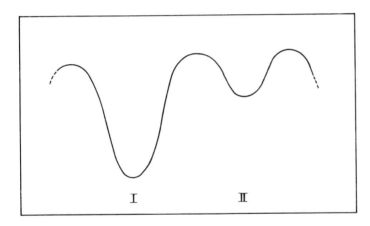

Fig. 14.3. *Light-curve of an EB-type (β-Lyrae) eclipsing binary*

ranging between 0.4 and 200 days, although short periods (around 24 hours) are most frequent (Table 14.2). Most often, the two stars are unequal in size (Fig. 14.3).

In most contact binaries, the components are small, so eclipses occur frequently. They are variables of class EW (the 'W-Ursae Majoris stars'), with periods that are always less than 1 day, and sometimes even less than 1 hour (Table 14.3). In general, the two stars are very similar, and the secondary minimum closely resembles the primary minimum (Fig. 14.4).

14.1.2 Stellar evolution

The intrinsic magnitude of a star is primarily a function of its mass, and of the type of nuclear reactions that are occurring in its core. These reactions change over the course of time, and the star undergoes overall changes in brightness. In general,

Table 14.2. *Some well-known EB-type eclipsing variables**

Star	Period	Magnitude			Radius		Sep. AB	Total mass
		M	m I	m II	A	B		
β Lyr	$12^d21^h47^m$	3.34	4.34	3.84	11.10	4.43	55.51	13.79
η Ori	$7^d23^h47^m$	3.14	3.35	3.23				24.9
u Her	$2^d01^h13^m$	4.83	5.51	5.09	6	6	13	10.1
μ_1 Sco	$1^d10^h34^m$	3.02	3.30	3.13	5.6	6.2	13	22.5

* Units as Table 14.1

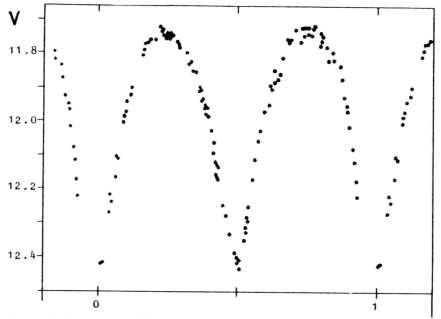

Fig. 14.4. *Light-curve of NSV 4070, an EW-type (W-Ursa Majoris) eclipsing binary. (One-metre reflector, Pic du Midi, Toulouse Observatory photometer, 1983 December. A. Figer, J. F. Leborgne, M. Dumont.)*

these changes are long-term, but the star also passes through several periods when it undergoes more rapid variations. Hot, luminous gas may be ejected or, conversely, dark absorbing material. In both cases the star is described as being an 'eruptive' variable. The variations in magnitude may arise as a consequence of cyclic variations in the diameter of the star: it is then known as a 'pulsating' variable.

After their birth, stars are initially unstable. They remain irregular eruptives until their energy production becomes stabilized, which occurs when the nuclear reactions

Table 14.3. *Some EW-type eclipsing variables**

Star	Period	Magnitude			Radius		Sep. AB	Total mass
		M	m I	m II	A	B		
W UMa	$8^h00^m23^s$	7.88	8.63	8.54	1.09	0.80	2.58	2.06
V566 Oph	$9^h49^m53^s$	7.45	7.95	7.90				
FZ Ori	$9^h35^m59^s$	10.61	11.02	10.95				
LO And	$9^h07^m49^s$	11.20	11.75	11.70				
i Boo	$6^h25^m37^s$	5.85	6.44	6.39	1		1.63	1.44

* Units as Table 14.1

turning hydrogen into helium become their dominant energy source. The star then enters the most stable, and longest, period of its existence.

The accumulation of helium at the core of the star steadily raises the central temperature, producing an increase in the star's diameter. It eventually precipitates the first of the various crises that occur.

A graphic representation of a star's life is given by the track that it follows on the Hertzsprung–Russell (H-R) diagram. The position and path of the star are related to its initial mass, which is the dominant factor in stellar evolution (Fig. 14.5).

The various phases of instability occur at different periods of a star's life. There are several types of pulsating and eruptive stars, and each class is relatively localized on the H-R diagram (Fig. 14.6). In other words, variables belonging to a specific class are at the same stage of their evolution. The Sun was undoubtedly an irregular eruptive some 4.6 thousand million (4.6×10^9) years ago; and it will doubtless become a Mira-type variable in another 7 thousand million years. At present, the chronological sequence of all the different phases of a star's lifetime is not perfectly established, but it forms the subject of extensive research, and will doubtless be fully understood in the next few decades.

14.1.3 Eruptive variables

14.1.3.1 Young variables
Stars are born in giant clouds of interstellar gas. At the beginning of its existence, a star is still unstable and is surrounded by the remains of the cloud from which it was born. Young stars are always irregular, eruptive variables, and are associated with nebulosity (Table 14.4). They do not all have the same characteristics:

- stars of type Ina are very hot (O, A, or B spectra);
- stars of type Inb are cooler (spectra of type F to M).

Many nebular variables have a specific form of spectrum, resembling that of T Tauri. This spectrum resembles that of the Sun's chromosphere, but with an exceptional abundance of lithium. These stars, denoted InT, are in the course of

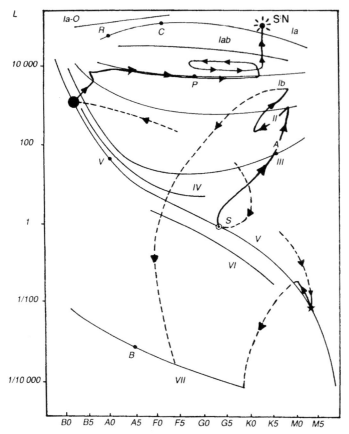

Fig. 14.5. *Evolutionary paths of three stars on the H–R diagram: • star of 10 solar masses; ⊙ the Sun; ⋆ star of 0.25 solar mass. The various luminosity classes are marked, luminosity L itself being plotted on the y-axis, that of the Sun being equal to 1.*
Ia-O: hypergiants; Ia, Iab, and Ib: supergiants; II: bright giants; III: giants; IV: sub-giants; V: main sequence; VI: sub-dwarfs; VII: white dwarfs.
A few individual stars are plotted: R = Rigel; C = Canopus; P = Polaris; V = Vega; A = Aldebaran; B = Sirius B.

formation. Atoms of lithium are not stable at high temperatures; their considerable abundance therefore indicates that the temperature inside the star is moderate, and that nuclear reactions may not yet have started.

Several stars of the FU Orionis type are known that have suddenly gained 5 to 7 magnitudes, and subsequently stabilized at the higher magnitude. They are consistently associated with nebulosity, and are undoubtedly very young stars.

14.1.3.2 Red eruptives
It is thought nowadays that between 10 and 20 % of red stars are variables. Red dwarfs, in particular, are often variable. They sometimes undergo sudden, irregular

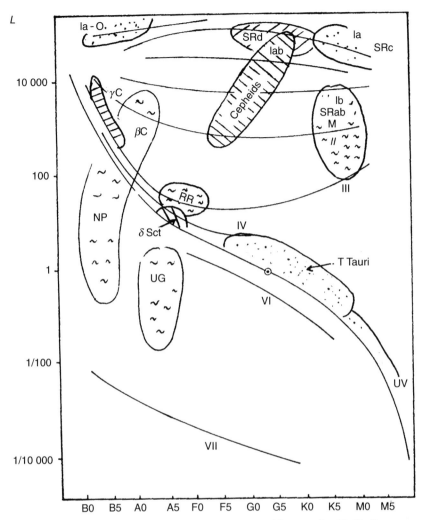

Fig. 14.6. *The position of some classes of variables on the H–R diagram.*
Dotted areas: young eruptives (ages less than 20 million years);
Diagonal hatching: middle-aged pulsating stars (ages between 15 and 1000 million years); Tildes: old stars.
PN indicates the central stars of planetary nebulae. The other abbreviations (SR, M, γC, βC, UG, UV, etc.) indicate classes of variables described in this chapter.

flares, the existence of which was established by Van Maanen in 1940. Over a period reckoned in seconds, the star gains 1 to 2 magnitudes. It subsequently declines, and returns to its normal magnitude over an interval that is measured in minutes. The phenomenon is analogous to the chromospheric eruptions seen in the atmosphere of the Sun, and elsewhere, and the amount of energy released is the same. Such a flare is insignificant when compared with the Sun's overall brightness, but very

Table 14.4. *Some young variable stars*

Star	Magnitude		Spectrum	Class	Associated nebulosity
	max.	min.			
R Mon	9.3	14.0	A–F	InT	NGC 2261
T Tau	9.6	13.5	G5	InT	NGC 1555
T Ori	9.5	12.6	B8–A3	Ina	NGC 1976
FU Ori	9.7	16.5	F2 I	FU	B35
V1057 Cyg	10	16	A1 IV	FU	NGC 7000
PU Vul	8.5	16	F–M	FU ?	

significant on a red dwarf, which has a luminosity that is 1000 or 10 000 times less. UV Ceti is the best-known flare star. These stars are therefore known as UV Ceti-type variables.

Variables of the BY Draconis type are far less spectacular; their variations are generally less than 0.1 magnitude. It is thought that their surfaces are of uneven brightness, and that the variations are the result of the stars' own rotations.

14.1.3.3 *R Coronae Borealis type variables*

R Coronae Borealis is normally at magnitude 5.8. At unpredictable intervals its brightness declines. The minima may be very deep, sometimes reaching below magnitude 14. After a lapse of several months, the star finally regains its normal brightness. In decline, the star shows absorption lines of carbon, which appears to be extremely abundant. Some 30 of these stars are known. The brightest include SU Tau, which is normally of magnitude 9.7 (at maximum), and RY Sgr (magnitude 6.5). The explanation of these variations is still incomplete.

14.1.3.4 *Novae*

Occasionally, a new star suddenly appears in the sky. After several days, it begins to decline, and the star disappears several months later. In fact, the star is not truly new, but was extremely faint before the outburst. Several types of novae have been observed:

- Fast novae (type Na) (Fig. 14.7): The star gains more than 10 magnitudes in a few hours. The decline begins almost immediately, and lasts several months, until the star reverts to its initial brightness.
- Slow novae (type Nb) take longer to reach maximum, where they may remain for several months, sometimes with frequent minor fluctuations, before fading slowly.
- Recurrent novae (type Nr) have undergone more than one explosion. In general, they change rapidly, but their amplitude is smaller. Table 14.5 gives details of the six recurrent novae known to date.

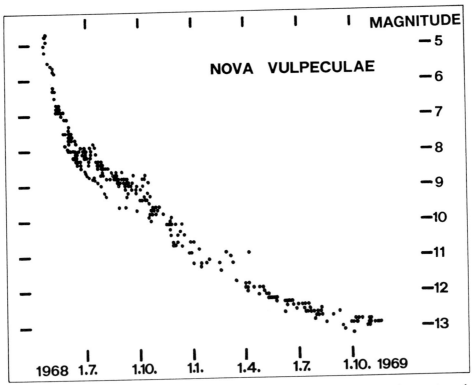

Fig. 14.7. *Light-curve of Nova Vulpeculae 1968 (LV Vul).* (*From observations by M. Dumont, R. Lukas, R. Germann and K. Locher, published in* Orion, *1970.*)

Table 14.5. *Recurrent novae*

Star	Known outbursts	Magnitude	
		max.	ampl.
T CrB	2	1.8	9
RS Oph	4	4.5	7.8
T Pyx	5	7.0	7.5
VY Aqr	2	8.5	8.5
U Sco	4	8.8	10.2
V616 Mon	2	11.3	8.5

Novae are known to be close binary systems. The interpretation of the events has undergone significant changes in the last 20 years. It is now thought that one of the components fills its Roche lobe, whilst the other is a tiny, white dwarf. There is a mass transfer from the component that fills its lobe towards the white dwarf. The gas builds up in a layer on the outside of the small star, until eventually the

Table 14.6. *Some dwarf novae*

Star	Magnitude		Average interval (d)	Sub-class	Binary period
	max.	min.			
WX Hyi	9.6	15.0	14	U Gem	1^{h}49^m
X Leo	11.5	15.5	23	U Gem	
Z Cam	10.4	14.9	23	Z Cam	6^{h}57^m
SS Cyg	8.2	12.5	50	U Gem	6^{h}38^m
SS Aur	10.5	15.3	55	U Gem	4^{h}20^m
U Gem	8.9	15.1	102	U Gem	4^{h}19^m

pressure and temperature become high enough for nuclear reactions to occur. The sudden onset gives rise to the outburst of energy and light that is emitted by the nova over a period of several weeks.

The greatest amplitude known for a nova is that observed for Nova Cygni 1975, which was magnitude 1.8 at maximum. Before its outburst on 1975 August 29, the star was not visible on the Palomar Sky Survey plate, which reached down to magnitude 21.

14.1.3.5 Dwarf novae

Dwarf novae are stars that are intrinsically faint, and which frequently, and fairly regularly, undergo sudden outbursts (Table 14.6). They rise by 4 to 5 magnitudes over a few hours, and then decline more slowly back to normal light. Again, they are binaries, and the mechanism has something in common with the brighter novae. In this case, the gas lost from the larger star forms a disk around the compact object. By a mechanism that is still poorly understood, this disk occasionally brightens. After the outburst, the system returns to its normal state, but the mass transfer soon causes yet another outburst to occur.

Dwarf novae of the U Geminorum type have distinct, well-separated outbursts, between which they are at minimum. Stars of the Z Camelopardalis type, on the other hand, are even less regular. They may sometimes remain at an intermediate light, at 'standstill' – known as 'stillstand' in North America – accompanied by rapid, minor fluctuations.

14.1.3.6 Symbiotic stars

The symbiotic (or 'Z-Andromedae') stars are binaries, both components of which are unstable. One star is a red giant, semiregular or Mira type (see the section on pulsating stars), and the other is a very hot, tiny star, which is often eruptive. The combination of these two variations gives rise to a peculiar light-curve. Examination of the spectra of these stars has enabled their nature to be understood (Table 14.7).

Table 14.7. *The five best-known symbiotic stars*

Star	Magnitude	
	max.	min.
Z And	8.0	12.4
CH Cyg	6.7	8.7
CI Cyg	9.0	13.1
AG Dra	9.1	11.2
AG Peg	6.0	9.4

14.1.3.7 The γ Cas stars

These are hot variables with Be spectra (i.e., with emission lines). They are probably very young and still slightly eruptive. They have characteristically fast rotation, which broadens the spectral lines. At the equator, the rotation velocity is only slightly less than escape velocity; if there is a slight eruption, a cloud of hydrogen escapes. The type star, γ Cas is the brightest of the class, but it includes other well-known stars, such as Pleione (also known as BU Tau) in the Pleiades cluster. Frequent ejections of material have created a shell around Pleione.

The brightness variations in these stars are small, being less than 0.5 magnitude, but from time to time absorption of light by the shell may cause a deeper decline.

14.1.3.8 Peculiar (eruptive?) variables

There are some stars that do not fit any of the preceding classes. They are stars that are unique. Some are bright and may be observed with small telescopes, so they deserve to be mentioned here. Table 14.8 gives details of some of these objects. The short description is obviously woefully inadequate to describe the behaviour of each object. A whole tome could be written about each of these stars.

14.1.3.9 Supernovae

The explosion of a supernova is certainly the most spectacular event that a variable-star observer can follow. Although the majority of stars end their lives as dense white dwarfs, there are certain stars that undergo a violent death, the mechanisms governing which have not yet been fully explained. The explosion of a star – we are dealing with the explosion of the whole star, which disappears catastrophically – is far more violent than any other stellar phenomenon. Quasars are the only relatively small objects that have intrinsic luminosities greater than those of supernovae. In our own galaxy, which contains 150 thousand million stars (1.5×10^{11}), the last visible supernova was observed in 1604 by Kepler. Nowadays, several supernovae are discovered every year in other galaxies, at distances of several million parsecs, so the objects are generally quite invisible before they explode. Supernova 1987A in the Large Magellanic Cloud was the first object for which the precursor could be identified. This was extremely important for supernova studies, even though it proved not to be a typical supernova.

Table 14.8. *Some peculiar variables*

Star	Magnitude		Notes
	max.	min.	
RT Ser	10.5	[16[a]	Very slow nova; decline lasting > 50 yrs
RR Tel	7	15.5	Semiregular; ejecting gas since 1944
S Dor	8.6	11.7	Resembles P Cyg; in the LMC
P Cyg	3	6	Exceptional eruptive supergiant; absolute mag. = −11.9 at maximum
η Car	−1	8	Slow variations over several centuries; young star cluster?
V389 Cyg	5.50	5.69	Pulsating, with changing period
AE Aqr	10.4	12.0	Slightly resembles UV Cet and U Gem types
V348 Sgr	10.6	17	Unique spectrum; slightly resembles an R CrB star
V605 Aql	11	20	Carbon-rich; normally very faint
V1343 Aql	13.0	14.5	Binary; emitting high-speed jets
FG Sge	9	14	Star turning into a red giant

[a] [16 indicates that the variable was fainter than magnitude 16.

Observations of extragalactic supernovae have allowed some progress to be made in studying these objects. Two major types are distinguished:

- Type I supernovae: old stars of Population II and moderate in mass;
- Type II supernovae: hot, massive, young stars, rich in heavy elements (Population I).

When the explosion occurs, the absolute magnitude may attain −20, and the supernova is sometimes as bright as the whole of the rest of the galaxy in which it occurs. The amplitude of the explosion is not known accurately: it is of the order of 14 magnitudes for Type II, or even up to 22 magnitudes for Type I, which are three times as luminous as Type II. (Before their explosion, Type I supernovae are much fainter than those that will become Type II supernovae.)

After the explosion, the star normally fades rapidly (approximately 2 magnitudes per month), and disappears. A nebula formed from the stellar debris appears some decades later. A small, very compact, central object may remain, as in the case of the Crab Nebula (the 1054 supernova). The youngest supernova remnant known in the Galaxy is the radio source Cas A. Strangely, the explosion, which should have occurred around 1667, was not observed!

Supernova explosions are highly significant. Heavy elements (oxygen, carbon, nitrogen, etc.) which are formed by nuclear reactions within the cores of stars are expelled by the explosion and enrich the surrounding space. Stars that form millions or thousands of millions of years later in the region will contain these elements, and

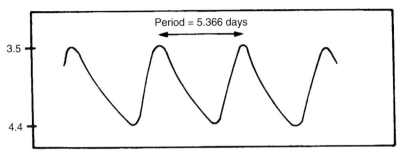

Fig. 14.8. *Schematic light-curve of δ Cephei*

solid planets may exist around these stars. On their surfaces, carbon compounds may arise. All living creatures, including the reader and the authors, are carbon-based beings, whose very existence is the result of the explosion of supernovae at least 6–8 thousand million (6–8 × 10⁹) years ago.

14.1.4 Pulsating variables

A star pulsates at several stages of its life; stars of different compositions encounter various stages during which they are subject to more or less regular pulsations. In effect, the different types of pulsating variables correspond to different ages, masses or chemical compositions.

14.1.4.1 Cepheids

Historically, δ Cephei was the first star to be recognized as pulsating. Its variations were discovered in 1784 by Goodricke. Its light-curve (Fig. 14.8) recurs with such regularity that at first it was thought to be an eclipsing variable. However, there is no form of eclipse that would reproduce such a light-curve. Subsequently, it was discovered that there were variations in radial velocity and temperature that were correlated with the variations in brightness. These suggested that the mechanism was one of pulsation. The star alternately expands and contracts. When it expands, it appears to approach us, and when it contracts it appears to recede. The surface temperature varies during the course of this cycle. The intrinsic luminosity E_0 of a star may be expressed, to a first approximation, by an equation of the type:

$$E_0 = kR^2T^4$$

where R is the stellar radius, T the temperature, and k a constant that depends on the unit chosen. Variations in T are more effective in causing changes in the brightness than those in R, because T enters the equation as the fourth power. For δ Cep and most pulsating variables, the brightness is a maximum when T is a maximum; the star is then in the process of expanding (maximum radial velocity, and intermediate diameter, Fig. 14.9).

The pulsation of a star somewhat resembles the oscillation of a pendulum: the extremely regular and stable period is mainly a function of the mass of the star.

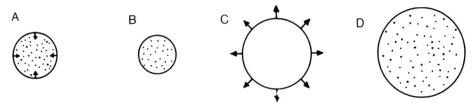

Fig. 14.9. *Cepheid pulsation:*
A: the star contracts; temperature and magnitude are lowest;
B: minimum diameter; temperature and magnitude increase;
C: maximum velocity of expansion, temperature and magnitude;
D: maximum diameter; temperature and magnitude have declined considerably.

Once initiated, the oscillation may continue for a very long time; it would require enormous forces to modify the oscillation over a short period of time.

'Normal' stars obey a mass–luminosity law that has been very well established for stars on the Main Sequence in the H–R diagram. Similarly, the oscillations of a Cepheid closely follow a mass-period relationship. We can see, therefore, that there should be a period–luminosity relationship as well. This was discovered in 1912 by Henrietta Leavitt for the Cepheids in the Small Magellanic Cloud. This relationship is of major significance in measuring distances. Once the period of a Cepheid is known, we can deduce its absolute magnitude M; but the apparent magnitude m is linked to M and the distance d by:

$$m - M = 5(\log d - 1).$$

The magnitude m is given by observation; once M is known, the distance of the Cepheid may be deduced. Cepheids are giants, which are visible at great distances. Once a Cepheid is detected in a cluster or in a nearby galaxy, the distance of the object is immediately known.

Unfortunately, Cepheids – and pulsating stars in general – are not 'normal' stars on the Main Sequence. We must not expect all pulsating stars to follow the period-luminosity relationship faithfully. Some pulsating stars do not obey any law of this sort, and even among the Cepheids, we have to distinguish between three sorts of objects each with its own period–luminosity relationship.

Cepheids of type Cδ are Population-I stars; they are supergiants, whose absolute magnitudes may reach -6 (more than $10\,000$ times the luminosity of the Sun). Their periods range from 2–50 days (Table 14.9). In this class, the period–luminosity relation is:

$$M_V = -3 \times 10 \log P + 1.70(B - V) - 2.37,$$

where M_V is the absolute visual magnitude, P is the period expressed in days and $(B - V)$ the colour index. The logarithms are to base 10. [Colour index is defined as the difference in magnitude at two given wavelengths. In the UBV system, these are defined by the effective wavelengths of the two pass-bands, *see* p. 1046.]

Table 14.9. *Some well-known Cδ stars*

Star	Observed mag.		Period (d)	Spectrum	M^*	d^*
	max.	min.				
δ Cep	3.48	4.37	5.3663	F5 to G2	−3.6	260
TT Aql	7.0	8.3	13.7546	F6 to G5	−5.1	2630
T Mon	5.6	6.7	27.0205	F7 to K1	−4.7	1150
Y Oph	5.9	6.5	17.1241	F5 to G0	−3.8	870
FF Aql	5.0	5.5	4.4710	F5 to F8	−3.0	400
DT Cyg	5.5	6.4	2.4991	F7 to G5	−2.7	435
α UMi	1.92	2.07	3.9698	F7	−3.2	106

* M is the absolute magnitude and d is the distance expressed in parsecs. Note that Polaris is a small-amplitude Cepheid.

Cepheids of type CW, similar to W Virginis, are old stars of Population II. They are found in the galactic halo and in globular clusters. Their range of periods is even wider: 1–60 days. CW stars are fainter than Cδ stars: for the same period they are about 2 magnitudes fainter. They have a different period–luminosity relationship that is less well-established.

For a long time, dwarf Cepheids were confused with the RR-Lyrae type stars that are described next. They are designated RRs. One of the most striking is CY Aqr (Fig. 14.10), which varies between magnitude 10.45 to 11.20 in $1^h27^m57.714^s$. In 20 minutes, the brightness of the star doubles. Several periods may be observed in a single night.

Dwarf Cepheids are much less spectacular objects than Cδ and CW stars. Their absolute magnitudes range between +1 and +5; they are probably Population-I stars.

14.1.4.2 RR-Lyrae class stars

Because they were first discovered in globular clusters, these stars were called cluster variables. They are very numerous; more than 7000 are now known. The brightest – and one of the closest – is the type star RR Lyrae itself, which varies between magnitude 7.0 and 8.1 in $13^h36^m12^s$. Their periods are always short: between 5^h15^m and 32^h21^m.

RR-Lyrae stars are Population-II stars. Their diameter is 4–5 times that of the Sun, but they have only half the mass. The absolute magnitude is approximately +0.5, and appears to be more or less independent of period: these stars have no period–luminosity relationship.

14.1.4.3 Double periods and period changes

In general, pulsation of a star involves only the outermost layer. However, it is possible for several layers, at different depths, to be the site of oscillations that produce pulsations with different periods. It is quite common for a pulsating star

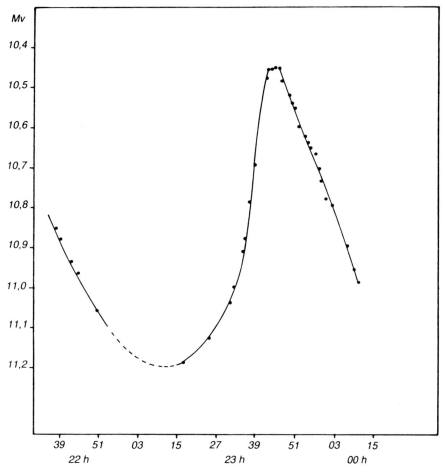

Fig. 14.10. *Light-curve of CY Aqr obtained on 1982 August 25–26, at the Jungfraujoch Observatory, using a photoelectric photometer on a 760-mm reflector.*

to have two different periods. We then find a light-curve of period P_1, which progressively alters from one cycle to the next (Fig. 14.11). After several tens of periods P_1, the curve repeats its original behaviour: calculating the secondary period P_2 is a problem in simple arithmetic.

This cyclic variation in the light-curve caused by a second period is known as the Blazhko effect, which is commonly found in RR-Lyrae stars, but which also exists in some Cepheids. In particular, the Cepheid CO Aur, which is well-known to amateurs, has two periods: $P_1 = 1.7841$ days and $P_2 = 1.4255$ days. We can see that $5 \times P_2 \approx 4 \times P_1 \approx 7.1364$ days; and $5P_2 - 4P_1 = 12^{\mathrm{m}}49^{\mathrm{s}}$. The light-curve of CO Aur therefore repeats after 7 days 3 hours. This is what is known as the 'modulation period'.

Progressive or sudden changes in period are sometimes observed. They are always small, but because the effect is cumulative from one period to the next, they are

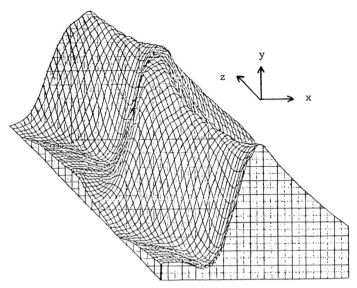

Fig. 14.11. *The Blazhko effect in the light-curve of RR Lyr (after S. Kuchto, GEOS Note NC 408); x = phase; y = magnitude; z = time. A different light-curve is obtained for every value of z.*

easy to detect. Such changes are observed in all the pulsating classes. The Cepheid T Mon, for example, has had several changes in its period, which oscillates between 27.0228 days and 27.0053 days (a variation of 25 minutes).

14.1.4.4 Variables of class M, similar to Mira Ceti

These are giants, very red, with long periods and large amplitudes (Table 14.10). Periods lie between 80 and 1000 days – which is why the stars are often called 'Long-Period Variables' (LPVs) – and are frequently subject to changes or irregularities. The amplitude is considerable, and ranges between 2.5 and 11 magnitudes. Long-period variables are the easiest for amateurs to observe, because errors in the estimates are generally negligible in comparison with the star's actual variation (Fig. 14.12).

These are relatively cool stars, with spectra of class M, C or S. They are generally immediately recognizable amongst other stars because they are so red. Their $(B-V)$ index is exceptionally high at maximum, and may exceed $+5$. They are surrounded by an expanding gaseous envelope. The existence of pulsations is definite, and is doubtless accompanied by eruptive phenomena. The exact mechanism behind the variations is still poorly understood.

Mira-type variables are Population-I stars, distinctly older than the Sun. The Sun will probably expand, become a long-period variable, and then, suddenly, become unstable and expel its outermost layers, which will form a planetary nebula surrounding it. More than 6000 stars belonging to this class are known.

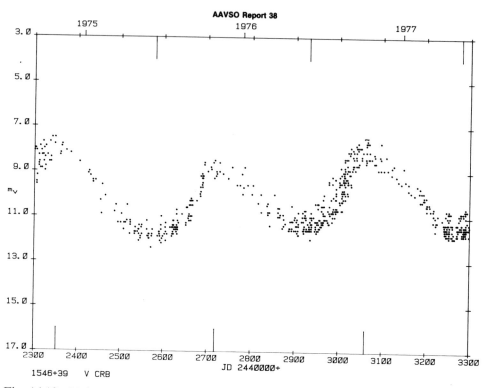

Fig. 14.12. *Light-curve of the Mira-type star V CrB, as observed by members of the AAVSO.*

Table 14.10. *Some of the best-known M-type variables*

Star	Magnitude		Period (d)	Spectrum
	max.	min.		
R Boo	6.2	13.1	223	M3e to M8e III
R Leo	4.4	11.3	310	M6e to M9e III
o Cet	2.0	10.1	332	M5e to M9e III
R Ser	5.7	14.4	357	M6e to M8e III
T Cep	5.2	11.3	388	M5e to M9e III
R Gem	6.0	14.0	370	S2e to S8e
χ Cyg	3.3	14.2	408	S6e to S10e

14.1.4.5 Semiregular variables

Mira-type stars are relatively regular, but there are very similar stars – red giants of spectral classes M, C or S – which are much less regular. These are the semiregular (SR) variables, which actually consist of various types of object.

The SRa sub-group closely resembles the Mira type, but with smaller amplitudes. The SRb stars generally have several superimposed periods, and show distinct irregularity. SRc stars are red supergiants ($M = -6$), but are very irregular: α Her, α Sco (Antares), and α Ori (Betelgeuse) belong to this class. The amplitude is generally less than 1.5 magnitudes.

Similar stars are also found that are completely irregular: they are divided into classes Lb (red giants with S, C or M spectra) and Lc (red supergiants).

There are also somewhat hotter, semiregular pulsating stars (with spectra between F and K): these are the SRd sub-group, in which it is difficult to detect a period, and also the RV Tauri stars, which resemble slightly irregular Population-II Cepheids. The period–luminosity relationship for the RV Tauri stars is essentially the same as that for CW stars.

14.1.4.6 Low-amplitude pulsating variables

Amateur variable-star observers may be interested in three other classes, which are just within their capabilities:

- The δ Scuti stars. These are very young, Population-I stars, which closely resemble the dwarf Cepheids (RRs), but which have a much smaller amplitude, always less than 0.2 magnitude.
- The β CMa stars (also known as β Cep stars, because both objects belong to the same class). The amplitude and period are both small. The radial velocity varies very considerably, but the stars' expansion does not occur at the same photometric phase as with the δ Scuti or RRs stars. The majority of these stars have two simultaneous periods, which differ only slightly. They are white stars with B0 to B3 spectra.
- The $α^2$ CVn stars are known as 'magnetic variables', because of their intense magnetic fields, revealed by the Zeeman effect in their spectra, which changes in parallel with the variations in light. These stars have an abnormal chemical composition and, in particular, an extremely high abundance of metals. It is possible that the variation (with an amplitude less than 0.1 magnitude) is not caused by pulsation, but by the stars' rotation.

14.1.5 Variable-star catalogues and nomenclature

The standard reference catalogue for variables is the 4th edition of the *General Catalogue of Variable Stars* (GCVS), which consists of four main volumes (Kholopov *et al.*, 1985–7). It contains approximately 28 500 variables (in our galaxy). In addition, a large proportion of the 14 810 stars that are listed in the *New Catalogue of Suspected Variable Stars* (NSV), published in 1982 (Kolopov *et al.*, 1982), are also variable.

The development of photoelectric photometry in the last 20 years has enabled many low-amplitude variables to be discovered. The number of known δ Scuti stars (for example) has increased considerably. Large numbers of 'microvariables' have been discovered by photometry. The best photoelectric measurements have an accuracy of around 0.003 magnitude. In general, a star is considered to be variable

Table 14.11. *Overall percentages of variables*

Category	%
Eclipsing	17.6
Pulsating	66.2
Eruptive	9.7
Cataclysmic	2.2
Rotating	0.9
X-ray	0.2
Other (peculiar, etc.)	1.8
Unknown	1.5

Table 14.12. *Pulsating stars*

Category	%
Mira	30.9
Cepheids	4.3
RV Tauri	0.7
RR Lyr	32.4
Semiregulars	17.9
Irregulars (L)	12.0
β CMa + δ Sct	1.6

when a series of 10 independent, accurate measurements, show a typical scatter greater than 0.02 magnitude.

Tables 14.11 to 14.13, taken from data in the 4th edition of the GCVS (1985–7), illustrate the approximate numbers of different types of variable. These figures are not exact, because some variables may belong to more than one class, and individual classifications are subject to change.

For 150 years, the same terminology has been used to denote variable stars. The first variable discovered in a constellation is designated by the letter R, followed by the name of the constellation (R And, R Cyg, etc.). The second is designated S, the third T ...to Z. Nine are therefore known by single letters. Then double letters are used: RR, RS, ..., RZ, SS, ST, ..., SZ, etc. up to ZZ. This covers another 54 stars. Then AA, AB, ..., AZ, BB, ..., to QZ are used, the letter J being omitted. This gives a total of 334 letter designations. If the constellation contains more variable stars (there are well over 4000 in Sagittarius!), subsequent objects are designated by V335, V336, etc.

To take an example, in the 4th edition of the GCVS we find that the last variable to be discovered in Ophiuchus is the star V2127 Oph, a Mira type, which varies between 14.5 and 19.5 magnitude, with a period of 331 days. Other discoveries

Table 14.13. *Eruptive & cataclysmic stars*

Category	%
R CrB	1.1
Novae	6.1
Symbiotic	1.4
γ Cas	3.2
Irregulars (I, FU Ori)	42.6
UV (flare stars)	33.7
Dwarf novae	9.8
Supernovae	0.2

have been made since the publication of that particular volume. Note that the designation of variable stars is the last area where the arbitrary division of the sky into constellations is still used.

14.2 Visual observation of variable stars

14.2.1 Visual estimation of magnitudes

Observing a variable star means determining the magnitude. There are five common methods of measuring magnitudes (mag.):

- Visual observations, which we will describe here, and which have an accuracy of approximately 0.1 mag.;
- Photographic measurements, where the probable error is around 0.2 mag.;
- Use of a visual photometer, which allows the accuracy to be improved (± 0.05 mag.). Such photometers are gradually disappearing in favour of photoelectric photometers;
- Photoelectric measurements, which have a far higher accuracy: 0.02 mag. with amateur equipment, 0.003 mag. with more complex photometers and accurate reduction taking atmospheric effects into account (*see* Chap. 19);
- In observatories, CCD detectors are frequently used. These allow quantitative measurements of stars' brightnesses to be made. Intrinsically, these detectors have good accuracy (0.03 mag.), but reduction of the measurements is still currently slightly less sophisticated than photoelectric methods, so the final accuracy is not yet comparable with that obtained with photoelectric photometers.

14.2.1.1 The Argelander method

The most effective method of visually estimating the magnitude of a star was introduced in 1840 by Argelander. Two comparison stars close to the variable V, are selected. One, A, is slightly brighter than the variable, and the other, B, slightly fainter. By interpolation, the magnitude of V is estimated with respect to A and B.

First, the difference in brightness between A and V is estimated. The differences in brightness are expressed in steps, which are defined as follows:

One step   If, at first glance, A and V appear equal, but after close examination, it appears that, apart from rare moments, A is slightly brighter than V, A is said to be one step brighter. This is recorded as: A(1)V.

Two steps   If A and V first appear equal, but then, rapidly and without hesitation, it becomes obvious that A is brighter than V, this is recorded as: A(2)V.

Three steps   If a slight difference in brightness is obvious at first glance, then A is 3 steps brighter than V: A(3)V.

Four steps   A distinct difference in brightness, visible immediately, implies a difference of 4 steps: A(4)V.

Five steps   Five steps indicates a major difference in the magnitudes of A and V. A and B should be chosen so that the difference in magnitude should, if possible, be less than 5 steps.

Beyond 5 steps, the method rapidly loses its accuracy. It is extremely difficult to define, in an accurate and reproducible way, what might be meant by 6, 7, 8 … steps.

After comparing A and V, the same method is used to compare V and B, thus obtaining an overall expression of the form:

$$A(4)V(2)B.$$

[Note that, in fact, these two estimates should actually be independent, and written in the form 'A(4)V; V(2)B'. – Trans.]

The magnitude of V is calculated by simple relative proportions. Call the magnitudes of A, B, and V, m_A, m_B, m_V, respectively. The comparison $A(\alpha)V(\beta)B$ allows us to write:

$$m_V = m_A + [\alpha/(\alpha + \beta)](m_B - m_A).$$

For example: $m_A = 9.85$; $m_B = 10.40$; $\alpha = 4$ steps; $\beta = 2$ steps. For A(4)V(2)B, $m_V = 9.85(4/6) \times 0.55 = 10.22$.

Note that the Argelander method does not attribute any specific value to a 'step', which is defined as the smallest difference in brightness that the eye is able to distinguish. In practice, it has been found that the value of a step, expressed in magnitudes, depends on observing conditions and the observer's experience. A beginner's step is often close to 0.3 mag., but it may become as small as 0.04 mag. under excellent conditions and with a highly experienced observer.

14.2.1.2 *Recording the observations*

Observation of a variable is a simple scientific task. This should be carried out and recorded with great care. The most common method of recording observations is as follows (Table 14.4):

Each observational report should include the year (frequently the observations are analysed several years later), the name of the observer, and the observing site.

Column 1   The date. Do not forget that the date changes at 00:00 (UT). Some organisations prefer the date to be expressed in Julian Days (*see* the next section).

Table 14.14.

1986 – Observer: DMT					Site: F-78 Versailles	
(1) Date	(2) UT	(3) Star	(4) Inst.	(5) Est.	(6) *m*	(7) Wt
Apr. 30	00:16	R CrB	B 50 × 10	A(4)V(4)B	6.06	II 3
Jun. 28	23:00	S CrB	G 80 × 73	11.8	[11.8	II 1

Column 2 The time in Universal Time (UT), to an accuracy of 1 minute. For some very fast variables (periods less than 2 hours, and flare stars), times to a half-minute are preferable.

Column 3 The name of the variable.

Column 4 The instrument used for the comparison: NE: naked eye; G: refractor (for **OG**); B: binoculars; R: reflector then the diameter expressed in millimetres and the magnification. (So, B 50 × 10 indicates 50-mm binoculars magnifying 10 times. G 80 × 73, a 80-mm refractor magnifying 73 times.)

Column 5 The estimate. This is the most important piece of information, which cannot be omitted under any pretext. The person reducing the observations must be able to determine which stars were used as comparisons, with no possibility of error. In the first of the two examples shown in Table 14.14, A and B are marked on the chart used by all observers. If observers use a chart that is not well-known, or a chart of their own, then this should be submitted at the same time as the report. In the second case, the observer did not succeed in seeing S CrB, which was too faint. Column 5 therefore contains the faintest comparison that could be seen, here a star of mag. 11.8.

Column 6 The magnitude *m*, calculated from the estimate. In some cases, the magnitude of the comparison stars is not known, or is subject to doubt, so the magnitude of the variable cannot be calculated immediately. In such a situation, the estimate must be preserved at all costs! The expression [11.8 indicates that the variable was fainter than mag. 11.8.

Column 7 Class of the observation and sky conditions (wt):
I: estimate considered to be good by the observer
II: average quality estimate
III: doubtful estimate
1: clear sky
2: even haze or mist
3: clouds
Other codes may be included here, such as M (for Moonlight), T (for twilight or dawn), etc. Details of sky conditions are useful in judging the weight that may be given to a particular observation.

[Note that the various observing organisations may require data to be submitted in slightly different ways, which may mean that different information must be recorded. Some groups (such as the AAVSO) do not record the actual estimates, only the deduced magnitude. In general, however, these are the sort of data that are required. – Trans.]

14.2.1.3 Julian Day

It is not always easy to tell immediately the interval of time between two dates (for example, how many days are there between 1945 October 8 and 1962 July 7?). A very simple method consists of giving days consecutive numbers. The idea goes back to Joseph Scaliger in 1583, who named this method 'Julian Days' after his father, Julius Scaliger (there is no link with Julius Caesar!). On this scale, the origin (day number 1) is 4713 BC January 1, which corresponds (according to Joseph Scaliger!) to the date of the Creation. A few anti-establishment progressives nowadays believe that the Universe is somewhat older!

Each Julian Day begins at midday (UT). The relationship is therefore:

1984 January 01, 12:00 UT = JD 2445 701.0
1984 January 01, 20:00 UT = JD 2445 701.33
1987 January 19, 23:38 UT = JD 2446 815.47.

Most yearbooks and handbooks give the Julian Days for each day of the year. It is easy to obtain the Julian Day for any given date, using a programmable calculator (*see*, for example, Meeus, 1978).

14.2.1.4 Variable-star charts

To identify a variable star and chose suitably convenient comparison stars, detailed charts are essential. Charts that show the whole sky on just one or two sheets may be decorative, but are quite unsuitable. Atlases that contain 8–20 charts are useful for the brightest variables, but are insufficient for the observation of fainter stars. A simple pair of binoculars enables far more stars to be seen than given in any normal atlas.

Specialized charts produced by the variable-star organisations are therefore essential. Figure 14.13 shows a chart of this sort. It is a chart used for R CrB, a star that varies between mag. 5.8 and 14.8. When the star is bright, it may be seen with the naked eye, or very easily with a pair of binoculars; Chart A is then the one to use. Comparisons A and B, used in our example estimate of 1986 April 30 (*see* Table 14.4), are marked by the arrows (A= π CrB at mag. 5.60; B is mag. 6.52). When the star fades below mag. 7, Chart B is better. Below mag. 10, Chart C is required.

14.2.2 Errors and inaccuracies to be avoided

14.2.2.1 Choice of comparison stars

The comparison stars should be chosen carefully. The difference in magnitudes, their angular distance from the variable, and their colour should meet certain criteria:

- An essential principle of the Argelander method is that in comparing the variable with the stars A and B, there should not be a very great difference

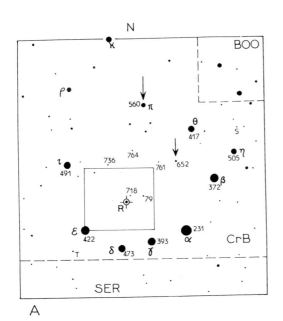

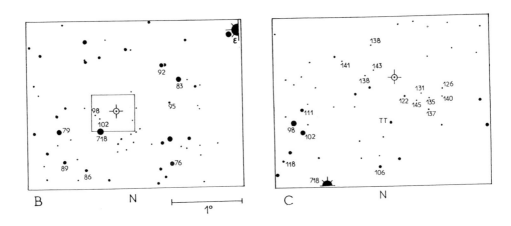

Fig. 14.13. *Set of AAVSO charts for visual estimates of R CrB. (Original charts reduced by a factor of 4/3.)*

in magnitude. In practice, $(m_B - m_A)$ should never exceed 1.5 magnitudes. Ideally,

$$0.5 \leq m_B - m_A \leq 0.8.$$

An $m_B - m_A$ difference of less than 0.5 mag. requires very careful, and difficult, observation. With a difference of 2 mag., on the other hand, it becomes very difficult to decide on the number of steps. The variable will systematically be estimated as too bright because it is easier for the eye to distinguish differences in brightness when the amount of light received is fairly low.

- The comparison stars should not be too far from one another. If possible, the angular distance between them (as seen through the astronomical instrument) should be less than 10°. Large magnifications are therefore not generally advisable.
- The stars to be compared should have the same colour. Atmospheric extinction is (mainly) a function of a body's height above the horizon and of its colour. If two stars of different colours are compared, the estimate is likely to vary, depending on the daily motion, which causes the star to rise to culmination and then set. In addition, the response of the retina is not independent of the colour of the light source, and may also vary from one night to be next.

14.2.2.2 *Choice of instrument*

The first thing that a variable-star observer requires from an astronomical instrument is a wide field. The most useful instruments therefore have fairly low f-ratios:

50-mm binoculars, magnification 7 to 10;
Short-focus refractors, 80 to 100-mm diameter; f/6; minimum magnification $\times 12$;
Newtonian reflector, f/5.

Ideally, an observer would have all three instruments.

Let D be the diameter of the objective, d that of the eye, and m_0 the limiting magnitude attained by the naked eye. The theoretical limiting magnitude of any instrument is then:

$$m = 2.5 \log(D^2/d^2) + m_0.$$

In principle, the limiting magnitude is reached with a magnification slightly less than that required for optimum resolution (which is equal to the diameter of the objective expressed in millimetres). The influence of the magnification on the limiting magnitude is not particularly great when the sky is completely dark. On the other hand, in urban areas the sky is light; increasing the magnification weakens the sky background, and increases the contrast between it and the stars.

With a pair of binoculars, the limiting magnitude is about 0.8 magnitudes higher than the theoretical limit. (The magnification is low; the prisms absorb a lot of light; and because binoculars are hand-held, they are difficult to hold perfectly steady.) Under good conditions, the naked eye can see down to 6th magnitude; but magnitude estimates are most accurate between magnitudes 2.5 and 5.0. The eye is bad at detecting small differences in brightness if the stars are too bright. It is possible to consider that the role of a telescope is to reduce the numerical magnitude of stars and that the aim is to ensure that the variable being estimated should have an apparent magnitude, through the telescope, of 2.5–5.0. Table 14.12 summarizes the situation.

With an 80-mm refractor, for example, stars with actual magnitudes of between 8.1 and 10.6 will have the same appearance as stars of magnitude 2.5–5.0 seen by the naked eye. Accurate observations may therefore be obtained with an 80-mm refractor, over this range of magnitudes (8.1–10.6). It should be noted that measurements of bright stars will be more accurate with small instruments. With

Table 14.15. *Range over which particular instruments may be used for visual magnitude estimates*

Instrument	Limiting magnitude	Range of greatest accuracy
Naked eye	6.0	2.5 to 5.0
B 30 mm	8.7	5.2 to 7.7
B 50 mm	9.8	6.3 to 8.8
G 50 mm	10.6	7.1 to 9.6
G 60 mm	11.0	7.5 to 10.0
G 80 mm	11.6	8.1 to 10.6
R 115 mm	12.4	8.9 to 11.4
R 200 mm	13.6	10.1 to 12.6
R 320 mm	14.6	11.1 to 13.6

a 7th-magnitude star, for example, a pair of 30-mm or 50-mm binoculars will give better results than an 80-mm refractor, and far better than a 200-mm reflector. An amateur who has just a pair of binoculars is perfectly adequately equipped to observe variables brighter than 8th magnitude. An observer who has 50-mm binoculars, a short-focus, 80-mm refractor, and a 200-mm reflector may cover a magnitude range of about 6.3–12.6. This means that several thousand variable stars are accessible. Any amateur who is less favorably placed, having perhaps just an 80-mm telescope, can gain extra precision by stopping down the telescope, using diaphragms with apertures of 20 and 40 mm that may be placed in front of the objective.

The main instrumental problem to be avoided is vignetting. Figure 14.14 shows a refractor, where the bundle of rays is cut off by an internal diaphragm that has been incorrectly calculated. The fully illuminated field is smaller than the eyepiece's field diaphragm. Point X receives light from the whole of the object glass, but point Y is illuminated by only part of the objective. It will be seen that if the images of stars being compared fall at X and Y, the estimate must be inaccurate. Variable-star observers need to know what is hidden away inside their refractors! If there is any doubt, it is possible to compare the two stars by bringing them successively to the centre of the field. Such a technique involves a slight loss of accuracy; it is, however, essential when the angular distance between the stars exceeds half of the field-diameter. Never compare a star at the centre of the field with one at the edge. In general, images are best, and true points, at the centre of the field; this is certainly not the case at the edge.

After several years of observation, it becomes obvious that the number of results obtained (i.e., the number of good light-curves) is strictly linked to the number of visual estimates that are made each year. In trying to get an even better return, the nature of one's equipment becomes of fundamental importance. A pair of binoculars is extremely efficient: between 20 and 50 visual estimates may be made in an hour. A short-focus refractor is still very effective, but pointing a reflector at a faint star that

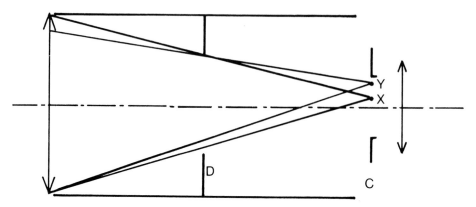

Fig. 14.14. *Vignetting caused by an incorrectly calculated aperture in a telescope's diaphragm D.*

is difficult to find obviously takes some time. With a properly adjusted equatorial mounting, it is possible to reduce the time required to find variables by marking their coordinates on the setting circles. The majority of variable-star observers prefer altazimuth mountings, which, in general, are much more practical. Estimates using the Argelander method should be made quickly, so diurnal motion does not have time to become a nuisance. Equatorially mounted telescopes in a dome are poorly suited for the observation of variables. The time required to move the dome reduces considerably the amount of work that may be carried out in the course of a night.

A driven, equatorial mount is useful for following a rapidly changing variable over a period of several hours. The observer locks the reflector on the star and makes a visual estimate every 3–5 minutes. Another object may be observed using a second telescope, between making two estimates of the rapid variable.

14.2.2.3 Position-angle error

The sensitivity of the retina is not uniform, so it is essential to use the same area of the retina when comparing stars. When the stars are fairly close to one another, there is a tendency to observe them both simultaneously: their images fall on different areas of the retina, and position-angle errors are inevitable. The true method is to observe them both alternately using the same part of the retina (not necessarily the centre!).

The central area of the retina has excellent spatial resolution, but it is not particularly sensitive to faint illumination. A few degrees away from the centre, the retina is much more sensitive, although the resolving power drops rapidly. (It is not possible to read the left-hand page of a book when looking at the right-hand page!). It is not possible to see faint stars, close to the limiting magnitude, with the centre of the retina. To see a faint star, it is necessary to look 'to one side', so that its image falls on the more sensitive outer area of the retina. Such 'averted vision' enables one to gain about 1.5 magnitudes (some areas of the retina are thus about 4 times as sensitive as the central area).

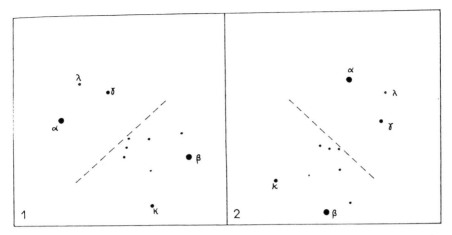

Fig. 14.15. *Position-angle error: the rotation of the field as seen in an altazimuth telescope may cause seasonal variation in magnitude estimates.*

Position-angle error may cause apparently seasonal variations. Consider a variable-star observer who makes an estimate of a star in Orion at the same time each night. At the beginning of October, Orion is in the East (Fig. 14.15-1). At the end of February, it is in the West (Fig. 14.15-2), and the field has rotated by about 80°, changing the position-angle error.

A simple method of revealing position-angle error is to observe the same variable under different conditions:

- with a refractor (where the image is inverted) and with binoculars (which have an erect image);
- if the variable is close to the zenith, it may be observed with binoculars if one lays on one's back; try two different positions: feet towards the east (and head to the west), and the reverse.

In both cases, the position-angle error will be obvious!

Figure 14.16 shows several observations of V Boo, made by nine individuals in 1971 July. The systematic shift in the observations made with binoculars with respect to those made with a refractor suggest that position-angle error probably affects all the observers alike.

There are two different suggestions for combating position-angle error:

- some observers try to avoid it at the time of observation by making two estimates under different conditions: with erect image and inverted image, for example, or else use two pairs of comparison stars with different orientations in the field;
- other observers, on the other hand, take pains to ensure that they always observe under identical conditions. The position-angle error is thus stable, and becomes a systematic error, which may be corrected when the mea-

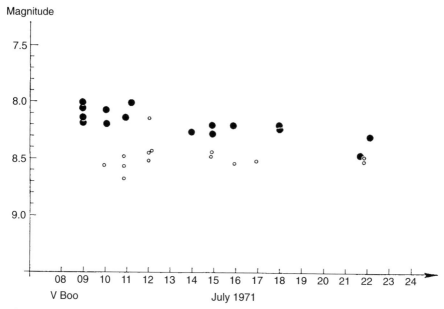

Fig. 14.16. *Observations of V Boo made in 1971 July by M. Vidalie, B. Legras, M. Du-mont, J. L. Duranceau, S. Roux, H. Ateb, F. Kachoukh, F. Ben Brahim and J. P. Gouy. Note the systematic shift between observations made with binoculars and those with a telescope with an inverted image:* • *telescopic estimates;* ○ *binocular estimates.*

surements are analyzed by comparison with estimates of the same variable made by others.

Statistically, both methods can be defended. In the former, it is important not to take the mean of the two estimates, but to submit both series of measurements, clearly stating the different methods used.

14.2.2.4 Bias
Errors caused by bias are undoubtedly the worst. Amateurs who know (or who think they know!) the magnitude of a star before observing have a distinct tendency to make estimates with the expected result. Several experiments have shown that if beginners are told the result of an estimate before making one themselves, their own estimates give the same result – even if the suggestion was deliberately erroneous!

A variable-star observer, even if very experienced, can never be completely free of bias. All that one can do is to try to minimize the effects by taking some precautions:

- Never use ephemerides that predict the minimum or maximum of a star that one intends to observe several hours later. This precaution presents a major problem in the observation of eclipsing variables of type EA, because observation of an EA star in eclipse is far more useful than observations of the same star at maximum; if no ephemerides are consulted, most estimates will fall at maximum where the star has a constant brightness.

- Do not consult observations made the night before, especially in the case of slowly changing variables (Mira, SR stars, etc.).
- Make as many observations as possible during a night so that none of them are remembered. An observer who makes 10 to 20 estimates a night is never able to 'forget' observations from one night to another. Someone who makes 80 observations cannot possibly remember each of the comparisons. It should also be noted that it is easier to remember a specific magnitude than an estimate of the type A(3)V(2)B. It is therefore sound practice to calculate the magnitudes some 8–10 days later. Someone with an excellent memory might be able to memorize 40 or 50 comparisons; it is then advisable to adopt the decimalized Argelander method (*see* next section).
- If observing in a group, never discuss estimates of the variables being observed during the course of the working session.

14.2.2.5 Decimalizing Argelander steps

After several hundred visual estimates, beginners start to have a good idea of the apparent difference corresponding to 1, 2, 3, 4 and 5 steps. Then one fine night, they hesitate for a long time between choosing 3 or 4 steps. Why choose? It suffices to write down 3.5. Half-steps thus appear, and estimates of the type A(2.5)V(3.5)B or C(2)V(4.5)D increasingly find their way into the observing notebooks. After thousands of observations, an observer may hesitate between choosing 3 and 3.5, and a true decimal series appears: 3, 3.1, 3.2, 3.3, etc.

Such decimalization has no drawbacks; on the contrary it offers several advantages:

- The apparent loss of time and waste of paper required to write down decimal fractions are largely offset by removing any hesitation that might otherwise have occurred during observation. It may take 10–15 seconds to decide on whether a step should be recorded as 3 or 4.
- Theoretically the decimalization allows differences in magnitude to be expressed to an accuracy of one hundredth of a magnitude. It is well known that such an accuracy is superfluous for visual estimates, but it is better to be too accurate rather than to introduce deliberate, and significant, rounding errors. The mean step of an average observer is 0.1 magnitude. If an observer hesitates between choosing 3 or 4 steps, there is a risk of introducing an error of 0.05 magnitude, if one value or the other is selected. This error is in addition to any other errors that may be present.

Under rare, very favourable conditions, the accuracy of visual estimates may reach 0.06 magnitude. It is obvious that under such circumstances it is essential for the decimal method to be used. Any observers, who have never decimalized their estimates, and who then try to determine the accuracy of their measurements, will, perforce, obtain a value of 0.1 magnitude (in favourable cases), because this value is precisely the step that the observers have established. In fixing this step at 0.01 magnitude, it is possible to study the real accuracy of visual estimates and subject them to standard statistical methods.

Decimalizing the steps is an effective way of avoiding bias. It is impossible to remember sixty-odd comparisons similar to A(3.2)V(4.7)B. Such decimalization is practically essential when the same variable is observed several times during a single night.

[It should be noted that many experienced observers consider that decimalization of estimates is extremely unwise, given that it tends to suggest a spurious accuracy that certainly cannot be attained in practice. – Trans.]

14.2.2.6 Atmospheric absorption

Consideration of atmospheric extinction is essential when photoelectric measurements are being made. For visual observations, its effects should be borne in mind under two sets of circumstances:

- if the stars that are being compared have different colours (already mentioned in Sect. 14.2.2.1);
- if a very bright star is being observed with the naked eye, and is being compared with stars (necessarily bright) that are at considerable angular distances from it. For example, the star α Ori (Betelgeuse) is a red supergiant that varies between magnitudes 0.4 and 1.4; it is observed by some amateurs, and suitable comparison stars – Capella, Aldebaran, Pollux, etc. – are well away from it on the sky. On the other hand, it is impossible to use Rigel (B8 spectrum), Sirius (A1 – and far too bright!), or Procyon (F5), which are much hotter than Betelgeuse (M2).

In comparing Capella or Aldebaran with Betelgeuse, there is no way of avoiding different zenith distances, and thus different atmospheric extinctions. Betelgeuse is a low-amplitude variable; under these conditions, visual estimates that do not take atmospheric extinction into account are only fit for the nearest waste-paper basket.

14.2.2.7 Some heresies to be avoided

Some charts are inadequate, and certain parts of the sky are poor in bright stars. It may therefore happen that only a single comparison star is available. It is then tempting to measure the brightness of the variable solely with reference to this star. This is a practice to be strongly discouraged, because it assumes that the value of a step is known in tenths of a magnitude. Unfortunately, in rare individual cases, it is the only solution. Accuracy then suffers, becoming ±0.25 mag. in most cases. Under such circumstances, do not hesitate to use (whenever possible) a comparison star with a suitable magnitude, even if the magnitude is not shown on the chart, or even if there is a risk that the colour differs from that of the variable. By consulting appropriate catalogues it will be possible to determine the magnitude of the star subsequently.

What should never be done, is to try to make an extrapolation such as A(3)B(2)V, hoping to deduce the magnitude of the variable from this. Given the properties of the eye, such an attempt is incapable of giving a result anywhere near the true situation. Table 14.16 illustrates this. [This objection is not, of course, valid if a true fractional estimate (*see* p. 830) is made instead. – Trans.]

The failure of this extrapolation is caused by the fact that the field of a 50-mm refractor at a magnification of ×24 is not at all suitable for the observation of R CrB

Table 14.16. *Errors introduced by the extrapolation method*

Date	UT	Instrument	Estimate	Mag.	Obs.
1971 Jly 02	22:47	G 30 × 4	5.6(5.5)V(7)6.52	6.00	DMT
1971 Jly 10	00:55	G 50 × 24	V(3)7.2(1)7.9	5.1	FS
1971 Jly 10	01:03	G 50 × 24	V(2)7.2(4)7.9	6.8	FBB
1971 Jly 15	22:15	G 30 × 4	5.60(5.5)V(4.5)6.52	6.11	DMT

The star observed was R CrB.

at maximum, when the brightness of the variable is steady. Note that the 30-mm refractor was being used with a magnification that was less than the equipupilary magnification (*see* p. 889). This does not pose any real problems; the aperture being used was doubtless closer to 23 mm. (DMT never having measured the diameter of his pupil in darkness!)

In submitting observations, no estimates should be included where there is a possibility of there having been an error of identification. Such cases are, unfortunately, to be found in every publication … but probably double-star measurements suffer most from this scandalous state of affairs. A double-star worker observes a star 10 to 15 times a decade, whereas a variable-star observer makes 10 to 15 estimates a month, and soon comes to recognize the field without risk of error. It may also be noted that an observer who knows all his charts by heart (which is quite common) can gain a valuable amount of time.

Observation of a variable should be made with great care. The quality of the observations and the corresponding results are more important than the mere quantity of observations made. Everything should be done to ensure that the optimum conditions apply. The eye should be fully dark-adapted (with 10 minutes in full darkness in most cases, or 20 after watching television or reading brightly illuminated paper). A very low level of illumination should be used to write down observations: an ordinary pocket torch is far too bright. The lamp may be shaded with a neutral or red filter; or the lamp may be under-run – used at a lower voltage than normally.

Attention to comfort is important. Using binoculars, for example, three positions are best:

- lying down with a cushion underneath the head;
- sitting in a garden chair;
- kneeling on a cushion, with the elbows resting on top of a balcony or other support.

With experience, the amateur will be able to avoid the normal problems: dew on objectives or eyepieces (use blotting-paper as a dew-cap); fatigue (it is better to get an early night than to make poor measurements); wind, which means that charts and observing notebooks must be firmly held (the Aletsch Glacier in Switzerland is harbouring several charts and observing notebooks!); cold, which is almost always a problem when it is a fine observing night. (This introduces the problem of gloves; can you write legibly with gloves?).

The Argelander method that we have described is the most effective, but there are others that are very similar. Two of them are frequently used:

- When comparing V with stars A and B, one may systematically describe the difference between A and B as 10. The estimate will then always be of the general form: A(10)V(0)B; A(9)V(1)B, A(8)V(2)B, ... A(1)V(9)B, A(0)V(10)B. Again, there is the possibility of decimalizing the steps: A(3.4)V(6.6)B. The only advantage of this method is a very slight gain in time; when noting down an estimate of the type A(α)V(β)B, it is sufficient to write down A(α)V, because we know that $\beta = 10 - \alpha$. [Again, caution should be exercised, because many experienced observers believe that the basic method – known as Pickering's decimal method – is unsound. Experiments have shown that the human eye is actually unable to distinguish the difference between (say) A(1)V(9)B and A(1)V(8)B, whatever the magnitude interval between A and B. – Trans.]
- Similarly, some observers (beginners) assume that the difference between A and B is known. If, from the chart, $m_B - m_A = 0.8$ (for example), one says $m_B - m_A = 8$, and the estimate will be of the form A(α)V(β)B, where $\alpha + \beta = 8$. This may also be decimalized.

These two methods give the same accuracy as the Argelander method, but do not provide any information about the difference ($m_B - m_A$), nor about the value of the step, expressed in tenths of a magnitude. In some, relatively frequent cases, the chart contains errors. In general, the sum ($\alpha + \beta$) obtained by the Argelander method picks up these easily. We shall return to the questions of correcting and improving charts in the next chapter.

We strongly advise amateurs interested in variables to adopt the Argelander method. After years of practice and thousands of observations, small personal alterations may perhaps be added; but with experience, it will always be possible to know which changes are of positive benefit. All the advice we have given in this section aims at minimizing random errors. Systematic errors may be eliminated in reducing the observations.

14.3 Reduction of visual estimates

After several months of observation, an observer will have a set of estimates from which various results may be obtained: first the light-curve; then this curve may be analysed to obtain the period, the amplitude, the type of variable, etc. The first concern should be to determine the accuracy of the observations, to judge how much confidence one can have in the light-curve that has been obtained, so we need to correct any systematic errors that may distort the curve.

14.3.1 Systematic errors

14.3.1.1 Analysis and correction of systematic errors

A perfect visual observation (which assumes that the magnitude m found by the observer corresponds exactly to what was seen) is, a priori, a function of several variables:

$$m = f(m_0, p, i, h, q)$$

where m_0 is the true magnitude of the star, p the observer's personal equation, i the instrumental equation, h the altitude of the star above the horizon, and q the quality of the seeing. We will assume for the time being that the chart is accurate.

If the sky is free from cloud, the quality of the seeing q has little effect on the final result. A change of site may alter the accuracy, but will not produce a systematic effect. After months of observing in an urban area, it is quite disconcerting to be up in the mountains, with a very dark sky. For two or three nights estimates will be less accurate, but things soon return to normal. A similar sort of effect occurs on returning to an urban, light-polluted sky. We may assume that the sky-background level does not introduce a systematic effect.

The effect of h is negligible close to the zenith or when the stars being compared are of the same colour. If no attention is paid to colours, apparent seasonal effects may appear. It is always possible to evaluate the influence of h on results, but the best way is always to bear the two following precepts in mind:

    (i)   Only compare stars of the same colour.
    (ii) Do not observe less than 20° from the horizon.

If these two conditions are observed, h will not introduce any systematic error.

Although it would therefore appear possible to avoid the effects of q and h, the values of p and i are, by contrast, difficult to assess. By definition,

$$m = f[m_0, (p, i)]$$

and there will be a different light-curve for each pair of values (p, i). The details occurring in each curve may well be smaller than the shift of one curve with respect to another, and combining the different curves without due precautions on the same graph may lead to a loss of information or erroneous conclusions (Fig. 14.17).

An individual observer can study the effect of i by making systematic observations of several variables, using two different instruments (a refractor and binoculars, say). On the other hand, study of p is only possible if observations made by several observers (say ten) are available; hence the value of variable-star observers concentrating on specific sets of stars.

If several light-curves of the same star are available, obtained by several observers (or observer–instrument pairs) over a specific period, then it is possible to correct the systematic errors. We begin by drawing the light-curve for each individual observer. We can read the amplitude of the variation from each of the curves; let a_0 be the mean of these amplitudes. Each curve is then scaled, so that its amplitude is equal to a_0. The axis of each rescaled curve is a straight line parallel to the time axis (Fig. 14.18), laid through the average obtained from the rough curve.

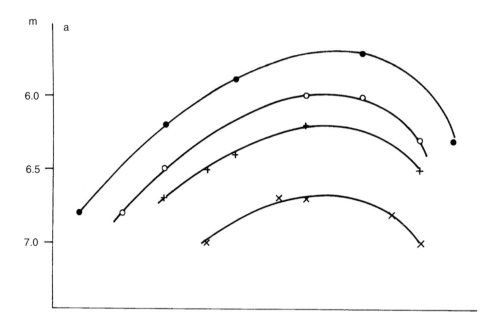

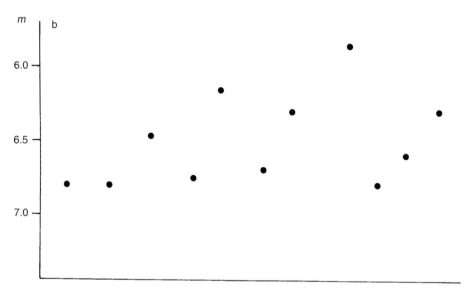

Fig. 14.17. *Systematic error. a : four observers have obtained the same light-curve, but their estimates show a systematic shift. This systematic error must be corrected before daily means are taken. Means taken without any such correction give the light-curve shown in b, where the information is lost. It is even possible to believe, mistakenly, that the magnitude was increasing over the time covered by the last three measurements.*

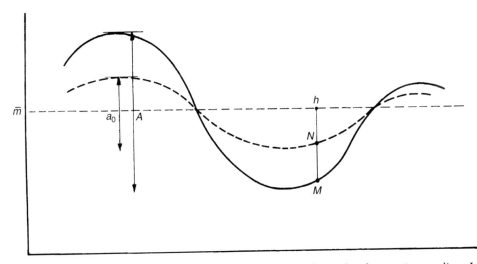

Fig. 14.18. *One observer has obtained the light-curve shown by the continuous line. Its amplitude A is twice that of the mean amplitude a_0 obtained by the group of observers. A scaling factor of $1/2$ is applied about $\bar{m}$ (the mean magnitude of the curve). For every point M on the curve, a point N is derived, such that $xN/xM = 1/2$.*

At this stage, we obtain a set of light-curves that have essentially the same shape, but which are shifted with respect to one another by an amount that may be as much as several tenths of a magnitude. It now remains to apply a translation to each curve so that its average value corresponds to the mean of all the averages.

Note 1 This reduction is not as arbitrary as it may seem. It depends on the fact – which has frequently been noted – that the preliminary light-curves of two observers may be derived from one another by means of rescaling and a translation.

Note 2 There are several ways of defining the average value of a function. The person carrying out the analysis needs to make an intelligent choice. If one can be sure that there are no aberrant estimates, the mean of the extreme magnitude will suffice.

Note 3 If the star has been observed several times by all the observers simultaneously, there is no need to calculate the mean values. The points with a common value on the abscissa will serve to determine the translations required by the curves.

14.3.1.2 The causes of systematic errors

The errors are of physiological and psychological origin. It is not always possible to explain them fully. Among the probable causes, we may note:

- Position-angle error (already discussed in Sect. 14.2.2.3). One striking result of this error is the systematic warping of a light-curve, caused by a change of instrument, and this is illustrated in Fig. 14.19. The star concerned has a large amplitude, and when it was faint, the observer used an 80-mm refractor to make estimates. The star brightened, and after May 3, the observer switched to 50-mm binoculars. The position angle changed abruptly, and the light-curve showed a sudden jump. Because the observer always changes instruments at about the same magnitude, the jump is systematic. Such errors may be found in many published light-curves.

- If the variable is of an ordinary colour, it is very easy to find comparison stars of the same colour. On the other hand, Mira-type stars are extremely red, and one is unlikely to find comparison stars with the same colour. The problems that we have mentioned then occur: differential extinction effects; seasonal, and even systematic, variations caused by the different red sensitivities of different observers. The limits of the visible spectrum (the red and violet ends) are not the same for all observers.

- The Carnevali effect (named after the professional astronomer who began as an amateur and who correctly investigated this effect in 1975). The Carnevali effect is any error that systematically affects the distribution of estimates over a particular magnitude interval (A–B).

  Assume that a star fades steadily from A to B (Fig. 14.20a). (For ease of discussion, assume that A and B are the designations of the comparison stars and their magnitudes.) If the interval A–B is divided into ten, the star remains the same amount of time in each division, and an observer who makes regular estimates will obtain equal numbers in each division (Fig. 14.20b).

  The most common form of the Carnevali effect consists of refusing to accept that the variable is equal in magnitude to one of the comparisons, in this case: V and A, and then V and B. Observers tend to prolong the process of estimation for 2 or 3 minutes until they believe that one of the two stars is brighter. The distribution of estimates over the interval is not even; it has two bumps (Fig. 14.20d) and the light-curve does not have the same shape (Fig. 14.20c).

  Other variants of the Carnevali effect exist, but they are rare. The case shown in Fig. 14.20c is very common, and perfectly explains the similarity observed between one observer and another when a moderate-amplitude variable always lies between the same two stars A and B.

14.3.2 Random errors

Random errors (by definition) introduce randomly occurring errors into the results. It is impossible to correct such errors, and the true information is disguised in an irreversible fashion.

One effective method of reducing or detecting random errors that arise with slowly changing stars, for example, is to observe the same star two or three times, independently, during the course of a night. One major error is rare; two occurring with the same star during the same night is a highly unlikely situation – if the

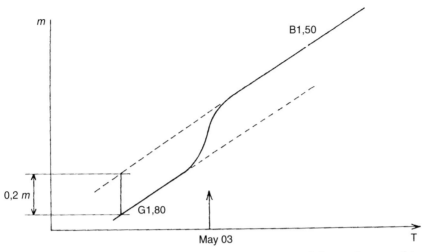

Fig. 14.19. *The sudden jump in the light-curve caused by a change of instrument. Before May 3, the observer was using an 80-mm refractor; after that date he used 50-mm binoculars. Observations made with these two instruments have a systematic difference of 0.2 mag. (say), which will cause an alteration in the curve around May 3. A change in the instrument used causes a distinct change in the curve.*

estimates are truly independent, which is easy enough to arrange if a lot of estimates are being made and they are decimalized.

By way of example, DMT, who made nearly 3000 estimates in 1982, detected two random errors of 0.5 magnitude during reduction of the observations. The quality of the observations and the nature of the error of any estimate may be determined by analysing the statistical distribution of the estimates.

When true measurements are made, study of instrumental methods frequently enables one to gain an idea of the probable (random) error of each measurement. Visual observations of variables are not measurements (even though the term is sometimes used), but estimates, and no analysis of instrumental factors can give an indication of the probable error. The only way of estimating the accuracy of visual observations is therefore to carry out a statistical study of a large number of observations. Various investigations of this sort have been carried out over the last 20 years or so. The first was that by J. Lecacheux (Lecacheux, 1970). This study involved 610 observations of HR Del (Nova Delphini 1967). Lecacheux showed that the distribution of estimates by a good observer about the mean curve had a standard deviation $\sigma = 0.08$ mag. Naturally, this value varies from one observer to another. When σ is known, it is possible to calculate the probability that an estimate will differ from a mean value by an amount that is greater than ϵ. If ϵ is expressed as a function of σ, we obtain the values shown in Table 14.17.

$\bar{m}$ is the mean about which the estimates m are distributed. In general, estimates that differ by more than 3σ are considered to be random errors and are eliminated. The standard deviation σ varies as $1/\sqrt{n}$, and the accuracy of results increases as

813

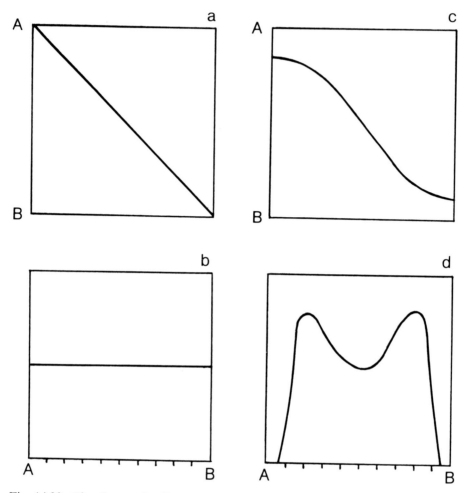

Fig. 14.20. *The Carnevali effect (see text)*

$\sqrt{n}$, the square root of the number of observations. Note that the standard deviation is calculated from

$$\sigma = \frac{\sqrt{\sum_{i=1}^{n}(m_i - \bar{m})^2}}{n}$$

where $(m_i - \bar{m})$ is the deviation from the mean of the i-th estimate in a series of n estimates.

Such a statistical investigation must be carried out after the correction of random errors, of course.

14.3.3 Smoothing the curves

After having corrected systematic errors and removed rogue estimates, a scattered series of points remains, forming a light-curve that may be neither complete nor

814

Table 14.17. *Error in estimates*

ϵ	Probability that $\| m - \bar{m} \| < \epsilon$
0	0
0.2 σ	0.159
0.5 σ	0.383
0.75 σ	0.547
1.0 σ	0.683
1.5 σ	0.866
2.0 σ	0.954
2.5 σ	0.988
3.0 σ	0.997

smooth. There should be a good density of points throughout the period of observation. This is why it is important not to observe too many stars. With a large-amplitude star, the errors in the estimates are small, or even negligible, in comparison with the star's variation. A light-curve may therefore be obtained by combining points from several observers. If the amplitude is low (less than 2 magnitudes), the observational errors are no longer negligible, and a statistical treatment is required. Such analysis is only possible if the observer has made enough estimates of the same star.

A hundred observations scattered over thirty-odd stars is of little interest. The same number of observations of two or three stars, would, on their own, enable two or three usable light-curves to be obtained. Some compromise has to be found between a single, excellent curve, ten good curves, twenty mediocre ones, or several hundred isolated points. An active observer makes more than 1000 observations per year. An effective observer makes, on average, more than 40 estimates per star per year.

The light-curve will be more or less smooth, and some of the variations will be real, whereas others will be caused by observational errors that the previous steps have not eliminated. It is frequently of interest to remove these irregularities by smoothing the curve.

- One rough method consists of laying a line through the points, to give a close fit without attempting any calculation.
- When the nature of the curve is known, or when one decides in advance what the curve should be (a straight line, a parabola, a sinusoid, etc.), a least-squares method may be used to obtain the best fit. Such a method is fairly simple to apply to variables that have short, regular periods.
- A simple, but effective method for any type of curve is to divide the period of observation into small, equal intervals, in which the star's variation is considered to be negligible. Several observations – the more the better – are required in each interval. The average is then calculated for each interval.

Table 14.18. *Chart magnitude errors*

Star	Chart magnitude	Photoelectric magnitude	Chart error
1	7.4	7.30	+0.10
2	9.0	9.36	−0.36
3	9.8	10.01	−0.21
4	11.0	10.71	+0.29
5	11.6	11.84	−0.24
6	12.4	12.61	−0.21
7	13.1	13.01	+0.09

This example illustrates the errors that may occur in magnitudes given on a chart, which in this case was for S Cep.

When these means are plotted, they give a much smoother curve than the original.

The choice of interval depends on the speed at which the variable changes and on the number of observations available. Smoothing is generally good when there are more than ten estimates per interval.

14.3.4 Correction of comparison-star sequences

We have already seen that habitual changes of instrument may introduce distortions into a light-curve. An error in the magnitude of a comparison star may also distort the curve. Old charts were prepared using inaccurate photographic methods and major errors are frequently present. Table 14.18 illustrates this sort of error. These are stars near S Cep that are frequently used to estimate the magnitude of the variable. The magnitudes of these stars were measured photoelectrically in 1985 July, using the 76-cm reflector at the Jungfraujoch Observatory. The errors are generally larger than the accuracy expected from visual estimates. In addition, the error is not systematic, and is not caused (as some people maintain) by the difference between visual magnitudes and the V magnitudes obtained using a photometric system. (Otherwise we would have to accept that the error in the magnitude of star 4 was as much as 0.5 mag.!)

Photoelectric-system V magnitudes are precisely defined. Visual magnitudes, in contrast, have never been defined; they date back to a time when the eye was the only instrument available to 'measure' stellar magnitudes. The only defensible position is to decide that photoelectric V magnitudes should be used (without alteration!) as reference for visual estimates.

Strangely, the people who advocate a mysterious shift between the two types of magnitude only invoke this shift below a certain magnitude. Previously, Polaris was thought to be of magnitude 2. Then it was used as a standard star, with a magnitude of 2 by definition. We now know that it varies slightly about magnitude

Magnitude

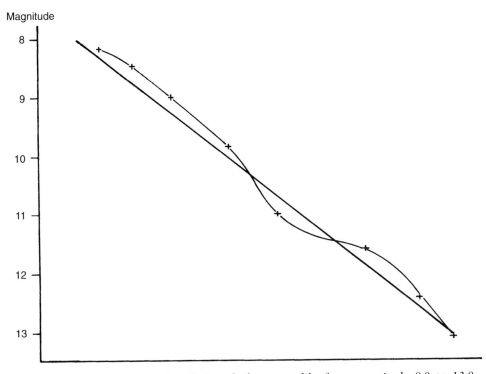

Fig. 14.21. *Let us assume that S Cep declines steadily from magnitude 8.0 to 13.0. An observer using the chart described, and making perfect estimates will derive the sinuous curve shown. In addition, the decline would – systematically – appear to be slower after magnitude 11. This is the reason for the humps that occur on the light-curves of Mira-type stars!*

V = 2.00. Yet the photometric V magnitude was defined to replace the earlier visual magnitude m_v.

The same people also advance a similar argument about colours, maintaining that for red stars the difference is particularly noticeable. We know, however, that no two eyes have the same response at the two extremes of the visible spectrum. Moreover, there is no red photometric standard. It is easy to see that the argument is neither particularly daring, nor verifiable.

If the observer is certain that the sequence is a photoelectric one, then the following discussion is not relevant. Unfortunately, nearly all the charts used for variable-star work are not of photoelectric origin, and errors abound. The major ones are generally noticeable to the observer. Taking the S Cep sequence for example, a close look will show that the interval between stars (3) and (4) is smaller than predicted. Figure 14.21 shows the sort of deviation introduced into the light-curve by the errors in the comparisons on the chart for S Cep.

When a major error is suspected, do not hesitate to modify the chart to bring it into line with visual impressions, using the following method. To simplify the explanation, we assume that the sequence consists of just three stars, A, B, and C. After having made several tens of observations with the Argelander method, we will have a set of comparisons such as: A(1)V(5)B, A(1.3)V(5.1)B, A(3.2)V(3.4)B, ..., B(1)V(4.5)C, B(1.5)V(3)C.

Using arithmetic means, we calculate the difference in brightness between A and B, and between B and C, expressed in steps. An expression of the form A(α)B(β)C is obtained. If u is the value of a step expressed in magnitudes; a, b, and c the magnitudes of A, B, and C, respectively on the chart, and m_A, m_B, and m_C the magnitudes of the same stars as they appear to the observer, then:

$$m_B = m_A + \alpha u; \quad m_C = m_B + \beta u = m_A + (\alpha + \beta)u.$$

We then set $\Delta = (a - m_A)^2 + (b - m_B)^2 + (c - m_C)^2$.

Adjustment of one's personal sequence by the method of least squares consists of calculating m_A and u such that Δ is a minimum. This algorithm is practically indispensable in obtaining good light-curves for large-amplitude stars that do not have photoelectric sequences.

14.3.5 Combining observations

Let us assume that P is the period of a variable, and that the star has been observed over several weeks or, in a more general case, over an interval much longer than P. Combining the observations involves deriving a composite curve over an interval equal to P.

For example, let us assume that the star TT Aql, with a period of 13.7546 days has been observed between 1982 June 17 and September 25. To combine the observations, we choose the cycle at the centre of the observational period: July 28 to August 11. We start by plotting the observations obtained between July 28 and August 11. Observations made shortly afterwards (August 12–16) are shifted by one period: an estimate made on August 14.9257 is plotted at the same point on the diagram as one made on (14.9257 − 13.7546 August) = 1.1711 August, which comes within the interval July 28 to August 11. An estimate made an any date T is shifted by an integral number of periods: $T - nP$, so that it falls within the same interval. Observations made earlier than July 28 are shifted by a similar procedure, by adding an integral number of periods.

The mean light-curve for the observational season is thus obtained. On a plot (Fig. 14.22), it is customary to show magnitude on the ordinate (y-axis) and time on the abscissa (x-axis). Time may be expressed as a function of the phase of the variation: with a pulsating star, phase $\varphi = 0$ corresponds to maximum; with an eclipsing star, phase $\varphi = 0$ is defined as the centre of primary minimum.

Combining observations like this is valid only if the period is completely constant throughout the observational period.

14.3.6 Heliocentric correction

Study of rapidly changing variables (RR-Lyrae, eclipsing stars, etc.) requires an accurate knowledge of the time of each estimate. It is also necessary to take account

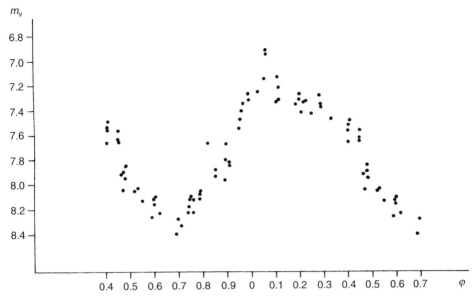

Fig. 14.22. *Combined light-curve from 70 visual estimates of the Cepheid TT Aql made by DMT between 1982 June 17 and September 25. The star varied between magnitudes 6.9 and 8.4. The maximum appears slightly later than suggested by the ephemeris given in the GCVS (Kholopov, 1985) for $\varphi = 0$: JD 24 372 36.10 + 13.7546n. But the difference is small, $\varphi = 0.06$, or 19^h45^m. This star warrants continued monitoring.*

of the heliocentric correction, so that the Earth's position in its orbit around the Sun does not introduce any shift in the timing that might otherwise be attributed erroneously to the variable. Light takes slightly more than 8 minutes to reach the Earth from the Sun; such a shift may be significant as soon as the timing of a particular event needs to be known to an accuracy better than half-an-hour.

Ephemerides of variable stars are always heliocentric, i.e., apply to a fictitious observer at the centre of the Sun. The heliocentric correction h must be added algebraically to the time (UT) of any observation to obtain the heliocentric time. Calculating h is moderately complicated:

$$h = -0.138\,612R[\cos\theta\cos\alpha\cos\delta + \sin\theta(\sin\epsilon\sin\delta + \cos\epsilon\cos\delta\sin\alpha)],$$

where h is expressed in hours, R is the Earth–Sun distance in astronomical units, θ is the longitude of the Sun, α and β the equatorial coordinates of the star, and ϵ the obliquity of the ecliptic. If great accuracy is required, R, i, α, β, and ϵ must be considered as variables, which explains the complexity of the calculation.

14.4 The major variable-star organisations

Visual observations of variables are an extremely valuable source of information. It is pointless for an amateur to accumulate hundreds or thousands of estimates and

819

keep them tucked away where they will never see the light of day. All amateurs
should join one of the specialized variable-star organisations, which will provide
detailed guidance in observing, and which will collect, analyse and publish the results. There are numerous small groups, often several in one country, but in the
English-speaking countries there are three, major international organisations: the
American Association of Variable Star Observers (AAVSO), the British Astronomical Association's Variable Star Section (BAA VSS), and the Variable Star Section
of the Royal Astronomical Society of New Zealand (RASNZ VSS). Note that despite their names, all of these groups are truly international and receive estimates
from observers in many different countries. Details are also given of three, rep-
resentative, European groups – the Association Française d'Observateurs d'Etoiles
Variables (AFOEV); the Bedeckungsveränderlichen Beobachter der Schweizerischen
Astronomischen Gesellschaft (BBSAG); and the Groupe Européen d'Observations
Stellaires (GEOS) – and the Variable Star Observers League in Japan (VSOLJ).
[This, and subsequent sections have been modified and amplified to include details
of the major English-speaking organisations and the VSOLJ. The descriptions of
specific groups of stars and their observation remain as in the original. – Trans.]

14.4.1 American Association of Variable Star Observers (AAVSO)

The AAVSO is the largest variable-star observers' association. It has already
amassed several million observations since its formation in 1911, under the aegis of
Harvard Observatory. Several hundred observers in many countries regularly submit
observations; and the annual rate is over 500 000. The observations are too many to
be published individually. They are stored in computer files and provided to those
requiring them in machine-readable or paper form. Review papers are published,
based on the observations received and, particularly for Mira stars, the light-curves,
where each individual observation is represented by a single point. This chapter
contains three such curves (Figs. 14.12, 14.23, and 14.24). These show:

- the relatively large amount of scatter;
- the variation between one cycle and another;
- the gaps when the object is in conjunction with the Sun.

The AAVSO publishes general scientific papers on variable stars in its *Journal*,
and issues a large number of charts for visual work. These charts contain about
2000 stars, and one is reproduced in Fig. 14.25.

14.4.2 Association Française des Observateurs d'Etoiles Variables (AFOEV)

This association was founded in 1922 by A. Brun, and has aims very similar
to those of the AAVSO. It currently receives about 50 000 observations per year.
This relatively small number means that all the most significant observations may be
published regularly. These amount to about 80 % of the total. The only observations
that are eliminated are those which are obviously in error or of no value, such as
those where an observer has not seen the star at a certain magnitude, but where it

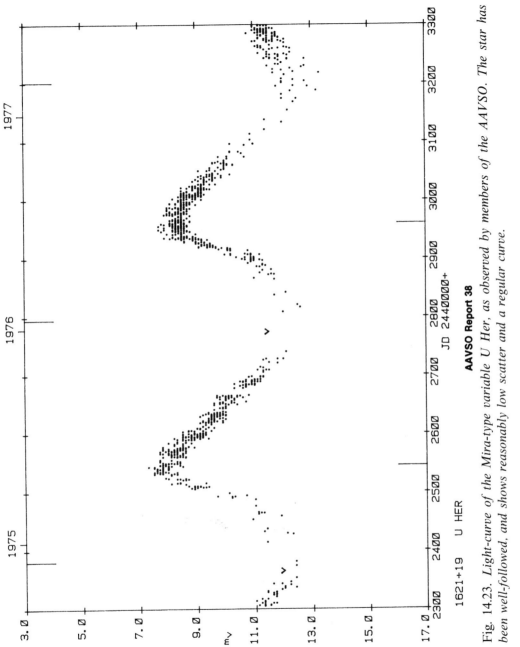

AAVSO Report 38

Fig. 14.23. *Light-curve of the Mira-type variable U Her, as observed by members of the AAVSO. The star has been well-followed, and shows reasonably low scatter and a regular curve.*

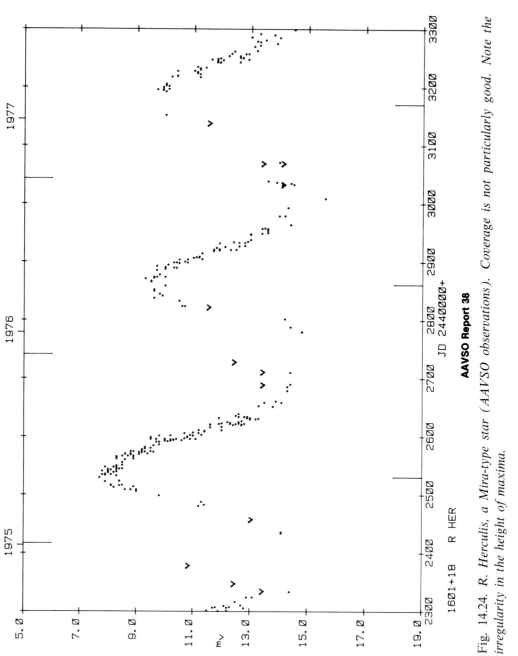

AAVSO Report 38

Fig. 14.24. R. Herculis, a Mira-type star (*AAVSO observations*). Coverage is not particularly good. Note the irregularity in the height of maxima.

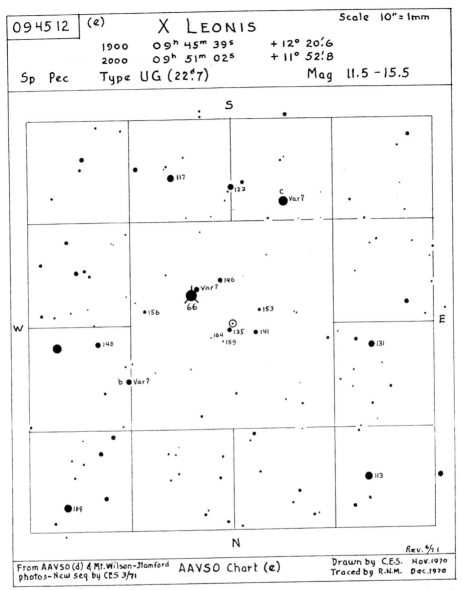

Fig. 14.25. *AAVSO chart for X Leonis (type UG). The faintest magnitudes shown would only be visible with a 600-mm telescope. (Original reduced by a factor of 3/2.)*

has been estimated at a lower magnitude by another observer. Figure 14.26 shows an extract from one of the quarterly *Bulletins*, which contain these observations, together with general articles about variables. A bimonthly newsletter *La gazette des étoile variables* concentrates on news about the behaviour of eruptives, novae, etc., as well as other information of interest to observers.

The observations of the AFOEV are stored on computer and copies are for-

21h 59m

JJ 2446000 +

TW Peg (cont.)

VED	699.3	7.2
	704.4	7.2

2201+34 SV Peg

KOV	682.5	8.7
VED	703.4	7.9

MPG

LEG	671.5	9.9
	684.4	10.0

2201+33 RZ Peg

FJH	613.5	9.4
PQR	625.5	9.8
SCW	627.5	9.9
FJH	636.4	10.6
	646.5	11.1
PQR	648.6	10.4
WAL	649.4	10.2
SCW	651.5	10.2
ZAL	658.6	10.9
FJH	660.5	11.5
SCW	676.4	10.8
GUN	676.5	10.7
ZAL	678.6	11.1
KOV	682.5	11.2
FJH	683.4	11.7
CMG	683.5	11.4
PQR	684.5	11.4
	702.4	11.6
FJH	702.4	12.0
VED	703.4	11.3
MZR	703.4	12.0

2202+28 AZ Peg

KOV	682.5	9.0

2207+14 RS Peg

FJH	623.5	13.8
	639.4	14.3
	645.4	14.5
SCW	651.5	<14.0
FJH	653.5	14.4
	673.4	14.5
GUN	676.5	<14.2
FJH	680.4	14.6
CMG	683.5	14.5
KPG	684.4	14.5
FJH	686.4	14.5
	699.4	14.3

2209+12 RU Peg

FRH	610.49	10.7
PRZ	613.45	:11.2
VER	613.46	11.6
SCW	614.44	12.2
VER	615.45	12.5
GUN	615.56	12.5
PPS	616.40	12.7
FRH	616.43	14.5
PRZ	616.50	12.3
VED	616.51	12.5
PPS	618.40	12.5
PPZ	619.43	12.7
	621.44	12.5
VER	621.45	12.5
FRH	621.47	12.5
PRZ	621.50	12.5
PPS	622.40	12.7
FRH	622.43	12.5
VER	622.45	12.6
SCW	622.47	12.8
FJH	623.54	12.7
PRZ	624.48	12.6
FRH	624.49	12.5
GUN	624.55	12.5
SCW	625.51	12.8
VER	626.44	12.6

RU Peg (cont.)

GUN	649.44	12.5
VER	649.44	12.7
SCW	649.44	12.8
FJH	649.46	12.7
ARN	649.56	12.8
VER	650.40	12.7
FRH	650.43	12.5
RIP	650.43	12.8
PRZ	650.45	12.6
VER	651.40	12.7
RIP	651.41	12.8
SCW	651.47	12.7
PQR	651.50	12.8
ARN	651.54	13.1
	651.56	13.1
PPS	652.40	12.3
PRZ	652.40	12.7
FRH	652.41	12.6
RIP	652.41	12.8
	653.41	12.8
FJH	653.46	12.8
PPS	654.40	12.5
PRZ	654.41	12.5
RIP	654.49	12.8
PRZ	655.38	12.5
FRH	655.40	12.5
RIP	655.41	12.8
VER	655.43	12.7
PRZ	656.38	12.7
FRH	656.39	12.6
PPS	656.40	12.5
RIP	656.41	12.8
VER	656.41	12.7
MZR	656.60	12.7
PRZ	657.37	12.6
FRH	657.40	12.5
VER	657.40	12.7
RIP	657.41	12.8
	658.40	12.8
PPS	658.40	12.5
FJH	658.44	12.8
RIP	659.41	12.8
ZAL	659.60	12.7

RU Peg (cont.)

ARN	679.44	:12.8
ZAL	679.50	12.5
PRZ	680.33	12.5
VER	680.33	12.6
ZAL	680.50	12.8
VER	681.35	12.6
ARN	681.45	12.8
GUN	681.48	12.5
VER	682.39	12.7
PPS	682.40	12.7
ARN	682.46	12.9
VER	683.36	12.6
PRZ	683.36	12.6
FJH	683.40	12.7
PPS	683.40	12.6
VED	683.42	12.5
CMG	683.50	12.9
KPG	683.50	12.9
PRZ	684.32	12.5
VER	684.36	12.7
VED	684.38	12.5
RIP	684.39	12.7
ARN	684.44	12.8
FJH	685.38	12.7
PQR	685.46	:12.5
PPS	686.40	12.6
PRZ	687.32	12.5
RIP	687.36	12.7
PPS	687.40	12.5
VER	688.32	12.7
PPS	688.40	12.6
	689.40	12.6
FJH	690.35	12.7
PPS	690.40	12.7
VER	691.34	12.6
PRZ	692.31	12.5
VER	693.35	12.6
GUN	693.38	12.5
VED	693.38	12.5
PPS	693.40	12.5
PRZ	694.33	12.4
VED	694.34	12.5
VER	695.31	12.7

SV Cep (cont.)

BJK	677.34	11.2
	679.31	11.2
	683.34	11.2
	685.35	11.2
	686.32	11.2
	687.31	11.3
	688.31	11.3
	692.29	11.4
	693.41	11.4
	698.28	11.2

2219+55 RW Cep

PPS	614.4	7.4
FID	614.5	7.0
	618.5	7.1
PPS	622.5	7.5
	628.4	7.5
	639.4	7.5
FID	639.4	7.0
KOV	639.5	7.5
PPS	643.4	7.6
	645.4	7.5
REI	645.5	7.1
PIR	652.4	7.0
PPS	656.4	7.6
FID	665.4	7.0
PPS	668.4	7.5
	673.4	7.5
	675.4	7.5
	682.4	7.6
	687.4	7.3

2221+29 RV Peg

FJH	643.5	14.9
SCW	651.5	:13.7
FJH	653.5	14.0
	666.4	12.6
SCW	676.4	11.4
FJH	682.4	10.3

Fig. 14.26. *Extract from the quarterly bulletin of the AFOEV. The individual estimates are reduced, stored and edited using the computer at Strasbourg Observatory.*

warded to the AAVSO, where they are incorporated into that organisation's archive. Information that is extracted in the short-term is mainly concerned with the dates of maxima and minima of periodic variables and outbursts of eruptives, but the archives may also be used for longer-term studies. Light-curves are produced, and some examples are given in Fig. 14.27.

The AFOEV's programme includes only stars with amplitudes of at least 1 magnitude. With such objects, simple archiving, with no additional reduction of the estimates allows fairly satisfactory light-curves to be obtained, despite the scatter among the observations. Stars with smaller amplitudes may be studied by visual methods, but they need to be subjected to statistical analysis, which is not, in general, carried out by the AFOEV.

The AFOEV publishes charts for about 500 variables. The magnitudes used are the same as those employed by the AAVSO, but the presentation is different. The format is smaller, and comparison stars are given letter designations to make it easier to record the estimates that are made. Examples of these charts are shown in Fig. 14.28. AFOEV members are able to obtain copies of these charts and also AAVSO ones for which there are no AFOEV charts.

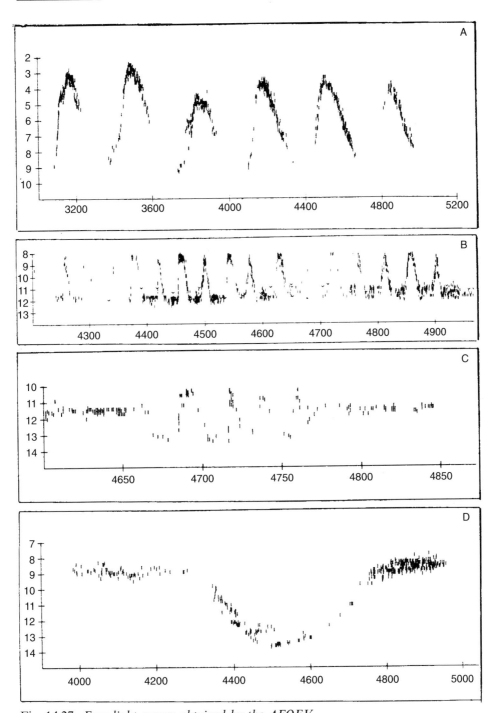

Fig. 14.27. Four light-curves obtained by the AFOEV.
A: light-curve of Mira Ceti (1977–81); B: light-curve of the dwarf nova SS Cygni (1980–81); C: light-curve of another dwarf nova, Z Camelopardalis (1981 January to August); D: light-curve of PU Vulpeculae: this is an unusual slow nova which has been constant since 1981.

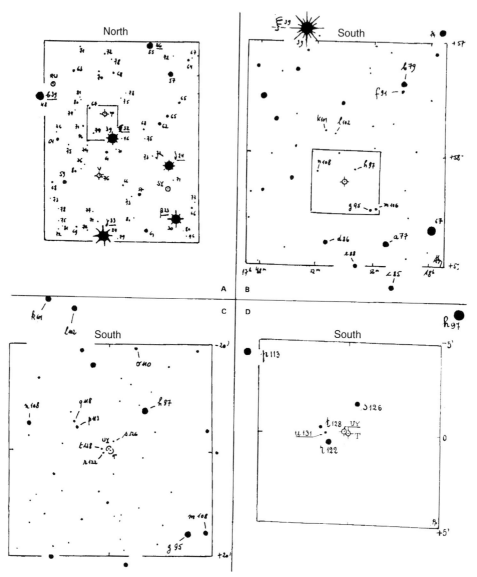

Fig. 14.28. *AFOEV charts for T Dra, a Mira-type variable.*
A chart: for binocular observation, North is at top;
B chart: for a short refractor or good finder, South is at top;
C chart: for telescopes up to 150 mm in diameter;
D chart: to be used when the variable is at minimum, with a telescope between 200
and 300 mm in diameter.
(Original reduced by a factor of 2.)

The AFOEV does not observe short-period variables (Cepheids, RR-Lyrae stars, etc.) that require intense observational coverage, which is rarely available with amateurs. On the other hand, Mira stars, semiregulars, R Coronae Borealis stars, novae and eruptives are well-observed. A few remarks on the two major types of stars covered by the AFOEV are in order.

14.4.2.1 Mira stars and other red giants

We have described earlier some aspects of the red-giant variables, which are extremely numerous. Among these, the majority are Mira stars. Of these, 80 % are accessible, at maximum, with amateur instruments, so the field to be covered is immense. Their long periods and slow variation mean that it is not advisable to observe a given star more frequently than once every 10 days. An active amateur could follow hundreds of stars, without devoting an excessive amount of time to the work. Given that their amplitudes are also very considerable, it is easy to see why these stars form a major part of many amateurs' programmes.

It should be noted that many semiregular stars – even if we only include those with amplitudes that are greater than 1 magnitude – are visible throughout their cycles in binoculars, or even with the naked eye. Examples are: μ Cep, g Her, R Lyr, etc. Observers with the minimum equipment are happy to follow these stars, whose semiregular nature means that they are somewhat unpredictable, which reinforces interest in them. Some may undergo long periods when there is little detectable variation, at least as far as visual observations are concerned, and this may be found rather disappointing.

The more or less red colour of these stars, for all it may make them attractive as individual objects, is a distinct disadvantage when it comes to making accurate visual observations. The scatter in the uncorrected estimates from several observers may amount to more than 0.3 mag. The AAVSO light-curves shown here (Figs. 14.12, 14.23, and 14.24 show this very well. Luckily, the amplitude is such (at least as far as the Mira stars are concerned) that the curves retain a degree of scientific value.

As may be seen from the light-curves reproduced here, although the curves of Mira stars are, broadly speaking, periodic, they are unpredictable to the extent that the epochs and exact values of maxima or minima do vary, as does the general shape of the curve. A long-term record of these variations is therefore a useful factor in trying to gain a better understanding of these stars, even though the physical causes escape us at present. It is possible that the real value of these observations will only come after decades (or even centuries) of observation. The patient accumulation of such data, which would be irretrievably lost if no one bothered with it, is the main aim of the amateurs who follow these objects.

With fainter stars, where only the maxima are detectable with amateur instruments, and which are, of course, even more numerous, it should at least be possible to determine the dates of maxima, even if the whole of the light-curve cannot be followed. Keeping the ephemerides up-to-date does provide information about which stars may be observable at any given time by observers who may have more powerful equipment, such as photoelectric photometers. In the long term, it also allows us to determine whether variations in the period, which are possible indicators of evolution, occur. A few, rare cases are known, but many more doubtless await discovery.

This is, however, no reason for not trying to obtain the complete light-curve for the brightest stars, with the aim of studying their shape and also of determining the times of minima. The periods obtained from the intervals between minima are of more physical significance than those obtained from the maxima, and therefore tend to show less random fluctuations. Because they are only attainable for a smaller number of stars, however, the chances of finding significant, long-term evolutionary changes are smaller. Such evolutionary changes will, in any case, eventually be detectable in the periods determined from the maxima.

The necessity for complete, careful observations of the light-curves of certain Mira stars has recently been reinforced by requests arising from the HIPPARCOS mission. For the satellite to carry out properly the mission's task of accurate astrometric measurements, it is essential to know the approximate magnitude of stars that are imaged. With large-amplitude, and somewhat unpredictable stars such as the Mira-type variables, this information can only be obtained by visual monitoring, which is why amateurs are undertaking this work.

Another example of the short-term scientific implications of estimates made by amateurs, we may mention the investigations that have shown a correlation between the OH maser emission from certain Mira stars and anomalies in the visual light-curves determined by amateurs.

Those who are discouraged by the idea that their observations will only serve to increase the size of existing archives, and will not prove of use until the distant future, can therefore still become involved with scientific programmes with the prospect of more immediate results.

14.4.2.2 Eruptives

The various types of eruptive variable stars that we have described earlier are represented by far fewer individual stars than pulsating or eclipsing types. This is basically because of they are intrinsically faint, which means that only the closest may be seen in amateur-sized telescopes. The mechanisms occurring in these eruptives are still poorly understood, and their classification is not well-established and is often revised. This means that frequent observations, well spread out in time, are required.

In general these stars are not red, so the accuracy of visual observations is higher than with the Mira stars. Here again, however, the amplitude of the variations is such that a very high degree of accuracy is unnecessary, so the raw observations are normally used. In fact, for many of the eruptives, the magnitudes determined are not particularly important, but what is significant is the knowledge of whether a star is, or is not, in outburst.

Some eruptive light-curves have been given earlier. Although observations of some specific classes, with very few members, are of particular interest (the R Coronae Borealis stars, for example), most eruptives are of the U Geminorum type, sometimes known as dwarf novae. This has three main sub-classes:

- SS Cygni stars, which vary between a quiescent state and outbursts, which occur at irregular intervals. [Although individual stars are characterised by specific mean periods. – Trans.] The length of the outbursts and the shape of the curves show a moderate degree of variation.

- SU Ursae Majoris stars, which in addition to the normal type of variation, occasionally undergo supermaxima, which are brighter and longer than normal outbursts.
- Z Camelopardalis stars, which sometimes remain for long periods part way between maximum and minimum.

Figures 14.27B and 14.27C show the prototypes of the first and third of these sub-classes.

The unpredictable nature of the outbursts means that observation of these stars contains an element of surprise for amateurs. The rapidity of the rise during an outburst means that it is advisable to examine such stars every night if at all possible.

We should, however, mention that these stars, which are not very numerous for observers with small- or medium-sized telescopes, are very well covered, so that observers with such telescopes, apart from the pleasure of observing, are unlikely to make any real scientific contribution of their own. Anyone with a large instrument, with a good site, can concentrate on less 'popular' stars, and can make a significant and useful contribution to our observational knowledge of these stars, which remain poorly understood. We should stress that a lot of patience is required, because some of these stars may remain for years without an outburst.

14.4.3 British Astronomical Association, Variable Star Section

The Variable Star Section of the British Astronomical Association produced its first report in 1891, and has continued to observe variables ever since. Since its inception, it has encouraged observers to follow specific stars, rather than accepting observations of any stars that members happen to observe. Long, continuous runs of data on many stars are therefore available. Although initially concentrating on LPVs (i.e., Mira stars), the programme soon broadened to include eruptive stars, and an extensive series of analytical reports on many different classes of variables and individual stars have been published, primarily in the BAA's *Journal*. Approximately 100 000 observations are received every year. Small, but active, sub-groups carry out photoelectric photometry and observe eclipsing binaries.

A major part of the BAA VSS current activity involves monitoring known, and suspected, recurrent eruptives, such as old novae, etc. In addition, there is close collaboration with the U.K. Nova and Supernova Patrol, which, despite its name, is international in its membership. This is coordinated by Guy Hurst, editor of *The Astronomer* magazine. The Nova and Supernova Patrol, which is carried out both visually and photographically, and also the recurrent objects' programme have had a large number of successes to their credit. One participant in the Nova Search Programme, Mike Collins, has made a phenomenal number of discoveries of new, suspected variables, as well as confirming variation in many previously suspected objects. This has been achieved by painstaking photographic work and meticulous checking of the images obtained.

The Early Warning scheme run by Guy Hurst has become the primary means of contact between amateur and professional variable-star astronomers worldwide. It is particularly noted for its rapid, accurate confirmation of suspected objects or outbursts. It operates an extremely fast scheme for notifying observers (and even providing them with charts) by means of electronic mail, which has often resulted in confirmation of a new object being obtained within hours of discovery. More details of this programme are given later in Sects. 15.3 and 15.3.5 (pp. 858 & 869).

The methods employed by the BAA VSS differ slightly from those used by many variable-star organisations. The charts, for example, do not carry the magnitudes of comparison stars. These are given letter designations, and the sequences are given separately, thus minimizing the effects of bias. Similarly, all estimates are reported in full, i.e., with comparison-star designations and numerical steps, not simply as deduced magnitudes. (This does in fact mean that the Section has more data to handle, with the resulting problems.) In addition, many observers use a method of estimating known as the 'fractional method'. Here, the observer estimates the magnitude of the variable as a relative fraction of the interval between two comparison stars. 'One-third of the interval from A to B' would therefore be written as: A(1)V(2)B. Note that this is a single estimate, not two independent (Argelander) step estimates. This method has been found to be particularly suitable for beginners and to give results comparable in accuracy to the Argelander method when the sequences are satisfactory, with moderately regularly spaced comparison-star magnitudes. (More advanced observers are encouraged to use other methods when sequences are less satisfactory.)

The Section is actively computerizing all its current records, and is also beginning to make inroads into the data (some of which are available in printed form) that have accumulated over the past 100 years. Portions of this large database of well over 2 000 000 observations are made available on request to both amateur and professional researchers.

One particular development may be mentioned, and this is the construction by J. Ells of an automatic photoelectric telescope. This computer-controlled telescope is the answer to all armchair variable-star observers' dreams! Once programmed, the telescope will carry out a sequence of observations, unattended, for hours on end, or even throughout the night. These may be either repeated observations of one variable and its comparison stars (as with an eclipsing binary, for example) or observations of different stars. The telescope was, in fact, more productive than any single, BAA VSS visual observer in 1990, with over 5000 observations. Despite Mr Ells' untimely death, the telescope has now been installed on a new site, and is again in operation. A full description may be found in Ells & Ells, 1989 & 1990.

The BAA Variable Star Section actively cooperates with the Variable Star Section of the New Zealand Astronomical Society (which is described later), exchanging estimates of certain variables at low declinations on either side of the equator to improve observational coverage, and enable more satisfactory analyses to be made.

14.4.4 Bedeckungsveränderlichen Beobachter der Schweizerischen Astronomischen Gesellschaft (BBSAG)

The BBSAG is a small group of observers, primarily in German-speaking Switzerland, which has specialized in the observation of eclipsing binaries. Although it has a small membership, this group has gained an excellent reputation, and it is certainly the most active group when it comes to eclipsing binaries. It has two main aims: the search for new eclipsing stars, primarily by observing stars suspected of variability and listed in the NSV catalogue; and the determination of the times of minima (both primary and secondary) of known eclipsing variables. This latter work provides a check on the published ephemerides and reveals any changes of period or motion in binary systems that may have occurred.

14.4.5 Groupe Européen d'Observations Stellaires (GEOS)

GEOS was born in 1973 with the amalgamation of several small, European groups based in France, Belgium, Spain and Italy. Its observations are mainly concentrated on variable stars, but the group is also interested in artificial satellites, the mutual phenomena of the satellites of Jupiter, occultations by minor planets, and photoelectric photometry. GEOS has about 200 members, about half of whom observe variables, and report some 80 000–100 000 visual observations per year (to which about one thousand photoelectric observations may be added). Since its foundation, the group has accumulated about 2 000 000 visual observations.

GEOS gives priority to variables, where observations appear to offer the possibility of rapidly discovering something new. Most estimates are therefore concentrated on little-known variables, which are mainly of two types:

- Faint variables, always below 10th magnitude, and because of this, frequently neglected by amateurs. Observations of them require telescopes of 200-mm aperture or greater.
- Low-amplitude variables, with amplitudes below 1 magnitude, which therefore require careful statistical analysis ... and plenty of observers. There are still bright stars (magnitudes 6–8) whose variations, of a few-tenths of a magnitude, remain undetected. An amplitude of 0.5 mag. is certainly detectable by a good variable-star observer; but discovery becomes less and less likely as the amplitude decreases. In practice, it is impossible to be certain of the variability of a star by visual methods when the amplitude is less than 0.2 mag.

14.4.5.1 Reduction and publication of observations
Unlike the majority of other variable-star organisations, the estimates made by members of GEOS are not centralized. Every month, each observer submits a detailed report: number of observations, number of nights, stars observed, and number of estimates per star. Once or twice a year, the group circulates a call for observations, which contains a list of variables, each one of which is of interest to one of the members (who has previously requested information about it). All the observations

of one specific star are then forwarded to the person who has undertaken to reduce the estimates of that star. That person analyses the observations and prepares a final report. GEOS thus publishes specific papers and not lists of observations.

In practice, GEOS publishes:

- the monthly report of the work carried out by the group (variable stars, minor-planet occultations, research in progress, articles, events ...);
- notes, restricted to members of the group. A note may contain interesting interim results, which may affect a current project and spur observers to follow a star that is of particular interest;
- *GEOS Circulars*, published in English and distributed more widely. These publications always contain a result that merits publication in the astronomical community;
- basic information for beginners;
- technical notes, which describe methods of observation and reduction, the programme of priority stars to be covered and the cumulative totals for previous years;
- popular articles in various European journals, and scientific contributions, notably in the IBVS (the *Information Bulletin on Variable Stars*, published by IAU Commission 27.)

In 1985, for example, GEOS published 44 notes and four *GEOS Circulars*.

The decentralized organisation of GEOS avoids a large secretarial force (handling 100 000 observations per year) and allows all the members to participate in the overall work. One objection might be that the 'Request for Observations' does not cover all the stars observed, and that a certain number of estimates thus remain dormant with their originators. This number of unused observations is very low. Obviously any variable-star observer who has accumulated a lot of observations of one particular star will, sooner or later, undertake to analyse the results – if only to justify their own observations. By the time 10 years have elapsed all the observations have been analysed.

14.4.5.2 Slow, low-amplitude variables

By virtue of their very nature, this type of variable is not particularly enthralling, and is not recommended for beginners. These stars are, however, actively observed by GEOS, precisely because they are not observed anywhere else, and because there are still many bright, low-amplitude stars, visible in binoculars, which are under-observed.

Observation of a faint, low-amplitude variable requires great care. If the variations are slow, it is reasonable to assume that the star does not vary during one night, and two or three independent estimates are made each observing session. Analysis of the estimates, after a campaign lasting at least 5 years, consists of determining the class of variation. The first aim of the analyst is to search for any periodicity. It may be noted that many stars have been (temporarily) classed as irregulars because they have been insufficiently observed.

14.4.5.3 Unknown, rapidly changing variables

An amateur who tackles a new variable always begins by looking for rapid variations. With this in mind, the star is observed five or six times a night. After several tens of nights it is possible to arrive at a definite opinion and if the star does not vary over the course of a night, then it is considered to be a slow variable.

If it does obviously vary during the course of a night, the second stage consists of searching for a possible period. Two situations may arise:

- The possible period appears to be shorter than a night. During the course of a night, the brightness may rise, and then fall. This gives a scattering of points (Fig. 14.29), to which one can attempt to fit a periodic curve. Short-period variables are often EW stars, or pulsating stars of the RR or δ Sct classes with quite low amplitudes. The scatter among the measurements is annoying and observations over one night will only give an approximate idea of the period. Observation of subsequent maxima will enable this period to be determined more accurately, and a preliminary ephemeris to be established. The $(O - C)$ values between the observed and the calculated times of maxima will enable the period to be progressively refined.

- The period is longer than one night. Let us assume (Fig. 14.30) that two maxima have been observed on two consecutive nights. The interval between these two maxima is therefore a multiple of the period. If this interval is D, then the true period will be D, or $D/2$, $D/3$, ... In this case we have to prepare an ephemeris for each of the possible periods and continue observing. The best ephemeris will give the lowest $(O - C)$ values and the true period. If two of the possible sub-multiples of D are still in contention, then composite light-curves for each of the periods must be prepared. The standard deviation of the estimates relative to each of the curves is then calculated. The smallest standard deviation will indicate the period to be retained.

There are very effective mathematical algorithms for extracting a period from a light-curve affected by the considerable 'noise' arising from random errors that are relatively large when compared with the amplitude. One of these methods consists of deriving a periodogram; this algorithm cannot be tackled by hand, but is quite feasible on a computer.

Let us assume that the period definitely lies somewhere between P_1 and P_2. All the observations are combined for period P_1 and a standard deviation σ_1 obtained. Then we take the period $(P_1 + \epsilon)$ and recalculate $\sigma_{1+\epsilon}$. Overall, we obtain the function $\sigma = f(P)$ between P_1 and P_2.

Possible periods appear as peaks in the curve. The successive calculations for P_1, $P_{1+\epsilon}$, $P_{1+2\epsilon}$, ..., are generally extremely numerous, and impractical with manual methods.

If we arrive at the conclusion that the star does definitely show periodic variations, we still have to determine whether we are dealing with a pulsating star or an eclipsing binary. Some low-amplitude EW stars have light-curves that definitely resemble certain RR or δ Sct stars. If there is still doubt, note that pulsating stars are generally close to minimum, whereas with an EW, on the other hand, it is more

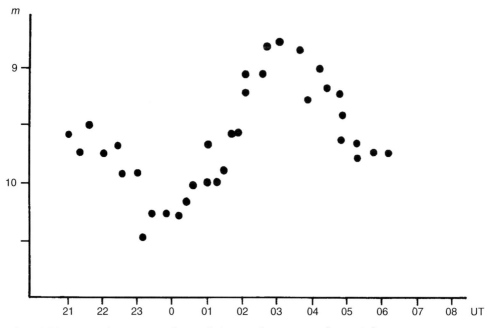

Fig. 14.29. *Typical estimates obtained during the course of one night*

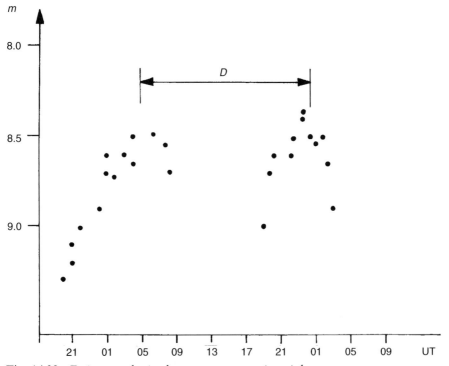

Fig. 14.30. *Estimates obtained over two successive nights*

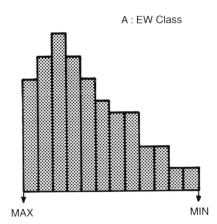

A : EW Class

MAX   MIN

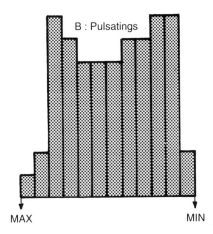

B : Pulsatings

MAX   MIN

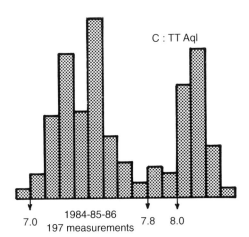

C : TT Aql

7.0   1984-85-86   7.8   8.0
       197 measurements

Fig. 14.31. *Histograms of numbers of observations versus magnitude. If the estimates were made at random intervals, the histograms for an EW eclipsing star and for a pulsating object are different. An EW star is more frequently near maximum. Naturally, the light-curves, and thus the histograms, differ from one star to another. Diagram C shows the histogram for TT Aql, obtained by DMT. The 'hole in the centre' is too deep. It is typical of a Carnevali effect (p. 812), caused by the fact that comparison C has a magnitude of 7.8 (B = 6.92; C = 7.80; D = 8.40).*

frequently close to maximum (Fig 14.31). If the spectral class of the star is known, this may resolve the doubts. RR-Lyrae stars have A, F or G spectra – as do the δ-Sct stars – whereas EW stars cover a wide range of spectral classes, from B to K. In general, the two components of an EW star have the same temperature, so the (B − V) index does not vary significantly. Pulsating stars, on the contrary, have indices that definitely vary. A few two-colour measurements always enable a final decision to be made.

If it is decided that the star is eclipsing, it must be remembered that the period is twice as long as the observed variation, because we see the eclipse of A by B and then of B by A in succession. Stars of class EA sometimes have secondary minima that are unobservable (because their amplitudes are too low), but the light-curve of an EA star cannot be confused with that of a pulsating variable.

14.4.5.4 Observation of pulsating stars (Cepheids and RR-Lyrae stars)
Observation of a well-known pulsating star is interesting routine work, which enables the period to be checked against the ephemeris. Analysis of the observations is carried out at the end of each observing season. All the observations are combined, using the (known) period. This gives a well-defined, mean light-curve and, in particular, the (mean) time of maximum enables $(O - C)$ values to be measured. If these are not negligible, it is often an indication of a slight period change. The time of maximum is given in the published ephemerides, either those in the GCVS or in the annual list issued by the Krakow Observatory in Poland.

In making observations, it is important to determine the time of maximum. One should therefore aim to secure as many maxima as possible, particularly for RR-Lyrae stars, because one is looking for evidence of a Blazhko effect, which indicates double periodicity (Fig. 14.11, p. 791). The asymmetry of a light-curve is the ratio between the time taken for the rise and the period. The value of this asymmetry gives a method of discriminating between the two families of RR-Lyrae stars. The RRab stars have a very pronounced asymmetry, whereas the RRc have almost sinusoidal curves. These are slightly hotter stars whose mode of pulsation is different from that of the RRab stars.

14.4.5.5 Observation of eclipsing variables
Observation of eclipsing variables is easier than that of pulsating stars because of the stability and symmetry of the light-curve. It is useful to obtain a full light-curve, but the most important thing is to carefully determine the time of minimum. Direct examination of the light-curve does not enable the precise time of minimum to be determined. The accuracy may be improved by using what is known as the 'tracing-paper method'. This method may be used because the light-curve is absolutely symmetrical about the time of minimum.

Each observation is plotted in the normal way (with magnitude on the y-axis and time on the x-axis). In chosing the scale for these axes, it helps to ensure that the two branches of the light-curve around minimum make angles of about $45°$ with the axes (Fig. 14.32). The points are copied onto a sheet of tracing paper, together with the x-axis (the time axis). The tracing paper is turned over and the two x-axes are aligned. This gives us two sets of points: the original points and the traced set, seen from the back. Then we find the best fit, by sliding the tracing paper along the x-axis. As we slide the tracing paper along, we see the two sets of points combine to give a very regular curve. Once we have found the best position, where the two sets of points form a single, regular curve, we can read off the x-value of the central point of the chord joining a point to its counterpart. This value indicates the time of minimum:

$$t_{m} = (t_{p} + T_{p'})/2,$$

where t_p and $t_{p'}$ are the relative times of point p and its counterpart p'.

When analysing observations of eclipsing stars, variable-star observers should remember that EW stars have a primary minimum that closely resembles the secondary minimum. In some cases it is impossible to discriminate between them. Such confusion is rarely likely with EA stars. It is more likely, in fact, that

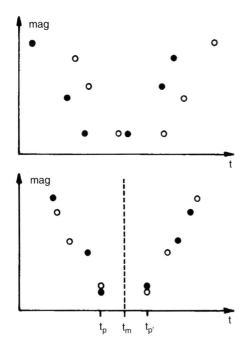

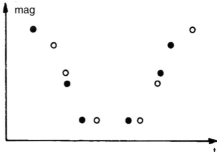

Fig. 14.32. *The 'tracing-paper' method. Successive positions are shown, with the tracing paper being moved from right to left. • original points (on the plot); ○ corresponding points (on the tracing paper)*

their secondary minima will be of very low amplitude, and thus be undetectable visually.

There is a relationship between the amplitude Δ of primary minimum (Min I) and that of secondary minimum (Min II). This follows from Pogson's magnitude scale. As the reader will realise after a moment's reflection, it is impossible for the amplitudes of both Min I and Min II to exceed 0.75 mag. We find:

if Δ(Min I) = 1 magnitude, Δ(Min II)≤ 0.55 magnitude;
if Δ(Min I) = 2 magnitudes, Δ(Min II)≤ 0.19 magnitude;
if Δ(Min I) = 3 magnitudes, Δ(Min II)≤ 0.07 magnitude.

For RW Tau, for example, Δ(Min I) = 4.49. The amplitude of secondary minimum must therefore be less than 0.18 [sic., note, however, that recent observational data give an amplitude of 3.61 mag. for Min I, and 0.11 mag. for Min II. – Trans.]

Eclipsing variables, even well-known ones, should be observed regularly when the secondary minima are detectable. In general, Min I is conventionally regarded as phase $\varphi = 0$. Min II therefore falls at some, normally constant, phase φ'. If φ' varies, this is because the relative orbit of the component that is producing the primary eclipse is rotating in space (Fig. 14.33). Such a rotation is caused by the presence of a third, unobserved, body. It may require several years (or decades!) to detect a slight shift in φ', but this is the sort of research that amateurs can carry out.

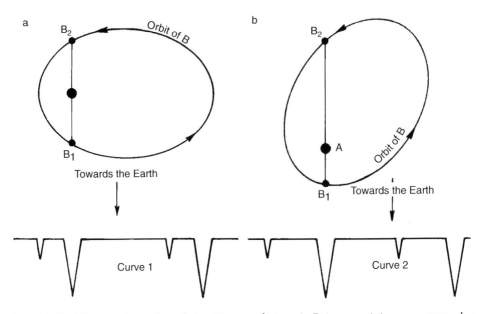

Fig. 14.33. *The relative orbit of star B around star A. Primary minimum occurs when B is at B_1, and secondary minimum when it is as B_2. There is a greater interval between B_1 and B_2 than there is between B_2 and B_1, giving rise to light-curve 1, where secondary minimum is shortly before primary minimum. If apsidal rotation has occurred, the situation is as shown in b: secondary minimum has moved closer to the preceding primary minimum.*

14.4.5.6 Searching for new variables

Searching for new variables forms a considerable part of GEOS' activities. It requires great patience and long experience in the art of estimating magnitudes. There are two principal methods, the first consisting of photographing a large number of regions of the sky with a refractor or long telescopic lens. Some months or years later, the same photographs are repeated. Comparison of the two photographs may enable new variables to be discovered. Some thirty years ago, Roger Weber discovered more than 100 variables by this method. [Note that a very successful, far more intensive method of searching and checking is carried out by the U.K. Nova Search programme, described in Sects. 14.4.3 and 15.3. – Trans.]

The second method consists of selecting stars from the NSV, and this is the one mainly adopted by GEOS. A star is then included in the programme and followed by several observers for several years, until some definite conclusion is reached. If variation is confirmed, a photoelectric campaign is organised to confirm the visual observations. The photoelectric light-curve is then published in the IBVS.

All (active) variable-star observers have noted that stars within the fields of variables that they are following sometimes appear too bright, too faint, or not constant. There is then a very strong temptation to follow this anomaly. This method of discovering new variables is not very productive. If the star appears brighter or fainter than indicated on the chart, this is generally a chart error. (Errors of 0.5 mag. are common on earlier charts.) Stars that do not appear constant are more interesting, but are difficult to detect. An active observer (with more than 50 observations per night) does not lose time carefully examining all the other stars in a particular field .... In 1982, for example, variations in the star HD 156860 were noted by Michel Dumont, during regular observations of the eclipsing star U Oph. The amplitude was close to 0.6 mag. and was thus sufficient to call attention to itself. Recent publications give HD 156860 as V2113 Oph.

Another idea is to monitor 'high risk' stars, i.e., those stars with spectra that strongly suggest that they may be variable. Amateurs have no means of determining such anomalies themselves, but most catalogues give the spectral class of stars. Several years ago, GEOS started a campaign of systematic observation of red stars. Several tens of red, supposedly constant, stars are regularly followed; a number of them are obviously variable. The choice of red stars is not arbitrary; probably more than 10% of stars with M-class spectra are variable. Stars covered by the programme lie between 6th and 10th magnitude.

Finally, the secret aim of every variable-star observer is to discover a nova or supernova. The appropriate search methods are described in Sects. 15.3 and 15.5, respectively.

14.4.6 *The Variable Star Observers League in Japan (VSOLJ)*

Japanese observers have been particularly active for many years in certain fields of variable-star research, notably the search for novae, where they have enjoyed considerable success. There have been three principal organisations: the Japan Astronomical Study Association (JASA), the Research Group for Variable Stars in Japan (NHK – Nippon Henkosei Kenkyukai), and the Japan Amateur Photoelectric Observers Association. These groups actively cooperate with various other variable-star organisations and, increasingly, with professional astronomers. All classes of variables are observed, but, as with other groups, considerable emphasis is placed upon monitoring of cataclysmic (i.e., eruptive) variables.

Around 1987, the Variable Star Observers League in Japan was formed by combining these three groups of variable-star observers. The principal aims were to publish a single *Bulletin* devoted to variable stars, and to establish a common archive of Japanese variable-star data. Both of these ends have been achieved and in a very short time (roughly a decade) approximately 1 000 000 observations have been entered into a computer archive. This database is being used to prepare and publish long-term light-curves of various classes of variable.

14.4.7 *The Variable Star Section of the Royal Astronomical Society of New Zealand (RASNZ VSS)*

This group is the leading variable-star organisation in the Southern Hemisphere, with observers in a number of countries. Its programme includes all types of variables, with the exception of short-period eclipsing stars. The principal objects observed, however, are cataclysmic binaries, and such observations are much in demand by professional astronomers. A number of major papers on such stars have been published in professional journals such as *Astronomy & Astrophysics* and *Monthly Notices of the Royal Astronomical Society*.

Although large-amplitude Mira and semiregular stars are observed, they have a low general priority apart from a few unusual objects, and certain southern stars about which little is known. The section also operates a nova search programme, which has had a number of very significant successes, including the discovery of novae in the LMC, previously unknown variables, and even comets. The section exchanges observations of variables with low northern declinations for estimates of stars with southern declinations that have been obtained by the BAA VSS, thus ensuring better coverage than each organisation on its own could achieve. There is also active collaboration with the AAVSO.

The section issues charts for all the stars that its members are monitoring. Its methods are similar to those of the BAA VSS, which is one reason why actual estimates may be freely exchanged between the two groups without a loss of scientific content.

14.5 Some variables with which to start

Finally, here are seven examples of spectacular variable stars that readers might like to use to get them started. If they find that they want to continue, we strongly advise them to get in contact with one of the organisations that we have mentioned [or their own national group], who will be able to suggest observational programmes that will keep them busy for several centuries!

14.5.1 R Leonis

This is a Mira-type star, which varies between mag. 5.4 and 10.5 with a period of 313 days [extreme range listed in GCVS is mag. 4.4 to 11.3 – Trans.] It may be observed with binoculars when bright and with a small, 60–80-mm refractor when near minimum (Fig. 14.34). One estimate every 10 days is adequate.

14.5.2 SS Cygni

This is a U-Gem type variable. It is generally at mag. 12, but its outbursts are sudden, and quite frequent (about every 50 days), when it rises by about 4 magnitudes in a few hours. It should be observed two or three times a night, but the frequency should be increased when an outburst occurs, or when it begins its decline. At maximum a very small refractor is quite adequate, but a 100-mm aperture is recommended for following it at minimum. For beginners, SS Cygni is an excellent exercise in finding a faint object (Fig. 14.35).

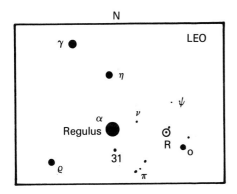

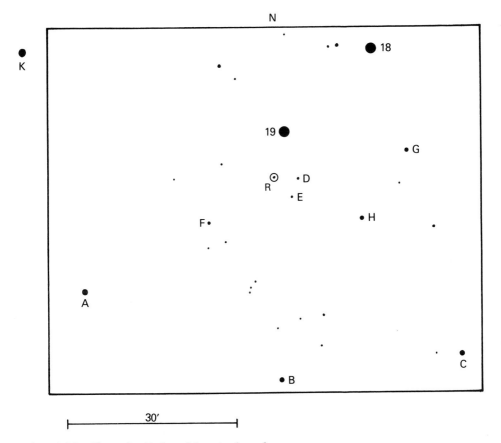

Fig. 14.34. *Chart for R Leo. Magnitudes of stars:*
18 Leo = 5.68; B = 7.81; E = 9.6; 19 Leo = 6.44; C = 8.6; F = 9.7; A = 7.60;
D = 9.0; G = 10.1; H = 10.4; K is a magnitude 6.6 star close to X Leo (see
Fig. 14.25)

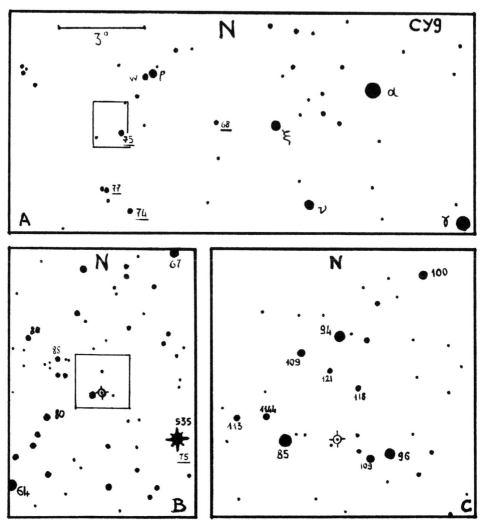

Fig. 14.35. *Charts for SS Cygni*

14.5.3 TT Aquilae

This is a classical, Cδ Cepheid, which varies between mag. 7.0 and 8.1 with a period of 13.7546 days. It is easy to find, and may be followed throughout its cycle with binoculars. It should be observed three or four times a night (every 2 hours, for example) (Fig. 14.36).

14.5.4 CY Aquarii

CY Aquarii is a dwarf Cepheid (RRs) which varies between mag. 10.4 and 11.2, with a period of $1^h27^m53^s$. It is one of the fastest pulsating variables. With dark

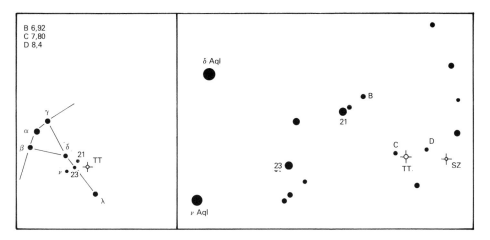

Fig. 14.36. *Chart for TT Aql. The Cepheid SZ Aql is nearby (mag. 7.92–9.26, P =* 17.137 939 *days)*

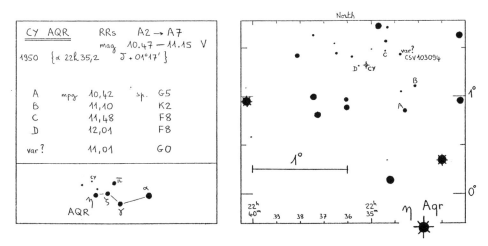

Fig. 14.37. *The field of CY Aqr (GEOS chart)*

skies, a 80–100-mm refractor suffices, but under urban condition a larger aperture is required (Fig. 14.37).

The star should be followed without a break for more than 90 minutes, with one estimate every 2–3 minutes. Never let more than 5 minutes go by without an estimate! Here is the ephemeris, in heliocentric Julian Days, for CY Aqr:

JD hel. $2445\,207.494 + 0.061\,038\,29n$,

where n is the number of epochs since a photoelectric minimum observed in 1982 August (GEOS RR 7).

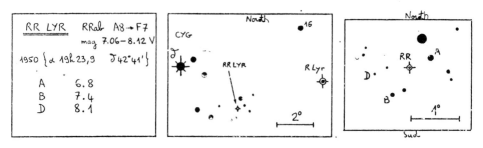

Fig. 14.38. *Chart for RR Lyr (GEOS)*

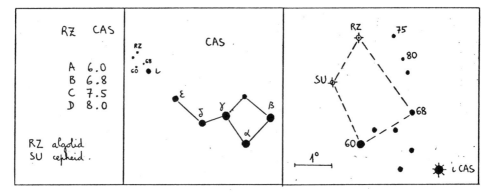

Fig. 14.39. *Chart for RZ Cas; SU is a low-amplitude Cepheid. (GEOS chart)*

14.5.5 RR Lyrae

This is the brightest of this class of variables. It varies between mag. 7.1 and 7.9 in $13^h36^m15^s$. It is very well placed during summer nights, which are, however, shorter than the period. A night centred on maximum should be chosen. RR Lyr should be observed with binoculars, with one estimate every 10–15 minutes (Fig. 14.38). The ephemeris is:

JD hel. $2442\,995.446 + 0.566\,839n$.

14.5.6 RZ Cassiopeiae

This is undoubtedly one of the best-known EA eclipsing variables. It varies between mag. 6.3 and 7.9 in $1^d04^h41^m09.5^s$. The decline is rapid, lasting about 2 hours. It does not remain at minimum, beginning to rise immediately, so that in another 2 hours it has regained its normal brightness. The secondary minimum, with an amplitude of 0.05 mag. is not detectable visually.

The star should be observed with binoculars. Observations should begin slightly

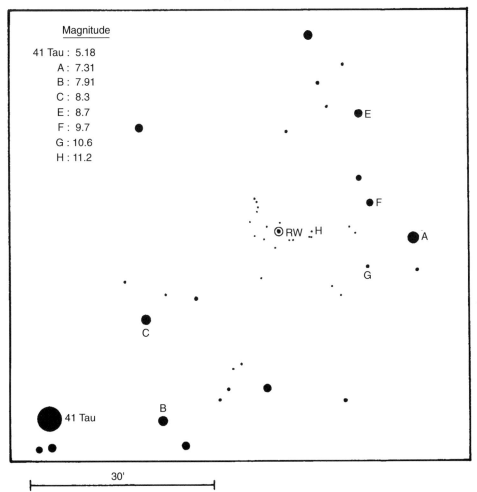

Fig. 14.40. *Chart for RW Tau*

earlier than 2 hours before predicted minimum, with one estimate every 5 minutes (Fig. 14.39). The ephemeris is:

JD hel. $2439\,673.1302 + 1.195\,248\,92n$.

14.5.7 RW Tauri

This is an EA-type eclipsing variable, with a large amplitude. It varies between mag. 8.0 and 11.5 in $2^d18^h27^m08^s$. There is a total eclipse of a hot (B8) star by a large, cooler (K0) star. It is advisable to use a 100-mm instrument to follow the star during eclipse (Fig. 14.40). One estimate should be made every quarter of an hour at the beginning of the decline; one every 5 minutes below 10th mag. until totality. The ephemeris is:

JD hel. 2446 005.386 + 2.768 8396n.

Use this ephemeris to choose nights for observing RW Tau. If the star is observed without using the ephemeris, 9 times out of 10 no eclipse will occur.

15 Deep sky, novae and supernovae

P. Martinez

15.1 General

It is an unfortunate fact that as soon as we consider objects beyond the Solar System, we find that they are generally very faint. Amateurs are therefore at a distinct disadvantage when compared with professionals, because studying such objects requires the largest possible apertures to collect the maximum amount of light. In the case of photography (or for any detector other than the eye), a particularly stable and accurate mounting that will allow long exposures is also required.

In addition, distant objects do not normally change with time: the shape of a galaxy and the spectra of most stars do not alter on a human timescale. They are thus far more suitable for study with large professional instruments than with the small telescopes available to amateurs.

The objects where amateurs have a role to play are the exceptions to this general rule. The most important are double stars and (in particular) variables. In the latter case, the sheer numbers, and the rapid changes that may occur require intensive work, which only amateurs are in a position to provide. Then there are nova- and supernova-searches, which involve monitoring large areas of sky; a task in which amateurs can undoubtedly compete with professionals. The observation of double stars and variables requires specific techniques, which are described in other chapters. The study of novae and supernovae has more in common with deep-sky 'observation', a field that is dear to many amateurs, so these objects are discussed in this chapter.

Disregarding the Moon, distant objects are, paradoxically, the most spectacular to look at and to photograph. This has caused amateurs to become increasingly interested in them in recent years. This interest has much in common with the yearning for larger and larger diameters that is found among telescope makers.

The various types of amateur activity involving objects beyond the Solar System may be summarized as:

- 'sight-seeing':

    * wide-field, naked-eye or photographic observation of the constellations; identification of stars from an atlas or else drawing a personal chart. Naked-eye identification of the constellations is described in numerous books for beginners and will not be discussed here. Photography is covered in Sect. 15.2;

    * observation of faint objects (nebulae, clusters, galaxies, etc.) with binoculars or a telescope; photographing these types of object with

847

telephoto lenses, a Schmidt camera, or at telescopic prime focus (*see* Sect. 15.4.4.3).

- 'educational' work: this consists of repeating work carried out by professional astronomers, but with simpler equipment. This type of work provides personal satisfaction in confirming the earlier results. Two main activities may be mentioned:
  - * spectroscopy of stars and nebulae. The methods are described in Chap. 18;
  - * colorimetry of stars. When associated with magnitude determinations, such work may lead to the 'rediscovery' of the Hertzsprung–Russell diagram (*see* Sect. 15.6).

- useful observations:
  - * double stars and variable stars; see Chaps. 13 and 14;
  - * searching for novae (*see* Sect. 15.3);
  - * searching for supernovae (*see* Sect. 15.5).

15.2 Constellation photography

We use this term to mean photography of the sky using standard, wide-angle, or short telephoto, lenses. Such lenses give fields some tens of degrees across, sufficient to cover the area of one or more average constellations. Table 15.1 shows the fields given by various lenses with common focal lengths for an image size of 24×36 mm, the standard 35-mm format.

15.2.1 Films

The choice of film depends on the type of photographs that are required. For black-and-white work, the ideal is hypersensitized Kodak Technical Pan 2415. When the sky background is poor, however, or if the lens is very fast ($F/D < 2$), exposure times cannot exceed a few minutes. Untreated film may then be used: either TP-2415 if high resolution is required or the object is relatively bright (an open cluster or the brightest stars in a constellation), or else a faster film such as Kodak T-Max 400 (which has replaced Kodak Tri-X) if the object is diffuse or faint (nebulosity in the Milky Way, etc.). Amateurs who are not equipped to hypersensitize films and who want to make exposures exceeding a few tens of minutes should consider Kodak 103a films, which are formulated to counteract the Schwarzschild effect. The problem with these films is that their resolution is worse than T-Max 400, and far inferior to that of TP-2415.

Taking photographs with a filter designed to select a particular range of wavelengths is interesting work. The filter (a Wratten gelatine filter is very suitable) merely has to be held in front of the lens. (A photograph in red light, using a Wratten 29 filter, for example, will show nebulae that emit light in the Hα line at 656.3 nm.) Black-and-white film is used for this, but because of the amount of light-loss produced by the filter, only 103a or hypersensitized films may be used.

Table 15.1. *Fields covered by various lenses on 24 × 36-mm format*

Focal length (mm)	Field
17	93° × 70°
20	84° × 62°
24	74° × 53°
28	65° × 46°
35	54° × 38°
50	40° × 27°
85	24° × 16°
100	20° × 14°
135	15° × 10°
200	10° × 7°
300	6.8° × 4.6°
400	5.2° × 3.5°
500	4.0° × 2.7°

(The choice is between 103aE (*see* Fig. 15.1) and 103aF for photography in the red, because 103aO is not red-sensitive.) TP-2415 is particularly suitable for photography in the red region, because it has extended spectral response at long wavelengths.

With colour films, the situation is changing rapidly, because of the intense competition between the various manufacturers, which has resulted in spectacular progress since the early 1980s. At present, the film best suited to constellation photography is Agfachrome 1000 developed at 3000 ISO; it has dethroned Fujichrome 400, which was the best choice for a number of years. Hypersensitization of colour films is less successful than with black-and-white emulsions. (Its gain is limited to 2–3 times, and there is a significant loss of contrast.) Because of the speed of lenses used for wide-field photography, it is not generally worthwhile hypersensitizing colour emulsions.

15.2.2 Exposure times and guiding

To record very faint objects, exposure times should be as long as possible. They are, however, limited by two factors: fogging caused by background sky illumination and, with guided exposures, by observer fatigue. [The limiting magnitude attained may be determined by photographing an area with well-established sequence of magnitudes. In the Northern Hemisphere this is should be the North Polar Sequence (Sect. 15.8, p. 940). – Trans.]

The advantage of very fast lenses is that they allow relatively short exposure times. By way of example, under a clear sky (at sea level, but without stray light; the Milky Way being clearly visible and with a fair degree of contrast), a 20-minute

Fig. 15.1. *An unusual view of the constellation of Orion (compare with Fig. 15.6), photographed in red light (103aE film and Wratten 29 filter) by J. Sylvain. The hydrogen emission (Hα) regions are particularly noticeable). f/1.9, 50-mm lens, exposure 50 minutes.*

exposure with an f/2 lens gives a sky background of unit density on hypersensitized Kodak TP-2415 film. Although the negative appears very dense, a background density of unity indicates optimum exposure for TP-2415: the signal/noise ratio is a maximum, which means that these conditions are best for detecting faint objects. Naturally, optimum exposure is a function of the film, the aperture of the lens, and the brightness of the sky background. It is advisable for everyone to carry out their own tests to determine what it should be, given the material being used and the quality of the site.

The simplest photographic method consists of fixing the camera onto a very stable tripod, and leaving the shutter open, unattended, on a B setting. Because the camera is pointing in a fixed direction and the sky appears to rotate around the Earth, each star moves during the exposure and leaves a trial on the film. Such photographs are informative (they clearly indicate the apparent rotation of the sky, especially when the camera is pointed towards one of the celestial poles), and frequently very beautiful. Scientifically, their interest is more limited; however it is worth noting that it may be a convenient method of photographing meteors or artificial satellites. One specific instance is that of geostationary satellites, which require a fixed-position exposure, because they are themselves essentially stationary with respect to the Earth.

If the stars are to appear as points rather than trails, then the motion of the sky must be followed. For anyone with an equatorially mounted telescope, the camera is simply mounted with its optical axis parallel to that of the telescope. The latter, with a graduated eyepiece, may then be used for guiding. The mounting must be an equatorial: even though it may be possible to follow a star's apparent motion across the sky using the two motions of an altazimuth telescope, there is still an unavoidable rotation of the field.

For preference, the equatorial mounting should be driven. If the motor is particularly stable and accurate, and if the polar axis is properly aligned, it is even possible to dispense with guiding through the eyepiece. On the other hand, even if the mounting is not driven, it is possible to track the sky manually, using the slow motion in right ascension, provided the latter is sufficiently sensitive in operation. Note that manual guiding is possible only with a lens of short focal length. Such a method cannot be used when making photographs at a telescope's prime focus. Nevertheless, any manual guiding is irregular, and it is essential to determine the extent of any deviations from the ideal mean motion that may arise, to ensure that these variations are acceptable. Guiding errors may be evaluated by determining how long unguided diurnal rotation takes to move a star by a comparable amount. In general, an experienced observer using a good-quality mounting manages to limit errors in guiding to the amount of motion that a star would show in 1 second of time. Are such errors acceptable? The answer is given by Table 15.2, which gives the angular error corresponding to a trail of 30 μm on the film (which corresponds to the photographic resolution), as a function of focal length, and also the time taken by a star on the celestial equator to cover this distance. A tolerance of 30 μm corresponds to what is obtained using films such as Tri-X, the 103a emulsions, or fast colour films. With TP-2415, which has a much better resolution, the errors should be restricted to about 10 μm. The corresponding angular motions and times

Table 15.2. *Permissible guiding errors for given focal-length lenses*

Focal length (mm)	30 μm		10 μm	
	Angle (″)	Time (s)	Angle (″)	Time (s)
17	364	24	121	8
20	309	21	103	7
24	258	17	86	6
28	221	15	74	5
35	177	12	59	3.9
50	124	8	41	2.8
85	73	5	24	1.6
100	62	4.1	21	1.4
135	46	3.0	15	1.0
200	31	2.1	10	0.7
300	21	1.4	7	0.46
400	15	1.0	5	0.34
500	12	0.8	4	0.28

This table shows the apparent angles corresponding to displacements of 30 and 10 μm on the emulsion, and the times taken by an object on the celestial equator to cover these angles. The 30-μm columns are to be used with Kodak 103a or similar, sensitive films; use the 10-μm columns for Kodak TP-2415. The maximum angle that a star may move off cross-hairs on a reticle may be determined for given lens and film types. When manual guiding is used, the value in the 'Time' column indicates the time taken by a star at the equator to move by the corresponding distance, when no corrections are made. Manual guiding becomes practically impossible when this time becomes less than 1 s. If the object being photographed is not on the equator, but at a declination δ, the tolerance is greater: all the values may then be divided by $\cos \delta$.

thus have to be divided by three: the values are given in the last two columns of the table.

Any amateur who does not have an equatorial mounting can construct one, which may be relatively light and simple, because it has to carry just a camera, and perhaps a guide telescope. The latter may not be necessary with careful construction, and if the alignment is sufficiently accurate for the guiding errors to be less than the values given in Table 15.2. Table 15.3 indicates the maximum errors that may be expected with inaccurate alignment, expressed as a function of exposure time and alignment error (the angle between the polar axis and the mounting's polar axis).

Figure 15.2 shows a simple mounting – sometimes known as the 'Scotch' or Haig

Table 15.3. *Image displacement relative to the film**

Alignment error (arcsec)	Displacement error (in arc-sec.) resulting from an exposure time of:				
	1 min	2 min	4 min	8 min	16 min
7	1.8	3.7	7	15	29
15	3.9	8	16	31	63
30	8	16	31	63	126
60	16	31	63	126	251
120	31	63	126	251	503

* Caused by alignment errors.

Mount – that acts as an equatorial table for holding a camera. A hinge functions as the polar axis. Many other designs are possible, and these are largely determined by the equipment to hand.

15.2.3 Subjects for wide-field photography

Most amateurs begin photographing the sky for the simple pleasure of collecting a series of photographs showing the different constellations. The region of the Milky Way is very spectacular: a short telephoto lens reveals a myriad stars, whereas a wide-angle lens with a field larger than the width of the Milky Way reveals it as a whitish band picked out with clusters and nebulae, flanked on either side by darker sky (Fig. 15.3). An extreme fish-eye or very wide-angle lens will show the Milky Way as a luminous band stretching right across the sky. Such methods are also able to reveal the large clouds of dark, obscuring dust that occur within the Milky Way, and which are often of very great extent. This type of wide-field photography is also able to show the zodiacal light (Fig. 15.4) and the gegenschein.

The major makes of photographic lenses (Canon, Minolta, Nikon, Olympus, etc.) offer two types of fish-eye lenses: those that produce a circular image, 180° in diameter, within the 24 × 36-mm format, and those where the image covers the whole of the 24 × 36-mm frame. The disadvantage of the former is that the image is small, because it is circular and its diameter cannot exceed 24 mm. On the other hand, the second type does not cover all of the sky, because the full 180° field applies to the diagonal of the 24 × 36-mm format (i.e., to 43 mm). For this reason two American observers, Dennis di Cicco and Roger Angel, devised a different scheme, where a fish-eye lens of the second type is fitted to a medium format 6 × 6-cm film back (which gives a usable area of 56 × 56 mm). This gives an image of the whole sky, 43 mm in diameter, which is far more detailed than that provided by the first type of fish-eye lens (*see* details and illustrations in Woolf, 1985 and di Ciccio, 1986). Note that this idea may be applied to any lens intended for a 24 × 36-mm format that is used for stellar photography, and especially for searching for specific objects. The

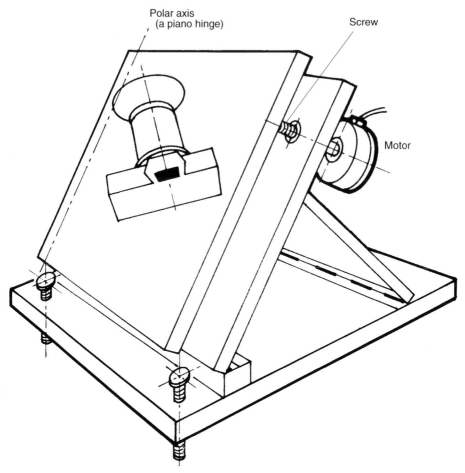

Fig. 15.2. *The 'Scotch' or Haig Mount: a simplified, equatorial mounting that will suffice for exposures of several minutes with a standard, photographic lens.*

use of a camera back that can accept an image 43 mm in diameter enables an area of sky to be covered that is 1.7 times as large as with a normal 24 × 36-mm body.

A cheaper method of photographing the whole sky than buying a fish-eye lens is to use a spherical convex mirror, with its axis pointing vertically (see Fig. 15.5). A camera is mounted above the mirror, pointing downwards. Light from the whole sky is then reflected to the camera. However, the image obtained is less detailed than with a true fish-eye, and the mounting is more cumbersome (*see*, for example, Schur, 1982).

Starting with a series of constellation photographs, it is possible to build up one's own personal sky atlas (*see* Fig. 15.6). Such a task is basically educational, because the atlases available commercially are generally quite adequate for amateur use. Under certain circumstances, however, notably with specialized search programmes, it is of value to have reference photographs taken under specific conditions.

Fig. 15.3. *The Southern Milky Way, with Comet Halley just above the horizon, photographed on 1986 March 14. A 24-mm lens was used at f/1.4; hypersensitized Kodak TP-2415 film; exposure 5 minutes. Photo: Akira Fujii.*

Fig. 15.4. *The zodiacal light, photographed just before dawn from the island of La Réunion in the Indian Ocean. Mercury (right on the horizon) and Jupiter are rising. f/2.8, 28-mm lens; original photograph on Fujichrome 400. Photo P. Martinez.*

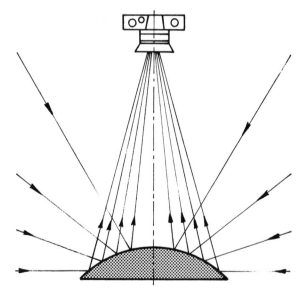

Fig. 15.5. *Arrangement allowing the whole of the sky to be photographed using a convex, spherical mirror and standard lens.*

Fig. 15.6. *Photograph of Orion taken by J. Mazel in 6 × 6 format, on hypersensitized Kodak TP-2415 film, with an f/4.5, 250-mm, Tessar lens. The field covered is 10° × 10°; such records may be used to form a photographic atlas of the sky.*

One can, for example, map a region of sky using a short telephoto lens, and use the images obtained as references for subsequent searches for comets, minor planets, or novae. If the searching is being carried out visually, the limiting magnitude shown on paper prints from the reference photographs should be comparable, or preferably slightly better than that given by the instrumentation (binoculars, small telescope, etc.) that is being used for the search. If the search is photographic, the patrol photographs should be taken under the same conditions as the reference frames (i.e., with the same film, exposure times, filters, etc.). Methods of searching for minor planets and comets are described in the chapters devoted to those objects, so just nova patrols are described in the following section.

857

15.3 Searching for novae

Novae are eruptive stars and are always close binary systems, where one component, a cool star (a red giant), loses material from its outer layers beyond a certain critical radius (the Roche radius) towards the nearby, white-dwarf companion. This leads to the formation of an accretion disk around the white dwarf, and sometimes to a 'hot spot' where the material encounters the disk. (This applies to the prenova state.) Eruptions of dwarf novae (or recurrent novae that show repeated explosions) of the SS Cygni, U Geminorum, Z Camelopardalis, and SU Ursae Majoris types, arise from brightening of the disk itself. 'Ordinary' novae are characterised by thermonuclear reactions that affect the outer layers of the white dwarf, onto which the material (hydrogen and helium) arriving from the cool star falls at supersonic speeds. In novae, the enormous quantity of energy released in a very short time during the explosion is accompanied by a significant ejection of material (of the order of 10^{-5} solar masses), at velocities of between a few hundred and a few thousand kilometres per second.

Unlike comets, novae do not take the names of their discoverers, but are instead initially designated by the genitive of the Latin name of the constellation in which they occur, followed by the year of the discovery (e.g., Nova Cygni 1975). If two novae appear in the same constellation in the same year, they are simply numbered in order of discovery (for example, Nova Vulpeculae 1968 No. 1 and Nova Vulpeculae 1968 No. 2). Subsequently, novae receive definitive, variable-star designations.

Strictly speaking, novae are variable stars, which are considered in detail in Chap. 14. With the exception of recurrent novae, however, nova outbursts are unexpected events. Their initial detection is therefore a matter for sky-search programmes, rather than involving variable-star observational techniques. This is why the methods for systematically searching for novae are discussed here.

15.3.1 Characteristics of novae

At outburst, the luminosity of a nova may be enormous. Sometimes it is impossible to find a star at its position on earlier photographs. This was the case for Nova Cygni 1975 (V1500 Cygni) which reached magnitude 2 at maximum; even the Palomar Sky Survey, an atlas that reaches down as far as magnitude 21, did not show a star at its position. This indicates that Nova Cygni 1975 rose by at least 19 magnitudes during its outburst. Such extreme amplitudes were, of course, the reason why ancient astronomers, believing that they were dealing with 'new' stars, called them novae.

The brightness of novae is not, however, sufficiently high for them to be detected in galaxies outside the Local Group with amateur equipment. Observers in the Southern Hemisphere can search for them in the Magellanic Clouds, where their magnitudes are still within reach. For example, a 10th-magnitude nova was discovered on 1986 December 13 by Robert McNaught, using an 85-mm lens and Tri-X film. The galaxies M31 and M33 may also be monitored, but any nova occurring in either of those systems cannot be expected to exceed magnitude 16, which means that searching must be carried out photographically. In the main, therefore, amateurs

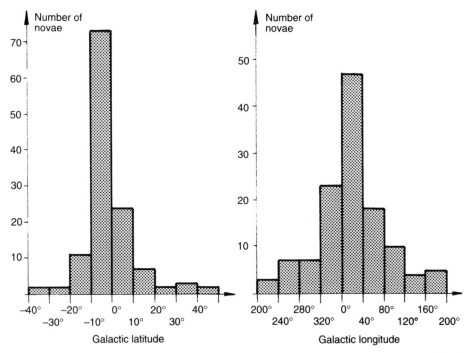

Fig. 15.7. *Distribution of 126 novae brighter than magnitude 10 as a function of galactic latitude and longitude. Sources: J. B. Sidgwick,* Observational Astronomy for Amateurs, *and IAU Telegrams.*

can expect to detect novae occurring in our Galaxy, particularly those stars near the Sun.

Naturally, the probability of a nova appearing in any particular direction is higher the greater the density of stars in that direction. This explains why most novae are found in constellations crossed by the Milky Way, and particularly in the direction of the galactic centre. Figure 15.7 shows the distribution of 126 novae with magnitudes above 10, as a function of galactic latitude and longitude. It clearly shows this concentration along the Milky Way.

The rise is very rapid; just a few hours may elapse between the time when a nova becomes detectable, and when it reaches maximum. The decline is relatively slow (lasting some weeks to some months), and generally shows various irregularities (oscillations, etc.). Figure 15.8 shows typical light-curves of two novae.

15.3.2 The value of discovering novae

A nova outburst is relatively rare (only about ten occur in any one year, *see* Table 15.5). It is therefore important that these events are studied in detail by professional astronomers. Again, the nova has to be detected, and this detection

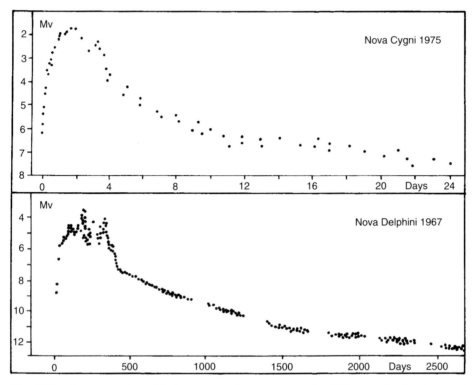

Fig. 15.8. *Two examples of light-curves of novae: note the difference in the time scale.* Top: *a fast nova, Nova Cygni 1975 (V1500 Cyg)*; bottom: *a slow nova, Nova Delphini 1967 (HR Del), whose light-curve showed numerous oscillations after maximum.*

should be as early as possible, so that the nova may be studied on its rise to maximum, or at least from maximum onwards.

Professional astronomers do not carry out any systematic sky patrols designed to detect novae. In most cases it is amateurs who first detect the event; an alert is then issued as quickly as possible so that professional observatories may study the star with more powerful instrumentation.

For an amateur, searching for novae has the attraction of being a search for something new. It is not essential to have sophisticated equipment, nor to be a very experienced astronomer. It can be a fascinating field, even for beginners.

15.3.3 Visual search methods

We have seen that novae may sometimes reach fairly bright magnitudes at maximum. It is therefore possible to carry out visual searches with the naked eye, with binoculars, or with a telescope. The naked eye is limited to about magnitude 6, whereas binoculars will reach magnitudes 9–10, and a small telescope magnitudes 11–12 (*see* Table 14.15, p. 801).

Table 15.4. *Frequency distribution of stars*[*]

m	Range	Number of stars per magnitude	Cumulative total
−1	−1.50 to −0.51	2	2
0	−0.50 to +0.49	7	9
+1	+0.50 to +1.49	13	22
+2	+1.50 to +2.49	71	93
+3	+2.50 to +3.49	192	285
+4	+3.50 to +4.49	625	910
+5	+4.50 to +5.49	1 963	2 873
+6	+5.50 to +6.49	5 606	8 479
+7	+6.50 to +7.49	15 565	24 044
+8	+7.50 to +8.05	21 225	45 269

[*]After Hirschfeld & Sinnott, 1985

The problem is not actually seeing the nova, but in identifying it as such. In checking a star field, it is essential to know whether every star was, or was not, visible on preceding nights. The larger the instrument used, the fainter and therefore the more numerous potentially detectable novae become, but the number of stars to be checked also increases. Table 15.4 shows the distribution of stars as a function of apparent magnitude m, and gives an indication of the problem that we face.

Beginners should choose a region of sky and compare this with a reference chart. This chart may be part of an ordinary commercially available atlas, one drawn by the observer during the first observing session, or a photograph (obtained by the methods described earlier). The reference chart should always be appropriate for the instrument being used; its limiting magnitude and scale should be such that all the stars seen during the course of the search may be identified unambiguously. Figure 15.9 shows the distribution of novae versus discovery magnitude.

The use of a chart while actually examining the sky, however, has two disadvantages. First, each star has to be identified on the chart, which means constantly switching backwards and forwards between the eyepiece and the chart, leading to a significant loss of time. Second, each examination of the chart affects the eyes' dark adaption, even if the chart is only weakly illuminated. This causes a loss of sensitivity to the faintest stars, or an additional loss of time, which may become unacceptable if several minutes have to elapse before the eyes re-adapt.

The time spent in examining each star is a fundamental criterion of the observer's efficiency; the longer it is, the fewer the stars that can be checked, and the lower the probability of finding a nova. Ideally, therefore, one should know the positions of all the stars in the field by heart; reference to the chart is then only necessary in doubtful cases, or if a suspect star is discovered. This is where each observer's experience and visual memory come into play. The British amateur astronomer George Alcock is reputed to know the whole sky down to magnitude 8; checking

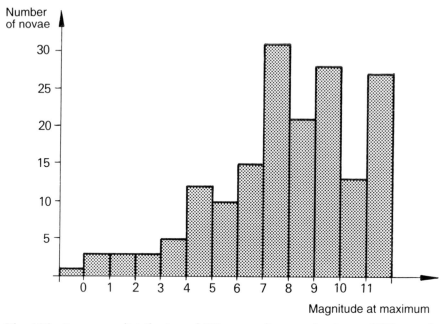

Fig. 15.9. *Frequency distribution of 172 novae discovered prior to 1986 as a function of their magnitude at the time of discovery. Sources: Sidgwick, J. B.,* Observational Astronomy for Amateurs *and IAU Telegrams.*

the sky with a simple pair of binoculars, he has become a renowned nova (and comet) searcher. It will be obvious that experience such as Alcock's cannot be gained overnight.

Another example that may be mentioned is that of the Californian amateur Peter Collins, who has also devoted several thousand hours to memorizing the sky down to magnitude 8. The region of the Milky Way took most of that time. On 1978 September 10, Collins found Nova Cygni 1978, using a pair of 7 × 50 binoculars, but he was anticipated by the Canadian, Warren Morrison (and followed some hours later by the Frenchman Michel Verdenet). Collins finally discovered 'his own' nova (Nova Vulpeculae 1984 No. 2), on 1984 December 22, after another 1500 hours searching with 11 × 80 binoculars, when it was at magnitude 5.5. It was then subsequently discovered independently by another observer, R. Hess. There is a story attached to this discovery: on the night it was made, Collins was at a Christmas party, which he left when the sky cleared. It was just as well he did! It is worth noting that discoverers of novae – and this applies equally to those who find supernovae or comets – are dedicated observers, and do not make discoveries by chance. There is an important moral here for observers! [Collins' most recent discovery is of Nova Cygni 1992. – Trans.]

Because a nova may appear suddenly, it is not a waste of time checking the same area every 4 or 5 days. The best advice for a beginner is not to try to cover too large an area: choose a region that is sufficiently restricted to be swept in a single session, using a chart, and systematically check the same area on every occasion. Visual memory is more effective than is generally supposed, and reference to the chart will rapidly become unnecessary; it is then that the beginner can consider increasing the area covered.

Even if they are not specifically trying to discover novae, variable-star observers become extremely experienced: observing the same variables regularly, they come to know the fields by heart. If a nova (or a comet) appears close to a variable, there is every likelihood that it will be detected immediately by one of the observers studying the star. Searching for novae and variable-star observation are perfectly compatible activities, the tedious nature of the former being offset to some degree by the short-term results provided by the latter.

A good example of the complementary nature of variable stars and novae is given by Michel Verdenet, who is a well-known French amateur astronomer and variable-star observer. On 1978 September 10, Verdenet discovered Nova Cygni 1978, only half a degree from the variable SS Cygni, which he had been observing for 11 years: the presence of an additional 6th-magnitude star in a field that was perfectly familiar to him could not go unnoticed. After checking his discovery, Verdenet immediately alerted professional astronomers, enabling spectra of the nova to be obtained that same night from the Observatoire de Haute Provence.

15.3.4 Photographic search methods

A visual observer examines every star and immediately determines whether it is a new star by comparing what is seen at the eyepiece with a chart or with the memorized field. A photographer examines each star shown on a photograph, determining if it is new by comparison with a reference photograph taken under similar conditions at an earlier date.

The major problem with photography is the fact that checking occurs at a later time: if examination of the photographs taken one night takes place during the next day, for example, one has to wait until the following night (if it is fine) to obtain confirmation of any suspect star with a telescope. This problem may be minimized by adopting two rules:

- systematically obtain two photographs of every field: if a suspect object appears on both, there is every likelihood that it is a star, not a film defect;
- check photographs the following day at the latest. If a nova remains undiscovered on an unchecked photograph for three weeks, either someone else will have discovered it in the meantime, or else it will be too late to start any useful observational work.

The great advantage of photography is that the time required to monitor the sky is much shorter than with visual methods, for equal numbers of stars – despite the fact that time has to be spent in obtaining the photographs and later in checking them. Photographic searching for novae appears to be more efficient than visual searching; it requires less experience, but greater care.

Table 15.5. *Novae and suspected novae: 1980–92 May*

Desig.	Discovery date	Mag.	Discoverer	*	Notes
(SS LMi)	1980 Apr.	15.9	Alksnis, Zacs	P	UG star?
V4065 Sgr	1980 Oct. 28	9.0	Honda	A	
(V1760 Cyg)	1980 Nov. 29	11.8	Honda	A	LPV
(Sct)	1981 Jan. 18	8	Branchett	A	doubtful
V693 CrA	1981 Apr. 02	7.0	Honda	A	
LMC 1981	1981 Sep. 30	12	Wischnjewsky	P	
V1370 Aql	1982 Jan. 27	6	Honda	A	
V4077 Sgr	1982 Oct. 04	8.0	Honda	A	
GQ Mus	1983 Jan. 18	7.2	Liller	P	
V4121 Sgr	1983 Feb. 13	9.5	Wakuda	A	
MU Ser	1983 Feb. 22	7.7	Wakuda	A	1985 Oct.
GW Lib	1983 Aug. 10	9	Gonzalez	P	
UW Tri	1983 Sep. 11	15	Kurochkin	P	
V341 Nor	1983 Sep. 19	9.4	Liller	P	
PW Vul	1984 Jly. 27	6.4	Wakuda	A	
V4092 Sgr	1984 Sep. 25	10.5	Liller	P	
V1378 Aql	1984 Dec. 02	10	Honda	A	
QU Vul	1984 Dec. 18	5.6	Collins	A	
(V344 Nor)	1985 Jan. 26	10.5	Liller	P	LPV?
(V345 Nor)	1985 May 28	9.0	Liller	P	Z And
V960 Sco	1985 Sep. 24	10.5	Liller	P	
(V840 Cen)	1986 Jan. 03	7.5	Liller	P	Z And
V1819 Cyg	1986 Aug. 04	8.7	Wakuda	A	
Sgr	1986 Oct. 28	10.4	McNaught	P	
V842 Cen	1986 Nov. 22	4.6	McNaught	P	
(Lac)	1986 Nov. 22	8	Honda	A	doubtful
OS And	1986 Dec. 05	6.3	Suzuki	A	
SMC 1986	1986 Oct. 05	16	McNaught, Garradd	P	1986 Dec.
V827 Her	1987 Jan. 25	7.5	Sugano, Honda	A	
BW Cir	1987 Feb.	16.9	Makino	P	X-ray Nova
V4135 Sgr	1987 May 18	10	McNaught	P	
LMC 1987	1987 Sep. 21	9.5	Garradd	P	
QV Vul	1987 Nov. 15	7.3	Beckmann, Collins	A	
PQ And	1988 Mar. 21	10.0	McAdam	A	
LMC 1988-1	1988 Mar. 21	11.4	Garradd	P	
V2214 Oph	1988 Apr. 10	8.5	Wakuda	A	
LMC 1988-2	1988 Oct. 12	10.3	Garradd	P	
V977 Sco	1989 Aug. 17	9.0	Liller	P	
V443 Sct	1989 Sep. 20	10.5	Wild	P	

Table 15.5. (cont.) *Novae and suspected novae: 1980–92 May*

Desig.	Discovery date	Mag.	Discoverer	*	Notes
LMC 1990-1	1990 Jan. 16	11.5	Garradd	P	
LMC 1990-2	1990 Feb. 14	11.2	Liller	P	
Vel	1990 Feb. 22		Sunyaev	P	X-ray nova
Sgr	1990 Feb. 23	8.0	Liller	P	
(Pav)	1990 Jly. 21	14.5	Wischnjewsky	P	U Gem
(For)	1990 Oct. 27	12.5	Liller	P	U Gem
Mus	1991 Jan. 12	17	Della Valle, Jarvis	P	
Her	1991 Mar. 24	5.4	Sugano, Alcock	A	
Cen	1991 Apr. 02	8.7	Liller	P	
Oph No.1	1991 Apr. 11	10	Camilleri	A	
Oph No.2	1991 Apr. 11	9.3	Camilleri	A	
Sgr	1991 Jly. 29	8.5	Camilleri	A	
LMC 1991	1991 Apr. 18	12.3	Liller	P	
Sct	1991 Aug. 30	10.5	Camilleri	A	
Pup	1991 Dec. 27	6.4	Camilleri	A	
Sgr	1992 Feb. 13	7.0	Liller, Camilleri	A	
Cyg	1992 Feb. 19	4.3	Collins	A	
Sco	1992 May 22	8.2	Camilleri	A	

* A/P = amateur / professional
LMC = Large Magellanic Cloud; SMC = Small Magellanic Cloud; objects in parentheses proved not to be novae; the discovery date is the date of the first visual or photographic observation that raised the alert.

15.3.4.1 Making an exposure

We require a wide-field, guided, exposure, which may be obtained by mounting the camera piggy-back on an equatorially mounted telescope. Because of the short focal lengths employed, however, it is sometimes possible to avoid having to guide visually, if the mounting is of good quality and is accurately adjusted. An amateur astronomer who aims to carry out a sky patrol is advised to construct an accurate equatorial mounting that may be left in position after careful alignment. This will avoid any necessity for guiding wide-field photographs. This gain is by no means negligible, especially because a sky patrol may require several hours of photography each night.

Some amateurs like to use colour reversal film, because checking is less stressful when using a positive image rather than a negative one. Colour films suffer from some disadvantages, however: their relatively high cost when a large number of photographs have to be taken each night; and the time required for development, which may be incompatible with the essential requirement for rapid checking. Our own preference is for Kodak TP-2415 black-and-white film, which may be developed immediately after the session has finished. Hypersensitized TP-2415 is the best choice, or failing that, the more recent T-Max 400, unless one is prepared to accept a higher limiting magnitude than the equipment is actually capable of reaching. Many nova searchers, particularly the formidable Japanese observers, have used the traditional Tri-X film (*see* Fig. 15.10); but this situation is likely to change very quickly.

The choice of photographic lens will depend on the field to be covered, and the required limiting magnitude. For equal exposure times, most current standard and short telephoto lenses reach approximately the same limiting magnitude, whatever their focal lengths. (This is simply because, for a comparable cost, all such lenses have about the same diameter.) The lower the limiting magnitude, however, the larger the number of stars that will be recorded and, in the area of the Milky Way, these soon reach spectacular numbers. Beyond a certain stellar density, photographs become difficult to check; it is therefore essential to choose a focal length that will ensure that the negative will be easy to examine. On the other hand, long-focus lenses cover smaller fields, which means that more photographs have to be taken to cover the same area of the sky. The choice of focal length therefore depends on the limiting magnitude desired. For example, a 135-mm lens appears to be needed to detect 10th-magnitude stars in the Milky Way, even using such a fine-grained film as Kodak TP-2415. Observers will need to determine the best solution for themselves, taking account of the equipment that they have, and the methods of checking to be used.

Once a lens has been selected, there is an optimum exposure time to reach any given magnitude. This may be determined after a few trials. It will be found that achieving faint magnitudes is costly in time. To gain one magnitude, it will not only be necessary to increase the exposure by a factor of three – experience shows that trebled exposure times are required to lower the limiting magnitude by one whole magnitude – but also to increase the focal length so that the photograph will remain usable. Twice the focal length means that four times the number of photographs have to be taken to cover the same area of sky, so the overall time for the exposures is multiplied by twelve!

Medium format cameras (i.e., those with 4.5 × 6-cm, 6 × 6-cm, or 6 × 7-cm formats) are of benefit because they cover wider fields for a given focal length. Their lenses are, however, generally slower than those available for the 24 × 36-mm format, requiring longer exposures. One gain is offset by another loss. We may mention, however, that the Hasselblad Series F lenses, which fit the Hasselblad 2000 camera back, are of excellent quality and very fast. The best of this range are the 110-mm f/2, the 150-mm f/2.8, and the 250-mm f/4 lenses. Their main disadvantage is their relatively high price. Kodak TP-2415 film exists in 6 × 6-cm format, and is known as TP-6415. Using 6 × 6-cm film does, however, mean a larger hypersensitization chamber and appropriately sized equipment for carrying out checking of the images.

It should also be noted that, in general, the time taken to check the photographs will be considerably longer than that required for the exposure. There is probably little point in attempting to reduce the latter but, on the contrary, one should be prepared to increase the number of photographs by choosing a longer focal-length lens, if this is likely to reduce the time taken for checking.

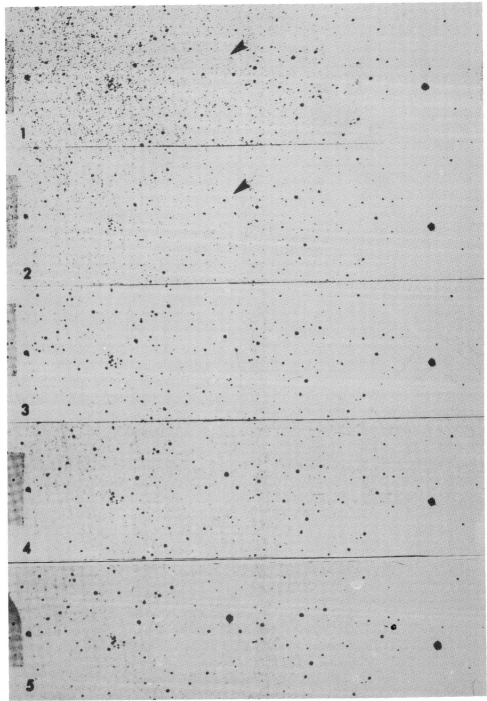

Fig. 15.10. *A series of photographs of the outburst of Nova Cygni 1975 (V1500 Cyg) on 1975 August 30. These photographs were obtained accidentally by Ben Mayer, using Tri-X film with a 35-mm lens at f/3.5.*

15.3.4.2 Checking

The problem with searching for novae is that it is essential to compare every single point on the photograph with those on a reference plate. Nothing distinguishes a nova from an ordinary star. Someone hunting for Earth-grazing minor planets can expect them to leave slight trails on the negative, which distinguishes them from the point images of stars. Similarly, the slightly fuzzy appearance of cometary images are different from those of stars. There is nothing so distinctive about novae. The ideal solution is to have a blink comparator. Amateurs interested in searching for objects such as novae, comets, etc., using photography are advised to build such an instrument, following the advice given in Chap. 16, for example. Even though construction of such a device may appear difficult and take a long time, it will swiftly repay the effort by reducing the time required to examine each photograph.

When using a blink comparator, the reference plate must naturally be a negative of the same region of sky taken with the same equipment, the same film and the same exposure time. It is therefore advisable to obtain a series of reference frames of the areas being covered and to store these carefully.

For an amateur who does not have a blink comparator, one solution is to examine the negative under a linen-counter or similar magnifier, or an eyepiece, in front of an illuminated screen. The best sort of reference material is then a paper print from an earlier photograph of the field. This should be obtained as follows: photograph the field with the film that will be used for the patrol, but with 3 to 4 times the exposure. Using an enlarger, print the negative onto sheet film such as Kodak Kodalith. (On this film the stars will appear white on a black background.) Then use this intermediate positive to obtain a contrasty, large-sized paper print. On this print the stars will appear as dark points against a light background, as on the negative film, which will help comparison. Care needs to be taken in the printing process to ensure that all the stars down to the chosen magnitude limit are definitely present.

Checking fields with a magnifier takes much longer than examining the same negative with a blink comparator. It is, however, much faster than visual searching, simply because it is easier to work in daylight at a desk than in the dark in a dome. There is also no problem with the eye having to re-adapt each time the chart has been consulted.

15.3.5 The areas to be monitored

The techniques that we have described show that the amount of time required for nova searching is roughly proportional to the number of stars examined. However, it is highly unlikely that a nova will be one of the stars normally visible (if it were, it would be extremely bright at maximum!). To simplify the argument, let us assume that all stars have the same absolute magnitude. If, at outburst, a nova gains 10 magnitudes, it is 10 000 times as bright as formerly, and it is thus visible at 100 times the distance. From this we can deduce that the volume of space in which we are able to see a nova is one million times larger than the volume in which we are able to see ordinary stars with the same equipment. Looking along the galactic plane, we can assume that stars are evenly distributed in space and so, for every normal star visible, there are one million in the same direction that might give rise to a detectable nova. On the other hand, when we observe outside the galactic plane, the proportion of distant stars is much lower, because the galactic disk is relatively thin. This is why most novae appear close to the plane of the Milky Way; and this is also why it is not worthwhile searching for novae away from the galactic plane: to have the same probability of discovering a nova, a much larger number of stars must be examined, which will require a far greater amount of time.

Whichever method is chosen, the time required is proportional to the number of stars being examined. Because the time available to any given observer is limited, the area of the sky that may be monitored is smaller, the lower the limiting magnitude. For example, the whole of the Milky Way may be scanned with the naked eye for stars down to magnitude 5, a single constellation with binoculars down to magnitude 7, or a more limited area with a telescope down to magnitude 10. The same applies to photographic methods.

In any given direction, the probability of finding a nova is proportional to the number of stars that are checked, whatever limiting magnitude is set. Nevertheless, any observer who chooses to cover a large area, but only down to a relatively high magnitude, will only see very bright novae, i.e., the novae that will have been discovered previously by hundreds of other observers. For this reason, it would seem to be more effective to search a limited area of sky, but down to as faint a magnitude as possible. Under these circumstances, a single observer cannot cover the whole of the Milky Way, but it is also pointless for several observers to cover the same area. It therefore seems sensible for nova searchers to work in a coordinated group, each person being responsible for a portion of the Milky Way, and the area covered being adjusted according to the amount of time available to each observer.

There is only one major, coordinated group of this sort: the United Kingdom Nova–Supernova Search Programme, led by Guy Hurst, and a collaboration between the magazine *The Astronomer* and the Variable Star Section of the British Astronomical Association (*see* p. 1146 for address). [Despite its name, it is international in its membership and coverage – Trans.] This group has been functioning for a large number of years, and has divided the Milky Way into 120 areas, down to declination $-20°$. Each member of the network is allocated eight areas, four in the summer sky and four in the winter, and these are monitored as frequently as possible, either visually down to magnitude 8, or photographically. Each field corresponds to the area of sky covered by a 135-mm camera lens. Apart from the discovery of novae, this search programme has also revealed completely new variables, and confirmed the variability of other objects listed in the NSV – the catalogue of suspected variables (Kukarkin *et al.*, 1982). Its observers are increasingly linked by electronic mail, and are able to receive almost instantaneous alerts and even charts by this method.

15.3.6 *Suitable times for searching*

As with any work on faint objects, it is preferable to search for novae when the Moon is absent. The Moon does not completely prevent searching, however. To reach the normal limiting magnitude visually, it suffices to use an instrument that is larger in diameter than that normally employed (a small telescope instead of binoculars, for example). In photography, the use of a telephoto lens with a longer focal length and greater f/ratio will give the same result. With both of these methods the area covered will be smaller, and searching will take longer. It is also possible to use the usual equipment, but be content with a higher limiting magnitude. (In photography this will be inevitable, because the greater luminosity of the sky background will demand shorter exposure times.) A visual observer can take advantage of times when the Moon is present to learn new star fields. The Moon is always a nuisance, but it does mean that few people are observing the sky when it is visible; if a nova appears at that time, there is less chance that it will have already have been discovered by another observer.

For the same reason, it is worth checking areas of the Milky Way that are visible in the later part of the night. Unlike comets, novae are not more likely to appear in the morning sky; but the majority of observers work only during the earlier part of the night, so anyone finding a nova in the morning sky is more likely to be the first to spot it.

15.3.7 Detecting a nova

When a 'suspect' is noted, either visually or photographically, it is essential to ensure that it is really a nova before announcing the discovery. If the object is at the limit of detection, it may well be a variable star, the amplitude of which, even if only low, may mean that it is detectable for a period of just a few days. It is helpful to check the positions of all known variables in the field covered and to mark these on the reference chart or exposure.

Another pitfall is created by the hundreds of minor planets that may be seen in amateur instruments. This is a more difficult problem, because these objects move, so it is necessary to obtain ephemerides for every minor planet above the search's limiting magnitude. Naturally, the majority are seen close to the ecliptic.

Only after having confirmed that the star that has been found is not a known variable, a minor planet, or a film defect, should one attempt to announce the discovery to the International Astronomical Union's Central Bureau for Astronomical Telegrams. It is undoubtedly advisable to go via the intermediary of a professional observatory, or of an experienced amateur group, such as that coordinated by Guy Hurst, to obtain previous confirmation. One final precaution is to ensure that it is not a nova that has already been announced in the IAU's *Circulars* and *Telegrams*.

15.3.8 The observation of novae

Amateur astronomers are not restricted to just searching for novae. The observation of known novae is of considerable interest. These are announced in the IAU telegrams, and by alerts issued by various national organisations, such as Guy Hurst's team or the AAVSO. The information provided generally includes the position of the nova, the date of discovery and the estimated magnitude on that date.

Fig. 15.11. *Nova Vulpeculae 1984 No. 1, photographed on 1984 August 30 by Roger Chanal, using a 210-mm reflector at f/6.9; film: hypersensitized TP-2415; Wratten 12 filter; exposure: 1 hour.*

Naturally, observations should begin as early as possible after the announcement of the discovery (Fig. 15.11).

There are two options open: magnitude determinations and spectroscopy. Determining the magnitude of a nova is made in exactly the same way as with any other variable star: using the Argelander (or other visual) method (p. 795), or with a photoelectric photometer (Chap. 19) – for those fortunate enough to possess one. The major difficulty is the absence of calibrated comparison stars in the field of the nova. Atlases such as that by A. Brun or by the AAVSO are then of value in providing comparison stars. For magnitudes below 9.5, it will be necessary to have recourse to the atlases of Selected Areas, such as those published by Vehrenberg and Brun (bearing in mind that these give photographic magnitudes and not visual ones). Because of the rapid changes in a nova, it is important to make estimates every night, or even more frequently around maximum.

Spectroscopy of a nova is identical to that of any other star, and the reader is referred to Chap. 18 on this subject. Observation of the spectrum is, however, only possible with amateur equipment if the star is fairly bright. In addition, once the discovery of a nova has been announced, spectra are obtained by professional observatories, with a quality that is far superior to that obtainable by amateurs.

15.3.9 Flash stars

Monitoring the sky to detect flash stars may be considered as an extension of nova searching. At the beginning of the 1970s, γ-ray bursts were accidentally discovered originating in various areas of the sky. The duration of these events is between a fraction of a second and a few minutes; and it is possible that they are accompanied by visible-light outbursts, detectable with amateur means. Obviously, careful observation of such an event would be of major interest. To this end, four Californian amateurs have fitted their telescopes (between 300 and 350 mm in diameter) with an automatic detection system, based on a photoelectric photometer. A rota is arranged whereby, every fine night, two of these telescopes are pointed at one of the fields where γ-ray bursts are suspected to occur. The photometer is capable of detecting the appearance of any source brighter than 6th-magnitude, over a field 13 arc-minutes in diameter. Comparison between the two telescopes that are operating enables any local, undesirable, events – such as the passage of an aircraft, a satellite, or a meteor across the field – to be eliminated. To date, this search programme has not detected an event, but it would be interesting if other amateur groups were to join in this project (Schwartz, 1986).

15.4 Nebulae, clusters and galaxies

15.4.1 Star clusters

Clusters are of two types: open (or galactic) clusters, and globular clusters. Through a telescope an open cluster appears as a concentration of a few tens to a few hundreds of stars. It is generally clearly resolved, and may be of any shape. A globular cluster, on the other hand, contains several thousand stars, and has a definite, spherical shape with the density decreasing from the centre outwards. In general, however, the stars at the outer edge may be resolved. In fact, the distinction between open and globular clusters is not simply one of form; they are groups of stars with completely different origins and characteristics.

15.4.1.1 Open clusters

Open clusters consist of stars that were born more or less simultaneously from a single primordial nebula lying within a galaxy. This is why they are found close to one another in space. Each of the stars has its own proper motion, however, which depends on the gravitational forces exerted by the other stars. Two opposing effects occur: the gravitational forces between the stars in the cluster tend to hold them together, whereas the tidal forces exerted by the rest of the galaxy tend to disperse them. The cluster is stable only if its density is greater than 0.1 star per cubic parsec. The Pleiades cluster (Fig. 15.12) is just at this limit, while the Hyades cluster, with a density that is only one-quarter of this value, is in the process of dispersing. This explains why the clusters that we observe are relatively young: older clusters have been gradually eroded, and their stars are now dispersed and intermingled with other stars in the Galaxy.

Open clusters are concentrated along the galactic plane or, more precisely, in the

Fig. 15.12. *The Pleiades open cluster: note the nebulosity surrounding the stars. 200-mm reflector, f/6.0; exposure: 1 hour on Tri-X. Photo: B. Fouquet.*

spiral arms. This is why most are visible in the region of the Milky Way. Some, like the Pleiades, appear away from the galactic plane because they are close to us: their angular distance from the galactic plane is considerable even though their actual distance from it is small in comparison with the dimensions of the Galaxy.

A cluster luminosity function is obtained by counting the number of stars in magnitude steps. This enables us to distinguish true clusters, where the stars are close together and therefore at more or less similar distances from the Sun, from fields of stars that are at different distances and therefore not physically connected, but which may, by chance, appear to form a concentration. Generally, the luminosity function of a cluster peaks and then declines, whereas that for any ordinary field of stars shows a continuous increase in numbers towards fainter magnitudes (*see* Fig. 15.13).

The classification of clusters that is most familiar to amateurs is that proposed by Shapley; the different classes are distinguished as a function of the number and concentration of stars:

c : very loose and irregular;	f : fairly rich;
d : loose and poor;	g : very rich and concentrated.
e : moderately rich.	

This classification has now been replaced by one devised by R. J. Trumpler, who introduced a code consisting three symbols: a Roman numeral for the degree of

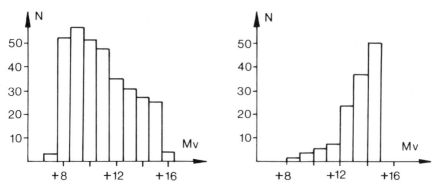

Fig. 15.13. Left: *the luminosity function of an open cluster. The shape of the distribution is caused by the fact that all the stars are at approximately the same distance from us.* Right: *the luminosity function of an arbitrary field of stars. The number increases towards fainter magnitudes, indicating a random distribution of distances. Even though the stars may appear to be more numerous in a particular direction, they do not form a true cluster.*

concentration; an Arabic numeral indicating the range of stellar brightness; and finally a lower-case letter for the cluster's richness. The details are as follows:

- Concentration:

  I   Detached; strong concentration towards centre
  II   Detached; weak concentration towards centre
  III   Detached; no concentration towards centre
  IV   Not well detached from surrounding star field

- Range in brightness:

  1   Small range in brightness
  2   Moderate range in brightness
  3   Large range in brightness

- Richness:

  p   Poor (less than 50 stars)
  m   Moderately rich (50 to 100 stars)
  r   Rich (more than 100 stars).

For example, the clusters h and χ Persei are of type I3r in Trumpler's classification.

15.4.1.2 Globular clusters

Globular clusters are the oldest objects yet detected in the Galaxy: their ages are estimated at about ten thousand million (10^{10}) years. Unlike open clusters, which mainly consist of young, hot stars, globular clusters are rich in cool stars. They contain numerous red giants, which are stars at an advanced evolutionary stage (Fig. 15.14).

Globular clusters are not concentrated towards the galactic plane, but instead

Fig. 15.14. *The finest globular cluster in the sky, ω Centauri. 300-mm reflector, f/6.0; film: hypersensitized TP-2415; exposure: 20 minutes. Photo: P. Martinez.*

form a spherical halo around the galactic centre. Numerous globular clusters have been detected in neighbouring galaxies: more than 300 in the Andromeda Galaxy, and about 6000 around M87 in Virgo.

The classification of globular clusters was also proposed by Shapley. It consists of 12 classes, where class 1 indicates the most concentrated cluster, and class 12 the least concentrated.

15.4.2 Nebulae

Nebulae are clouds of dust and gas. Some consist of material lying in the galactic plane: these are clouds of primordial gas that have not yet condensed into stars, and are generally enriched in heavy elements that have been introduced by the explosion of old stars. Other nebulae consist of shells of gas ejected by certain stars; these are the planetary nebulae and supernova remnants.

Planetary nebulae were given their name by Sir William Herschel in 1785, be-

Fig. 15.15. *One of the most famous dark nebulae, the 'Horsehead' in Orion. 406-mm reflector, F = 2135 mm; film: 103aE with a red filter; exposure: 75 minutes. Photo D. Cardoën.*

cause of their distinct, circular form, which resembles a planetary disk. Amateur astronomers tend to call all other bright nebulae 'diffuse nebulae', but this term includes different sorts of object. The following description differentiates between the various classes on the basis of the process by which they were formed.

15.4.2.1 Dark nebulae

These are vast clouds of dust, which probably consist of grains of graphite some $10\,\mu m$ in diameter that have been formed in the cool atmospheres of red giants. These clouds are not illuminated by nearby stars. They are completely opaque, and are only visible because they conceal the stars that lie behind them (Fig. 15.15).

Although the density of material in these clouds is extremely low, they are so large that light passing through them is scattered and absorbed to a very considerable extent. (Absorptions of more than 50 magnitudes have been determined in some cases.) Scattering of light is more significant at short wavelengths than at longer ones, so any stars that are visible through these clouds appear fainter and reddened.

Some very large dark nebulae may be detected with the naked eye. The most famous is the 'Coalsack' in the constellation of Crux, visible from the Southern Hemisphere. The Great Rift, which divides the Milky Way in Cygnus and Aquila, is also caused by an enormous dust cloud.

These nebulae are concentrated along the galactic plane; it is because of them that we are unable to see the galactic centre. It is also thought that they obscured a supernova explosion that occurred in 1668 (± 8 years), which produced the powerful radio source Cassiopeia A.

15.4.2.2 Reflection nebulae
When a nebula exists close to a star it reflects some of the light from the latter, and may therefore appear bright against the sky background. These nebulae contain more gas than dust. It is estimated that on average there are 10 grammes of dust for every kilogramme of gas, all within a volume of one million cubic kilometres. The dust scatters light far more than the gas, however, and is therefore responsible for the visibility of reflection nebulae, just as its absorption causes dark nebulae.

The spectrum of a reflection nebula is naturally the same as that of the star illuminating it, but because scattering is greater at short wavelengths, the nebula may appear blue by comparison with the star. For just the same reason, sunlight scattered in the Earth's atmosphere makes the sky appear blue, whereas when the Sun is low it appears red, because the considerable thickness of air that it has to traverse preferentially scatters short-wavelength light. One of the best-known reflection nebulae is that surrounding the stars in the Pleiades cluster (Fig 15.12).

15.4.2.3 Emission nebulae
When an atom is excited, either by a photon or by collision with another atom, it may lose one of its electrons; it is then said to be ionised. Subsequently, the ionised atom may capture an electron, which releases its excess energy in the form of radiation. Because electrons bound to an atom may occupy only specific, discrete energy levels, the wavelength of radiation emitted by an electron in jumping from one energy level to another, lower one, may only take specific values, depending on the element concerned. This gives rise to specific emission lines in the spectrum.

H II regions   A hot star emits a large flux of ultraviolet light. When a gaseous nebula lies close to such a star, its atoms are ionized by the ultraviolet radiation, and may then re-emit light by the process just described.

The state of an atom is indicated by a Roman numeral: I when the atom is neutral (not ionized); II when it is singly ionized (having lost one electron); III when doubly ionized (with the loss of two electrons); etc. Hydrogen has only one electron, so it can exist in the H I and H II states only. Similarly, doubly ionized oxygen is designated O III, etc.

Nebulae are rich in hydrogen, which is the most abundant element in the universe; nebulae exposed to ultraviolet radiation from nearby stars are therefore rich in ionized hydrogen H II, hence their name of H II regions. Their spectra show characteristic hydrogen lines, known as the 'Balmer series', in the visible, particularly the Hα line in the red at 656.3 nm; this line is responsible for the typical red colour of H II regions. Naturally, the spectrum also includes a certain number of lines characteristic of other elements (see Chap. 18).

H II regions are stellar nurseries: the nebulae fragment into clumps of gas, which condense, giving rise to stars. These young, very hot stars frequently illuminate the remnants of the nebulae that surround them (Fig. 15.16).

The H II regions in our Galaxy lie within the spiral arms, and the most famous is M42 in Orion. It is possible to detect H II regions in nearby spiral galaxies using amateur equipment (*see* Fig. 15.20). The Large Magellanic Cloud contains a giant, very spectacular, H II region, known as the 'Tarantula Nebula'.

Planetary nebulae Planetary nebulae generally have very small angular diameters, and often have a strongly symmetrical, sharply defined, circular form. (This is in contrast to diffuse nebulae, which are irregular in shape, and the outer edges of which appear to fade gradually into the background sky.) Many planetary nebulae appear as rings, like M57, the famous 'Ring Nebula' in Lyra, indicating that the density is greater at the edge than around the central star (Fig. 15.17).

These objects arise when, at the end of their lives and having exhausted their supplies of hydrogen, the outer layers of stars expand and cool, while the inner core collapses and heats up until helium fusion begins. Some stars go so far as to eject their outer layers, creating the expanding shells that we see as planetary nebulae. The central star that remains is a hot star of spectral class W or O, which radiates a large flux of ultraviolet light and thus excites the nebula. If the star is very hot (100 000 K), it emits nearly all its radiation in the ultraviolet, and may be extremely faint in the visible region. The nebula produces an emission spectrum, which naturally contains the Balmer series of hydrogen lines, as well as lines caused by O III (doubly ionized oxygen), N II (singly ionized nitrogen), Ne III and Ne IV (neon that has lost two or three electrons), etc. In general, planetary nebulae appear greenish to the eye, because of the O III doublet at 500.7 and 4959 nm.

Planetary nebulae are classified using a scheme proposed by Vorontsov-Velyaminov:

1 stellar appearance

2a smooth disk, brighter towards centre

2b smooth disk, uniform brightness

2c smooth disk, traces of ring structure

3a irregular disk, very irregular brightness distribution

3b irregular disk, traces of ring structure

4 ring structure

5 irregular shape similar to a diffuse nebula

6 anomalous shape

M57 is classed as type 4, for example, while M27 (the Dumbbell) is 3a.

Fig. 15.16. *An example of an H II region: M8 in Sagittarius. Lichtenknecker camera, 190-mm aperture, 760-mm focal length; hypersensitized TP-2415 film; 20-minute exposure. Photo: C. Viladrich.*

Fig. 15.17. *The Helix planetary nebula. Exposure: 30 minutes on hypersensitized TP-2415 film, using a Schmidt camera, 300-mm focal length, f/1.5.*
Photo: Société Populaire de Poitiers.

Supernova remnants (SNR) When a supernova explodes, it ejects large amounts of material at high velocities, and this forms a rapidly expanding shell, which is a sort of 'super planetary nebula'. Any central star that may remain after the explosion soon becomes incapable of exciting the gas. The kinetic energy of the ejected material is sufficiently high, however, for collisions with interstellar matter to cause it to glow (Fig. 15.18). This is the case with the Cygnus Loop, which is the still-visible remnant of a supernova that has long disappeared.

M1 (the Crab Nebula) in Taurus (*see* Fig. 5.36) is an unusual type of supernova remnant, undoubtedly because of its extreme youth. (The explosion that gave rise to it was observed by the Chinese in 1054.) As well as the classical line spectrum of an emission nebula, there is a superimposed continuum spectrum emitted by spiralling electrons through the synchrotron effect. In addition, the parent star has become a pulsar.

Herbig–Haro objects These objects, which are relatively poorly known by amateurs, were discovered in 1950 by the two astronomers whose names they bear, but were

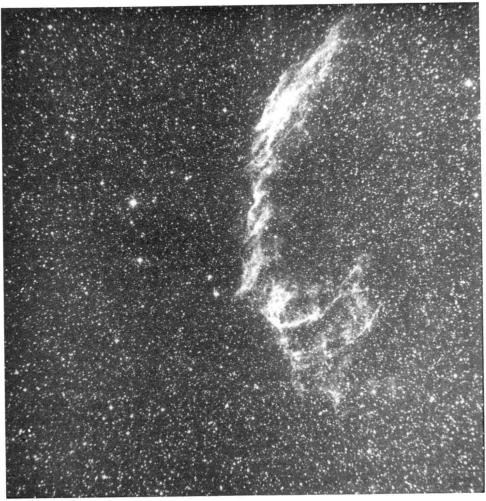

Fig. 15.18. *NGC 6992-5, a supernova remnant, taken using a 406-mm, Newtonian reflector, on 103aF film; exposure: 40 minutes. Photo: D. Cardoën.*

not explained until 1976. They are collision-excited emission nebulae that are not related to any visible star. In fact, they are caused by a strong stellar wind that is produced by a star hidden in a dark nebula. The dust in the nebula prevents us from seeing the parent star, but the collision between the stellar wind and the nebular material causes the radiation that we see. Herbig–Haro objects are too faint to be seen visually, even with large telescopes.

15.4.3 Galaxies

Galaxies are gravitationally bound concentrations of stars (several hundred thousand million in the case of giant systems), gas and dust. Although when seen through small telescopes they resemble nebulae, they are replicas of our own Galaxy. In those that are closest to us, sufficiently powerful telescopes are able to detect star clusters, H II regions, variable stars, occasional novae and supernovae, and sometimes the characteristic structure of the galaxy concerned (i.e., the nucleus and the spiral arms).

15.4.3.1 Classification of galaxies

In 1926, the American astronomer Edwin Hubble classified galaxies in three major categories according to their shape:

E   elliptical galaxies. These are galaxies that, apart from their shape, characteristically lack gas and consist of old stars. They are sub-divided into eight sub-classes designated E0 to E7 on the basis of their apparent flattening. (If a and b are, respectively, the apparent major and minor axes, then the nearest integer value to $10(a - b)/a$ gives the sub-class. For example, if $a = 50$, and $b = 35$, the galaxy would be classified as E3.) Galaxies of class E0 appear spherical, and those classified E7 are the most flattened.

S and SB   spiral galaxies. These consist of a central nucleus and spiral arms that encircle the nucleus. Sometimes these arms appear to arise from the ends of a straight bar that crosses the nucleus: the galaxy is then called a 'barred spiral', and designated SB. The S and SB classes are subdivided into three classes: a, b, and c. Types Sa and SBa have prominent central bulges and tightly wound arms; Sc and SBc, on the other hand, have loose spiral arms and inconspicuous nuclei (Fig. 15.19). Sb and SBb are intermediate between the two extremes. In spiral galaxies the nucleus consists mainly of old stars, whereas the arms are rich in young stars and H II regions, in which new stars are being born.

Ir   irregular galaxies, which cannot be classified in one of the classes just described.

The S0 category was introduced in 1936 as a designation for lenticular galaxies similar in shape to spiral galaxies, which may contain clouds of dust, but are also rich in old stars. Hubble's classification was extended subsequently by Sandage and then by de Vaucouleurs; but it would require too much space to describe these modifications in detail here.

Other classification systems were proposed in the second half of the 20th century, either based on the shape of galaxies (Vorontsov-Velyaminov), or on their luminosity (Morgan and Mayall; Van den Bergh). Catalogues have also been compiled of specific categories of galaxy; Markarian galaxies, which emit a lot of ultraviolet radiation; compact galaxies (by Zwicky); interacting galaxies (by Vorontsov-Velyaminov); etc.

15.4.3.2 Distribution of galaxies

A glance at an atlas shows that few galaxies are to be seen in the region of the Milky Way. This is not caused by a lack of galaxies in that region, but simply because they are hidden by the clouds of dust that exist in the plane of our own Galaxy.

Away from the Milky Way, the distribution of bright galaxies is not homogeneous:

Fig. 15.19. *The Whirlpool Galaxy, Messier 51 (NGC 5194), an Sc-type galaxy. Licht-enknecker Flat-Field Camera, 190-mm aperture, 760-mm focal length. Exposure: 40 minutes on hypersensitized Kodak TP-2415. Photo: C. Viladrich.*

some constellations contain particularly large numbers of them (Virgo, Leo, Coma Berenices, Ursa Major, etc.). Galaxies are in fact concentrated into clusters, which are themselves grouped into superclusters. Recent studies appear to show that superclusters are the largest structures that exist; on a still larger scale, the universe appears to be perfectly isotropic.

Our Galaxy belongs to the Local Group, which contains some thirty members, the most spectacular of which are the Magellanic Clouds (satellites of our Galaxy), the

883

Andromeda Galaxy, M31, with its two satellites, M32 and NGC 205 (*see* Fig 15.22), and M33 in Triangulum (Fig. 15.20).

The nearest cluster is the Virgo Cluster; it contains more than 200 galaxies, many of which are visible with amateur-sized instruments. It covers an area of sky that is about 12° across.

15.4.4 Observation of diffuse objects

15.4.4.1 Location and detection

The first criterion that comes to mind when considering how difficult it may be to observe a faint object is its magnitude. It is essential to remember, however, that the magnitude listed in catalogues corresponds to the light emitted by the whole of the object concerned; the apparent area of such an extended object may vary considerably. If the area is large, the light may be so diffuse that the galaxy is difficult to see. An extreme example of this is M33 in Triangulum, which has a visual magnitude of 5.8 (it is one of the brightest galaxies), but which covers an area of 60′ × 40′ (Fig. 15.20). M33 is relatively difficult to detect with a small telescope, whereas 10th-magnitude galaxies that are only a few minutes of arc across pose no problem.

As a result, the concept of surface brightness has been devised, i.e., the luminosity per unit area. This is a better indication of the ease with which an object may be observed. Catalogues generally compare galaxies to ellipses, for which they give the values of the major and minor axes (which we will denote A and B, respectively), expressed in minutes of arc. If m is the galaxy's magnitude, its surface brightness M, expressed in magnitudes per square arc-minute, may be calculated easily from:

$$M = m + 2.5 \log(A \cdot B) - 0.26.$$

For example, although large, M33 has a surface brightness of $M = 14.0$, whereas a neighbouring small galaxy, NGC 672 (magnitude 10.9; size 3.5 × 2.0 arc-minutes), has a value of $M = 12.8$; the latter is thus easier to see!

The equation just given may also be applied to objects with spherical symmetry, such as globular clusters; it then suffices to take $A = B =$ diameter of the cluster. It may also be extended to nebulae for which catalogues often give approximate dimensions. Because these objects are irregular, however, the result is only approximate.

It must also be remembered that surface brightness is an average value that would apply only if the objects had an overall, uniform, luminosity distribution, or if their luminosity always varied in the same way between the centre and edge. But this is not the case. Sa galaxies, for example, have a central bulge that is far more luminous than the arms. Yet the size quoted in catalogues is that of the whole galaxy, not that of the central bulge. The nucleus of the galaxy will therefore be brighter than the surface brightness would lead one to expect, whereas the arms will generally be invisible in a small telescope. On the other hand, Sc galaxies (such as M33) have a luminosity that is distributed more evenly between nucleus and arms. Surface brightness is therefore a better indicator of their visibility.

Another factor that determines the visibility of an object is its contrast relative

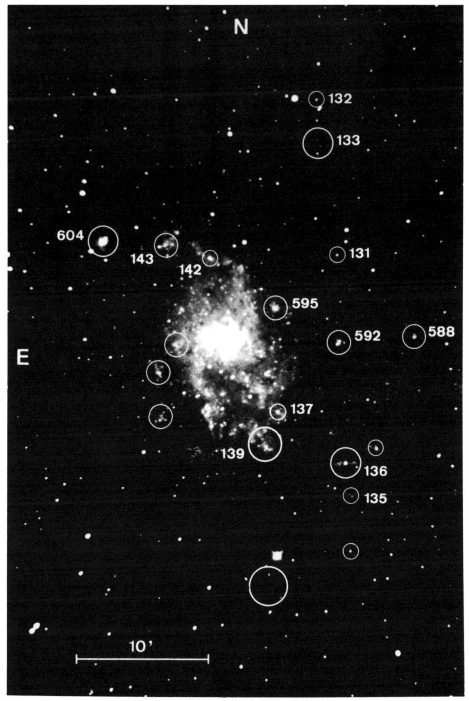

Fig. 15.20. *The spiral galaxy M33 in Triangulum. Newtonian reflector, 260-mm aperture, f/4.6; exposure 45 minutes on hypersensitized TP-2415 film. The circles indicate the H II regions scattered throughout the galaxy. Photo: S. Bertorello.*

to the background sky. A spiral galaxy seen side-on has relatively sharp edges, and therefore has a fairly high contrast (Fig. 15.21). On the other hand, an identical galaxy, seen face-on, has arms whose luminosity decreases from the centre towards the edge. Although the latter is far more impressive when photographed with a large telescope than a galaxy seen side-on, is likely to be far more difficult to see visually.

Stray light (urban lighting or moonlight) tends to increase the brightness of the sky background. Because of the resulting loss of contrast, the visibility of diffuse objects is rapidly effected. The neighbourhood of large towns and cities must therefore be avoided. For some years it has been possible to buy filters that are designed to remove wavelengths characteristic of urban lighting (primarily sodium and mercury vapour lines). The results are sometimes very spectacular, but it remains preferable to find an observing site that is well away from sources of light pollution. This requirement for having the darkest possible sky background also means that the observation of faint objects is impossible when the Moon is present, except in emergency, such as when a supernova erupts, for example.

Planetary nebulae generally have small angular diameters (about 1 arc-minute or less) and their images have sharp edges. They are therefore often fairly easy objects to see, even when their apparent magnitude is low. With low magnifications, some planetary nebulae, whose diameters do not exceed a few arc-seconds, appear stellar. They may, however, be distinguished from neighbouring stars by the fact that they tend not to scintillate, because their apparent diameters are greater than those of point sources. (For the same reason, a planet scintillates much less than a star to the naked eye.)

Identification of small, bright planetary nebulae may be made easier by the use of a prism (or a grating), placed between the eyepiece and the eye. The prism disperses light according to its wavelength. Stars show extended images in the form of a small streak of light running from violet to red. Planetary nebulae, on the other hand, have a distinct emission-line spectrum, which appears as a series of individual images, each of which corresponds to an emission line, and thus to a different colour. This technique may be applied to any emission nebulae. If the angular dimensions are too great, however – which is often the case with H II regions – the individual monochromatic images are superimposed on one another, and it is more difficult to distinguish the spectrum from a continuous stellar spectrum.

Reflection nebulae are more difficult to observe. They are always close to the bright star that is the source of their light, but which also dazzles the observer. The visibility may be improved by interposing a small mask, designed to hide the offending star, in the focal plane of the eyepiece. The presence of a dark nebula may be detected easily when the bright background shows a sharp edge. Normally the edges of diffuse nebulae and star-clouds are fairly indistinct.

Open clusters, on the other hand, are often difficult to identify when they are set against a rich background. This is definitely the case with the numerous clusters that occur throughout the region of the Milky Way.

To locate a diffuse object in a small telescope, it is advisable to plot the object's position on a chart beforehand, and to identify the field stars. It is generally easier to locate the surrounding stars than it is to find the object itself. In addition, they

Fig. 15.21. *The Sc-type galaxy NGC 253, in Sculptor. 406-mm reflector; hypersensitized TP-2415 film; exposure: 30 minutes. Photo: D. Cardoën.*

Fig. 15.22. *M31 in Andromeda, with its two satellite galaxies M32 and NGC 205. Schmidt telescope, 165-mm aperture, 394-mm focal length; exposure: 30 minutes on Tri-X. Photo: B. Fouquet.*

enable one to identify its position unambiguously. If the telescope is not provided with setting circles it is only too easy to point the telescope slightly to one side (even with circles, errors are still possible), and in a small telescope, one galaxy looks much like another.

15.4.4.2 *Visual observation*

Observation of faint objects requires the largest possible telescope. Newtonian reflectors are without doubt best for this field of study, because they combine large aperture with moderate cost. Deep-sky observation has even spawned a new type of telescope: the Dobsonian (named after the American amateur who invented the design). This is, in effect, the fastest and lightest Newtonian possible. The mirrors should be good, but optical quality is considered secondary to aperture. The mounting is a simplified altazimuth. It is impossible to take photographs with such an instrument, but its light-grasp and convenient handling make it very effective for visual observation of diffuse objects.

In a few cases, where the object is relatively bright and requires good resolution (such as a bright planetary nebula, a compact open cluster, or a globular cluster), use of a refractor may be of advantage.

In general, it is best to use the lowest possible magnification, i.e., the equipupilary magnification (*see below*), especially when the object is extended (as with diffuse nebulae). On the other hand, it is easiest to study stars (open clusters, for example) by using the magnification giving the best resolution ($D/2$, where D is measured in mm), which provides a better contrast relative to the sky background. An interesting experiment is to count the number of stars visible in a small area of sky with equipupilary magnification: then use the higher, optimum-resolution magnification and repeat the count for the same area. It will be found that the latter allows the detection of faint stars that were not visible with the equipupilary magnification. The explanation is simple: for magnifications below the resolving magnification, the eye sees stars as point sources, and their brightness is independent of magnification; the brightness of sky background, on the other hand, decreases with increasing magnification. For stars, the greatest contrast therefore occurs with the magnification that gives optimum resolution. For objects that are almost point sources (small planetary nebulae, H II regions in nearby galaxies, etc.) one needs to find the most suitable magnification, which will be between the two extremes.

We may wonder how useful amateur visual observations of faint objects are. The most honest response is to recognize that the principal gain is simply the observer's own pleasure in seeing these beautiful objects through the eyepiece. Apart from studying cluster variables [RR-Lyrae stars – Trans.], nebular variables, or supernovae, this sort of observation is of no value in advancing astronomy. It does, however, have the great advantage that it teaches the observer's eye to detect fine, very faint detail that is low in contrast. Such training is invaluable when it comes to observing a comet. One can also make drawings of nebulae (Fig. 15.23), galaxies or star-fields (open clusters). Again, the value of this is discovered when, at a later date, one attempts to draw a comet. The advantage of using a nebula rather than a comet for this preliminary training is that the observer can subsequently check the work by comparing the drawing with a photograph taken with a large telescope. The techniques required for drawing faint objects have been fully described in the chapter on comets, so they will not be discussed here.

Equipupilary magnification   By means of geometrical optics it is possible to show that all the light collected by the objective of a telescope leaves the eyepiece through a circular aperture, known as the exit pupil. Its diameter d may be derived from $M = D/d$, where D is the diameter of the objective and M the magnification employed. Obviously d decreases as M increases. It is essential that d should not exceed the diameter of the observer's pupil, otherwise a portion of the light collected by the telescope cannot enter the eye and is lost. The lowest possible magnification, known as the equipupilary magnification (because the exit pupil is equal to the eye's entrance pupil), is $M_{eq} = D/7$, where D is expressed in millimetres.

15.4.4.3 Faint-object photography

Here we are not concerned with wide-field photography (which does not present many problems and which has been discussed in Sect. 15.2), but with photography of small, faint objects, which are photographed at the telescopic prime focus (Fig. 15.24) or with a camera that has a focal length of several hundred millimetres.

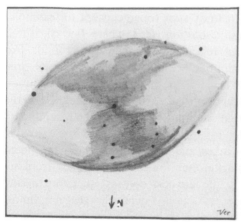

Fig. 15.23. *The planetary nebula M27 drawn (*left*) by M. Verdenet (530-mm aperture reflector, magnification 250×, and photographed (*right*) by R. Chanal (410-mm reflector, f/4.8; exposure 15 minutes on hypersensitized TP-2415 film).*

The problems encountered in all astronomical photography (mounting of the optics, focussing techniques, vibration, development of the film, printing, etc.) will not be considered here for want of space. The reader is referred to the various specialized books that cover these aspects of the work. It may be useful, however, to discuss two specific techniques employed in photographing faint objects that amateurs should master to obtain good results: hypersensitization of films and methods of guiding.

Films and hypersensitization For black-and-white work, the situation is relatively clear: Kodak TP-2415 is the obvious solution because of its fine-grain images, its high contrast (when it is developed in D19b or HC 110 at low dilution), and its extended colour sensitivity. In addition, this film is now widely available from ordinary commercial sources. The problem is that TP-2415's sensitivity, which is only moderate at short exposures, is very low over the long exposure times that are required to photograph faint objects. This is a problem that affects all, or almost all, photographic emulsions and which is known either as the Schwarzschild effect or as reciprocity failure.

With TP-2415, the most effective remedy is to hypersensitize the film with forming gas. This treatment consists of soaking the film in the gas (a mixture of hydrogen – usually 8 % – and nitrogen) whilst heating it. The efficiency of the process depends on the duration, temperature, pressure, and flow rate (or the frequency with which the gas is changed). Everyone has their own solution: by way of example we soak the film for 24 hours at 60°C, at a pressure of 1 bar, replacing the gas every 6 hours.

The equipment required naturally includes a gas- and light-tight chamber, within which the film itself is normally wound onto a spiral from a developing tank. The chamber has to be evacuated before introducing the gas, so a suitable pump is

Fig. 15.24. *The H II region surrounding the star η Carinae. 200-mm reflector, f/6.0; exposure: 45 minutes on hypersensitized TP-2415. Photo: P. Martinez and N. Née.*

Fig. 15.25. *The Lumicon hypersensitizing equipment. Note the chamber, fitted with a pressure gauge and a thermometer, the bottle of gas, and the manual vacuum pump. The small box contains the temperature control.*

required, and a pressure gauge. The chamber is surrounded by an electrical heating element. Some people prefer to use resistance heating inside the chamber itself to bake the film, but this does run the slight risk of producing a spark in the hydrogen atmosphere. The temperature needs to be checked with a thermometer, but use of a temperature control will avoid the necessity for constant attention throughout the soaking period. There are various complete commercial designs, such as the Lumicon equipment (Fig. 15.25). Alternatively, one can make up one's own form of 'pressure cooker'. The gas may be purchased from specialist firms. Failing that, it is worth heating the film in a vacuum, which will remove oxygen from the film that otherwise has a deleterious effect on sensitivity during long exposures.

In the absence of any means of carrying out sensitometry, it is possible to check the effectiveness of the hypersensitization by examining the density of the background fogging of the film. (Cut a small piece of treated film and develop it without exposing it to any light.) The fogging should have a density of 0.3; if it is less, treatment should be extended. The test film may be compared with a Wratten gelatine filter with a density of 0.3.

One major question is that of storing treated film. It is accepted nowadays that the gain from hypersensitization disappears quickly. This is true if one wants to gain the maximum benefit. Trials that we have made with TP-2415 show that about 20 %

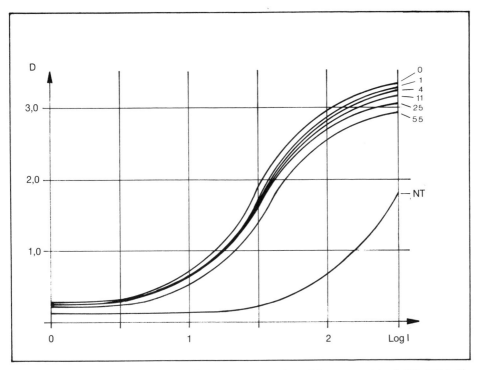

Fig. 15.26. *Characteristic curves for various samples of hypersensitized TP-2415 film. The faster the film, the farther its curve lies towards the left of the diagram (the weaker the degree of illumination required to obtain a given density). The numbers against each curve indicate the storage time (in days) at 20°C, between the film's treatment and use; the curve marked NT is that of untreated film. Note the significant gain attained by hypersensitization, and the good storage properties of TP-2415.*

of the gain from hypersensitization is lost in the first 24 hours, but that the decrease then tends to stabilize (Fig. 15.26). There is an overall loss of about 30% after about 1.5 months. But treatment gave an initial gain in sensitivity of 10 times, for a 40-minute exposure. Trials have also shown that the loss of sensitivity is reduced if the treated film is stored at low temperature (−5°C in our case) and in a vacuum. By contrast, storing the film at atmospheric pressure at the same temperature gave a worse result, probably because water vapour condensed on the emulsion.

We would strongly recommend all amateurs who are interested in photographing faint objects to hypersensitize their films. Alternatives do exist, but they are far less satisfactory. The best consists of using Kodak 103a emulsions, which are partially corrected for reciprocity effect. These emulsions have only mediocre resolution, however, which is not comparable with the fine grain of TP-2415; in addition they are generally much more difficult to obtain. Other common films may be used, such as the old stand-by Tri-X, or its replacement T-max 400, which is to be preferred because it gives far better definition. Using TP-2415 that has not been hypersensitized is unfortunately possible only with very fast optics, such as those in Schmidt cameras.

The greatest detection efficiency (the best signal-to-noise ratio) with TP-2415 is obtained when the exposure is such that the background sky fogging has a density of unity, which may be checked by comparing the developed negative with a Wratten neutral-density filter with that value. Do not worry if the film appears very dense (it only transmits 10 % of incident light). Even with hypersensitized TP-2415, such a density is difficult to attain. For example, with a relatively dark sky, 30 minutes were needed using f/2 optics. It is not possible to attain a background sky density of unity with Newtonian telescopes, unless there is light pollution, when it would be better not to take the photograph at all.

With colour films, the rapid progress made by various manufacturers in the last decade means that any advice is likely to become out-of-date very rapidly. Among the reversal films available at the end of the 1980s, the most interesting for astronomical photography are Agfachrome 1000 and Fujichrome 1600. Curiously, hypersensitization with forming gas does not give good results with colour-reversal films. There is a significant loss of contrast for a relatively low gain in sensitivity. The best way of combating reciprocity failure is to cool the film. One method of doing this is to modify a camera body so that a metal chamber filled with dry ice may be pressed against the back of the film. Apart from the problems of ensuring that the device is light-tight, some way has to be found to prevent condensation from forming on the emulsion surface. Dry air or nitrogen may be blown across it, or the emulsion may be placed in contact with a block of transparent acrylic plastic. (The low thermal conductivity of the plastic ensures that the opposite face, through which the light enters, will remain warm enough to prevent the formation of condensation. Care must be taken, however, because the focus will be shifted outward as the light crosses the acrylic plastic.) A more ambitious scheme is build a small vacuum chamber, with the film on one side and a transparent window on the other. A number of amateurs have made up such devices, and the results have been commensurate with the amount of effort that has been expended. The Celestron company in the U.S.A. also markets a cold camera of this sort.

Some photographers prefer to use fast negative films (such as Kodacolor VR 400, Konica SR 400 and SR 1600), which appear to behave better than reversal films when hypersensitized (Iburg, 1987). [Three comparatively new films, Ektar 1000, Kodacolor Gold 1600, and Ektapress 1600 also give good results, with the first seeming to give the best compromise between speed and grain size (Quirk, 1991). – Trans.] Another possibility, which increases contrast, is to develop reversal films as negatives.

Colour photography provides attractive images, but it is not suitable for obtaining scientifically useful results, especially when applied by amateurs. It must be borne in mind that the colours reproduced are often far from the reality. A colour film consists of three different layers, each sensitive to a single colour: blue, green, or red. Each layer has a different reciprocity effect, so the colour balance of the film is falsified with long exposures. In addition, colour films are designed to give realistic images of ordinary subjects, where the lighting is sunlight, incandescent light, or flash – i.e., lighting with a continuous spectrum. If an emission nebula is photographed,

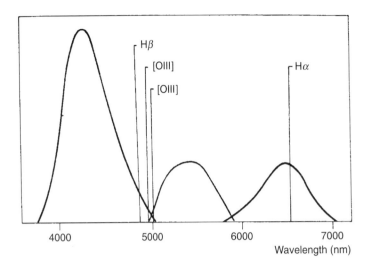

Fig. 15.27. *Spectral sensitivity of the three sensitive layers in a colour film, compared with the main, nebular emission lines.*

the Hα line will affect the red-sensitive emulsion, but the final photograph will show a red colour that is determined by whichever red dye is used by Kodak, Fuji, etc., but which will have no resemblance to that of the Hα line at 656.3 nm. Worse: the green lines of O III and Hβ, both close to 500 nm, lie at a point on the spectrum that is at the very edge of the sensitivity of the blue and green emulsions, and where colour films have practically no response (Fig. 15.27).

Guiding It is essential to have a driven, equatorial mounting for taking long-exposure photographs at the focus of a telescope or even with a telephoto lens. It is not enough to have just a drive, however, because drift occurs for various reasons, and the observer needs to constantly monitor the accuracy of the tracking and make corrections in both right ascension and declination whenever necessary. Corrections are made using a variable frequency generator (when the drive is by means of a synchronous motor), a correction motor, or a mechanical slow motion.

We shall not discuss the different methods of moving the telescope; these are directly related to the instrument's design. We shall confine our attention to the equipment and methods that are required for the observer to detect guiding errors.

An absolutely essential piece of equipment is an eyepiece with an illuminated reticle. There are two types: the graticule may consist of cross-hairs, or of lines engraved on a glass plate. Both are held at the focal plane of the eyepiece. Cemented cross-hairs are very fragile; great care needs to be taken not to damage them when handling the eyepiece. The glass plate has the advantage that it may be specially engraved. (It may, for example, have a graduated scale that may be used for making measurements.) It does, however, involve an additional loss of light and (above all), the reflections it creates and the dust that may collect on its surfaces may interfere with the visibility of the guide star.

The focal length of guiding eyepieces is frequently around 12 mm, which is very

Fig. 15.28. *The Rosette Nebula, taken with a Schmidt camera, 300-mm aperture, 620-mm focal length; film: Kodak TP-2415 hypersensitized with forming gas; Wratten 25, red filter. Photo: M. Leroux, J. F. Boyer and D. Albanèse.*

suitable. Choose one with the finest possible engraved lines. To obtain accurate guiding it is necessary to make every effort to bisect the diffraction disk of the star with the reticle's cross-hairs throughout the exposure. If the lines are too thick, they will interfere with the visibility of the star and will cause deceased accuracy. Double cross-hairs or a glass plate on which several lines are engraved in both directions are very useful, because they provide an image scale. This gives one some idea of any accidental deviations or, when tracking a comet, enables one to offset the guide star by a previously calculated amount at specific, regular intervals to compensate for the comet's motion. Avoid graticules with large numbers of unnecessary circles or rectangular grids: the glare from them affects the eye and tends to mask faint stars, even when the illumination is at its lowest setting. The manufacturers claim that the shift will not be detectable on the film if the star remains within certain specific limits. This applies only if the focal length of the photographic instrument is very small in comparison with that of the guide telescope (using a camera body mounted piggy-back, for example). In photographing at prime focus, which is what interests us, it is essential that the star stays on the cross-hairs, and this is, moreover, more effective. It is a good habit to orient the cross-hairs so that they are parallel to the right-ascension and declination axes, and also to check the direction of motion of the star when a control button is pressed, or when the slow motion is turned in a specific direction. Naturally this should be done before the exposure is started.

The eyepiece needs to be illuminated for the cross-hairs to be visible. Anyone who claims to be able to do without illumination is not doing serious work: if the sky background is so light that the cross-hairs appear as dark silhouettes, it is far better to pack up and go to bed rather than trying to photograph faint objects. As regards the trick of defocussing, so that the guide star appears as a small disk against which the cross-hairs are visible, it is not very effective. For one thing, you need to be extremely lucky to find a star within the field that remains bright enough to be defocussed in this way; for another, guiding is less accurate, because it is extremely difficult to judge when the cross-hairs bisect the false disk exactly. Such a procedure is acceptable only when guiding a wide-field exposure with a long-focus telescope.

It must be possible to vary the strength of the illumination, so that it may be matched to the brightness of the guide star that is being used. It should be red so that it does not affect dark adaption. In addition, because the guide star normally appears more-or-less white, it may be more easily distinguished against the cross-hairs.

There are three possible ways of using a guiding eyepiece during an exposure. Commercial beam splitters are available that use a semi-reflecting plate ahead of the camera body (Figs. 15.29 and 15.30). The plate reflects about 20 % of the incident light over the whole field towards the guiding eyepiece, in which one sees exactly the same image as that being recorded by the film. This device is the easiest to fit and is the one most widely used. It is very convenient for checking the field of view before the exposure starts, but does have some disadvantages. First, the film receives only 80 % of the light gathered by the objective. Second, the eyepiece is centred on the optical axis of the telescope and cannot normally be displaced in the image plane. When a galaxy is being photographed, for example, the image on the cross-hairs is that of the galaxy; but it is not possible to guide accurately on a diffuse image. If a star is centred, the galaxy moves off-axis; if it is too far from the nearest suitable star, its image may be affected by instrumental coma, or it may even be out of the field altogether.

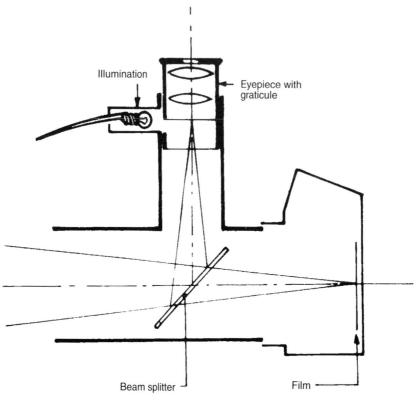

Fig. 15.29. *The principle behind the use of a beam splitter.*

A variation of the previous scheme is to insert a ring immediately in front of the camera body. Inside this ring is a prism or a small mirror, which lies at the edge of the field and reflects a small portion of the image formed by the objective into the guiding eyepiece (Fig. 15.31). If the field used for guiding is sufficiently far from the centre it does not interfere with the photography. (If it is too far, however, if may be affected by coma and vignetting.) The mechanical arrangements are such that the prism may be rotated 360° around the optical axis and a guide star may be located anywhere within this ring of sky. This is not always an easy operation, because the area of the ring is quite small, so it may be difficult to find a suitable guide star. In addition, the eyepiece, which is at 90° to the optical axis, turns with the prism, which may result in some acrobatics being required to reach it. After a guide star has been located it is necessary to check that the orientation of the 24 × 36-mm format within the field is still satisfactory (because the camera body turns with the rotatable ring), to ensure that the focus has not been accidentally altered, and to re-orient the graticule so that the cross-hairs are parallel to the right-ascension and declination axes.

The solution that is heaviest, most difficult, but yet the most effective, is to use two telescopes in parallel on a single mounting (Fig. 15.32); one is used for photography,

Fig. 15.30. *A Clavé beam splitter*

the other for guiding. It should be possible to offset the latter by a few degrees to find a suitable guide star near the field being photographed. Once the object being photographed and the guide star are centred within their respective telescopes there should be no flexure whatsoever between the two instruments. To benefit from the fine grain of films such as TP-2415, guiding needs to be precise. In theory the focal length of the guide telescope should be at least twice that of the photographic instrument. This is often difficult to achieve, and a good-quality Barlow lens will be found to be extremely useful.

15.4.4.4 Checking the photographs

Any astronomical photograph is a record that is worth examining in detail, even if the initial reason for taking it was purely aesthetic. The accidental discovery of some unexpected object (such as a comet, a minor planet, a nova, an eruptive star, etc.) is a possibility that should not be ignored. Naturally, the probability that the comet of every amateur's dreams has been caught by the photograph that has just been taken is very low; but it is obvious that the probability of discovery is zero unless the photographer takes a few minutes to examine the plate carefully.

We cannot stress enough to amateurs the necessity of examining every photographic plate, even if they appear perfectly ordinary at first glance. Rather than labouring the point, we will give just two examples.

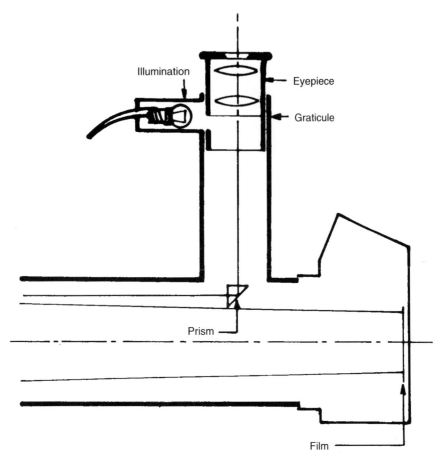

Fig. 15.31. *Arrangement for guiding on an off-axis star*

On 1983 December 23, Roger Chanal took a photograph of the Orion Nebula, M42, with his 210-mm f/7 reflector; the film was Fujichrome 400, and the exposure 1 hour (*see* Fig. 15.33). When examining the photograph, Chanal discovered a star of magnitude 13.0, which did not appear on earlier photographs. To obtain confirmation, he immediately alerted Jean-Claude Merlin, who passed the information to Guy Hurst. Investigations then found the star on other photographs, thus confirming the discovery. Then 'Chanal's object' was identified with the suspect variable NSV 2229, which had a visual magnitude of 18! The alert begun by Chanal enabled several observers to follow this strange star at the beginning of 1984; it appears to be a nebular variable of the FU Orionis type, or perhaps a star that has just been born (a Herbig–Haro object?). This story poses two questions:

- would the outburst of NSV 2229 have ever been detected if Roger Chanal had not bothered to check his photographs?
- 'Chanal's object' was magnitude 13 when detected, and would appear to have erupted at the end of spring 1983, or the beginning of summer, when

Fig. 15.32. *This 300-mm reflector has been fitted with a guiding reflector 150 mm in aperture, and 1260 mm in focal length. Adjustments allow the guide telescope to be pointed in a different direction to the main telescope ($\pm 4°$ in right ascension, and $\pm 3°$ in declination), enabling a bright star to be located for guiding. A photographic camera, visible beneath the main telescope, helps to balance the whole assembly.*

Orion was in conjunction with the Sun and thus unobservable. Between then and its discovery by Chanal, how many amateur astronomers had observed or photographed M42, and found nothing because they did not trouble to look? For the whole of this time the star was detectable by photography with any amateur telescope, and could have been seen visually in a 200-mm reflector.

The second example is the experience that occurred to the Californian amateur Ben Mayer, which he describes in the *Cambridge Astronomy Guide* (Mayer & Liller, 1985). After having spent the night of 1975 August 30 taking numerous photographs of the region of Cygnus, with the aim of detecting meteors, Ben Mayer developed the films the next morning and, not finding any meteors, threw them into the waste-bin.

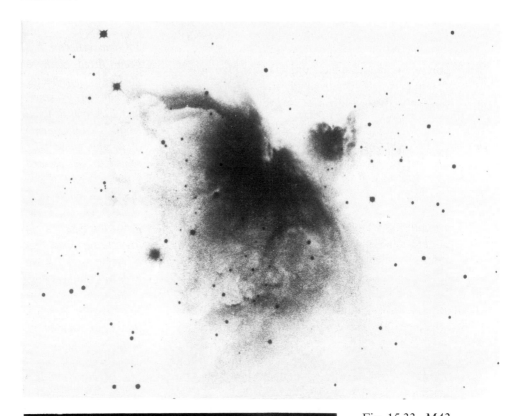

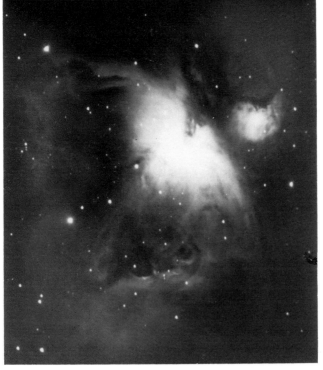

Fig. 15.33. M42, taken with a 410-mm reflector. Top: the discovery photograph for 'Chanal's Star' (indicated by the arrow), 1983 December 29. The original is on Fujichrome 400. On the second photograph (left), taken on 1985 August 14, the star is no longer visible. Photos: R. Chanal.

That evening he attended a star party in southern California, and heard that a nova had been discovered in Cygnus the previous night by a Japanese amateur. Extremely anxious about the fate of his films, he had to wait until the next day until he was able to phone home and make sure that the precious records were rescued. Luckily, the waste-bin had not been emptied, and Ben Mayer was able to recover the historic series of photographs showing the rise of Nova Cygni 1975 (*see* Fig. 15.10).

15.5 Searching for supernovae

Supernovae are distinguished from novae by the considerably higher amount of energy that is liberated when the eruption occurs (of the order of 10^{51} ergs, i.e., equivalent to the amount of energy radiated by the Sun over 9×10^9 years). There are two main classes of supernovae. Type I consists of old stars, low in heavy elements, and generally low in mass (Population II stars), which, like novae, are components of close binary systems. Mass transfer from the red giant to the white dwarf initially produces quiescent fusion reactions on the surface of the star, which has a degenerate core. When the mass of the dwarf approaches the critical value of 1.4 solar masses (the Chandrasekhar limit), carbon fusion occurs within the core, and the star becomes unstable and explodes. The progenitors of Type II supernovae are single, massive stars (minimum 8 solar masses), rich in heavy elements (Population-I stars) that have reached the end of their evolution and the gravitational-collapse stage. In both types the material ejected by the explosion is expelled with velocities that sometimes reach 20 000 km/s.

Supernovae are classed as variable stars and studied by the methods described in the preceding chapter. Their outbursts are always unexpected, however; their detection requires monitoring of galaxies. This is why the search for supernovae is discussed here.

Like comets, supernovae are designated by the year of discovery, followed by a capital letter, indicating the order of detection. But, unlike comets, supernovae are not given the names of their discoverers. The first supernova discovered in 1987 was 1987A, the second 1987B, etc. [When more than 26 supernovae are discovered in any one year, subsequent SNae are given designations such as 1991aa, 1991ab ...1991ba, 1991bb, etc. – Trans.]

15.5.1 The behaviour of supernovae

When a supernova explodes, the increase of light is far greater than that shown by a nova. Figure 15.34 shows the light-curves of various types of supernovae (Doggett & Branch, 1985). The interesting point is that a Type-I supernova explodes when it reaches a well-defined, critical mass. As a result, the absolute magnitude reached by Type-I supernovae at maximum is relatively consistent, estimated at about −19.5. The apparent magnitude of such a supernova therefore depends only on its distance. Supernovae of Type II are less consistent and, in general, are fainter; their average absolute magnitude is about −16.5.

The explosion of a supernova in our Galaxy, close to the Sun, is a very spectacular

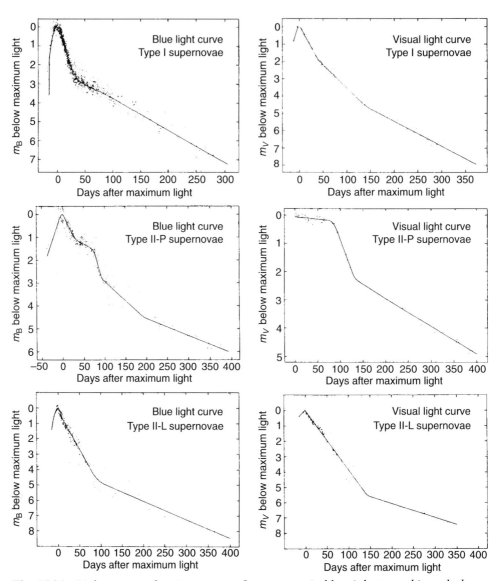

Fig. 15.34. *Light-curves of various types of supernova, in blue (photographic and pho-toelectric), or visible light (after Doggett & Branch, 1985).*

event, which has been observed only five times in the last ten centuries (*see* Fig. 15.36). If we accept that a galaxy produces one supernova every 30–100 years, it may appear surprising that so few events have been observed in the Galaxy. It is, however, probable that other supernovae have exploded in our Galaxy, but that the clouds of dust that lie in the galactic plane have prevented us from seeing them. Figure 15.35 shows the estimated positions of supernovae observed in our Galaxy, showing that these essentially occurred in the same region as the Sun.

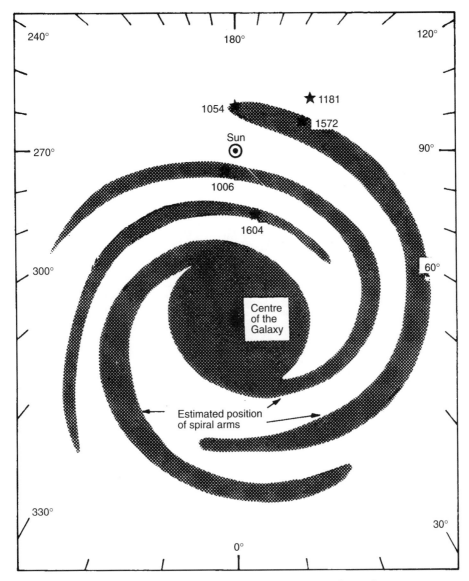

Fig. 15.35. *Estimated positions of supernovae observed in the Galaxy*

If a supernova were to explode in the neighbourhood of the Sun, it would be a highly spectacular event – let's hope that the star concerned is not too close, because it could have serious effects on life on Earth! No such event has been observed since 1604, and it is difficult to predict whether readers will see one during their lifetimes.

On the other hand, a supernova is a sufficiently brilliant event for it to be seen in a neighbouring galaxy, even with amateur equipment. The first extragalactic supernova was detected in the Andromeda Galaxy, M31, on 1885 August 17 by L. Gully, an

905

Fig. 15.36. *M1 in Taurus, the most famous supernova remnant. 200-mm, f/5.5 reflector; exposure 30 minutes on hypersensitized TP-2415. Photo: C. Viladrich.*

amateur astronomer and professor of mathematics at Rouen (de Vaucouleurs, 1985). It is listed in the catalogues as the variable star S And. This supernova attained 6th magnitude at maximum and was observed by thousands of astronomers, who were convinced that they were seeing the birth of a new star. Since the beginning of the 20th century, we have known that the 'Andromeda Nebula' is not a solar-type system in the stage of forming from a cloud of gas, but an entire galaxy.

Because of the rarity of galactic supernovae, we have to search for them in extragalactic systems. They appear to have relatively faint magnitudes because of the distances of these galaxies.

15.5.2 The reasons for searching for supernovae

The incentives are the same as with nova searching: the rarity of the events, the necessity of studying them as early as possible (before maximum if possible), and the attraction of making a discovery (*see* Table 15.6).

In addition to the interest that professional astronomers find in studying supernovae with the aim of understanding their nature, measurements of their apparent magnitude at the time of maximum enable one to deduce their distance, and thus the distance of their parent galaxy. Again, it is important to detect a supernova as early as possible, before its luminosity begins to decline.

Table 15.6. *Amateur discoveries and co-discoveries of supernovae*

Galaxy	Date	Discoverer	*	Amateur co-discoverer	Notes
NGC 224 (M31)	1885 Aug.	Gully	A		1
NGC 4374	1957 Apr.	Romano	A		2
NGC 5236 (M83)	1968 Jly.	Bennett	A		
NGC 4321 (M100)	1979 Apr.	Johnson	A		
NGC 6946	1980 Oct.	Wild	P	Bryan	
NGC 1316	1980 Nov.	Wiscniewski	P	Evans	
NGC 1532	1981 Feb.	Evans	A		
NGC 1316	1981 Mar.	Evans	A		
NGC 4490	1982 Apr.	Wild	P	Newton	3
NGC 4753	1983 Apr.	Okazaki	A	Evans	
NGC 5236 (M83)	1983 Jly.	Evans	A		
NGC 1448	1983 Oct.	Evans	A		
NGC 1365	1983 Nov.	Evans	A		
NGC 3169	1984 Mar.	Evans	A	Okazaki	
NGC 7184	1984 Jly.	Evans	A		
NGC 1559	1984 Jly.	Evans	A		
NGC 991	1984 Aug.	Evans	A		
NGC 3675	1984 Dec.	Ikeya	A		
NGC 4045	1985 Jan.	Horiguchi	A		
NGC 4451	1985 Mar.	Horiguchi	A		
NGC 1433	1985 Oct.	Evans	A		
NGC 3367	1986 Feb.	Evans	A		
NGC 5253	1986 Apr.	Evans	A	Greenwood	
NGC 5128	1986 May	Evans	A		4
NGC 1559	1986 Oct.	Evans	A		
LMC	1987 Feb.	Shelton	P	see note	5
NGC 5850	1987 Feb.	Evans	A		
NGC 2336	1987 Aug.	Patchick	A		
NGC 7606	1987 Dec.	Evans	A		
NGC 4579 (M58)	1988 Jan.	Ikeya	A	see note	6
NGC 4772	1988 Jan.	Taniguchi	A		
NGC 3627 (M66)	1989 Jan.	Evans	A		
NGC 150	1990 May	Evans	A		
NGC 5493	1990 Jun.	Evans	A		
NGC 4527	1991 Apr.	Knight	A	see note	7
NGC 2565	1992 Feb.	Buil	A		
NGC 4411B	1992 Jly.	Evans	A		
NGC 2082	1992 Sep.	Evans	A		

* A = Amateur; P = Professional
1. Supernova S And (de Vaucouleurs, 1985)
2. Co-discovered by Gates (professional)
3. Photographed by J. Newton, end of March 1982, but photograph not checked in time...
4. Uncertain: may be an H II region
5. 1987A: co-discovered by Duhalde (professional) & Jones (amateur)
6. Co-discovered by Evans & Horiguchi (amateurs) and Pollas (professional)
7. Co-discovered by Evans, Villi, Cortini, & Johnson (all amateurs)
Visual observers:
J. C. Bennett (South Africa), J. T. Bryan (U.S.A.), G. Cortini, (Italy), R. O. Evans (Australia), D. Greenwood (U.K.), L. Gully (France), K. Ikeya (Japan), G. Johnson (U.S.A.), M. Villi (Italy)
Astrophotographers:
S. Horiguchi (Japan), J. Newton (Canada), K. Okazaki (Japan)

There are a few professional supernova-search programmes, which are carried out with Schmidt telescopes. These programmes are, however, inadequate to cover the whole of the sky; in addition, the frequency at which each galaxy is covered (30 to 40 times a year, when the Moon is absent), is not enough to ensure that a possible supernova is discovered sufficiently early. So there is an important gap to be filled by amateurs.

Searching for supernovae requires more equipment than searching for novae, because their apparent magnitude is fainter. On the other hand, it is perhaps more attractive to some amateurs, because it may be combined with a programme of observing galaxies.

15.5.3 *The visual search method*

The basic technique consists of locating the galaxy to be checked, and determining if a new star has appeared in its vicinity (*see* Fig. 15.38). Be warned that frequently only the nucleus of the galaxy is visible, but that supernovae often occur in the outer regions (in the arms of spiral galaxies, for example); so a supernova may actually appear to be just outside its parent galaxy (*see* Figs. 15.39 and 15.40). In practice, it is not necessary to search for any possible supernova beyond some 10–15 arc-minutes from the centre of the galaxy, except in the case of a few nearby galaxies that have large apparent diameters. Because the area to be checked is relatively small, and galaxies are generally well away from the Milky Way, the number of stars to be checked is not very large. Unlike the situation with novae, there is no advantage in limiting the faintest magnitude of the stars that are checked. Quite the contrary. Because most supernovae are detected at magnitudes below 13 or 14, the fainter the magnitude observers can attain the better their chances of making a discovery. As a result, the largest possible telescope should be used, together with the optimum magnification (equal to the radius of the objective expressed in mm), which provides the best contrast between faint stars and the sky background.

The observation takes place in two stages: first, locating the galaxy, then searching for any possible intruder in the field. Naturally, the chance of finding a supernova is proportional to the number of galaxies that are checked; it is therefore necessary to devote the minimum amount of time to each one, which leads to the following considerations:

- The magnification recommended for obtaining the lowest possible limiting magnitude gives a relatively small field, about one-third of the diameter (and therefore one-ninth of the area) that is given by equipupilary magnification, which is the lowest magnification that can be used. But a small field tends to increase the amount of time spent on checking the galaxy, unless the area is extremely well-known. On the other hand, changing eyepieces between finding the galaxy and searching the field involves an additional loss of time, especially as a change of eyepiece usually involves refocussing. Observers will have to find their own personal compromises from experience.
- Although we have recommended the largest possible telescope, so that the

faintest possible magnitude may be attained, it will not be particularly easily manoeuvrable. Once again, a compromise has to be reached: a 150–200-mm reflector should suffice for examining nearby galaxies, and will allow rapid finding, but a 300-mm will be required for more distant galaxies.

Finding the galaxy may be done by using an atlas such as *Sky Atlas 2000.0*, etc. Such an atlas normally indicates stars down to a magnitude between approximately 6 and 8. When examining the field, however, it is essential to have a better reference that shows stars down to the telescope's limiting magnitude (at least). The few charts that are available have been compiled by supernova searchers; the main ones are:

- *Check a possible supernova*, a set of 80 photographs of galaxies that may be monitored. All the photographs were taken by Juhani Salmi from Finland, with his 305-mm Newtonian on 103aO film. They are published with black stars on a white background, in two volumes, each with 40 charts (Fig. 15.37 and 15.39).

- The charts prepared by Guy Hurst and the United Kingdom Nova-Supernova Search Programme. Including charts designed for supernova searching and the photometric sequences for galaxies where a supernova has been discovered recently, this series of charts numbers about 150 at present.

- The charts prepared by James Bryan and Ronald Buta, which only cover a dozen or so galaxies. These are 'professional' charts, which give, apart from a chart of the field, a description of the galaxy and a photographic, photometric sequence, and various references, as well as a history of supernovae previously detected in the galaxy.

- The series of photographic charts being prepared by the Argentine amateur astronomer Manuel Lopez-Alvarez, using a Schmidt camera. Lopez-Alvarez has also drawn up a list of galaxies that he considers to be worth monitoring, classified as a function of their distance, and therefore of the magnitude that a supernova within them may attain. (This list was published in *Pulsar*, 1986 September–October, p. 250).

- *Supernova Search Charts and Handbook*, by Gregg Thompson and James Bryan, Cambridge University Press, 1990. The 236 charts show more than 300 galaxies. The scale is $10''/\mathrm{mm}$, and the field covered by each chart is $0.5°$. The limiting magnitude lies between 14.5 and 16.0; the magnitudes of some comparison stars are given.

- To this list may be added all the photographs of galaxies that may be found in various publications, all of which may be used as reference material.

We would advise the beginner to supernova searching to obtain the charts that have been mentioned. The number of galaxies covered is not very large, however, and if all supernova searchers concentrate on those galaxies simply because charts of them exist, some galaxies will be monitored by dozens of amateurs, whilst others will be ignored. In addition, it is not necessarily the case that the charts are perfectly

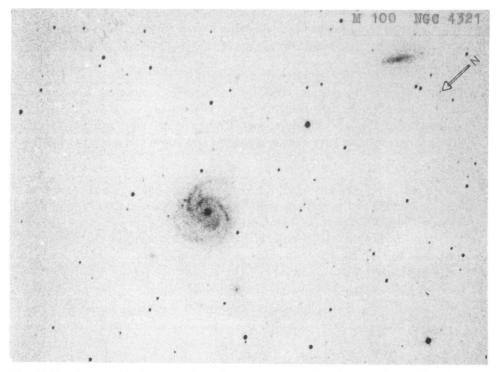

Fig. 15.37. *An example from an atlas of supernova-search reference plates. This photo-graph of M100 is from* Check a possible supernova, *by J. Salmi. Reproduction ratio: 0.9:1.*

suitable for any telescope (the limiting magnitudes may be too faint or too bright, for example).

It would seem desirable for any amateurs searching for supernovae to compile a set of reference charts for themselves. Two methods may be used:

- Make a drawing of the galaxy and the field stars as they appear through the eyepiece. Mark the orientation (the direction of North), and note down the scale, for example by drawing the limits of the field given by the particular eyepiece that will be used for the actual searching.
- Take a photograph of the galaxy at prime focus. The photograph should then be enlarged considerably onto contrasty paper. As was suggested for novae, it is a good idea to make an internegative, so that one can obtain black stars on a white background, which is easier to check. Care should be taken to ensure that the print shows stars slightly fainter than the visual limiting magnitude. This will prevent confusion later. On the other hand, the print should not be too contrasty: the galaxy should remain as half-tones. Otherwise the printed image may hide stars that would be distinctly visible through the eyepiece, under conditions when diffuse objects are not seen so easily by the eye. This problem occurs particularly when one makes

Fig. 15.38. *Supernova 1986O in NGC 1667, photographed on 1987 January 2. Newtonian telescope, 406-mm aperture; exposure 30 minutes on hypersensitized TP-2415 film. Photo: D. Cardoën.*

an internegative, so it may be as well to replace the usual type of material that is used (such as Kodalith) by a more conventional type of film.

Drawing the galaxy is the quickest method, and will be preferred by amateurs who are not used to taking photographs of faint objects. There are two disadvantages, however. First, a drawing is less accurate than a photograph, and second, one cannot go beyond the telescopic limiting magnitude. If the sky is more transparent (or, more likely, the observer has become more experienced) when a search is carried out, new stars at the limiting magnitude will be seen, and may wrongly be taken for supernovae.

Although reference charts are required at first, and as a check in doubtful cases, they do have the important disadvantages already mentioned with regard to novae: there is the loss of time, and problem of the eye being dazzled each time they are consulted. It is preferable for the observer to become familiar with the field of each galaxy that is being monitored.

One example to be emulated is that of the Australian amateur Bob Evans. Blessed with an exceptional visual memory, Evans has memorized hundreds of fields, which he is able to find and check in just a few minutes. He is thus able to check more than 100 galaxies every night (di Ciccio, 1984). The method certainly pays off, because Evans has discovered 25 supernovae by 1992 October – a figure that has probably been exceeded by now!

911

Fig. 15.39. *Photograph taken by Alan Young of the supernova 1986I in M99. 570-mm, f/4.7 reflector; exposure: 20 minutes on TP-2415. Below: a reference chart for M99, taken from Salmi's atlas. It is left to the reader to find the supernova!*

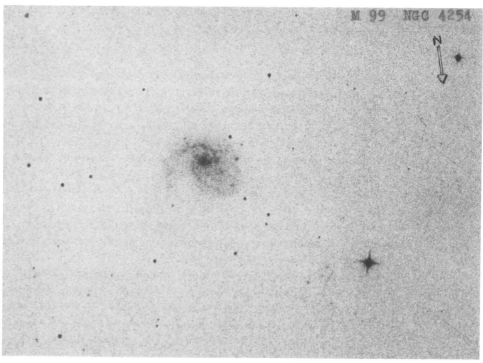

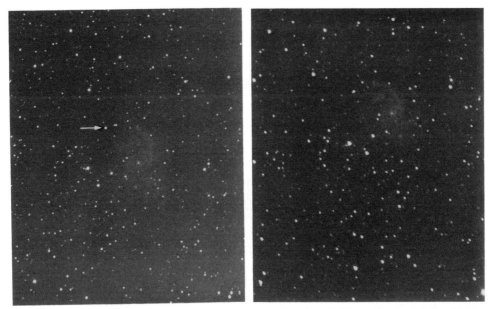

Fig. 15.40. Left: *the supernova 1980K discovered by P. Wild in NGC 6946 in 1980 October, photographed on 1980 November 12 by J. C. Merlin, 260-mm, f/6.0 reflector; exposure: 11 minutes on 103aE. This may be compared with the photograph* right: *NGC 6946 photographed by J. C. Merlin on 1980 September 3, just before the supernova exploded. (Same telescope, exposure 15 minutes on 103aF.)*

15.5.4 Photographic search method

The required limiting magnitude and the large number of stars above that limit mean that a long focal length has to be employed: a telephoto lens of 300–500 mm at least. It is simplest to take a photograph at the telescope's prime focus. The best film is hypersensitized Kodak TP-2415, or failing that, 103a. Exposure times should be at least 10 minutes.

This type of photography is much more difficult than that employed in searching for novae and the series of photographs to be taken is longer. Because of the long focal length used, the telescope (or the telephoto lens) has to be carried on a driven equatorial mounting. In addition, it is necessary to guide on a star. Because the photographic field is limited in size, the first available guide star may be outside the field and well away from the galaxy, which means that there is additional work in bringing the guide star into view in the guiding telescope while the galaxy remains centred in the photographic instrument. Counting the time taken in locating the field, the time required by an experienced worker in preparing for a photograph is about 5 minutes, to which has to be added twice the exposure time. We say 'twice' because it is imperative to take two exposures to ensure that a suspect star is not a film defect.

In searching for novae, a wide-field photograph may record thousands of stars, the visual examination of which may take several hours or tens of hours. But with supernovae a long-focus photograph only includes the field of one or two galaxies, which may be examined visually in just a few minutes. Searching for supernovae photographically is therefore less profitable. The big advantage is that photography allows one to reach fainter magnitudes than visually, provided the exposure times are more than just a few minutes. The photographic method is therefore recommended for relatively distant galaxies, where the supernovae are likely to be too faint for visual searches. It is particularly suitable for clusters of galaxies, where several galaxies may be captured on a single frame, and also for distant active galaxies (clumpy galaxies – *see below*).

Apart from the time taken to make the exposures, we also have to add the development time, and the time taken for checking. The latter, however, is fairly fast, because the number of stars to be checked is not very large. As with novae, the checking should be carried out as rapidly as possible after the photographs have been taken.

Given the relatively small number of stars to be checked on each photograph, it is not necessary to use a blink-comparator for checking. It will probably suffice to examine the negative with a loupe, comparing it against the reference chart. The latter may be one of the charts mentioned that have been published by supernova hunters, or a print made from a negative taken with the same telescope and the same film. It is useful if the reference photograph is given a slightly longer exposure to ensure that all the field stars that may subsequently be detected are readily visible on the print.

15.5.5 Galaxies to be monitored

There are several thousand galaxies accessible with amateur instruments. It is impossible to observe them all regularly, so a choice has to be made. Rather than suggesting that observation should be restricted to certain objects, we will discuss some criteria that may be used to make that choice.

15.5.5.1 Type of galaxy

Type I supernovae are observed in all classes of galaxy; Type II, on the other hand, appear preferentially in Sb, Sc, SBb, or SBc spiral galaxies. The reason is simple: the arms of these galaxies are more pronounced than in the Sa and SBa classes, and contain a large number of young stars. Ellipticals, by contrast, mainly consist of older stars. Type II supernovae are massive stars that have very short lifetimes; they are therefore found only in regions that contain young stars.

Sb, Sc, SBb, and SBc galaxies are thus the preferred targets, particularly those that are seen face-on. When a spiral galaxy is seen from the side, most of its stars are hidden by its dust clouds (*see* Fig 15.41). The problem is the same as with our Galaxy: a supernova exploding on the far side will be undetectable.

Another criterion is to observe galaxies that have already shown a considerable number of supernovae. There is nothing to say, however, that the next supernova will occur in one of the galaxies that have already seen several eruptions; on the other

Fig. 15.41. *NGC 891, a spiral galaxy seen edge-on; note the dust clouds obscuring the galactic plane. Newtonian, 310-mm reflector, 1900-mm focal length; exposure: 75 minutes on hypersensitized TP-2415 film. Photo: S.A.P.P.*

hand, you can be sure that they are better covered than the others. Table 15.7 gives a list of the most prolific galaxies; note the preponderance of spirals (Fig. 15.42).

15.5.5.2 Distance of the galaxy

Let us take a supernova that reaches an absolute magnitude of −19 at maximum; this means that it would have a magnitude of −19 at a distance of 10 parsecs (10 pc). Each time its distance is multiplied by 10, its luminosity is divided by 100; we thus have to add 5 to its apparent magnitude. If the supernova occurs at 100 Mpc, its apparent magnitude will be 16, which represents the visual limit of a 600-mm telescope.

Catalogues of galaxies do not generally give their distance, which cannot be measured directly, but instead list the radial velocity corresponding to the Doppler shift observed in their spectral lines. Except in the case of a few nearby galaxies, this shift is towards longer wavelengths, i.e., towards the red for visible wavelengths, hence the term 'redshift'. This means that all these galaxies are receding from us; in fact they are all receding from one another, the effect being caused by the expansion of the universe. In 1925, the American astronomer Edwin Hubble showed that the velocity of recession V of galaxies is proportional to their distance

915

Fig. 15.42. *NGC 6946, one of the galaxies that show most supernovae. 500-mm re-flector, 2500-mm focal length; exposure: 60 minutes on hypersensitized TP-2415 film. Photo: J. Dijon.*

Table 15.7. *Galaxies in which most supernovae have occurred*[*]

Galaxy	Type	Supernovae
NGC 2276	Sc	1962Q, 1968V, 1968W, 1993X
NGC 2841	Sb	1912A, 1957A, 1972R
NGC 3184	Sc	1921B, 1921C, 1937F
NGC 4254 (M99)	Sc	1967H, 1972Q, 1986I
NGC 4303 (M61)	SBc	1926A, 1961I, 1964F
NGC 4374	Sc	1957B, 1980I, 1991bg
NGC 4321 (M100)	Sc	1901B, 1914A, 1959E, 1979C
NGC 4725	S(B)b	1940B, 1969H, 1987E
NGC 5236 (M83)	Sc	1923A, 1950B, 1957D, 1968L, 1983N
NGC 5253	Irr	1895B, 1972E, 1986F
NGC 5457 (M101)	Sc	1909A, 1951H, 1970G
NGC 6946	Sc	1917A, 1939C, 1948B, 1968D, 1969P, 1980K

[*] List compiled by J. C. Merlin, revised by S. R. Dunlop

$D : V = H_0 \times D$. The constant of proportionality H_0, known as the Hubble constant, is still not known very accurately. In recent years, most astronomers have accepted a value of 50 km/s per Mpc, and we will use this value to be consistent with the absolute magnitudes indicated for supernovae. Some teams of researchers consider that H_0 is 75 km/s or even 100 km/s, which would result in the galaxies being 'closer'.

From the Hubble equation we can deduce that a galaxy at a distance of 100 Mpc has a radial velocity of 5000 km/s. If the telescope being used is not capable of reaching magnitudes fainter than 16, it would be advisable to select from the catalogues galaxies with radial velocities that are below 5000 km/s. Galaxies with velocities of 2000 km/s are 2.5 times as close (40 Mpc). If a supernova of absolute magnitude −19 occurs within them, its apparent magnitude will be 14 (because its luminosity will be 2.5 × 2.5 times greater, and a factor of 2.5 in luminosity means a difference of 1 magnitude), and may be detected with a telescope 250 mm in diameter. Conversely, using a telescope 600 mm in diameter on a galaxy with a radial velocity of 200 km/s will enable fainter supernovae to be detected (down to absolute magnitude −17). Readers may calculate other values for themselves.

15.5.5.3 Magnitude of the galaxy

A galaxy may be bright because it is close, or because is contains a large number of stars. Conversely, the region around our Milky Way system contains a number of insignificant dwarf galaxies. The distance of a galaxy is therefore a more fundamental criterion than its magnitude in deciding whether a supernova will be observable. For a given distance, however, the magnitude gives a good indication of the size of

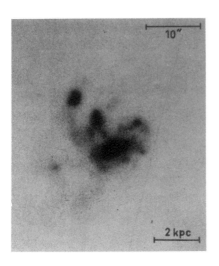

Fig. 15.43. *The galaxy Markarian 325, showing the 'clumpy' structure. Image-tube photograph taken with the 2-m reflector at the Pic du Midi Observatory.*

the galaxy, and the more stars it contains the more likely one is to find that one of them is a supernova. In practice, some supernova hunters restrict their searches to galaxies brighter than magnitude 12 or 13, which have the advantage of being shown in most atlases, and which are not too difficult to find in the sky.

15.5.5.4 Clumpy galaxies

The French astrophysicist Jean Heidmann has called the attention of both professional and amateur astronomers to the value of observing this class of galaxies (Heidmann, 1982). The 'clumps' are very active H II regions, where new stars are born (Fig. 15.43). The star-formation rate appears to be much higher than in the H II regions known in our own Galaxy or in nearby galaxies.

The spectra of these clumpy galaxies indicate that they are rich in hot, supergiant stars. This suggests that the frequency of supernovae in clumpy galaxies may be one every 2 or 3 years, i.e., 10 times the rate in a normal spiral galaxy.

Only five clumpy galaxies are definitely known, and their main characteristics are given in Table 15.8. They are all irregular Markarian galaxies. They have the disadvantage of being faint and of lying in fields where there are few bright stars, so they are difficult to find. The five charts shown in Fig. 15.44 should, however, enable them to be found on other photographs (such as Fig. 15.45).

Table 15.8. *Principal known, and possible^a clumpy galaxies*

	Coordinates (1950.0)		Mag.	Size (arc-sec.)	Distance (Mpc)*
	α	δ			
Mkr 7	$07^h22^m18.7^s$	$72°40'24''$	14.7	40×20	64
Mkr 8	$07^h23^m38.5^s$	$72°13'50''$	14.6	34×24	75
Mkr 296	$16^h01^m13.4^s$	$19°17'53''$	16.0	26×12	94
Mkr 297	$16^h03^m01.2^s$	$20°40'43''$	13.6	38×28	94
Mkr 325	$23^h25^m12.0^s$	$23°18'53''$	13.4	35×30	74
VV 523	$11^h54^m56.5^s$	$32°37'00''$	13.8	55×10	
VV 645	$09^h01^m48.9^s$	$14°47'47''$	14.3	17×10	

* Distances are given assuming: $H_0 = 50\,\text{km/s}$ per Mpc
^{a} The Vorontsov-Velyaminov galaxies, VV 523 and VV 645, are possible candidates that have yet to be confirmed as true clumpy galaxies.

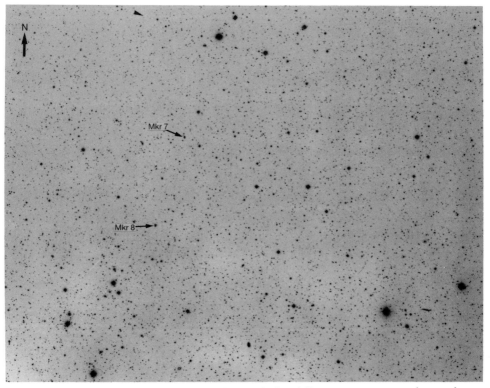

Fig. 15.44. *The five photographs on this and the following pages may be used to locate the galaxies listed in Table 15.8. These images are taken from the National Geographic Society/Palomar Observatory Sky Survey: limiting magnitude: 21; scale: 1 mm = 1 arc-minute.*

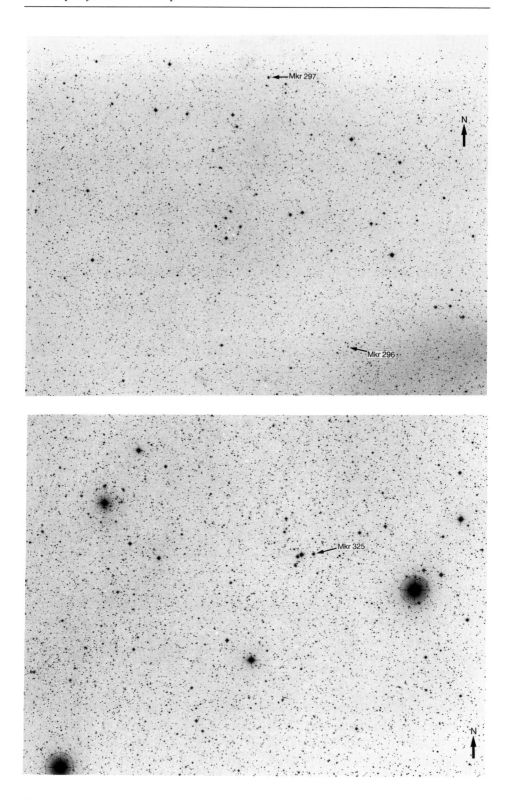

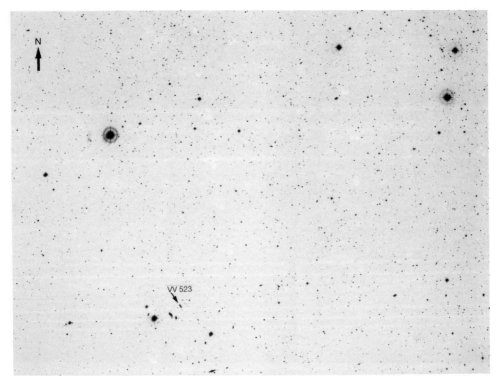

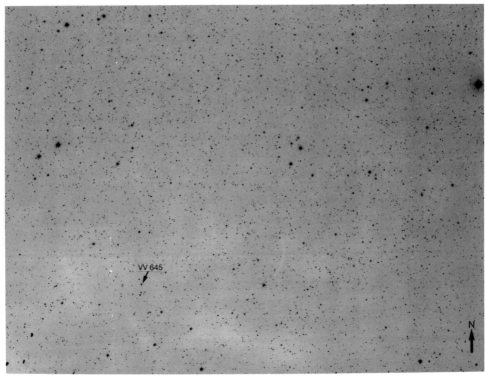

Fig. 15.45. *Markarian 297, photographed using a 410-mm reflector. Exposure: 30 minutes on hypersensitized TP-2415. Photo: R. Chanal.*

The other disadvantage is that all these galaxies are relatively distant: between 65 and 100 Mpc. A Type II supernova is therefore likely not to exceed magnitude 15–16. It would seem best to monitor these clumpy galaxies by photographic methods, which will allow fainter limiting magnitudes, especially as the small number of these galaxies known means that they may all be photographed in a single night's work.

The estimates that we have made of the magnitude of supernovae in these clumpy galaxies should not be viewed pessimistically. Recent radio observations appear to show that Markarian 292 was the site of a much more violent phenomenon a few years ago (Heidmann, 1986). This 'hypernova' was not observed when it should have been detectable in visible light, but it was undoubtedly within the range of many amateur telescopes. Its observation would have had great scientific value.

Research is currently taking place to identify new clumpy galaxies. The detection and study of supernovae that may occur in these galaxies is of the greatest interest in reaching an understanding of these phenomena.

15.5.5.5 Clusters of galaxies

In visual work, a not-inconsiderable amount of time is taken up by locating the galaxies. It is therefore of advantage to choose galaxies that are close to one another, and to decide on an order in which they are swept that will minimize movement of the telescope. Regions of the sky that are rich in galaxies (Ursa Major, the Virgo Cluster, etc.) are thus the best areas on which to concentrate.

15.5.6 Discovery and observation of a supernova

The procedure to be followed if a supernova is discovered is exactly the same as with a nova. It is first essential to ensure that it is not some other type of object: a known variable star, a passing minor planet, or a film defect. It is also advisable to obtain confirmation from an independent observer. If, after this, one is certain that a supernova has really been detected, its magnitude should be estimated immediately, and its exact position determined. In general, the positions of supernovae are defined with respect to the centre of the galaxy (for example, 100 arc-seconds south). The IAU also insists that accurate coordinates of the star should be given in right ascension and declination (Marsden, 1982). It is important that observations are made as soon as possible at wavelengths other than the visible (for example at radio wavelengths), where the observers are unable to use the galaxy as a reference.

After having determined the brightness and position of the supernova, the observer should report the discovery as quickly as possible, preferably by contacting the coordinator of an observational network, or the nearest professional observatory. Failing this, the Central Bureau for Astronomical Telegrams may be contacted directly, but be warned: the higher up the chain one reports a supernova, the less forgivable it is to make an error!

When a supernova has been discovered, the only work that amateurs can usefully carry out is to determine its magnitude. If possible, the comparison stars should be chosen from those in the field.

15.6 Photometry and colorimetry

15.6.1 Colours of stars

The colours of stars are linked to their temperature. The spectrum of light emitted by a star follows a curve, the maximum wavelength of which is inversely proportional to the surface temperature. For example, the cooler the star, the redder it appears.

In practice, to determine the colour of a star, its luminosity is measured in various spectral regions. The most widely used system is the UBV, which consists of three pass bands: U in the near ultraviolet; B in the blue, and V centred in the visible region (yellow). This system has been extended by defining other pass bands at longer wavelengths, but we shall not discuss them here (*see* Chap. 19, 'Photoelectric photometry').

If we measure the magnitude of the star in the B and V bands, the difference between these two values (denoted B − V) is known as the colour index. If it is positive (B, expressed as a magnitude, is greater than V), the star is faint at short wavelengths, and thus cool. If it is negative, the star is bright at short wavelengths, and thus hot. It will be seen that the algebraic value of (B − V) is a measure of the temperature of the star. The same reasoning applies to the (U − B) index.

15.6.2 Photographic photometry

The brighter a star, the denser and larger its image will be on a photographic negative. We may therefore use either the diameter of the image, or its density to derive an estimate of the magnitude of the star. Naturally, such measurements are not absolute; they depend on the film, on the optics used, and on the accuracy of the focussing. It is therefore absolutely essential that the image of the star being measured should be compared with those of comparison stars obtained under the same conditions and, if possible, on the same photograph.

Photographic photometry is not widely used, because it is less accurate than photoelectric photometry, and more onerous than visual photometry. It may, nevertheless, be of considerable value in the following cases:

- the object is too faint to be reached by a photometer or by visual observations;
- subsequent observations; this applies to novae, for example, that are found on photographs taken before their actual discovery;
- relative measurements obtained for a large number of stars situated in the same telescopic field (*see*, for example, the suggestions made in Sect. 15.6.4).

15.6.2.1 Measuring the diameter of images

In this case we make use of a phenomenon that is normally considered undesirable in photographic emulsions: an over-exposed image tends to be larger, thus affecting the photograph's resolution. The diameter of the image should be measured from the negative by using a small microscope fitted with an eyepiece micrometer. It has been found that there is a linear relationship between the diameter of the images and the magnitude of the stars that have been photographed. The scatter is relatively large, however, and one should not expect an accuracy greater than about one-tenth to several tenths of a magnitude.

15.6.2.2 Measuring densities

The image of a star on a negative frequently does not exceed a few hundredths of a millimetre in diameter, and its density varies considerably between the edge and the centre. To measure this density, we require an extremely fine beam of light (the diameter of which has to be less than that of the stellar image). This beam is passed through the image and its attenuation is measured using a photoelectric cell. The equipment capable of doing this is known as a microdensitometer (Fig. 15.46). It is a laboratory instrument that is highly accurate mechanically. Its construction is beyond the capacity of the vast majority of amateurs – and its price as well – and it is only found in a few observatories or laboratories.

The microdensitometer beam is scanned in a line across the stellar image; the resulting curve produced by the instrument therefore represents a cross-section of the image density. In addition, the densities cannot be measured properly unless the microdensitometer has been calibrated with a grey scale (such as that produced by Kodak, for example). Figure 15.47 shows a typical trace from a microdensitometer; it will be seen that the curve varies considerably because of the grain of the film, and that it is not immediately obvious what one ought to measure: we could choose the

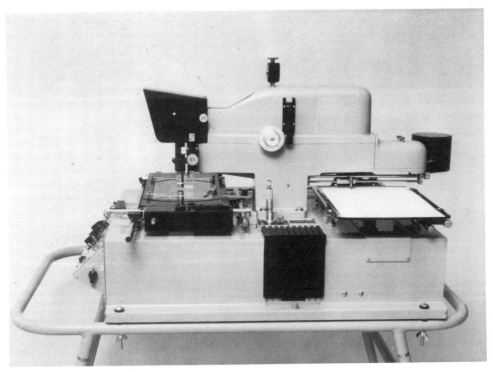

Fig. 15.46. *The microdensitometer at the Ecole Nationale Supérieure de l'Aéronautique et de l'Espace in France.* Left: *the moving plate carrier, where the flying spot 'reads' the plate;* right: *the plotting table.*

peak value or the width at half-height, or even the integral, i.e., the area defined by the curve. It is not surprising that the accuracy is somewhat limited (approximately 0.05–0.1 magnitude) despite the considerable technical resources being employed (*see* Lagerkvist, 1975).

One solution consists of defocussing the images when the photograph is taken: each star then appears as a uniform disk that does not require such a fine beam of light, or such mechanical accuracy. Averages may also be taken to eliminate the effects of the grain of the emulsion. The limiting magnitude that may be reached will be much higher, however, and the procedure cannot be used with a prediscovery photograph, which will – at least in principle – have been made with accurate focussing.

One area of photographic photometry with defocussed images that, as far as we know, has not been explored is that of estimating cometary magnitudes. This would be the equivalent of the Bobrovnikoff method (Volume 1, Sect. 8.5.3.1), but applied to photography. It would have the advantage of being objectively based, because it would be easy to measure the integrated signal of both the comet and of comparison stars. This method may perhaps be developed in the future, and perhaps by an amateur astronomer.

925

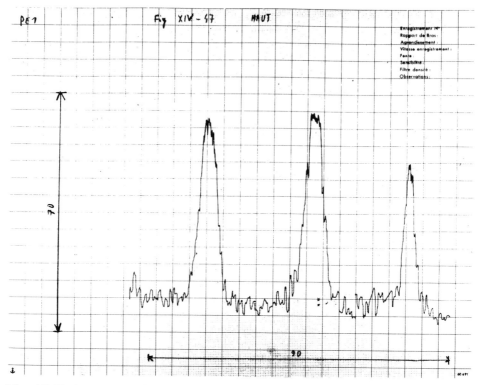

Fig. 15.47. *Density profiles of three stellar images, obtained by analysing a photo-graphic negative with a microdensitometer.*

15.6.3 The Hertzsprung–Russell diagram

We may plot the properties of stars on a diagram, each point representing one star: the abscissa is the star's temperature (or a colour index), and the ordinate is its luminosity, expressed in magnitudes. It is, of course, the star's intrinsic luminosity, i.e., its absolute magnitude. This may be calculated from the apparent magnitude, which is all that may be measured directly, using the distance of the star, from:

absolute mag. = apparent mag. − 5 × log(distance/10),

the distance being measured in parsecs. Absolute magnitude is the apparent magnitude that a star would have if it were at a distance of 10 pc.

The diagram thus obtained is known as the Hertzsprung–Russell diagram, after the names of the astronomers who first devised it (Fig. 15.48). We can see that stars are not distributed at random, but that most of them are found on what is known as the 'Main Sequence', which is where stars spend most of their lifetimes. The top of the Main Sequence is occupied by the most luminous and hottest stars; these are massive, young stars with very short lifetimes. At the end of its life, a star becomes a red giant, leaves the Main Sequence and moves across to join the red giants that lie

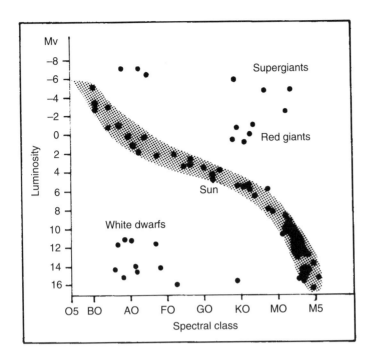

Fig. 15.48. *The Hertzsprung–Russell diagram*

Table 15.9. *Filter/film combinations for photographic photometry*

Film	Filter	Type	Transmission (%)	Exposure time (min)
103aO	Ultraviolet	UG 11	16	6
103aO	Blue	GG 385	24	4
103aF	Green	VG 5	26.5	4

at the top right-hand corner. More details about the Hertzsprung–Russell diagram may be found in nearly any book on astronomy.

15.6.4 The age of a star cluster

Determining the age of an open cluster is an exercise that may be carried out fairly easily. Its interest lies in the fact that it requires the use of photographic photometry and of the Hertzsprung–Russell diagram.

Two photographs of the cluster are taken, one in the B pass band, and the other in V. The exposure times should take account of the transmission coefficients of the different filters used. For example, combinations that are sometimes used are given in Table 15.9.

The focal length must be sufficiently long for the images of the stars to be clearly separated on the negative, and thus capable of being individually measured. (This is why, for an amateur experiment, one chooses an open cluster rather than a globular cluster.) It is also necessary to ensure that the accuracy of the guiding is comparable on both exposures.

Reduction consists of measuring the magnitude of each star in the two pass bands. By subtracting one value from the other, we then obtain the (B − V) colour index. We then only have to plot the values on the cluster's Hertzsprung–Russell diagram, using the (B − V) index on the abscissa, and a magnitude (the B magnitude, for example) on the ordinate. Because all the stars in the cluster may be considered as lying at the same distance from us, we do not have to calculate their absolute magnitudes. The scale on the ordinate is simply shifted by the value of $5 \times \log(D/10)$, but the shape of the diagram is correct. The other feature is that we assume that all the stars in the cluster were formed at the same time. Because all the stars have the same age, those that have masses above a certain value (which is a specific function of the age of the cluster), have left the Main Sequence and become red giants. Because luminosity is directly related to the mass of star, the whole of the top of the Main Sequence has disappeared. The colour index at the 'turn-off', where the Main Sequence turns into the red-giant branch, enables the age of the cluster to be estimated. The greater the colour index, the older the cluster.

15.7 Catalogues and atlases

There are a number of catalogues and atlases available to amateurs. Some cover the whole sky down to a specific magnitude limit; others are more specialized, and describe just a single category of objects. None of the references is comprehensive, and observers will find that they require more than one to cover all the aspects that they will require. To make the presentation simpler, we have classified the various publications into groups, starting with the most general and popular, and ending with the most specialized and complex.

15.7.1 General

Any beginner should have at least one publication in this category. Even experienced observers make frequent reference to this sort of work when they require simple, wide-field charts: for example, when determining which bright star within a constellation should be used as a starting point for finding a faint, neighbouring object.

- *Handbook of the Constellations*, by H. Vehrenberg and D. Blank, 1977. This work contains 55 charts with stars down to magnitude 6, and 1050 non-stellar objects, grouped by constellation. The interesting objects are described in a list that accompanies each chart.
- *Norton's Star Atlas*. The classic work, which has now reached its 18th edition (entitled *Norton's 2000.0*). It shows 600 non-stellar objects and all stars down to magnitude 6.5.
- *Burnham's Celestial Handbook*. Unlike the others, this is not an atlas, but describes the main objects of interest to amateur astronomers by constella-

tion (bright stars, doubles, variables, nebulae, clusters, galaxies, etc.). It is a major reference work, consisting of 2000 pages in three volumes.

15.7.2 Diffuse objects for amateur observation

Here we list some works giving details of the nebulae, clusters and galaxies that may be most easily observed. The charts included do not cover the whole sky, but just give the fields surrounding the objects concerned.

15.7.2.1 The Messier Catalogue

In the 18th century, the great French comet seeker, Charles Messier, compiled a catalogue of the diffuse objects that he knew existed (many of which he had discovered himself), so that he would not confuse them with a comet. This catalogue initially contained 103 objects: 32 galaxies, 28 globular clusters, 27 open clusters, 5 emission nebulae, 1 reflection nebula, 4 planetary nebulae, and 1 supernova remnant. The open cluster M45 (the Pleiades) is also associated with a reflection nebula, not counted in the list, as were some emission nebulae associated with open clusters. M102 and M101 are the same object; three numbers in the catalogue (M40, M73 and M91) are simple asterisms, recorded in error and generally omitted in modern versions. On the other hand, M104 to M110 have been added by later authors after Messier's death: the well-known galaxy M104, for example, was added by Flammarion.

As the first catalogue of non-stellar objects, Messier's catalogue contains a selection of those that are easiest to observe. For this reason, it remains popular with amateurs, who often mistakenly restrict themselves to its listing, when there are actually several tens of other objects, not catalogued by Messier, that are just as notable (not including those that are in the southern hemisphere).

We may mention that once a year, around March 21, 109 of the 110 objects in the Messier catalogue are visible in a single night; the exception is the globular cluster M30 in Capricornus, which is the only one hidden by daylight. This event has not passed unnoticed by amateurs in the United States and various groups organize 'observing marathons' aimed at finding all (or nearly all) the Messier objects in one night (Houston, 1979 & Harrington, 1985).

There are two books that are widely available and that describe all the Messier objects:

- *The Messier Album*, by John H. Mallas and Evered Kreimer, Sky Publishing Corp., 1978 and Cambridge University Press, 1979. This work contains a brief description of each object, accompanied by a finding chart, a photograph, and frequently a drawing. There is also a section of colour plates;
- *Messier's Nebulae and Star Clusters*, by Kenneth Glynn Jones, Cambridge University Press, 2nd edn, 1991. This is the definitive, English-language work on the Messier objects. The principal illustrations are drawings of the field and appearance of every object, but a large number of photographs are also included.

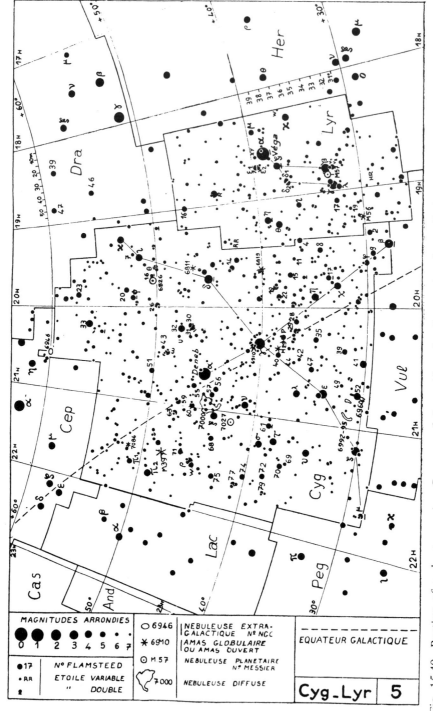

Fig. 15.49. Portion of a chart covering Cygnus and Lyra (from La Revue des Constellations 1963). Reproduced at the same scale as the original.

15.7.2.2 Other catalogues

- *Sarna Deep Sky Atlas*, Willmann-Bell, Richmond, VA, 1985. This comprises a set of 102 charts, drawn to the scale of 1 arc-minute per mm, giving the locations of 115 galaxies, 41 globular clusters, 56 open clusters, 25 planetary nebulae, and 17 diffuse nebulae.
- *Webb Society Deep-Sky Observer's Handbooks*, Enslow, New Jersey. This is a series of volumes (eight to date), containing very useful information:
  - volume 1: Double stars, 1979
  - volume 2: Planetary and gaseous nebulae, 1979;
  - volume 3: Open and globular clusters, 1980;
  - volume 4: Galaxies, 1981;
  - volume 5: Cluster of galaxies, 1982;
  - volume 6: Anonymous galaxies, 1987;
  - volume 7: The southern sky, 1987;
  - volume 8: Variable stars, 1990.

Volumes 2 to 6 are the ones of most interest here. Each contains a description of the object concerned, suggestions for observation, and a fairly comprehensive catalogue accompanied by drawings made at the eyepiece.

15.7.2.3 Photographic albums

Several works are both collections of magnificent photographs and a source of reference material for observers who require a catalogue of fields of galaxies or nebulae (Fig. 15.50). Among others, there are:

- *The Cambridge Deep-Sky Album*, Cambridge University Press, 1983, largely consisting of 126 colour photographs taken by the Canadian amateur Jack Newton.
- *Atlas of Deep-Sky Splendours*, by Hans Vehrenberg, Cambridge University Press, 1983. The photographs in this book, which were taken by the celebrated German amateur, and by professional astronomers, show more than 400 nebulae, clusters and galaxies.
- *The Hubble Atlas of Galaxies*, published by Sandage in 1962 after Hubble's death. This contains 208 photographs of galaxies taken by the latter using the 1.5-m and 2.5-m (60-inch and 100-inch) reflectors at Mount Wilson, and which illustrate his classification scheme.

15.7.3 Atlases

The works included here are atlases that cover the whole sky down to a limiting magnitude of between 7.5 and 9.5, and containing varying numbers of non-stellar objects. At least one is required by any amateur who observes regularly, but because each tends to be specialized, several may be considered desirable.

- *The AAVSO Variable Star Atlas*, AAVSO, Cambridge, Mass., 1990. Based on the Smithsonian Astrophysical Observatory Atlas, this atlas, for 1950 coordinates, covers the whole sky down to about magnitude 9.5 with 178

Fig. 15.50. *NGC 7000, the North America nebula in Cygnus. Schmidt telescope, 200-mm aperture, 300-mm focal length; exposure: 20 minutes in red light (Wratten 25 filter). Compare this photograph with the same region as shown on the charts on the following pages. Photo: S.A.P.P.*

charts. The scale is 11 mm per degree. Primarily designed for variable-star observers, it gives photometric sequences for all stars reaching magnitude 9.5 or above at maximum. It also gives the positions of about 200 double stars, and numerous non-stellar objects. Details of all the variable stars are given in an appendix (Fig. 15.51).

- *Atlas Borealis* (+90° to +30°), *Eclipticalis* (+30° to −30°), and *Australis* (−30° to −90°). Sold individually, these three atlases, prepared by Antonin Becvár, cover the whole sky down to magnitude 9. Non-stellar objects are not included on the charts. On the other hand, the stars are coloured in accordance with their real colours. The coordinates are for 1950, and the scale is 3 arc-minutes per mm; these atlases are of interest primarily to observers of stars and minor planets.
- *Sky Atlas 2000.0*, by Wil Tirion, Sky Publishing Corp., 1981. The first atlas with 2000.0 coordinates, 'Tirion' covers the sky in 28 charts with a scale of approximately 8 arc-minutes per mm. The stellar limiting magnitude is 8; *Sky Atlas 2000.0* does, however, show many diffuse objects, and colour coding in the 'de luxe' edition enables galaxies, nebulae and clusters to be differentiated. This is an atlas that is particularly suitable for observers of nebulae, clusters and galaxies, but which will also be useful for comet hunters (Fig. 15.52).
- *Uranometria 2000.0* by Wil Tirion, Barry Rappaport and George Lovi, Willmann-Bell, Richmond, VA, 1987 & 1988. In two volumes (northern and southern hemispheres), each containing 259 charts, this atlas has more than 330 000 stars down to magnitude 9.5 and 10 000 non-stellar objects; its scale is 1° = 18 mm (Fig. 15.53).

15.7.4 A photometric atlas

- *Atlas of Selected Areas*, by A. Brun and H. Vehrenberg. Just this one atlas is mentioned here. It contains 206 charts, 15 arc-minutes square, which are spaced at regular intervals across the sky. Each one shows stars down to 16th magnitude. The photographic magnitude of each star in the blue is given on the charts. This atlas is ideal for calibrating photographs, but caution must be exercised when using modern films such as Kodak TP-2415, which is more sensitive than older emulsions to red light.

15.7.5 Photographic atlases

These atlases are prints made from photographs; naturally no details are given on these prints, but transparent coordinate grids are provided, which enable objects to be identified from their positions. When there is a choice, it is better to choose black stars on a white background. Not only is the legibility better, but photocopies may be made on which the track of minor planets and comets may be plotted.

- *Atlas Falkauer*, by Hans Vehrenberg. This covers the whole sky, but the northern portion (which also covers the equatorial zone), with 303 charts,

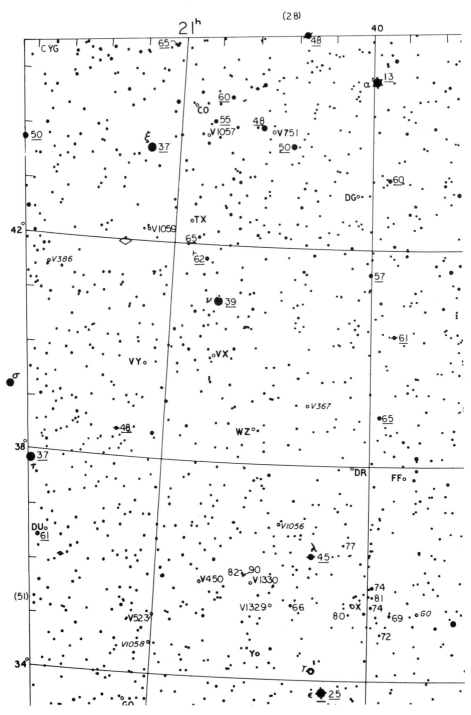

Fig. 15.51. *The area around Deneb and NGC 7000, from the AAVSO* Atlas *(1st edition). Partial reproduction at the same scale as the original.*

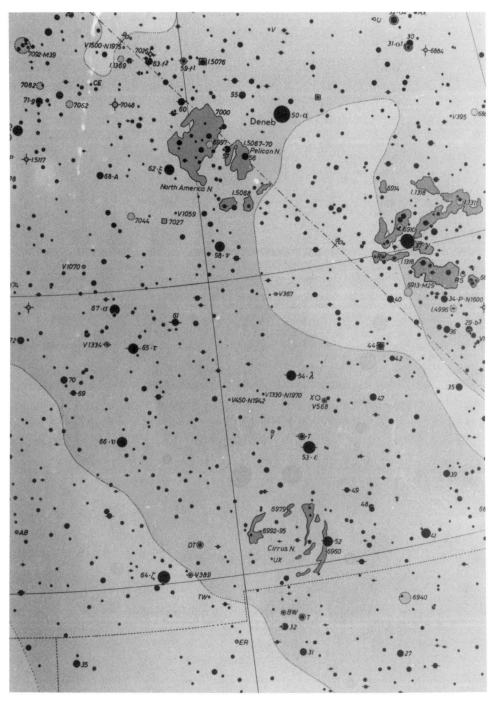

Fig. 15.52. *The area around Deneb and NGC 7000, as shown in* Sky Atlas 2000.0 *by Wil Tirion. Partial reproduction at the same scale as the original. On the original chart, the Milky Way, nebulae, clusters, and galaxies are colour-coded.*

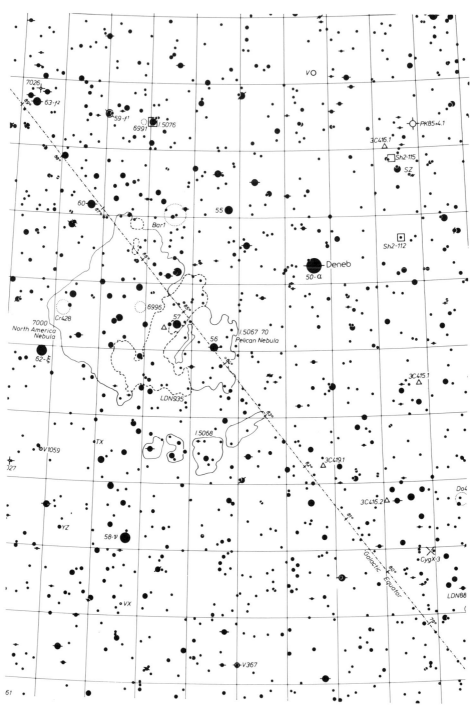

Fig. 15.53. *The area around Deneb and NGC 7000, as shown in* Uranometria 2000.0. *Partial reproduction at the same scale as the original. Reproduced by courtesy of Willmann-Bell Inc.*

and the southern portion, 161 charts, may be bought separately. The charts are 180 mm square, with a scale of 4 arc-minutes per mm; the limiting magnitude is about 13, and the coordinates are for 1950 (Fig. 15.54).

- *Atlas Stellarum*, by Hans Vehrenberg. This is a 'super-Falkauer', with charts 330 mm square, and a scale of 2 arc-minutes per mm. The northern portion consists of 315 charts and the southern 171. The limiting magnitude is about 14 (Fig. 15.55).

- *True Visual Magnitude Photographic Star Atlas*, by Popadopoulos and Scovill. This atlas was obtained using a filter which enabled the stars to be recorded with their true visual magnitudes. Apart from this, it resembles Vehrenberg's atlases (2 arc-minutes per mm, charts 11° square, limiting magnitude approximately 13.5), although slightly more expensive. It is available in three parts: northern (120 charts), equatorial (216 charts), and southern (120 charts).

- *Palomar Observatory Sky Survey* (POSS). This is the main working reference for professional astronomers, and was obtained with the Big Schmidt at Mount Palomar. Each region of the sky was photographed in blue and in red light. The scale is 1 arc-minute per mm and the limiting magnitude 21. A similar survey for the southern sky is the *ESO/SERC Survey* and a new version of the Palomar Survey is being made with modern emulsions. Because of their cost, these atlases are found only in professional observatories.

15.7.6 Catalogues

15.7.6.1 Stellar catalogues
These generally give the coordinates, proper motion, magnitudes, and spectral type of stars. The most widely used by amateurs are:

- *Fourth Fundamental Catalogue* (*Vierte Fundamental Katalog*) FK4, published in 1963. It gives the absolute positions of 1535 stars to 0.001 s in right ascension and 0.01 arc-second in declination. It is mainly used for astrometry.

- *Smithsonian Astrophysical Observatory Star Catalog* (SAOC), published in 1966; contains 259 000 stars for equinox 1950.0.

- *Sky Catalogue 2000.0*, Cambridge University Press, and Sky Publishing Corp, Cambridge MA. Volume 1, 1982, contains details of stars to magnitude 8.0. Coordinates are for 2000.0 and the catalogue contains all the objects plotted on *Sky Atlas 2000.0*.

15.7.6.2 Catalogues of non-stellar objects
The great classic remains the *New General Catalogue of Nebulae and Clusters of Stars* (NGC), which dates from 1888, and which was supplemented in 1895 and 1908 by the *Index Catalogue* (IC) and the *Second Index Catalogue*. These three publications contain 7840, 1529, and 5386 objects, respectively. The NGC has been reissued in the form of the *Revised New General Catalogue of Non-Stellar Astronomical Objects*

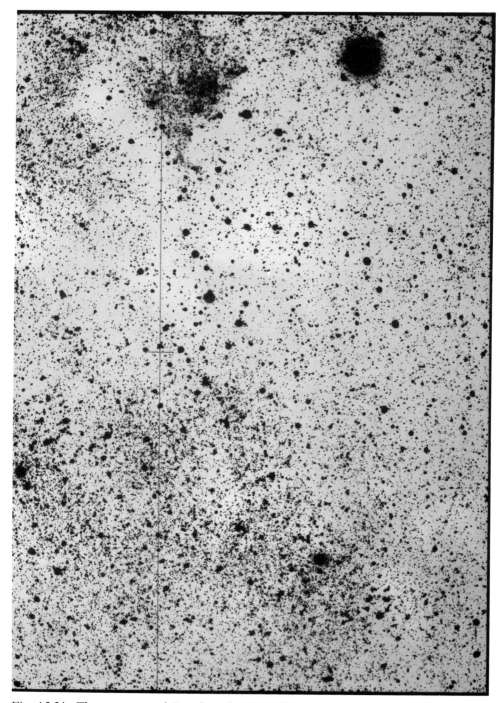

Fig. 15.54. *The area around Deneb and NGC 7000, as shown in* Atlas Falkauer *by H. Vehrenberg. Partial reproduction at the same scale as the original. Reproduced by courtesy of the late H. Vehrenberg.*

Fig. 15.55. *The area around Deneb and NGC 7000, as shown in* Atlas Stellarum. *Partial reproduction at the same scale as the original. The nebulosity is invisible on this plate, because the film used is not sensitive to the Hα line. Reproduced by courtesy of the late H. Vehrenberg.*

939

(RNGC), gathering together more than 8000 entries. There is one major defect with this, however; the positions are given for 1975.0, which means that some calculation is required to obtain the positions on any atlas used for finding objects. A complete revision of the NGC and RNGC is:

- *NGC 2000.0*, Cambridge University Press, and Sky Publishing Corp., Cambridge MA, 1988. This contains all 13 226 objects, with 2000.0 coordinates and modern magnitudes.

The most useful reference for amateurs is undoubtedly:

- *Sky Catalogue 2000.0*, Cambridge University Press, and Sky Publishing Corp, Cambridge MA. Volume 2, 1985. This gives 2000.0 coordinates for numerous double stars, all variables down to magnitude 9, as well as a list of suspected variables, 500 open clusters, 150 globular clusters, 300 bright nebulae, 150 dark nebulae, 600 planetary nebulae, and more than 2100 galaxies.

Another catalogue that may be mentioned is the *Uppsala General Catalogue of Galaxies*, which contains, with 1950 positions, about 13 000 galaxies, north of declination −2°. This is an invaluable reference for identifying faint background galaxies that may be found on photographs taken of other objects. There are also a large number of specialized catalogues used by professionals, many of which concern galaxies, but including some that cover clusters and nebulae.

15.8 The North Polar Sequence

The North Polar Sequence is a set of stars, centred on the North Celestial Pole [for 1900 – Trans.]. Like any sequence it is used to determine the magnitudes of stars, comets or other objects, or to calibrate experiments. It main interest comes from its position: the North Celestial Pole is visible throughout the year for observers in the northern hemisphere, and its altitude above the horizon is constant for any given site. In addition, the slow diurnal motion of stars close to the pole means that the sequence may be used to evaluate the performance of optics used for astrophotography.

The charts given here are a set, published by the American Association of Variable Star Observers (AAVSO), and reproduced with permission. The magnitudes shown are photovisual magnitudes. The original is reproduced here with a reduction of 1.56, which gives the following scales: 1 mm represents 220 arc-seconds on the first chart; 31 arc-seconds on the second; and 7.2 arc-seconds on the third.

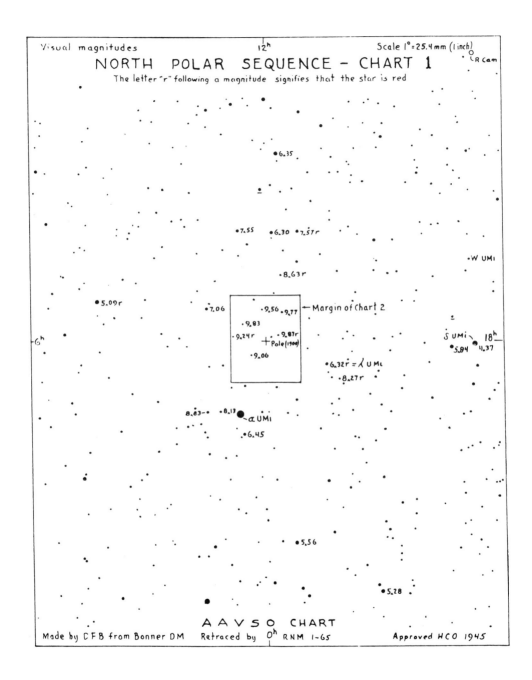

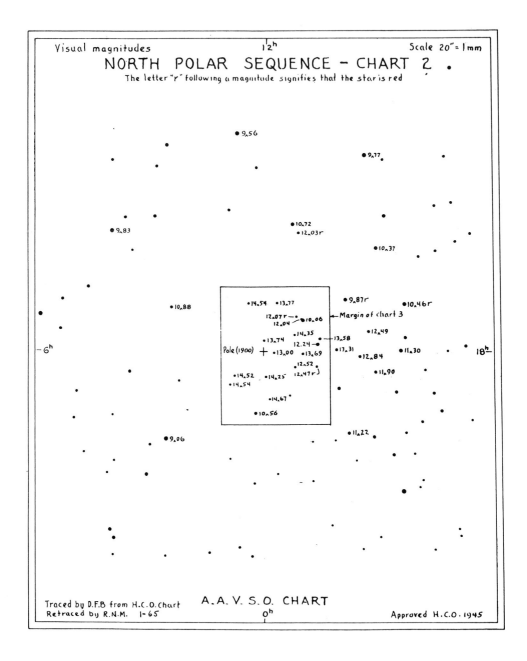

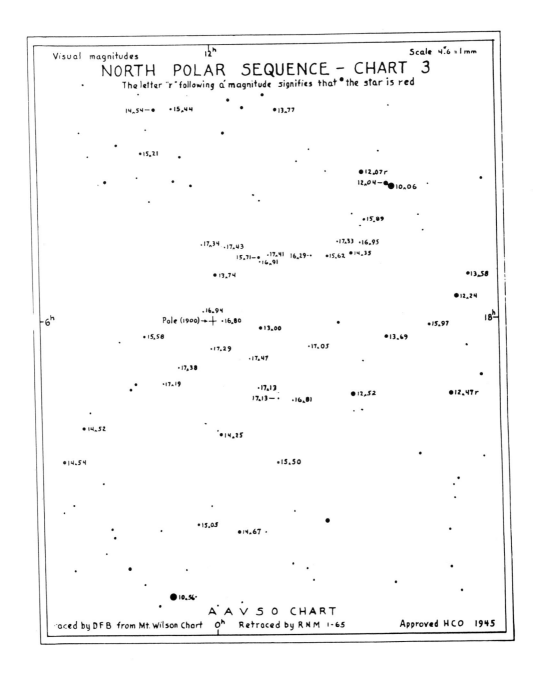

16   Plate comparators

J. P. Tafforin

All serious amateurs who have made one or more photographs find themselves confronted, just like professionals, with the problem of exploiting the scientific potential of the images. This chapter examines one specific way by which this may be attained: the use of a plate comparator.

The principle of a comparator is simple: it consists of superimposing the images of a reference frame and the photograph to be examined by means of a suitable optical system. It is possible to use either a subjective impression of depth (in a stereo comparator) or the alternation of the images (in a blink comparator). These systems, which are extensively used by professionals, are equally within the reach of a handy amateur who is prepared to take the trouble to construct a relatively complicated instrument. In what follows, we propose to discuss the overall principles that need to be taken into consideration, and to describe a few specific solutions that have been implemented by the authors.

16.1   Stereo comparators

This system uses the principles of binocular vision. Everyone knows that the brain processes the images received by the two eyes to provide depth information (i.e., a perception of distance). To do this it uses the difference in the information provided by each eye to 'reconstruct' the appearance of depth. This is why all objects 'at infinity' appear with no indication of their apparent distance: Jupiter appears just as far away as the Andromeda Galaxy, which itself seems no farther away than the Moon.

Informed readers will have realised that our brains use parallax to provide distance information, just as astronomers use it to calculate the distance of astronomical objects. Only the value of the 'baseline' is different: a few centimetres separate our eyes, but the length of a diameter of the Earth's orbit is approximately three hundred million (3×10^8) kilometres.

Let us assume that we have taken a photograph of part of the sky in which there is a comet. Nothing in the image indicates that the object is 'in front of' the background stars. But if we take another photograph of the same area of sky after an interval of time, we will obtain identical images as far as the background stars are concerned, but different for the comet, which has moved. We will have created a pseudo-parallax. If we simultaneously examine each of these two images with a different eye, our brains will create a pseudo-stereoscopic effect, and the comet will 'jump out at us'.

All we basically require is the arrangement shown in Fig. 16.1. The construction may be simply made in light plywood or in any other material that is easy to work, provided care is taken over the way in which the two plates are held. The use of slide mounts will be of considerable benefit in this respect. The light source may be simply ambient light or some form of artificial light, provided there is adequate ventilation in the latter case.

The construction may be simplified by making the distance between the two eyepieces a fixed amount; its value may be chosen to suit most individuals; i.e., somewhere between 60 and 70 mm. The slide mounts are located at the focus of the two eyepieces; the latter may quite simply be two simple lenses with a focal length of about 100 mm. The more demanding makers will want to provide a means of focussing, and will use achromatic eyepieces. At the base of the instrument there is a sheet of ground glass that diffuses the light entering the base. Two slits may be left so that the slide mounts may be inserted into guides that are lined with black velvet. The whole of the interior of the case should be painted matt black.

The two plates should be taken with the same objective, and the same guide star should have been used. If these two points have not been met, it will be almost impossible to obtain a suitable superimposition of the two images, and the device will not work. It would, however, be possible to modify this simple scheme and provide a means of adjusting the plates laterally, 'up-and-down', and to rotate one relative to the other. This would considerably complicate the device, especially when one considers that it only gives the appearance of depth. The stereo comparator has one major defect: it only reveals a displacement of an object and nothing else; this is why it is only used in searching for minor planets and comets [*see*, however Note 1, p. 1147].

Amateurs are frequently anxious to obtain the maximum information from their photographs: comets and minor planets, of course, but also novae, variable stars, meteors, etc. This was the reason for devising a system that would show unambiguously any alterations that had occurred on a photograph when it was checked against a reference plate. This is the blink microscope.

16.2 The blink microscope

The principle of the blink microscope is the same as that used in animated photography: two images, which only differ in slight detail, give the appearance of motion when viewed alternately at a sufficiently high rate. If a reference plate and another carrying the image of a recent nova (for example) are examined alternately, the presence of the new object will be immediately obvious by the fact that it blinks, whereas the images of 'ordinary' stars will remain unaltered, because they are present on both images. This assumes that the two plates appear perfectly superimposed, because otherwise the result will be impossible to interpret. It is therefore essential that the alignment should be very accurate. Note that in such a device the concept of depth, i.e., of binocular vision, is not used at all. This is why both images are brought to the same eyepiece. Naturally a binocular head may be provided – and this is frequently the case with laboratory microscopes – to make viewing more comfortable, but only a single image is examined at any one instant.

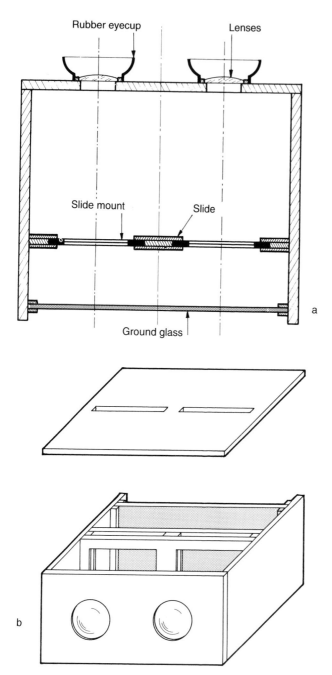

Fig. 16.1. *A simple stereo-comparator.*
a) Plan view. A division may be added to separate the two negatives. The rubber eye-cups are not obligatory.
b) Perspective view. The top cover has been removed to show the arrangement of the carriers. Note the two slots through which the slide-mounts may be introduced.

16.3 Building a blink microscope

The following section describes how a brave amateur may construct such a device, without, however, giving a specific 'recipe'. There are numerous possible solutions, and there is no point in trying to enforce one or other of them. We simply describe our own experiments and a device that we have built ourselves.

16.3.1 Optics

The general principle is that each image shall be presented alternately to the eyepiece, so we need to use two objective lenses and an optical system that allows the two images to be superimposed. One solution is to use a pair of binoculars (8 × 30, for example), from which one salvages the objectives and prisms, together with a commercial beam splitter, i.e., a semi-reflective plate. Note that lenses that have been specifically calculated to form images of objects at a short distance (i.e., lenses for macrophotography) are by far the best – they are also much more expensive.

Other arrangements are possible: Godillon (1971) and Jean Texereau (1947), in particular, describe systems that use prisms instead of the beam splitter. The example shown in Fig. 16.2 is a system that uses a beam splitter.

It is essential to be able to equalize the magnifications of the two lenses, so one of them needs to be adjustable, as does the prism (or mirror) feeding it. There are various means of making allowance for this adjustment; it should be remembered that it is done once and for all – or at least only rarely requires attention (if the mounting tends to go out of adjustment). So there is no need to provide a highly sophisticated arrangement. Anyone who is meticulous will perhaps want to provide a slide – or rather two slides, one for the lens and one for the prism/mirror – controlled by threaded rods that extend outside the optical box, and fitted with suitable knobs. If the threaded rods are 6 mm in diameter, standard potentiometer knobs – which are available at low cost from electronic shops – may be used.

In general, the mounting of the optical components needs to be accurate. The prism or central mirror, and the beam splitter need to be provided with accurate adjustments in each plane, for example, by using pairs of opposing screws (push-pull screws), or else by using springs. Attention must be given to the distances between the various components, and the adjustments designed taking into account the final overall dimensions, which are largely determined by the focal length of the lenses. Specific dimensions are given in Fig. 16.2 for this reason, and they show the values chosen to suit lenses 35 mm in diameter, and 130 mm in focal length, derived from a pair of 7 × 35 binoculars. Note that this instrument provided nearly all the optical components required. The optical path ends at a single eyepiece, the focal length of which is chosen to give the desired magnification. This should be sufficient for the grain of the film to be just detectable, but no higher, because the time taken to examine each photograph is shorter, the larger the field seen in the eyepiece at any given time. For a 6 × 6 format, a 10× magnification is comfortable.

Naturally, the whole of the interior of the optical box should be painted matt black. Finally we may mention that it may be desirable to stop down the lenses slightly to limit aberrations (to 15–20 mm, for example, for 35-mm diameter lenses).

947

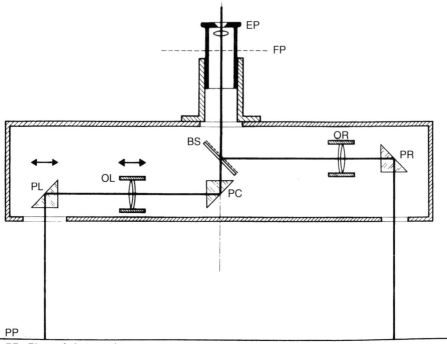

PP: Plane of photographs
PL: Prism (left)
OL: Objective (left)
PC: Prism (central)
BS: Beam splitter (the dotted line indicates the reflecting surface)
OR; Objective (right)
PR: Prism (right)
FP: Focal plane
EP; Eyepiece

Values for objectives with a diameter of 35 mm, and focal length 130 mm:
PP–PL: 160 mm; PL–OL: 100 mm; OL–PC: 100 mm; PC–BS: 40 mm; PP–PR: 200 mm;
PR–OR: 60 mm; OR–BS: 140 mm; BS–FP: 120 mm.

Fig. 16.2. *An example of the optical arrangement of a blink comparator that uses a beam splitter. All the prisms may be replaced by good-quality, surface-aluminized, plane mirrors. Note how the left-hand objective and prism may be adjusted, as well as the difference in height of the two optical paths.*

16.3.2 The blink device

So far, the optical box effectively combines the two images in a single light path and simultaneously superimposes them. The latter is, of course, essential in the process of positioning the images. But for the examination proper, each image should be presented singly to the observer's eye. We therefore need a device to blink the two paths.

The first method of doing this is to occult one image after the other inside the optical box. This may be carried out either in front of the objectives, or else between the latter and the beam splitter. In the latter case, the arrangement described by Texereau may be used: two disks with cut-out sectors mounted vertically on the optical axis, and rotating at a speed selected by the user, so that the occultation frequency is between 1 and 10 Hz. If it is preferred to occult the paths in front of the objectives, the two sector disks may be situated flat at the bottom of the box, together with all the gearing required to rotate them. Simple 'Meccano type' gears make it very simple to make up a suitable arrangement (Fig. 16.3). Because the two images must be superimposed, there needs to be a more complicated mechanical arrangement, with one of the idler gears capable of sliding on its shaft, being lifted by a lever and a fork. Rotation of the disks may be either by means of a knob or crank, or by using a small motor. This will require a reduction box. Readers choosing this method can use a low-voltage DC motor and make an electronic speed control (Fig. 16.4), which will make operation much more convenient. Whichever method is employed, every effort should be made to avoid vibration, which becomes very disturbing with prolonged observation.

One variation is to place the occulting disks between the source of illumination and the plates, i.e., outside the optical box. There are two advantages: increased accessibility of the occulting system, so that it is easier to arrange for it to be moved out of the way when both plates have to be illuminated simultaneously; and mechanical isolation of the motor and gearing and the optics, which is the best way of avoiding vibrations. Using this method it is preferable to provide artificial illumination (a bulb beneath each photograph), and to be prepared to use the blink microscope in a darkened room.

As may be seen, whatever arrangement is adopted it involves careful mechanical construction. There is a great temptation to forgo occulting the images and instead to 'switch them off' by controlling the illumination electronically. This solution suppresses all vibration at a stroke, but means that artificial illumination has to be used, and with it use in the dark.

The electronic version is shown in Fig. 16.5. The lifetime of the lamps is likely to be relatively short when they are rapidly switched on and off. To simplify the electronics and ensure greater safety, these lamps are low-voltage ones (12 V), and the amount of illumination provided is usually more than enough. If this were found to be a problem, the type of bulbs used in vehicles (12 V, 5 W or 21 W) would provide a solution, at the expense of having to increase the current available from the power supply. The 12 V, 120 mA or 250 mA bulbs used in the most powerful electric hand-lamps appear to be more than adequate. The light-box (Fig. 16.6) should be painted white, and divided into two compartments corresponding to the two plates. With the more powerful forms of illumination, forced ventilation will be required, using, for example, the small, 12 V, 50 Hz fans that may be bought from electronic suppliers. In any case, there should be adequate ventilation holes.

People who choose the mechanical method can also decide what type of illumination they will use. If artificial light is used, then ordinary, domestic-voltage lamps may be fitted, which will give ample illumination. Forced ventilation will then become imperative, however, as will two sets of components to vary the intensity of the

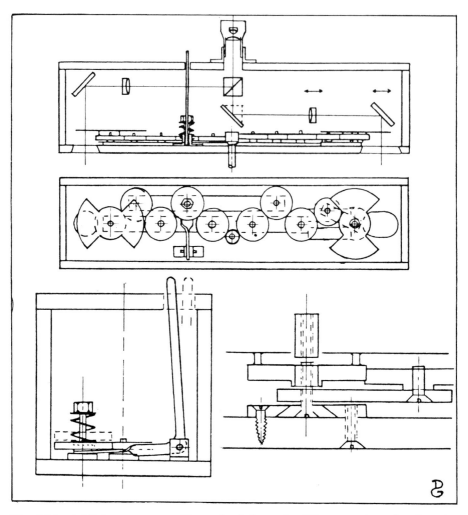

Fig. 16.3. *A blink comparator.* Top: *the light path.* (*The mountings for the mirrors, prisms and lenses are not shown.*) *Note the lever to the left of the eyepiece that lifts one gear, and the greater height of the right-hand gear-train caused by the adjustable slide.* Centre: *plan view of the occulting system and gear-train.* Bottom left: *the lever that lifts one of the gears.* Bottom right: *section of the slide that adjusts the position of the sector wheel. Note the two methods of fixing the slide rails* (*The various components are shown slightly separated for clarity.*)

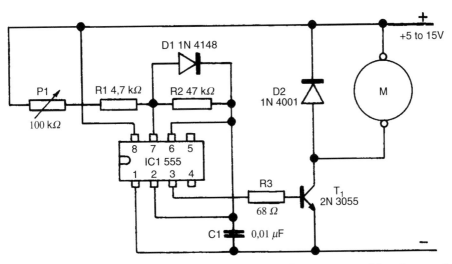

Fig. 16.4. *Speed control for an electric motor. This circuit enables the speed of a miniature motor of between 5 and 15 V DC to be altered. The values of P1 and R1 depend on the characteristics of the motor, but are not really critical. The values shown should be suitable in most cases. It is perfectly possible to find the correct values by trial and error, provided R1 is never zero. When the value of P1 is increased, the motor slows down, and conversely.*

Fig. 16.5. *(Opposite) – A power supply to achieve electronic 'blinking'. The apparently complex circuit is, in fact, fairly simple: sections Ⓐ and Ⓒ are identical; they enable the brightness of each lamp to be adjusted. In addition, the way in which they function is similar to the speed control described earlier. Section Ⓑ is an oscillator, with a square-wave output, the frequency of which may be adjusted by P2, R3 and R4 as given by the equation: $F = 1.44(C7)/[(R3+P2)+(2R4)]$. The values shown give a frequency range between approximately 0.5 Hz and 10 Hz. This signal is then injected into IC4 (section Ⓓ), which divides the frequency of the signal by 2 and gives a symmetrical output. This circuit also allows one to select which lamp is illuminated (or both) by means of the double-pole, 4-way switch S1a and S1b. Position A: blinking; B: both lamps illuminated; C: one illuminated, the other dark; D: as C, but reversed. The output from this circuit therefore consists of two identical frequency signals, which are injected into IC5 (section Ⓔ), at 13 and 1. This circuit adds the signals controlling the brightness of the lamps. The last part of the circuit consists of IC6, which acts as the first power output stage, and two power transistors that control the lamps L1 and L2. The power supply is very simple and requires no explanation, other than the fact that the transformer primary winding should be suitable for the mains input voltage (i.e., 220 V or 240 V). Note that the output may be any value between 5 and 12 V, depending on the voltage of the lamps being used.*

Component list:

IC1, IC2, IC3: NE 555	C1, C2, C3: 10 nF
IC4: 4013	C4: 1000 μF, 25 V
IC5: 4011	C5: 100 nF
IC6: 4049	T1, T2: 2N 3055
P1, P3: 100 kΩ	F: fuse 220 V (240 V); 1.5 A
R1, R5: 4.7 kΩ	I1: 220 V (240 V) switch
R2, R6: 47 kΩ	TR1: 220 V/12 V (240 V/12 V) transformer, 3 A
R3: 10 kΩ	IC8: rectifier bridge 4 A
R4: 22 kΩ	IC7: voltage regulator LN 7812
R7, R8: 68 Ω	S1: double-pole, 4-way switch
P2: 1 MΩ	D1, D2: 1N 4184
C6, C8: 0.01 μF	L1, L2: 12-V lamps, 250 mA
C7: 1 μF	minimum, 500 mA maximum

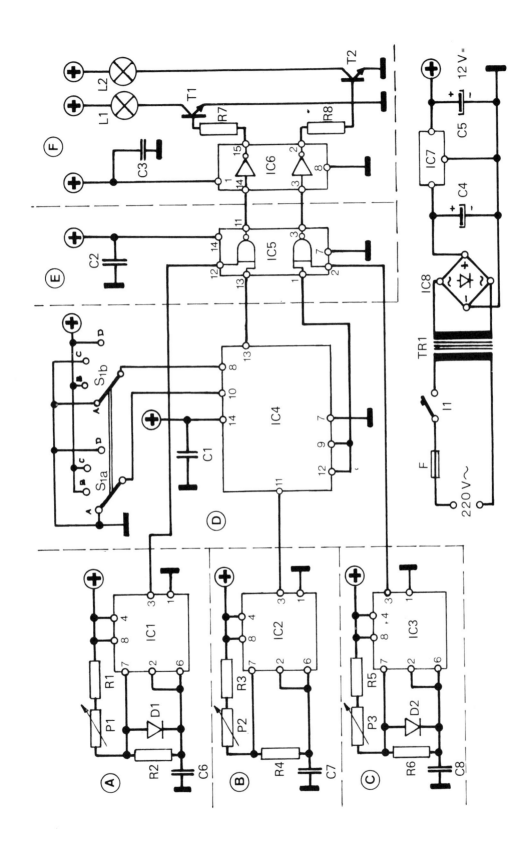

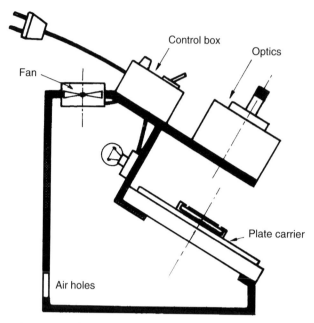

Fig. 16.6. *The light-box; the lamp is placed so that the light reaching the negatives is diffused. Note the position of the fan, and the direction of the flow of air. The box containing the electronics has been placed above the optics, but this is not critical.*

lamps (Fig. 16.7). It would seem more logical to rest content with ambient lighting, the light being fed to the two plates by mirrors that may be adjusted to equalize the illumination. This solution is not particularly simple, however, because it involves obtaining an evenly illuminated source of light. This might be a large sheet of white paper, illuminated by two lamps that are sufficiently bright, and about one metre from the equipment. On the other hand, the construction is simplicity itself. The mirrors may be anything suitable; ordinary looking glass is quite adequate.

16.3.3 The plate holder

This is the most critical part of the equipment. It needs to hold each plate directly beneath one of the lenses, and allow the images to be accurately superimposed. In theory, three motions are required: longitudinal movement of one plate, transverse movement of the other, and rotation of one of them. In our case, we decided to allow both plates to be rotated, thus extending the equipment's potential – enabling us to measure angles and the displacement of objects – but this is not obligatory.

It will be recalled that one of the two folding prisms (or mirrors) may be moved. In fact, it must be possible to vary the distance between the film plane and the objective, and also between the objective and the eyepiece. Because the plate-holder needs to be directly beneath the prism, the former's amount of travel needs to be increased accordingly. For example, if the plate-holder that moves longitudinally

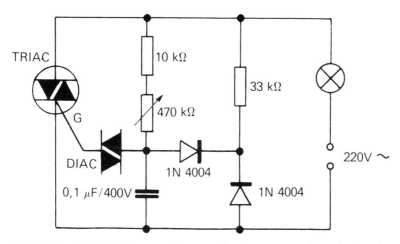

Fig. 16.7. *A 220-V dimmer. This standard circuit enables the brightness of a 100-W, 220-V lamp to be regulated. Two are therefore required (one for each lamp).*

has 20 mm of travel, the other one, which moves transversely, should have the same amount of travel, plus the 20–30 mm that corresponds to the prism's movement.

There are a number of mechanical solutions, so we will just mention a few basic principles:

- all the movements should be as free from play as possible;
- the movements should be slow;
- it must be possible to mount and remove the plates without disturbing the system;
- the major movements that allow the different parts of the plates to be examined must take place at right-angles to one another, and without affecting the alignment of the two plates;
- the equipment should be tilted at an angle (about 10–40°) to increase the observer's comfort.

It is best to build the device from metal, which does not help those amateurs who are not fully equipped to handle such material; in many cases it will be necessary to obtain help from someone with appropriate skills, which will increase the cost of the equipment. It is possible to use the strips of brass that are sold by model and hobby shops, as well as the dural sections used in furniture kits. We do not give precise dimensions here, because everything depends on the format chosen and on the equipment available, but Fig. 16.8 shows the principles that should be followed. In our case, only the two disks of aluminium and glass were made by professionals (a turner and a glazier). Particular attention was paid to the rigidity of the frame; simple glue and screws will not suffice here. Threaded rods run right across the frame and are held by nuts and large washers. Nevertheless, wood would probably work; here chipboard is probably best, because it has better dimensional stability than even marine plywood.

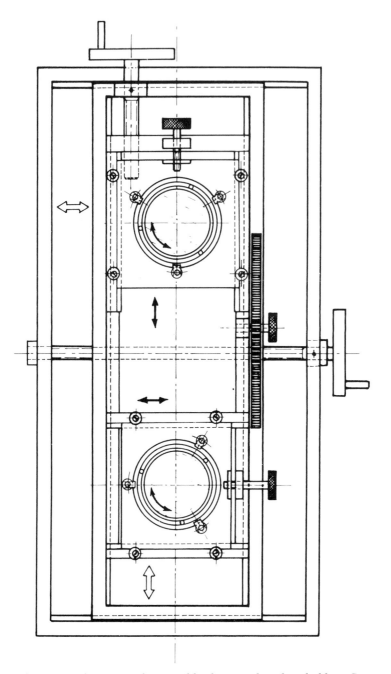

Fig. 16.8. *Plan view of a possible design of a plate-holder. Coarse motions are provided by long screws, fitted with handles. The rack shown provides transverse motion. Note that the frame's transverse travel is greater than that at right-angles, enabling adjustments to be made to suit the movable prism or mirror (see Sects. 16.3.1 and 16.3.3). The film carriers are not shown.*

The dural rails that permit the major movements (for examining different parts of the plates) slide on screws fitted with brass, domed nuts. Such a system is obviously inferior to a true ball, but it is adequate in practice, especially if the nuts are polished with very fine crocus paper. Each reference plane is defined by four screws, one of which may be adjusted in height to avoid a 'wobbly' mounting.

We come to the two rectangular metal frames. The first carries the two plate-holders, together with their adjusting mechanisms. The whole of this frame slides transversely within another frame, equally rigid, which itself slides within the outer structure. The latter motion may be provided in a different way, by making this outer frame slide on two tubes, or on a thick central rail. The latter arrangement is particularly useful if one has chosen mirror illumination, because the mirrors are then free from obstruction on each side of the centre (see the mountings used by Godillon or Texereau). We shall leave readers to decide on the solution that suits them best.

The plate-holders themselves should suit the format of the film being used. It might be worth considering whether the film may be held in strips. Some people prefer to mount each image in a slide mount (the format is then fixed). One could also adopt a mixed system. Whichever solution is adopted, the glass plate must be completely clean. This is why we chose to mount the plates in such a way that their top surfaces are completely free and accessible. This ensures that they may be thoroughly cleaned, using a very soft cloth, with which any dust may be swept to the edge, where there is nothing to retain it. The two metal disks rotate on three domed nuts and are retained by three brass feet. Rotation in the plane of the plate is controlled by three small ball bearings, spaced at 120° intervals (Fig. 16.9). By turning one with the fingers, there is a reduced, contrary rotation of the plate. Some form of clamp is needed to prevent accidental alterations.

It must be possible to position the plates themselves very accurately with respect to one another. It seems best to use a screw, threaded into the plate-carrier, with a knob at one end (Fig. 16.10). The gross positioning of the frame is not so critical and a rack-and-pinion arrangement appears quite suitable. If the design used includes a central rail or tube, it is even possible to make do with a pin that engages in holes spaced along it. If the whole instrument is inclined at an angle – as is desirable – the frame is held in position by gravity.

16.3.4 Using the blink microscope

After having switched on the lights and checked that the fan is working, the two negatives are placed in position and correctly oriented. Choosing two stars, the line joining them on one plate is aligned parallel with the same line on the other plate (both images being visible at the same time). This is done by rotating one of the disks. Then using the fine adjustments the two pairs of stars are superimposed. Naturally, the magnifications will have been previously equalized once and for all, and the focussing carried out with the greatest care.

The occulting device, or its equivalent, is then set in motion, and the frequency adjusted to the operator's liking. The results are then:

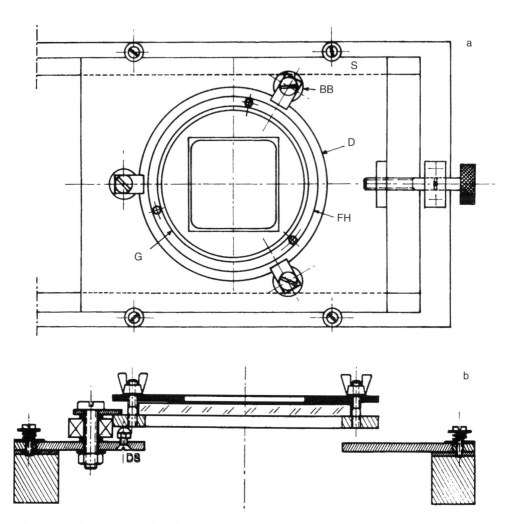

Fig. 16.9. *Plate-holder (detail)*

a) Plan view of one plate-holder. S: brass slides; BB: ball-bearing; D: disk of dural, OD = 120 mm, ID = 90 mm, thickness = 5 mm; FH: film-holder; G: glass plate, 5 mm thick. Note the brass washers on top of the slides, and the three screws that keep the glass disk in place. The size shown here is 6 × 6.

b) Section of one plate-holder. Note the spring-loaded washers that prevent the slides from rising, and also the dome-ended screw DS, on which the dural ring rotates. A smooth rotation of the film is achieved by turning one ball-bearing.

- a minor planet appears as a point that jumps backwards and forwards, or is elongated;
- a nova seems to blink in a single position;
- a variable star either blinks or seems to pulsate.

In general, any object present on just one of the plates seems to blink.

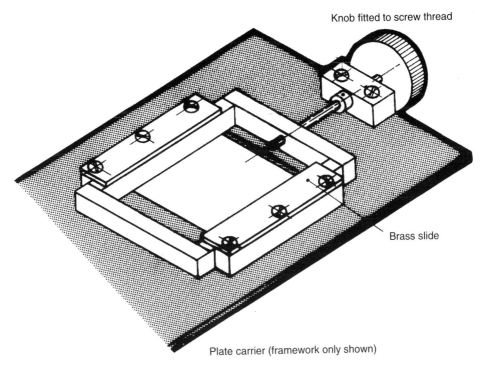

Knob fitted to screw thread

Brass slide

Plate carrier (framework only shown)

Fig. 16.10. *The basic principle used to provide motion of plate-holders. This arrangement should be adopted, whatever other form is used for the actual film holders. It is the only type that provides a sufficiently delicate motion. The slides and arrangement shown here are only schematic.*

When the equipment is first used, the operator will be amazed at the number of specks of dust that appear on the images, despite every precaution being taken to clean the glass and the film. Don't be discouraged: generally the magnification is sufficient to distinguish these from the true images. Sometimes, odd specks of silver may appear to be a real object, but this does not happen very often, and use of a proper microscope on them will soon remove any doubt.

16.3.5 Possible improvements

There are a number of improvements that may be made, but they require better engineering facilities than most amateurs possess. We may mention:

- measurement of the position of an object;
- direct read-out of the coordinates;
- a link to a computer, etc.

Any such system would be much more complicated and would demand a lot more time. In any case, is it really 'doing astronomy'?

16.4 Other systems and variants

Various authors have suggested ways of checking photographs. We may mention, amongst others, two proposed by Ben Mayer from California (Mayer & Liller, 1985):

- PROBLICOM (Projection Blink Comparator): This is a blink comparator that uses two slide projectors. The two projectors are mounted one above the other in a framework that allows the images to be accurately aligned (Fig. 16.11). The images are occulted by a sector wheel mounted between the two projectors. There is one particular feature that is interesting in that it allows one to work in a group, which might be useful for teaching purposes. In addition, the actual method by which the adjustments are obtained is very simple, and well within the scope of any amateur without many tools or equipment. On the other hand, two identical projectors are required, and the format is fixed (generally 24 × 36 mm). It is undoubtedly very suitable for an astronomical society or club.
- STEBLICOM (Stereo Blink Comparator): This is a greatly simplified blink comparator, which may also be used as a stereo comparator (Fig. 16.12). It is a device very similar to that already shown in Fig. 16.1, but with the addition of an arrangement to alternately switch the lamps on and off. It obviously uses an artificial light source. When both lamps are illuminated, it functions as a stereo comparator and things are seen in false relief. When the lamps are blinked, this effect disappears, but then it works as a blink comparator. One advantage is that this device is extremely simple to make (when compared with a blink comparator). Its disadvantages are that it does not give a very large magnification and is limited to a single format. In addition, blinking with binocular vision (each eye sees an image alternately) becomes uncomfortable after a while. An identical type of device, but even simpler in construction, has been suggested by P. Bourge, J. Dragesco and Y. Dargery (1977).

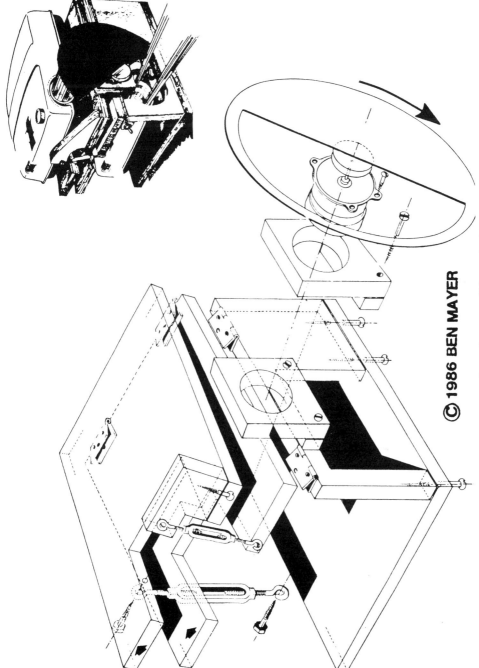

© 1986 BEN MAYER

Fig. 16.11. *Details of the* PROBLICOM *arrangement for using two slide projectors.*

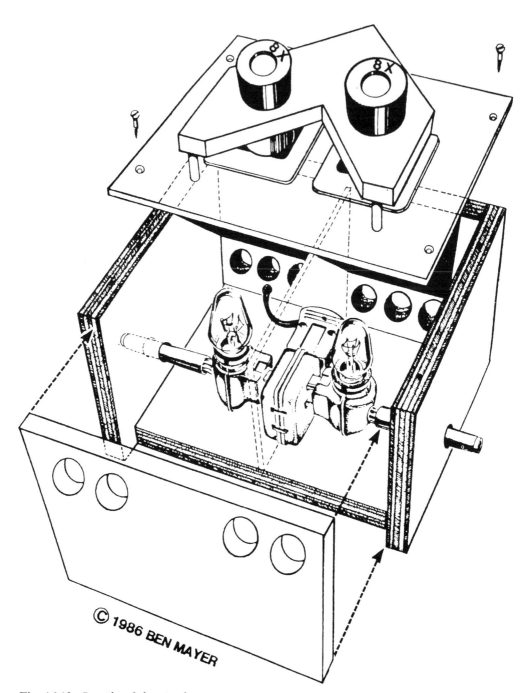

Fig. 16.12. *Details of the simple* STEBLICOM *stereo comparator.*

17 Astrometry

R. Chanal

17.1 Introduction

17.1.1 Archaeoastronomy

From the very dawn of humanity, people have watched the sky. This is shown by the designs carved or painted in prehistoric caves and shelters. At a very early stage mankind saw the march of stars across the sky and recognized the periodic return of some stars with changes of the seasons.

These simple observations, more statements of fact, were soon felt to be insufficient, and our ancestors started to erect structures, often made from the only material readily available: stone. They installed these stones, which were sometimes gigantic, so that their alignments indicated the rising or setting points of certain specific astronomical objects, especially those of the Sun.

The site of Stonehenge in Great Britain is one of the best-known examples. Specific alignments indicate the direction of sunrise at midsummer (when $\delta = +24°$) and at midwinter ($\delta = -24°$), as well as the direction of sunset on the same dates. Other alignments indicate the rising and setting positions of the Moon at its maximum and minimum declination ($\delta = +29°$ and $-29°$).

Such monuments bear witness to an accurate knowledge of the movements of the Sun and Moon, and are, in fact, the first instruments devised by Man to measure angles on the sky. Almost all over the world, different structures may be found: tumuli, muraghi, alignments of menhirs, and pyramids, and later, temples and observatories. They are silent witnesses to Man's efforts to measure the motion of objects in the sky. They are, in fact, the first astrometric instruments.

17.1.1.1 Ancient astrometry

Later, in the Hellenistic period, new methods appeared. People were able to take advantage of current technology, and we find them using portable instruments, such as the hemispherical sundial and the astrolabe, as well as more complex instruments such as the Antikythera mechanism, which appears to have been designed to calculate the position of the planets. It was then that the earliest catalogues of stars and of astronomical data appeared: the Almagest, for example, which was compiled by Ptolemy, is a very important catalogue of stars that dates from this period. Little progress was made in subsequent centuries, and the construction of devices such as armilliary spheres and quadrants brought little improvement in the accuracy of measurements.

17.1.1.2 Modern astrometry

Visual astrometry A new astrometry arose in the 16th century. Construction of improved instruments such as rotating and mural quadrants, enabled dedicated observers like Tycho Brahe to measure the positions of stars with an accuracy of about 1 minute of arc. He measured the positions of 1005 stars, which he brought together in a catalogue that was later used by Bayer to create his *Uranometria*.

The introduction of the telescope brought about a revolution by providing greatly increased accuracy. This technological breakthrough came in 1610, and optics remained the driving force behind astronomy for many years.

Photographic astrometry   Another major event occurred in 1816, when Niepce invented photography, although it was 1840 before the first astronomical photograph (of the Moon) was obtained by J. W. Draper. W. Bond photographed a star in 1850, but it was 1880 before photography of the sky became widespread, and gave rise to the first photographic atlases. This new technique provided a permanent record of the relative positions of all the stars and other objects that lay within the field of view at the time the exposure was made.

Before then, making a chart required measuring the time of meridian passage of stars, one after another, to determine their angular separation in right ascension (α), and measuring their difference in declination (δ), in the plane of the meridian, directly. All these values then had to be plotted onto paper; a long and difficult process that was subject to errors in all the various mathematical and graphical manipulations that were required.

With photography, everything is available, instantly. Naturally, the positions of certain reference stars must be known, when appropriate mathematical methods may be used to determine the position of any other star visible on the plate. This is known as 'differential astrometry' in contrast to 'fundamental astrometry' which determines positions with reference to the celestial equator and the vernal equinox (Υ).

Astrometry in space   We will say just a few words about this subject, significant work in which is currently being carried out by the HIPPARCHOS mission. [This satellite should have been placed in geostationary orbit, but because of a booster rocket malfunction during its launch by an Ariane rocket in 1989, is in an elliptical, intermediate orbit. Despite this, its mission has been highly successful. – Trans.] Its rotation causes a complete 360° band of sky (in fact, two bands of sky 70° apart) to be swept by its instruments. By altering the inclination of its axis of rotation, a new band of sky is examined, and so on. The first survey of the whole of the sky was covered in the relatively short time of about 2 years.

The positions obtained in this way are not affected by atmospheric refraction and absorption, and their accuracy is improved by 10 to 100 times in comparison with terrestrial measurements.

17.1.2 *The aims of astrometry*

17.1.2.1 *Astrometry and time-measurement*
Until recently, these two elements were inextricably linked in fundamental astrometry, because the rotation of the Earth on its axis (assumed to be uniform) was used

to determine the angular separation in right ascension of objects crossing the plane of the meridian.

It is immediately obvious that this regularity is questionable, not least because the length of the day varies according to the time of year, as shown by the equation of time. Other irregularities are revealed by the meridian passage of distant stars, and this led to the development of better and better clocks that indicated mean time. We shall not, however, discuss this in any greater detail here, but instead refer the reader to the sources given in the bibliography.

17.1.2.2 *The development of celestial atlases*

For a long time the sky was represented on globes, showing the constellations as envisaged by ancient peoples, and one such is that carried on his shoulders by the Farnese Atlas. The first charts appeared during the Renaissance, like that by Albrecht Dürer, which was printed in 1515.

The first modern representation of the sky (without mythological representations of the constellations) was that by Harding, which covered the sky from the North Pole down to declination −30° in 27 charts, and which appeared in 1822. In 1863, Argelander's 44 charts (derived from the Bonner Durchmusterung or BD catalogue) were published, showing 324 198 stars. It took Argelander and two assistants 7 years to measure all the positions. The charts and catalogue continue to be used today.

Another catalogue and its atlas were published by the Smithsonian Astrophysical Observatory in 1959. It is a compilation of older catalogues, converted to a consistent reference frame. It consists of 152 charts, 22° square, to a scale of 120 arc-seconds per mm, equinox 1950, and contains 258 997 stars down to about magnitude 10.

17.1.2.3 *Photographic atlases*

Strangely, photographic cartography began with that of the southern sky. This was carried out by David Gill in 1882–5, because there were no charts comparable with those available for the northern sky. This atlas contains some 455 000 stars down to magnitude 10.5 on 613 photographic plates. It is known as the CPD (*Cape Photographic Durchmusterung*).

The *Carte du Ciel*, a photographic atlas of the sky was begun in Paris in 1887. Observations were split between 18 observatories around the world. They came to an end in ... 1965. All the work of reducing the measurements was made without computers, of course. This cartographic work was never completed, because, in the meantime, a photographic telescope that was extremely powerful – both by virtue of its optics and the area covered – had been commissioned. This was the Big Schmidt at Palomar in California.

The project carried out with this instrument did not include measurement of stellar positions, and the atlas for the northern hemisphere was completed in a few years. It consists of 1870 photographic plates, with a quality only dreamed of by amateur astronomers, and is known as the 'Palomar Observatory Sky Survey' (POSS). The plates were taken with both blue and red filters, thus giving some indication of the spectral class of the stars that are recorded.

17.1.2.4 Objects for which positions may have to be determined

Unusual stars Here we include objects, found by observers, that are visible for a limited time, and where their coordinates are required so that they may be observed worldwide. Novae are an example, as well as some faint variable stars, which suddenly become accessible to smaller instruments.

Planets Thanks to Tycho Brahe's astrometric measurements, Kepler was able to formulate his famous laws. He succeeded in explaining the planetary positions that had been carefully recorded by his predecessor. Without that meticulous work, it would have been impossible for the discovery to have been made.

Comets It is from measurements of their positions on the sky at different times that the orbits of comets may be determined, and ephemerides prepared. For periodic comets, the orbital elements may vary as a result of possible encounters with the major planets, so their paths must be monitored at every return. Comet Halley was the subject of particularly comprehensive astrometry with the aim of refining its orbit before its encounter with the GIOTTO probe, and also of understanding the non-gravitational forces that it exhibits (in particular the reaction produced by the jets of gas that it emits).

Minor planets Like comets, the low mass of these objects (relative to other Solar-System bodies) makes them very subject to perturbations, and their low albedo means that their passage across the sky is not very prominent. Some of their orbits are poorly known, or unusual, and they require further study. Astrometric measurements provide the main source of data for the required calculations.

17.1.2.5 Determining the motions of stars

Proper motions of stars It is now widely recognized that nothing in the universe is static. It is nevertheless true that observers on the Earth have every reason for taking the centre of the body on which they are situated as being a reference point, and considering it as being at rest to enable the amount of motion of other bodies – even the most distant one, i.e., the stars in our Galaxy – to be determined.

The comparison of the positions of stars in old catalogues and modern observations has enabled their motions to be detected. This motion takes place in both α and δ, and is given in catalogues in the form $\mu\alpha$ and $\mu\delta$. Together these give the proper motion of a star. It is only the apparent motion, relative to two orthogonal axes, projected against the background sky. We shall have to take these proper motions into account in reducing astrometric measurements.

Definition of stellar parallax The Earth's motion in its orbit enables us to obtain a slightly different view of the stars, when, after six months, the Earth is on the opposite side of its orbit. Because of perspective, the nearer stars appear to move in ellipses against the background stars over the course of a year. The amplitude of this motion is inversely proportional to the distance of the star observed relative to the Sun, which is conventionally, and to simplify matters, considered as fixed. In fact, the Sun's proper motion, is that of just one star among many, and 'deforms' the circles (or ellipses) mentioned above. Because of this, it is possible to determine the direction in which it is itself moving.

Obviously the size of the small apparent orbits shown by stars enables their distances to be determined, because the value of the radius of the Earth's orbit is accurately known.

In 1838, the German astronomer Bessel was the first to determine the parallax of a star (61 Cygni), which amounted to 0.3 arc-seconds, i.e., 680 000 AU. It was a major success, because it was the first time that the true distance of the stars could be fully appreciated, something that previously had been only supposition.

17.1.3 Modern, ground-based astrometric methods

17.1.3.1 Transit instruments

Transit telescopes We will briefly describe these instruments: they are refractors that are only able to move in the plane of the meridian that runs through their site. During the course of the night, stars lying east of this plane cross it towards the west. The next night, the same stars cross it (on average) $23^h56^m04^s$ later. By determining the time between the passage of two stars, the angular distance between them (along planes parallel to the equatorial plane) may also be determined.

Transit telescopes have been perfected over the years, and the operation of some has now been completely automated. One such is the transit telescope at the Bordeaux Observatory, which has been used to obtain data for the HIPPARCHOS input catalogue (Fig. 17.1).

Other transit instruments Among other astrometric instruments, the photographic zenith telescope must be mentioned. This determines the latitude of a place. As its name indicates, it points towards the zenith and photographs stars that are carried across that area of the sky by the rotation of the Earth. It is an extremely accurate instrument, capable of measuring angles of 0.005 arc-seconds.

Another astrometric instrument that cannot be omitted is one developed by a great French astronomer, André Danjon: this is the impersonal 'astrolabe' named after him. This instrument is able to measure the latitude, and to determine the time of meridian passage of stars, whose declinations are as much as 30° greater or less than the latitude of its site. It essentially consists of a horizontal telescope which receives the light from a star via two different optical paths. One is direct, through an equilateral prism with horizontal edges, the other is reflected by a perfectly horizontal mirror of mercury before passing through the prism. The observer therefore sees two images of a star in the eyepiece. When the two images coincide, the star itself is exactly 30° from the zenith.

17.1.3.2 Photographic instruments

Astrometric telescopes Most astrometric instruments are refractors, astrolabes, or zenith telescopes; reflecting telescopes are rare. The reasons for this may be mentioned briefly. The optics of refractors give better images over the whole field than do those of reflectors, which normally have small focal ratios (F/D). There are, however, a few notable exceptions. The best known is undoubtedly the one

Fig. 17.1. *The transit telescope at Bordeaux Observatory*

commissioned in 1964 at Flagstaff in Arizona. It has a fork-mounting of generous dimensions, designed to give exceptional rigidity. Its optics consist of an f/10 mirror with a diameter of 1.55 m. Its focal length is thus 15.5 m. The optical path is folded like that of a Cassegrain, but with a plane mirror instead of the traditional hyperboloid. Photovisual magnitude 18 is reached with 10-minute exposures, and the telescope provides astrometric measurements with an accuracy of ±0.015 arcseconds.

Wide-field cameras To avoid the defects of telescopes with parabolic mirrors, which have relatively small distortion-free fields, Bernhard Schmidt devised a wide-field reflector that gave excellent images over the whole field. This type of telescope is now widespread. The one at Palomar, as we have said, obtained remarkable images of the northern sky.

Other optical designs have been suggested to obtain coma-free images, but only one has really been adopted: this is the Ritchey–Chrétien telescope. The focal ratio of this type of instrument generally lies between 7 and 8.

Fig. 17.2. *A plate-measuring machine at Toulouse Observatory*

17.1.3.3 Plate-measuring machines

Once a good photograph of stars has been obtained – with the accompanying difficulties that we shall describe later – it needs to be measured, i.e., the distances between the stars and two orthogonal reference axes, known conventionally as the *x*- and *y*-axes, need to be measured. These measurements must be made to a high degree of accuracy, if the whole information content of the plate is to be utilized.

Optical instruments have been designed to do this: they are very accurate indeed, and are generally maintained in constant-temperature environments to ensure dimensional stability. They are, in effect, metrology instruments (Fig. 17.2)

To obtain measurements in two perpendicular axes, the simplest method is one that is found in many machine tools. A table is able to move along the *x*-axis and carries a second table capable of moving along the *y*-axis. This enables a rectangular area (which may or may not be square) to be covered. Some measuring machines have this type of design. Instead of a cutting tool, they carry a microscope with cross-hairs, which may be positioned over the centre of the image of a star present on the negative, which is generally illuminated from below. The value for each axis

969

may then be read from a graduated scale or drum. The operation is repeated for all the reference stars as well as for the objects whose position is to be determined.

Another method of measuring the values on two perpendicular axes is to move the table carrying the eyepiece along one axis only, frequently the x-axis, note the value, and then turn the plate-holder by 90° to take the y-axis measurement. Obviously it is essential to check that the rotation is exactly 90°. Some observatories have this type of machine, which is generally simpler in construction.

The reader will doubtless note that we have not mentioned making measurements on paper prints, printed as negatives. In fact, given the accuracy required in their work, professional astronomers are unable to accept the distortions that occur when information is transferred from one type of emulsion base to another.

17.1.3.4 Reduction
Reduction of measurements made by automated transit telescopes Obviously when the motion of the refractor in declination is controlled automatically by an appropriate program, reduction of the measurements is also automatic. Generally the same computer carries out both of these functions, and the operator only supervises its operation.

Reduction of photographic plates In this instance, the operator has the task of providing the computer with the x and y values for all the reference stars, and for the objects whose positions in α and δ need to be determined. This is work that requires great care, to avoid entering inaccurate values, especially as the measurements (in mm) have to be taken to 3 or even 4 decimal places.

At some observatories, such as at Bordeaux, the measurements are entered directly, from digitized scales, once the operator has centred the reticle on the star, thus avoiding any transcription errors.

17.1.3.5 Star catalogues
We described previously how catalogues of stellar positions gradually increased in number. Some of those that are most frequently used are:

- The AGK3 catalogue (*Astronomische Gesellschaft Katalog*, 3rd edition), containing 183 000 stars, down to magnitude 11, which was derived from the old BD, already mentioned, improved by three later campaigns – 1890 + (= AGK1), 1921 + (AGK2), and 1950 + (= AGK3), making full use each time of the latest instrumental or photographic improvements.
- The *General Catalogue* (GC), published in 1936 by Benjamin Boss, with 33 342 stars, limiting magnitude approximately 7, is a compilation of 238 catalogues since 1755. Its equinox is 1950.
- The *SAO Catalogue*, published in 1965, is another catalogue produced from a compilation of older catalogues including the FK4, FK3, GC, AGK2, AGK1, Greenwich AC, Yale, Cape Zone, BD, and Me 4. It contains 258 997 stars of which 8712 are double and 499 variable, and goes down to about magnitude 11, although most of the stars included (113 737, or 44%) are between magnitudes 8 and 9. Stars of magnitude 11 amount to only

1.6 %. (These figures apply to visual magnitudes.) This catalogue consists of four volumes and, together with its associated charts, covers the whole sky. The data given in this catalogue have often been derived by differential astrometry. It was published by the Smithsonian Astrophysical Observatory, in Cambridge, Mass.

- The FK4 (*Vierte Fundamental Katalog*), published in 1963 by the Astronomisches Recheninstitut in Heidelberg, gives the positions of (only) 1535 stars, with magnitude below 7.6. The main difference between this catalogue and the preceding one is that the positions of the stars given here have been derived by fundamental astrometry. They are therefore absolute positions, corrected to the equinox of 1950. These serve as references in determining the positions of other stars. This catalogue is the main astrometric reference catalogue. The unfortunate thing about it is that, because of the small number of stars that is included, very wide fields have to be photographed if one wishes to have several stars on one plate.
- *Sky Catalog 2000.0* by Hirshfeld and Sinnott, published in 1982. This derives from NASA's Skymap project, Version 3. It contains 45 269 stars brighter than visual magnitude 8.05. This catalogue should be used in conjunction with *Sky Atlas 2000.0*, which contains all the stars mentioned.

There are, of course, many other catalogues, but these are very specialized works. By type of object, there are:

- for galaxies: RNGC, UGC, ...
- for double stars: ADS, Σ, ...
- for variable stars: GCVS, NSV, ...
- for planetary nebulae: PK, ...
- for quasars: 3C, 4C, ...
- for radio sources: PKS.

There are also others devoted to specific characteristics, such as photometric properties, parallaxes, radial velocities, etc. These catalogues are not commonly used by amateur astronomers, but it is as well to know that they exist, because one may want to refer to them occasionally.

17.1.3.6 Atlases

We have already said something about the development of atlases. The ones most commonly used are:

Atlases based on cartography

- Becvár's atlases (*Borealis, Eclipticalis*, and *Australis*), limiting magnitude 10 (80 charts).
- The *AAVSO Variable Star Atlas*, limiting magnitude 9.5 (178 charts), specifically intended for variable-star work.
- The *Handbook of the Constellations*, by Vehrenberg, limiting magnitude 6 (55 charts).

Photographic atlases

- The POSS, which we have already described, taken with the Palomar Schmidt for the northern hemisphere. It consists of 1872 plates for the sky north of −33°, limiting magnitudes approximately 21 (in blue) and 20 (in red). The SERC/ESO southern sky survey taken with Schmidts in Australia and Chile, has 660 plates of the southern sky, limiting magnitude 23. [A new northern survey, with a much lower limiting magnitude is currently in progress. – Trans.]

- Vehrenberg's *Atlas Stellarum*, limiting magnitude about 14: North, three volumes, 303 photographic plates; South, one volume, 161 photographic plates.

- *True Visual Magnitude Photographic Star Atlas* by Papadopoulos and Scovil, three volumes, limiting magnitude 13.5 approximately: North, 120 plates; Equatorial, 216 plates; South, 120 plates. This is notable in showing visual magnitudes.

- Vehrenberg's *Atlas Falkauer*, two volumes, limiting magnitude about 13: North, 300 plates; South 160 plates. Scale: 4 arc-minutes per mm. Very similar is the *Atlas Stellarum*, which has a larger scale (2 arc-minutes per mm) and a lower limiting magnitude.

- The *Atlas of Selected Areas* by Brun and Vehrenberg. This is a photographic atlas, converted into cartographic form. It has two specific features: first, it does not cover the whole sky, but zones spaced regularly across it, at varying scales, at declination intervals of approximately 15°. Each chart is given at two scales, one being a partial enlargement of the main chart. Second, it is a photometric atlas (with photographic magnitudes). Each star shown is labelled with a 4-figure number representing its magnitude (2 significant figures + 2 decimal places). This is a very useful work for calibrating photographic fields and determining the limiting magnitude of a photograph. To do this, two photographs are taken on the same night, with the same emulsion, the same exposure, and (of course) the same instrument. One is of a reference field, and the other of the field where one wants to determine stellar magnitudes.

There are many other very specialized photographic or cartographic atlases, which we will not discuss, not so much because there are so many of them, but because they are not directly relevant for astrometry.

17.2 Amateur astrometry

17.2.1 Preamble

Readers will have gathered from the previous sections the ways in which amateur astrometry may be applied. Obviously, these cannot include fundamental astrometry, because that has already been carried out, and uses means that are beyond the amateur's grasp. We are left with differential astrometry. This may be divided into two categories: visual astrometry and photographic astrometry. We will briefly give a couple of examples of these from the amateur point of view.

(i) Let us assume that one night, whilst generally 'sight-seeing' with your telescope, you discover a nova.

The first thing to do is to check that your discovery is actually correct. To be certain, you need to get out your charts and catalogues, and try to determine the position of the object that you have discovered. You can progressively narrow down the position to a specific point in your atlas, and you will then be able to give an approximate position, and quote an appropriate source.

But there is more to be done if you want to be more accurate: you need to draw a chart of the field and then interpret it (we shall see how this is done later). If you work properly, you can expect a positional accuracy that will leave little room for any doubt. This is the type of work that we call visual astrometry.

(ii) If you are an enthusiastic deep-sky photographer, you may, on checking a photograph – you should always check your own photographs, there are all sorts of odd things that you may find – discover a small 'trail' that is not in the same direction as your guiding. (There is nearly always some slight error, even though it may be very small.) You check that it is not dust on the negative or a flaw in the emulsion. If it is neither of those things it could be a minor planet (or a faint comet).

There is a strong chance that it is one of the objects given in the ephemerides (4256 for 1991). Never mind! It would be wise to report its position. It might be well away from opposition and its coordinates would be greatly appreciated, especially if it is one of the objects with a poorly known orbit. If it is a new object, then it is absolutely essential.

To come back to earth, however; if you want to obtain its position you have an unexpected bonus, because you have a record that probably shows stars that will serve as suitable reference points. We shall see later how this record may be utilized, but for now it suffices to point out that you need to carry out photographic differential astrometry. With a little care, you will be able to obtain a position accurate to better than 5 arc-seconds .

Before describing how to proceed in either of these situations, we will discuss the methods that are required.

17.2.2 Methods to be used

17.2.2.1 Observational methods

The first essential is to have an instrument, whether it be a refractor or a reflector, that has a driven equatorial mount. The proper conditions are essential in making drawings (and even more if you want to take photographs). It is well-nigh impossible to cope simultaneously with slewing the telescope with one hand, while holding a drawing pad, red light and pencil with the other. It is necessary to have at least a modicum of comfort and organisation.

It would be sensible, for example, to fit the drawing board with a faint red light. (There are various small, adjustable, clip-on, battery lamps that may be found and which are eminently suitable if fitted with a red bulb or screen.) The pencil and eraser may be tied to the board with a piece of string – it is easier to find a drawing board that you have dropped in the dark than it is to find a pencil – and the board should have a clip that firmly holds the paper.

It is also useful, if not indispensable, to have an eyepiece with illuminated cross-hairs.

For those interested in photography, we suggest that they read the works mentioned in the bibliography.

If you use a 24 × 36 camera body, it is best to have one that takes interchangeable viewfinders and which may be fitted with a magnifying eyepiece. Second-hand Practica VLC 2 bodies may be found quite cheaply or, if your means permit, you can buy a Pentax LX or a Nikon F2 or F3. But in any case, remove the focussing screen, which is of no use. You will obtain an aerial image which is much sharper than any given by a ground-glass screen.

17.2.2.2 Reference material

It helps to have reference material that is appropriate to the size and degree of sophistication of the instrumentation that one is using. We would suggest that a good basis would be to have *Sky Catalog 2000.0* and *Sky Atlas 2000.0*, which have the advantage of being homogeneous. All the stars in one may be found in the other.

If one is more fortunate, or more ambitious, one might acquire *Atlas Stellarum*, which, as we have said, goes down to magnitude 14. For those who are not familiar with handling photographic plates, it is not as easy to use, because it shows an extremely large number of stars, and beginners may find some difficulty in deciding where they are. A useful adjunct would be the *SAO Catalogue*, which is for the same epoch (1950). Using these two references it is possible to locate faint objects without any great problems.

17.2.2.3 Reduction methods

From a drawing made at the eyepiece If a very high degree of accuracy is not required, we may be satisfied with obtaining the coordinates by graphical interpolation, using the atlases and catalogues just mentioned.

If a higher accuracy is sought, it is possible to use an enlarger to project the drawing and make measurements on the projected image. We shall describe this technique, which takes longer, but is much more accurate, in more detail later.

From a photographic negative Here, there are several possibilities. The negative could be put into a measuring machine – as is done by professional astronomers. Unfortunately, such equipment is not readily available.

Some microscopes have stages that may be moved along two axes at right-angles to one another. Generally, however, the movement does not exceed 25 mm, which is a bit too low to measure a 24 × 36 photograph. It is possible to resort to trickery to increase the amount, but with a loss in accuracy. We may also add that in practice it is not easy to make measurements with this sort of equipment, because generally only part of the field is visible at a time, and it becomes difficult to locate the reference stars and the object being measured. In addition, the financial cost for such an instrument is quite high. If one wants to make measurements to 0.01 mm over the whole negative, it soon becomes prohibitive.

Those with considerable engineering experience can consider building a small measuring machine. This has been done by a number of amateurs, but it does require very accurate work (Everhart, 1982).

Again, there is the possibility of projecting the negative onto the base of a photographic enlarger. If one has a small workroom, this is probably the best method to adopt, subject to certain precautions, which we will discuss shortly.

17.2.2.4 *Calculations*

Calculation is one of the major factors involved in astrometry. A certain amount of familiarity with mathematical methods is required before tackling this part of the problem. Those who are strong in mathematics could devise a method of working from a study of works that deal with fundamental astronomy, but this will be quite complicated. We shall discuss the principles of this later.

We should point out that this solution, which does have the advantage of being very instructive, will soon become a chore if it has to be carried out frequently. It is therefore a very good idea to create a short program that may be used on a programmable calculator or a microcomputer.

Let us now examine one method in detail. We take this from a work published by Brian Marsden (1982a). It has subsequently been reproduced in a number of works on astrometry, including a more accessible source (Marsden, 1982b).

The first step is to determine the focal length of the instrument as accurately as possible. If the angular distance between two stars shown on a plate is known, the scale of the plate (in seconds of arc per millimetre, "/mm), and which we will call S, may be determined by measuring the separation (in millimetres) between the images.

The focal length of the instrument, expressed in millimetres, is given by:

$$L = 206\,265/S$$

(NB: This is only valid, of course, for photographs taken at prime focus. In any other instance one will obtain the effective focal length, including any associated optics.)

The right ascension and declination of the object to be measured (say a minor planet) are derived from those of reference stars, which should be at least three in number. We will take this as being the case, because it is the simplest to deal with.

Defining the parameters to be used, we have:

L: focal length of the instrument
A: right ascension of the centre of the photographic field
D: declination of the centre of the photographic field

For each of the three reference stars, which should, if possible, be evenly spaced around the minor planet, obtain the coordinates α and δ from a catalogue for the reference epoch. We then obtain the coordinates that apply at the time of the photograph by applying the corrections for $\mu\alpha$ and $\mu\delta$ (i.e., the annual proper motion), which are multiplied by the number of years that have elapsed between the reference epoch and the time of observation. These new values of α and δ are used in all subsequent calculations.

We now calculate the parameters H, ξ, and η for each reference star:

$$H = \sin\delta \sin D + \cos\delta \cos D \cos(\alpha - A).$$

Check that the value of H is approximately equal to 1.

We then have:

$$\xi = \cos\delta \sin(\alpha - A)/H,$$

$$\eta = [\sin\delta \cos D - \cos\delta \sin D \cos(\alpha - A)]/H.$$

Using a measuring machine, or by projecting the images onto the baseboard of an enlarger – we shall discuss the errors to be avoided later – we determine the values of x and y (i.e., the rectangular coordinates relative to arbitrary axes), for each of the reference stars. The north–south direction on the plate should be oriented as closely as possible parallel to the y-axis. The values of x and y should be expressed in the same units as those used for the focal length L (which we have specified in mm).

We then calculate:

$$\xi - x/L = ax + by + c$$

$$\eta - y/L = a'x + b'y + c'$$

where a, b, c, a', b', and c' are the unknown plate constants.

This procedure is repeated for each of the three reference stars, so we obtain three pairs of equations, from which we may obtain the values of the six constants by standard algebraic methods.

We then measure the values of x and y for the minor planet in the same way as those for the reference stars (checking to ensure that the negative has not been moved in the meantime). We will call these x_{ast} and y_{ast}. We now know the values of the plate constants, so we may derive:

$$\xi_{ast} = (x_{ast}/L) + ax + by + c$$

$$\eta_{ast} = (y_{ast}/L) + a'x + b'y + c',$$

which are the rectangular coordinates of the minor planet on the sky.

To obtain the astrometric coordinates of the minor planet on the sky, we then have to solve, in order, the following equations using the values of A and D, which, as we will recall, are the coordinates of the plate centre:

$$\Delta = \cos D - \eta_{ast} \sin D$$

$$\Gamma = \sqrt{(\xi_{ast}^2 + \Delta^2)}$$

and we obtain the equatorial coordinates of the minor planet from:

$$\alpha_{ast} = A + \arctan(\xi_{ast}/\Delta)$$

$$\delta_{ast} = \arctan[(\sin D + \eta_{ast} \cos D)/\Gamma].$$

This is the principle. If readers are familiar with the method of least squares, we strongly recommend that more than three reference stars are used; the accuracy of the results will be greatly increased.

There are various other methods that may be used, one of which is described in Acker and Jaschek (1986). The problem is obviously ideally suited to implementation on a microcomputer. [The original French work (Martinez, P. (ed), *'Astronomie, Le Guide de l'observateur, Tome 2*, pp. 920–6) included a BASIC reduction program developed by J. L. Heudier at CERGA. This has been omitted here. Readers are instead referred to the program and description given by Marche (1990). In addition, a recent book by Montenbruck & Pfleger (1990) and the accompanying disk of programs may be highly recommended. They include efficient PASCAL procedures and programs for astrometry (and many other astronomical calculations). – Trans.]

17.2.3 *Visual astrometry*

17.2.3.1 *Making drawings at the eyepiece*
Anyone already familiar with making drawings at the eyepiece will be at an advantage, because they know the main points of technique. We will describe them briefly, with specific reference to astrometry.

We assume that an eyepiece with cross-hairs is available and that one of the hairs is aligned north–south. For astrometry, we strongly recommend that when placing the stars that form the background, the eye is kept on the optical axis. Although there is a slight magnitude loss, there will be an increase in the accuracy of the stellar positions.

A sheet of paper should have been prepared beforehand, with a circle to represent the field of view, and a cross representing the cross-hairs. Use the most powerful eyepiece compatible with seeing the object concerned easily. If possible the cross-hairs should be set on a star that is fairly close to the object. This may be used to check that the drive is working properly, and that the reference stars did not move while the field was being drawn.

The positions of the brightest stars are then marked on the sheet, using polar coordinates, with the intersection of the cross-hairs as the origin, for example:

- one star is at 2/3 radius, at 30° in the northeastern quadrant,
- a second is at 1/10 radius, at 15° in the southeastern quadrant,
- a third is at 1/4 radius, at 25° in the southwestern quadrant,

and so on.

The fainter ones are then included, using the brighter ones already positioned as additional guides. The stars should be drawn as small circles, the size of which should be proportional to the brightness. This will help when reducing the drawing. (Don't lose time over this, though: we are not trying to carry out photometry.)

All the stars that are bright enough to be identified easily in the atlas should be

977

included. Anyone who is using the *Atlas Stellarum* could include very faint stars, but there is no point in going to far. Experience will soon prove the best guide as to what should be included.

The positions of all the stars are marked, without bothering about the position of the comet (if that is what the object happens to be). Once the background stars have been located – don't hesitate to alter the position of a star if it appears wrong – we can finally make the all-important observation and mark the location of the comet. (Note that we say 'mark the position' not 'draw'.) This position should be carefully noted, and marked on the chart by a cross. The time should also be recorded to the nearest second, if possible. Do not attempt to draw the comet, because this entails additional time, and any time-lag is detrimental to the accuracy of the marked position of the comet.

No alterations should be made to the drawing from now on, but we can add stars around the outside of the field. A larger eyepiece with cross-hairs should be used, and the cross-hairs correctly oriented by letting a star trail along the declination cross-hair. The cross-hairs should then be repositioned on the reference star.

The visible field is larger, and the new stars that have appeared should be marked, using precisely the same techniques as before. This process may be repeated again, if there is a real danger of becoming lost, so that there are adequate reference stars over a large field. When the wider-field eyepieces are used, only the brightest stars are added, but great care should be taken to ensure that the positions are as accurate as possible; this is very important at a later stage.

Once all this has been done, the position as given by the telescope's setting circles should be recorded. It is always useful as a reference. This ends the work to be carried out at the telescope. The rest may be done inside in the warm.

17.2.3.2 Reducing the drawing

We may have one of two basic aims: either we require a quick, approximate estimate of the position, or we want the most accurate position possible and are prepared to take some time over it. In the first case, it will suffice to identify the reference stars that have been marked with those on the appropriate chart in the atlas. This is easy if care was taken to mark on the drawing the cross-hair that was aligned north–south. A cross may be marked lightly in pencil on the chart at the comet's position, paying particular care to the distances between it and the various background stars. The coordinate grid for the atlas may be laid over the sheet, and the approximate position interpolated from the scale.

This quick procedure will only give an approximate estimate of the position. But we can do better.

Start by drawing a circle and two axes at 90° (to represent the field and the cross-hairs) on tracing paper at the same scale as the photographic plate (or the chart) from the atlas that is being used. This circle may be moved around over the chart to locate the area of sky that corresponds to the drawing made at the eyepiece.

At first, some difficulty may be experienced, especially with a photographic atlas that shows stars much fainter than those included on the drawing. With some perseverance, however, this usually becomes quite easy. The circle is located on the chart as accurately as possible, placing the intersection of the cross-hairs on the primary reference star that was used in making the drawing. We cannot afford any ambiguity over this, which is why a star (rather than any other object) was taken as the primary reference point. The tracing paper is turned about this point into the correct north–south orientation, and then fixed to the chart with adhesive tape. Next, using a fine needle, prick the tracing paper at the position of the comet. Don't hurry over this, because it is all-important. The comet's position should be shown by a *single* small hole.

Prick through the tracing paper at the location of each of the surrounding stars that have been previously identified from the reference catalogue, writing their designations alongside the holes in the tracing paper. The coordinates α and δ, and the values of $\mu\alpha$ and $\mu\delta$ for each star are noted down on a sheet of paper. About ten stars should be used. Why ten? Because if the reduction is done manually, it is likely to take some time, but if the number is much less than ten, the accuracy obtained will not be consonant with the amount of work required. Once again, everyone will have to find the most suitable figure for themselves, given the means at their disposal. But to return to the perforated tracing paper. It should be removed from the chart (not before having marked the direction of north), cut to a convenient size, placed between two slips of glass (to keep it flat) and inserted into the enlarger's negative holder.

An image of the tracing paper is then projected onto the enlarger's baseboard, choosing the magnification so that all the perforated 'stars' are included. Focussing may be done on the grain of the tracing paper, and the mark indicating the north–south line should be aligned parallel to one edge of the baseboard. We now have a large-scale chart of the area of sky, which – provided care has been taken in pricking the tracing paper accurately at the location of each star – is a faithful reproduction of the photographic atlas.

We now have a working copy, from which we can take the x and y measurements for each of the reference stars and of the comet. The reference axes are taken as being the left-hand edge and the bottom of the enlarger's baseboard. The Cartesian coordinates of each star (and of the comet) may be read off by using a graduated scale that has been turned into a T-square by the addition of a cross-bar, carefully aligned with the 0 marking. These measurements may be added to the values for α, δ, $\mu\alpha$, and $\mu\delta$ that we have already written down on our work-sheet. We then follow the procedure outlined in Sect. 17.2.2.4.

Using an enlarger gives an accuracy that is significantly better than that obtained by the interpolation method, provided care has been taken in locating the comet against the stars shown in the atlas, and in making very fine perforations in the tracing paper.

Experience shows that the accuracy obtained may be of the order of 1 minute of arc. This shows that visual astrometry is capable of obtaining a very reasonable degree of accuracy. In the next section we shall take this further by discussing photographic astrometry.

17.2.4 Photographic photometry

We have seen in the preceding section how the visual determination of a position depends on numerous factors, the most important of which is the quality of the drawing made at the eyepiece. Second is the accuracy with which the information on the drawing is transferred to the tracing paper. If we replace the eye with a camera, and take care when making the exposure, we will obtain an objective record, in which the observer's personal equation will not appear. As a result, we will obtain a considerable gain in accuracy.

17.2.4.1 Preparations

It is essential to make adequate preparations before making any astrometric exposures. It is useful to have a rough chart of the area being photographed. The first step is to interpolate from the ephemeris of the object (comet or minor planet), whose position is to be determined, for the expected time of observation.

Next, a small chart representing the field covered by the film should be prepared, showing the most important stars that will serve as reference points for finding the object. This chart is indispensable when the object being photographed is faint and cannot be seen in the telescope. (Not, it should be noted, through the camera's viewfinder, which generally allows only the very brightest stars to be seen.) If the object is visible in the eyepiece, finding it does not take long. If no field chart is to hand, however, a lot of time is lost in making sure where the telescope is pointing, even if it has proper setting circles.

17.2.4.2 Taking the photograph

The scale of the photograph will depend on the type of telescope being used. Wide-field instruments have a considerable advantage, because it is much easier to find reference stars, as well as a convenient guide star. We consider that a field-diameter of 1 degree is a practical minimum.

As far as the film is concerned, the choice should be made from high-contrast, fine-grain films. Note that an astrometric plate is not intended as a work of art, and that we are not concerned with recording delicate details in the tail of a comet. What we require is the greatest possible accuracy in the position of the central condensation. That is all that will be recorded. Preference should be given to a film like Kodak TP-2415, hypersensitized or not, depending on the magnitude to be reached and the resulting exposure time with the given instrument. This emulsion gives very sharp, high-contrast images, which are easy to measure subsequently.

Exposure times will vary, as we have said, according to the object being photographed. For example, using a Newtonian 410-mm reflector with a focal ratio of f/4.83, we made a 15-minute exposure on TP-2415 hypersensitized with forming gas, of Comet Halley on 1985 August 14, when it was about magnitude 14.5 (Fig. 17.3). Using the same equipment, the exposure was just 30 seconds in December, later the same year. It is worth giving this point some thought, before beginning the exposure, because it would be a great pity if the photograph were over-exposed. Measurements from it would be difficult: how can you determine the exact centre of the fuzzy head of a comet? The photograph might even be unusable.

Fig. 17.3. *An astrometric plate of Comet Halley taken on 1985 August 14, between 02:55 and 03:10 UT. 410-mm reflector, f/4.83; hypersensitized TP-2415 film. The comet is a tiny dot at the limit of visibility. Photo: R. Chanal.*

Before releasing the shutter, just two more points. Has the instrument been really sharply focussed? And are the camera and film truly at right-angles to the optical axis? The first point may be checked by careful knife-edge testing, using the methods described in specialized works on astrophotography. The second point is more difficult to ascertain. Only examination of earlier photographs will provide information about this. The stars at the edges of the field should all show equivalent aberrations relative to the centre of the field. If this is not the case, and if stars on the left and right (for example) appear different, the mounting of the camera body much be checked, because it is not perpendicular to the axis of the cone of light leaving the telescope. Deformed stars would be difficult to measure and the accuracy of the reduction would be affected.

Let us just say a quick word about guiding. (The techniques are fully described elsewhere.) It should be pointed out that, for astrometry, good guiding is vital for the quality of the photograph, and thus of the measurements. But this raises a dilemma: should the guiding be 'sidereal' or 'cometary' for an object that is moving? The answer to this is not easy. We believe that it is a mistake to make comparisons with the way that professionals go about this. They have very different means at their disposal.

What do we find in practice? Let us take a simple case: we want to measure the position of a relatively bright comet (magnitude 7–8), and we believe that an

Fig. 17.4. *An example of a faint, fast-moving object: Comet Sorrells, 1986n, photographed on 1986 December 1, between 20:15 and 20:35 UT. 410-mm reflector, hypersensitized TP-2415 film. Photo: R. Chanal.*

exposure of 5 minutes with good guiding will give a good image of the central condensation. Unfortunately the comet's motion is relatively fast.

If we use 'sidereal' guiding, the stars will be round, and therefore easy to measure, and the comet will appear as a small trail, the mid-point of which will be taken as the point to be measured, just as the time is taken as the mid-point of the exposure (Fig. 17.4). There is only one error that we might make, and that is in determining the centre of the trail, but this will be all-important for the overall accuracy. However, with care, we will not be far from the true centre of the trail. Generally, in this sort of situation, there will be no difficulty in finding a suitable guide star, assuming that we are using a telescope, and that guiding is carried out by deflecting a portion of the beam of light from the mirror to a guiding eyepiece.

Now let us assume that we want to guide on a comet. Amateurs generally use a beam splitter consisting of a semi-reflecting plate that diverts about 20 % of the light collected by the primary, the remaining 80 % being used for photography. With this equipment we have to guide on the photometric centre of the comet. But it will be very faint (20 % of the light from a comet of magnitude 7–8 in an amateur-sized telescope is not particularly bright), and in any case is likely to appear diffuse and low in contrast. Guiding under these circumstances is likely to be indecisive, especially if the comet is not strongly condensed, and the result will be an indistinct cometary image and the stars will appear as wavy trails, reflecting the intermittent guiding. It will be difficult to obtain satisfactory measurements of either the comet or the stars from such a photograph. After a considerable number of trials with greater or lesser success, we have practically abandoned this type of guiding for astrometric work. This does not mean, however, that it should always be rejected. We feel that it should be reserved for bright, highly condensed objects, or for amateurs who are equipped with plate-holders that allow them to track a comet but still using a guide star. Everyone will have their own preferences, depending on their degree of experience and the methods at their disposal.

When it is a question of minor planets, or of faint comets – for example, we currently have a programme involving minor planets of which the brightest are of magnitude 15 – the choice is made for us. It has to be 'sidereal' guiding.

If everything has been done properly, the reference stars will be perfectly round, and easy to measure, and the minor planet will be shown by a small, narrow, straight trail, which will not be too difficult to measure when it is projected with the enlarger.

There is one final point before releasing the shutter: record the time of the beginning of the exposure, which should be an integral number of minutes. The same applies to the end. These times should be given in UT, accurate to the nearest second. What is the use of obtaining an accurate position, if the corresponding time is not known accurately? None at all, and all the work will have been wasted, which would be a great shame. This is an important point that must not be underestimated. Don't take a photograph unless you are absolutely sure of the time.

17.2.4.3 *Developing the negative*
We want a contrasty negative, so choose a developer that gives a fairly high γ (2 or more), and develop according to the film manufacturer's instructions. For hypersensitized Kodak TP-2415, we have found that HC 110 developer, dilution D, for 10 minutes at 20°C gives good results. The contrast is good, and faint stars are not 'lost'.

17.2.4.4 *Identifying reference stars*
To help when working, we would advise making a paper print from the negative. It will be easier to identify stars on this than on the image given by the enlarger, and it will give a print that may be kept as an archive. Once the print has been made (on moderately hard paper, grade 4), we can lay a sheet of tracing paper on top and start work.

On a small piece of tracing paper or transparent sheet (overhead transparency foil works very well), draw a rectangular mask (for a 24 × 36 camera body) or a circular one (for a Schmidt) that represents the exact size of the field at the scale of the atlas being used. We made something similar to help us with visual astrometry. The size of this mask may either be calculated or taken approximately from the print.

Lay the coordinate grid over the chart, and locate the area photographed using the mask. Depending on the orientation of the camera body when the exposure was make, the rectangular mask may be more or less inclined to the north–south direction. Fix the mask to the coordinate grid with adhesive tape, and read off the coordinates of the centre of the field. Although this may only be done by interpolation, estimate it as accurately as possible.

From the reference catalogue, the SAOC or *Sky Catalog 2000.0*, try to identify as many of the stars shown on the print as possible. With each identification mark the tracing paper with the star's designation, and write down the values of α, δ, $\pm\mu\alpha$, and $\pm\mu\delta$ on a work sheet. Other information, such as magnitude, and spectrum may also sometimes be useful.

Depending of the area being photographed, there will be more or less reference stars. Although with visual astrometry we said that 10 stars was about the maximum, here there is everything to be gained from using as many as possible, because we will be obtaining far more accurate measurements of x and y, and it would be a pity not to use all the reference stars that are available. Obviously, we must be able to handle the calculations required for all these stars, but we have already discussed this point.

This completes the least rewarding part of astrometric work. If we are lucky, and have a large number of reference stars, we may expect to get a good reduction. Occasionally there are none suitable, and our work comes to an abrupt halt. Experience shows that with the equipment mentioned previously (i.e., with a field that is 42×63 arc-minutes in size), in 80 exposures, we had an average of 8 suitable reference stars per field. The minimum was 0 (on three occasions), and the maximum 16 (once).

17.2.4.5 *Measuring the values of x and y*

In what follows we assume that measurement of the x and y values is done using an enlarger, because we believe (wrongly perhaps) that few amateurs will undertake the construction of a measuring machine.

Make every effort to obtain the best possible image of the negative from the enlarger. If possible, fit a low-voltage, point-source lamp and condensers, which will give the sharpest possible definition, with sharp, black stars without fuzzy haloes. The enlarger should be fitted with a high-quality lens, which should have been checked by placing a Ronchi grating in the negative carrier. The lines should be completely parallel and straight. If necessary, stop down the aperture to eliminate edge defects. Do not use the lens at full aperture, but at about f/5.6 ± 1 stop.

Before making any measurements, a few essential checks need to be made. Is the projection axis perpendicular to the baseboard? This is by no means a foregone conclusion. To ensure that it is, draw a rectangle (a true rectangle, not a trapezoid or any other quadrilateral) on a piece of tracing paper (again!). This rectangle should be 23×35 mm for 24×36 negatives. Slip the tracing paper between two thin plates of glass, put it in the negative carrier, and project it onto the baseboard at a high magnification (about 10 $\times$). Measure the projected rectangle and adjust the position of the head or column (or both) until an accurate rectangle is obtained. This is very important; if this condition is not met, there will be distortions in the image and the measurements will be affected. After all the work that has been done so far, it would be a crime to skip this point.

Once everything is set up, we can start the measurements. By definition these

Table 17.1. *Astrometric measurements of Comet P/Halley*

Date	RA (1950)	Dec. (1950)	Residual	
			α	δ
1985 Aug. 14.12674	$5^h59^m56.25^s$	$+19°05'18.8''$	$-4.99''$	$+3.05''$
– 19.13472	$6^h02^m38.82^s$	$+19°09'45.7''$	$+2.64''$	$+4.27''$
– 20.11389	$6^h03^m08.90^s$	$+19°10'33.7''$	$-1.34''$	$+1.01''$
– 21.10278	$6^h03^m39.28^s$	$+19°11'26.4''$	$+1.53''$	$+1.99''$
– 22.11389	$6^h04^m09.61^s$	$+19°12'21.0''$	$+1.49''$	$+3.67''$
		mean residual	$-0.134''$	$+2.80''$

are in Cartesian coordinates and the *x*-axis must be precisely at right-angles to the *y*-axis. Exactly 90° not 89° or 91°.

To make the measurements accurately, we make use of the edge of the baseboard, against which we place a T-square consisting of a graduated rule to which a cross-piece has been cemented. It is important to check that the left-hand and bottom edges of the baseboard are truly perpendicular. If not, this must be corrected.

The whole of the field that has been photographed appears projected onto the baseboard. Using a pencil, mark the designation as it appears on the work sheet against each reference star and start to make the measurements. Don't forget to measure the object that is the subject of the whole exercise (Fig. 17.5).

Make at least three sets of measurements, taking great care to estimate them to a tenth of a millimetre (this is quite easy with some experience), and take the mean of the values obtained. Experience shows that if one works carefully, there is no advantage in making more than five sets of measurements. In practice, we restrict our measurements to that figure, and often to just four. We may now turn to the problem of reducing these values.

17.2.4.6 Accuracy of the results

If every care has been taken in the preceding steps, the results will be excellent; frequently the scatter will be about 2–3 seconds of arc, sometimes about 1 second, and occasionally below this. The accuracies quoted here are not estimated ones. They are values that have been duly confirmed by professional organisations, such as the IAU's Central Bureau for Astronomical Telegrams.

This shows that with equipment that is fairly modest, and could even be called primitive – especially the means by which the actual measurements are determined – very creditable results may be obtained. By way of example, Table 17.1 shows some measurements of P/Halley made by the author that were submitted to the IAU, and which were judged to be accurate and worth publication. The errors indicated are those determined by Dr Yeomans at the Jet Propulsion Laboratory in Pasadena, California. The magnitude of the comet was about 14.5 at the time.

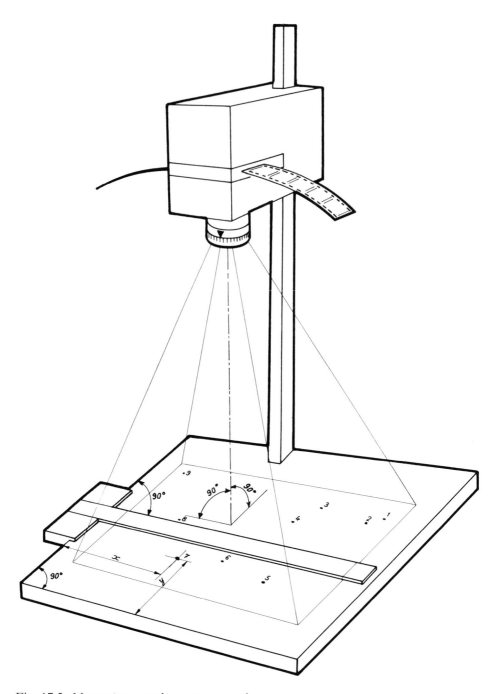

Fig. 17.5. *Measuring x and y, using an enlarger*

17.3 Why carry out astrometry?

17.3.1 Observational programme

In recent years in France, the RCP 639 (Research Programme Coordination) group, which was concerned with studying Comet Halley, set up a section for amateur astronomers. The impetus for this was given by the team at CERGA (Centre for Geodynamic and Astronomical Studies and Research), under the leadership of J. L. Heudier and R. Chemin.

A large number of amateurs were interested in photometry, estimating magnitudes, or in the comet's structure, which was studied by making drawings, and they made a major contribution to research on the object.

Astrometric measurements were also carried out in a systematic manner all along Comet Halley's path by a few French amateurs – they may literally be counted on the fingers of one hand – beginning as early as 1985 August. The campaign continued into 1987.

Comets Giacobini–Zinner and Crommelin were also selected objects, serving as preparation for Comet Halley. Other comets were also measured in the meantime during their apparitions, including Tiele, Hartley–Good, Wilson, Sorrells, as well as some older ones. It will be seen that, on the cometary side alone, there is plenty of work to be done.

Minor planets may also be an important field of activity for astrometric work. Obviously it is not a matter of measuring Ceres or Pallas, or other minor planets whose orbits are well-known. But there is a significant number of these small bodies in the Solar System, and some have not been observed very often, and some of the unnumbered object must be considered lost.

The *Ephemerides for Minor Planets*, which are published by the Russian Academy of Sciences, on behalf of the IAU, show some 120 of these small objects that require astrometric measurements. These are, of course, generally faint bodies, whose magnitudes at opposition are about 15 (at brightest). But there are plenty of amateur telescopes that are capable of reaching that magnitude, so the field is by no means barren.

Apart from such standard programmes, there is also the fact that any amateur involved in the discovery of a nova, a supernova, or a comet should be able to provide an astrometric position (even an approximate one) of the object in question.

17.3.2 Amateur/professional cooperation

Photographic astrometry in France began under the aegis of RCP 639. It is largely thanks to the team at CERGA that this work was started, and many professional astronomers gave up a considerable amount of time (and some their weekends) to get the programme going. A plate-reduction program was provided, which the persons involved were able to adapt to their own particular computing facilities.

What were the results of this programme on Comet Halley? Three observers contributed 18 measurements, which were checked by the Schmidt Telescope team; 13 of these received the ultimate accolade because they were published in the list of measurements in the *Minor Planet Circulars* (MPC). They were considered sufficiently accurate to be included. [Astrometry is also carried out by similar small numbers of amateurs in other countries, notably Japan and the United Kingdom. Again, professional advice and help has been available, including access to observatory measuring machines in some cases. The field is undoubtedly one where increased amateur involvement is highly desirable. – Trans.]

17.3.3 How should results be reported?

Anyone who wants to make significant contributions to any organisation (even on a more or less informal basis) needs to ensure that the results are presented in a standard format, which will make them easier to interpret and use. The information required includes:

- name of the photographer (or the person making the drawing);
- name of the person who carried out the measurements, if not the same;
- details of the observing site: address, longitude, latitude, altitude;
- type of instrument used: diameter, focal length or focal ratio; if necessary, optical layout and effective focal length;
- film used;
- filter used (if any);
- name of the object measured;
- date of observation: year, month, day (decimal);
- time of the beginning and end of the exposure (UT) or when the object was marked on the drawing;
- equinox used;
- photograph identification number;
- coordinates of the centre of the field: α, δ;
- reference catalogue;
- list of stars used in the reduction, each with: the arbitrary number on the plate, the catalogue designation, α, δ, proper motion ($\mu\alpha$, $\mu\delta$), x and y measurements;
- x and y measurements of the object concerned.

The report form may also include details such as magnitude and spectra of stars, from the same catalogue. These may help in any subsequent checks. As an example we show the report form that we devised for submitting our results (Fig. 17.6). This may serve as a model, that could perhaps be improved.

17.4 Conclusions

Before considering what may result from amateur astrometry, perhaps we should say a few words to prevent disappointment on the part of amateurs who might be in a hurry to obtain results.

Making an astrometric observation begins with the preparations for the observation and ends when the report form has been completed. In between these times

No	IDENTIFICATION	α 1950	δ 1950	Spec	MAG. B	MAG. V	PROPER MOTION $\mu\alpha$	PROPER MOTION $\mu\delta$	x	y

POSITION OF SUBJECT

REM

PLATE

INSTRUMENT   OF   FILTER   No

FILM

COORD OF CENTRE OF PLATE
α 1950 =   δ 1950 =

PHOTO TAKEN ON   AT   UT   OBJECT   DATE

NAME
ADDRESS
TEL.
LAT.
LONG.
ALT.

Fig. 17.6. *Sample report form for astrometric measurements*

there is a lot of work, and there must be constant attention to ensure that everything is carried out with as great an accuracy as possible, within the limits set by our relatively modest means. But this should not be taken as an excuse for slapdash work, which will only produce bad results.

There is no way of compensating for inaccuracy in measuring x and y values, and little hope of correcting incorrectly read α and δ coordinates. Similarly, if the camera back is misaligned with the telescope's optical axis, or the enlarger head is crooked, good results are impossible. They may only be obtained by scrupulous care at all stages of the process.

And what may be achieved? First, there is the personal satisfaction of knowing that with some perseverance, highly accurate work may be carried out. Second, there is the pleasure of collaborating, even if only on a modest scale, in obtaining data that will be added to those obtained by professional astronomers. Finally, there is the fact that one will be prepared to measure, without delay, the position of a nova or comet that one has discovered, or which one is trying to confirm.

18 Spectroscopy

O. Saint-Pé

The function of telescopes is to collect as much light as possible from any celestial object: they are, effectively, 'light-buckets'. Behind a telescope, – here generally termed an 'objective' – we may place a detector (a photographic plate, for example) that records the signal from the object being examined. But we could also place some form of an analyser in the light-path. As its name indicates, this enables us to derive specific information about the incoming radiation.

These three successive elements form what is generally known as an instrumental chain:

source	⇒	objective	⇒	analyser	⇒	detector
planet		refractor		filter		eye
nebula		reflector		grating		photographic plate

Although amateur astronomers frequently employ appropriate objectives and detectors, and become proficient in their use, analysers, on the other hand, tend to be neglected.

The two, principal analytical methods are polarimetry and spectroscopy. A distinction is often made between photometry and spectroscopy, but this is arbitrary, because both have the same primary aim: determining the quantity of light as a function of wavelength. Photometry is the measure of the quantity of light over a specific, small, wavelength interval, such as that selected by a filter, for example. The measurement is made by an appropriate detector (in general terms, a photometer). Spectroscopy aims to measure the quantity of light emitted at different wavelengths by the object being studied, so that its properties may be established.

This second method of investigation is the subject of this chapter and has been instrumental in making and confirming numerous astrophysical discoveries. It remains one of the most powerful techniques in this field.

18.1 Light

18.1.1 The Nature of Light

The majority of objects in the universe interact through electromagnetic radiation, the analysis of which enables us to determine numerous properties of the object being examined. This electromagnetic radiation may be considered as consisting of both an electromagnetic wave and a particle, a quantum of energy, known as a photon.

One of the main characteristics of waves is the wavelength λ, i.e., the distance between any specific point and the nearest point at which the wave has gone through a complete cycle (Fig.18.1). The value of λ is expressed in metres or sub-multiples of

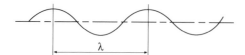

Fig. 18.1. *Wavelength* λ

a metre – for light, λ is often given in nm (nanometres), i.e., in thousand-millionths of a metre, 10^{-9} m).

For an object that is moving at a constant velocity v, the relationship between the distance covered x and the interval t is: $x = vt$. The same applies to electromagnetic radiation, which moves at a velocity c in vacuo (where $c = 300\,000$ km/s), and at a velocity that is less in any other medium, becoming $v = c/n$, where n is the refractive index of the medium concerned. For example, the refractive index of water is approximately 1.33; the velocity of light in water is thus: $v = 300\,000/1.33 = 225\,000$ km/s.

The time taken for light, in vacuo, to travel a distance equal to a single oscillation, known as the period T, is:

$$\lambda = cT,$$

where T is expressed in seconds or sub-multiples of seconds.

A wave may also be described in terms of its frequency v, which is equal to the reciprocal of the period:

$$v = 1/T.$$

It will be seen that an electromagnetic wave, e.g., light, may be described by one of the three properties: λ, T, or v. With radiation, however, we may also use the concept of a particle associated with the wave, such that for a given wave we have a given particle. This particle, the photon, is described by its energy E, which is linked to the properties of the associated wave by Planck's constant h ($h = 6.626 \times 10^{-34}$ J s):

$$E = hv = h/T = hc/\lambda.$$

The electromagnetic spectrum is subdivided into various regions, depending on its characteristic properties (Fig. 18.2). Although the term 'light' does not have a precise physical definition, it corresponds to radiation that is visible to the eye, together with the near-infrared and the near-ultraviolet. This part of the spectrum is also described as the 'optical region'.

On the ground, only certain parts of this overall spectrum may be detected, the remainder being blocked by the atmosphere, which acts as a filter (Fig. 18.3). In what follows, we are thus dealing with only a small portion of the electromagnetic spectrum, namely the visible (i.e., what is detectable with the human eye), the near-infrared, and the near-ultraviolet. Nevertheless, we can see the importance of observations from space that overcome the filtration effect of the atmosphere, and which may therefore observe objects in other wavelength regions (X-rays, infrared, etc.).

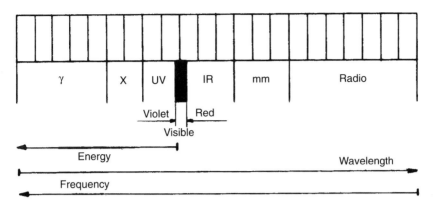

Fig. 18.2. *The electromagnetic spectrum*

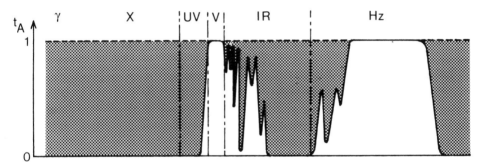

Fig. 18.3. *Atmospheric transmission as a function of wavelength*

There is obviously a qualitative relationship between colour and the values describing visible radiation. The term 'colour' is not, however, capable of defining a specific radiation exactly, because such a classification is highly subjective.

To carry out spectral studies, whether photometrically or spectroscopically, we need to ensure that the detector is sensitive to the radiation that we want to measure. We can determine this by studying the spectral response curve of the detector. If we use photographic film, the curve is determined by the manufacturer, although it may be modified by hypersensitization.

The spectral response of the eye is given in Fig. 18.4. From this we see that our 'natural detector' is sensitive to radiation between the violet and the red, with a maximum sensitivity in the yellow-green.

18.1.2 Continuum Radiation

Study of the radiation emitted by a heated solid, high-pressure gas, or fluid shows that emission occurs over a wide spectral range. The characteristics of this radiation vary as a function of the temperature of the source of emission. The hotter the

993

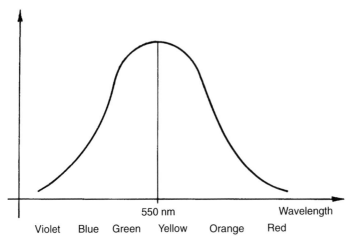

Fig. 18.4. *Spectral sensitivity of the eye*

latter, the greater its electromagnetic radiation, and the greater the emission at short wavelengths (at higher energies). An example is a heated bar of iron, the colour of which will change from red to orange and then to white if its temperature is raised sufficiently high.

A 'black body' is defined as an object that absorbs all electromagnetic radiation falling on its surface. It is possible to calculate the intensity of the energy radiated by such a body, and the emission curve always has the form shown in Fig. 18.5. This curve depends only on the temperature of the surface of the black body: the higher this is, the greater the peak emission and the farther this is displaced towards high energies, i.e., towards shorter wavelengths. The wavelength λ_{max} of the peak is given by Wien's Law:

$$T\lambda_{max} = \text{constant} = 3 \times 10^6$$

where the temperature T is expressed in Kelvin and the wavelength in nanometres.

In practice, the radiation from stars is assumed to be similar to that from a black body. For the Sun, we know that the peak emission is at 520 nm. From this we can deduce the temperature of the photosphere:

$$T = 3 \times 10^6/520 = 5770 \,\text{K}.$$

For Vega, on the other hand, with a surface temperature of 9200 K, the peak emission of electromagnetic radiation will be at a wavelength of:

$$\lambda = 3 \times 10^6/9200 = 330 \,\text{nm}.$$

18.1.3 Emission

With objects that consist of low-pressure gas, or with crystals, there is another mode of emission. As an example, let us take hydrogen, which consists of a nucleus orbited

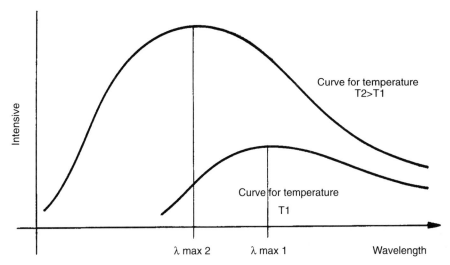

Fig. 18.5. *Black-body radiation curves*

by a single electron. This electron may occupy only specific, well-defined energy levels, which are identical for all hydrogen atoms.

At rest, this electron occupies the lowest level, or ground state. If it gains energy, no matter how, it jumps to a higher state, absorbing exactly the amount of energy that separates the two states. If this energy is carried in the form of light, the electron will only absorb photons whose wavelength corresponds exactly to the energy required for the transfer.

An electron that is at any level other than the ground state is liable to leave it spontaneously, dropping to a lower level; the difference in energy is thus liberated in the form of a photon, whose wavelength is defined by the amount of energy released ($E = hc/\lambda$).

Figure 18.6 shows the various levels for the hydrogen atom. It will be seen that the difference between the levels decreases with increasing distance from the ground state, and that there exists an upper limit, which corresponds to ionization, i.e., to the atom losing the electron.

By way of example, let us take an electron in the third level. It may (1) drop to the ground state, emitting a fairly energetic photon, or (2) to the second level, emitting a less energetic photon. If the electron is to jump to the fourth level (3), it needs to acquire energy. But if the electron gains too much energy (such that the latter exceeds the upper limit), it is lost by the nucleus (4); this produces ionization, resulting in a proton plus a free electron.

If low-pressure gaseous hydrogen is introduced into a flask and then excited (by an electrical discharge, for example), in each excited atom the electron will jump to one of the higher levels and drop back, emitting a photon, whose wavelength corresponds exactly to the transition. As the orbital structure is the same for all the hydrogen atoms, all the electrons jumping from one level (a) to another level (b) will emit photons at exactly the same wavelength. If the overall emission from the flask

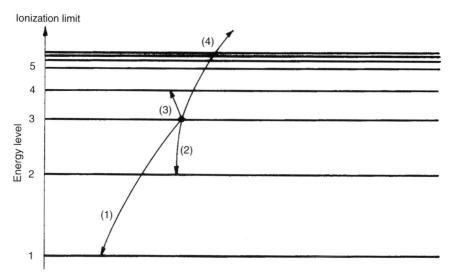

Fig. 18.6. *Electron levels in a hydrogen atom*

Table 18.1. *The first lines in the Balmer series of hydrogen*

Transition	Wavelength (nm)	Identification & colour of the emission
Level 3→2	656.2	Hα – red
Level 4→2	486.1	Hβ – turquoise-green
Level 5→2	434.0	Hγ – blue
Level 6→2	410.1	Hδ – blue

is considered, each transition will cause the emission of monochromatic radiation (because only a single colour is emitted). The sum of all the different emissions will give a discontinuous spectrum, also known as an emission-line spectrum. With hydrogen atoms, the difference between the ground state and the other levels is so great that the radiation emitted by a transition between any level and the ground state is in the ultraviolet; this series of lines is known as the Lyman series. The difference between level 2 and the other, higher levels is such that its various emission lines occur in the visible. They are therefore lines that we are able to see. This is the Balmer series, which is defined in Table 18.1. (*See also* p. 1037.)

With hydrogen, emission caused by transitions down to levels 3, 4, or 5 lies in the infrared. The radiation emitted by each of these transitions is known as an emission line.

The other chemical elements have the same sort of emission mechanism, except that the mechanism and the levels become more complex the greater the number of electrons in the atom. Similar mechanisms exist for molecules.

The explanation given here is obviously highly simplified, and although it is

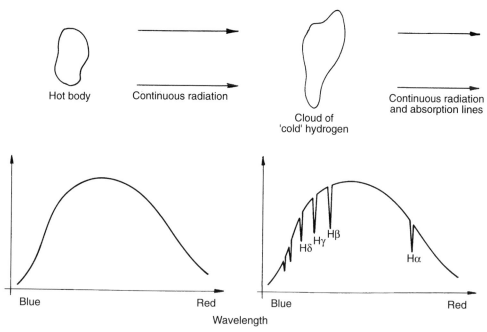

Fig. 18.7. *The formation of absorption lines when continuum radiation passes through a cloud of hydrogen.*

easy to understand, does neglect certain important points. One factor that may be mentioned is that a transition between one level and another is not automatically permitted, and does depend on properties that are not discussed here. (For more information, the interested reader should consult an introductory work on quantum mechanics or atomic physics.)

18.1.4 Absorption

The principles governing absorption are similar to those of emission but act 'in the opposite way'. If a cloud of hydrogen is exposed to a continuous spectrum of radiation, the hydrogen atoms will absorb energetic photons and move to higher levels. But they absorb only photons that correspond exactly to the energy required to make the transitions from one level to the next. To pass from level 2 to 3, for example, an electron has to absorb a photon with a wavelength of exactly 656.2 nm (Hα), because in dropping from level 3 to level 2, that is the radiation that is emitted.

After the continuum radiation has passed through the cloud of hydrogen, it will lack radiation at specific wavelengths that correspond to the emission lines discussed previously (Fig. 18.7). The same principle hold for all the other elements.

Once an electron has jumped to an upper level, it will obviously drop from the excited state down to a lower level, emitting a photon as it does so. It might be wondered why this re-emission does not compensate for the energy lost in absorption.

997

A moment's thought will show that the latter occurs only in the direction of the observer, whereas the former occurs in all directions, so it is unable to compensate for more than a small fraction of the energy lost to absorption.

18.1.5 Combined mechanisms

Obviously we may sometimes encounter emission or absorption lines superimposed on a continuum spectrum. Here are two examples:

A star

- a continuum (whose peak corresponds to the temperature) with absorption caused by elements in the photosphere;
- emission-line radiation (at radio, infrared, visible or X-ray wavelengths) caused by chromospheric eruptions and the corona.

A planetary nebula

- stellar radiation (continuum + absorption + emission) from the central star;
- emission-line radiation from the gas in the nebula, caused by absorption of radiation from the central star.

We shall return to these problems in the third part of this chapter.

18.2 Spectroscope design

18.2.1 Obtaining a spectrum

As already described, the aim of spectroscopy is to disperse electromagnetic radiation – which we will henceforth generally assume to be light – as a function of wavelength. Spectra may, of course, be observed in everyday life without having resort to spectroscopy: in rainbows (where sunlight is dispersed by rain drops or spray from a waterfall), when light falls on the foot of a glass or the stopper of a decanter, or when a streetlight is seen through a curtain with a very fine, parallel weave. In each of these cases, there has to be a physical object that disperses the incident light: this is called the dispersing element. We shall restrict discussion to the two types of dispersing element that are most easily used by amateurs: gratings and prisms, concentrating on the former, which are now most frequently used.

18.2.1.1 Gratings

In physical terms, a grating is a plane or spherical surface, ruled with a series of parallel, equidistant, straight lines. The number of lines is very high (between 100 and 4000 lines per mm in most gratings). We shall only discuss plane gratings. There are two forms of grating: transmission gratings and reflection gratings.

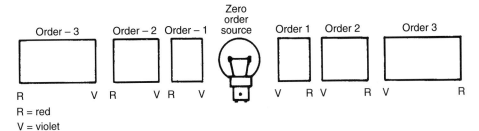

Fig. 18.8. *A schematic diagram of the way in which light from a lamp is affected by a grating.*

Transmission gratings In general terms, transmission gratings consist of a transparent support (of glass, acetate, etc.) on which parallel lines have been engraved, impressed or printed. The simplest example, as we have said, is a piece of fine, closely woven cloth. A grating may, in fact, be created very simply – but rather tediously – by drawing a series of fine parallel lines on white paper, and then photographing the result with a high-resolution, high-contrast film. If this is then reversed by contact printing onto another piece of film, we obtain a transmission grating.

If such a grating is held in front of the eye and a light source examined, we see (Fig. 18.8):

- the light source (because the grating is transparent). This image is known as the zero order;
- two identical spectra symmetrically situated to right and left of the zero order, and known as first order spectra (−1 and 1);
- farther away, two, fainter, additional spectra with greater dispersions (second order, −2 and 2);
- and so on with third and fourth orders, etc. The spectra become progressively fainter and have a greater dispersion.

In fact, if the light-source is faint, only orders 0, −1 and 1 are seen. Moreover, there is not an infinite series of orders: the more lines per mm, the smaller the number of orders. To see the effect just described a grating of between 150 and 600 lines/mm is the most suitable.

The principle of a grating is quite simple (Fig. 18.9). If a grating is illuminated with a beam of parallel light (this last condition is required to make the most of the properties of the dispersing element), the incident light is diffracted by the parallel lines in such a way that several spectra (orders) are formed, and within each spectrum the dispersion is greater the longer the wavelength (red is deviated more than blue).

The dispersion is given by a simple equation. If the incident light is perpendicular to the plane of the grating, we have:

$$\sin \theta_2 = nk\lambda,$$

where θ_2 is the inclination of the exit beam with respect to the normal to the grating,

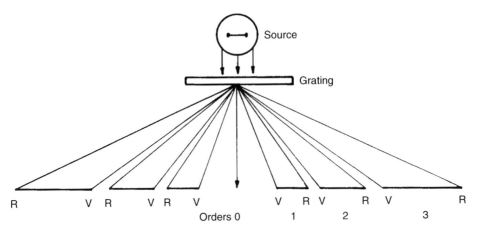

Fig. 18.9. *The formation of the various spectra shown in Fig. 18.8*

Fig. 18.10. *Section of a blazed grating*

n is the number of lines/mm, λ is the wavelength, and k is the order of the spectrum. Obviously, for a given grating and order, the value of the angle of deviation is a simple function of the wavelength, i.e., of the colour.

This equation shows that as the order increases, the angle increases, and that for a given order, the longer the wavelength, the greater the deviation. For example, with a grating having 600 lines/mm and the first order, the deviation is 22.2° for a wavelength of 630 nm, and 14.6° for a wavelength of 420 nm.

Cheap gratings (costing perhaps $10.00) are available, made by photographic reproduction on gelatine, and although their quality may be variable, it may be quite acceptable. Other versions, produced by holography or engraving, are of excellent quality, but very expensive.

Reflection gratings Instead of a transparent support, a reflecting one is used on which the lines that produce the diffraction are ruled. In general, the lines are ruled in such a way that the profile of the grating appears as a series of steps (Fig. 18.10); the grating is described as being 'blazed'. (The same profile is also used, very rarely, on transmission gratings.) The great advantage that this offers is that most of the light does not fall in the zero order, as it does with ordinary, unblazed gratings, but in another order, and at another wavelength, which may be chosen when the grating is designed.

The important factor is the angle φ formed by the basic triangular shape of the profile. If λ_m is the wavelength at which maximum energy is required, and k_m is the required order, we have:

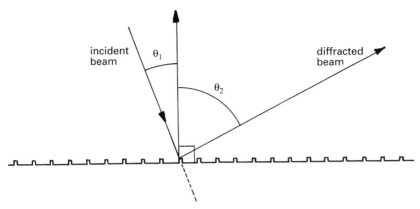

Fig. 18.11. *Dispersion by a reflecting grating*

$2 \sin \varphi = k_m \lambda_m n$ (n = number of lines/mm).

The equation describing the dispersion of a reflecting grating is similar to that for a transmission grating:

$\sin \theta_2 - \sin \theta_1 = nk\lambda,$

where θ_1 is the angle of incidence, and θ_2 is the dispersion angle, defined as shown in Fig. 18.11.

In general, this type of grating is expensive, because it is produced mechanically, using a diamond and a ruling engine. The lines are engraved in an extremely accurate fashion on a metallic support, called a matrix, which is then used to reproduce other gratings on aluminized glass plates.

18.2.1.2 Prisms
Historically, prisms were the first type of dispersing element to be used. They consist of triangular blocks of glass. After entering the glass, the light (again, initially parallel) undergoes greater refraction the shorter the wavelength. At the exit face therefore the red has undergone less dispersion than the blue wavelengths (Fig. 18.12).

If the spectrum of a light-source is examined through a prism, it will be found that on turning the prism about an axis at right-angles to the triangular faces, the spectrum moves, first in one direction, shrinking as it does so, and then in the opposite direction, expanding again. The angle when the spectrum is smallest, just before reversing direction, is known as the angle of minimum deviation. At this position, the properties of a prism are optimum.

Prisms suffer from numerous disadvantages when compared with gratings:

- the maximum resolution is limited (we shall discuss this later);
- fairly high price, even though they cost less for small sizes than good gratings;
- non-linear spectrum: the lines are more closely spaced at red wavelengths than at blue;

1001

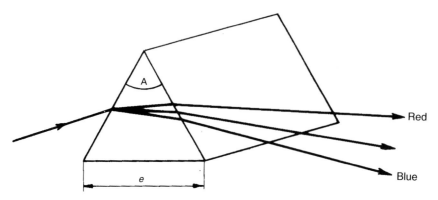

Fig. 18.12. *Dispersion in a prism*

- prisms are more difficult to use, because they have to be adjusted to give minimum deviation.

For a long time prisms had the advantage of giving a brighter spectrum, but since the appearance of blazed gratings this no longer applies. They are therefore suitable for faint-object spectroscopy if cost is an important factor.

18.2.1.3 Properties of dispersing elements
Resolution   This is a dispersing element's ability to separate wavelengths that are close together. There are several ways of defining resolution, one of which is: Two lines of the same intensity, at a wavelength of approximately λ and separated by $\Delta\lambda$, are considered to be resolved if the intensity minimum between them is less than 80 % of the peak value. The limiting resolution is then obviously the value of $\Delta\lambda$ that gives a minimum of 80 % (Fig. 18.13). Generally, the resolution is not specified as $\Delta\lambda$, but as $\lambda/\Delta\lambda$, because the resolution is dependent on wavelength.

Grating resolution: The true resolution of a grating is given by:

$$R = kN$$

where k is the order of the spectrum that is being used, and N is the total number of lines illuminated. We can therefore say: for a given order, the greater the number of lines, the better the resolution; or the higher the order, the finer the resolution. This agrees with the fact that the higher the order, the greater the dispersion of the spectrum.

For example, if a grating with 1200 lines/mm, measuring 40 mm square is fully illuminated, the resolution in the second order would be:

$$R = 2 \times 1200 \times 40 = 96\,000.$$

At a wavelength of 480 nm, it would be possible to separate two lines less than 5/1000 nm apart! We shall see later (p. 1013) that in fact this resolution is an upper limit that is never attained in spectroscopy.

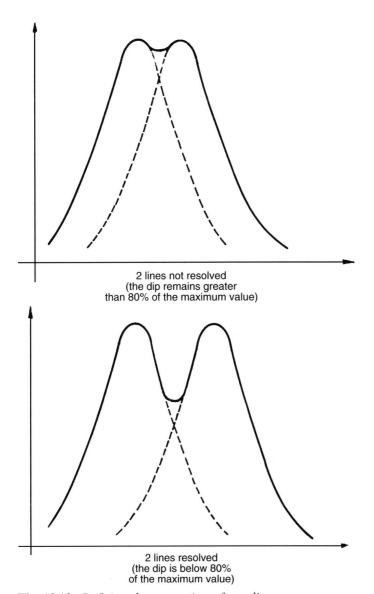

Fig. 18.13. *Defining the separation of two lines*

In addition, as we said earlier, it is not possible to obtain an infinite number of orders, and the more lines per mm, the lower the number of orders. In effect, very numerous lines and many orders are mutually exclusive. This means that the maximum resolution of a grating (for a given wavelength) does not depend on the number of lines per mm, but on its size:

$R_{\mathrm{max}} = 2a/\lambda$,

where a is the illuminated width of the grating.

We can therefore say that for equal size, a grating with 100 lines/mm will have the same resolution in the tenth order as a grating of 1000 lines/mm has in the first order. The latter cannot be used beyond the second order.

We may well ask why gratings with larger numbers of lines are used. The main reason is that when there are a large number of lines per mm, the spectra do not overlap, whereas with gratings that have lower numbers of lines, they do overlap, and the amount increases with increasing order.

Prism resolution: The resolution of a prism is given by:

$$R = e \times dn/d\lambda$$

where e is the width of the prism that is illuminated (in general, the base of the prism) and $dn/d\lambda$ is a factor describing the dispersion of light by the material involved. The larger this term, the greater the prism's dispersion, and the better the resolution.

For visual work, we obviously require a material that is very transparent, and which is obtainable in homogeneous blocks of sufficient size. In general, flint-glass or quartz are used.

Because the resolution is proportional to the illuminated base of the prism, it might seem that very flat prisms with a very large apex angle could be used. Unfortunately, beyond a certain value for this angle (about 60°) significant losses occur through reflection.

Light-gathering power   The light-gathering power may be considered as a measure of a spectroscope's ability to obtain spectra of faint objects.

The efficiency of gratings: The major disadvantage of poor-quality gratings is that they have a low efficiency. In the majority, only a few per cent of the incident light goes into the spectra, most passing straight through to form the zero-order image. On the other hand, with blazed gratings it is possible to 'concentrate' a large percentage (several tenths) of the light at a given wavelength into a specific order.

The efficiency of prisms: Amongst other factors, a prism's efficiency depends on the transparency of the material used for the prism. In general, it is very high (approximately 80 %).

Dispersion   Dispersion is less important that the two factors just mentioned, but the greater the dispersion in a spectroscope, the smaller the range that may be studied at any one time. This may be a disadvantage under certain circumstances.

Dispersion of gratings: The angular dispersion is given by:

$$d\theta_2/d\lambda = kn/\cos\theta_2,$$

where θ_2 is the deviation angle of the spectrum. From this we may derive the linear

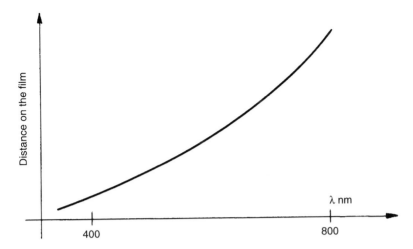

Fig. 18.14. *Dispersion curve for a grating*

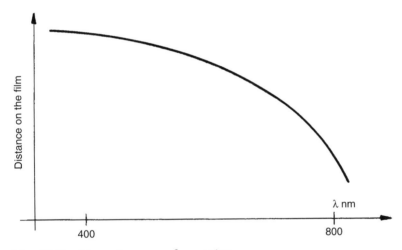

Fig. 18.15. *Dispersion curve for a prism*

dispersion on the emulsion that is obtained using an objective of focal length f:

$$\mathrm{d}x/\mathrm{d}\lambda = knf/\sqrt{1 - (kn\lambda)^2}.$$

The important point here is not the equation, but the fact that over short intervals of wavelength the dispersion may be represented as linear (Fig. 18.14). This is of considerable help in the reduction of spectra obtained with gratings.

Dispersion of prisms: With this type of dispersing element, on the other hand, the dispersion varies considerably with wavelength (Fig. 18.15).

1005

18.2.2 Observing the spectrum

Using a suitable dispersing element and the eye, we can observe different types of spectra, such as those described earlier, provided, of course, that the source is bright enough to give a spectrum detectable by the eye. A telescope may be placed behind the dispersing element to observe the spectrum (in which case we have a spectroscope). When the source is faint, or if we want to record the spectrum, then we can use a camera behind the dispersing element (when we have a spectrograph).

It is an interesting exercise to look at the spectrum of a streetlight that, like the sample of hydrogen that we discussed earlier, gives a line (non-continuous) spectrum, where the different 'colours' depend on the low-pressure gas inside the lamp. Here are some examples:

- Mercury-vapour lamp: these are the violet-white lights often used for public lighting. If their spectrum is observed with a grating (of 600 lines/mm, for example), spectra of order -1 and 1 may be seen (on either side of the image of the streetlight, which is the zero order), and which consist of a series of lines of light, two of which, in the violet and the green, are very bright (*see* p. 1040). As with the Balmer series of hydrogen each of the lines corresponds to a transition between two of the electron states of the mercury atom.

    It will also be seen that the shapes of the images in these lines are exactly the same as that of the source, which is normal, because each of the lines is a monochromatic image of the source.

- Most TV and radar masts, etc. are fitted with red, aircraft-warning lights. Access is difficult, so neon lamps are used because they have long lifetimes. Their spectra should be a series of lines in the orange and red, which correspond to the transitions in neon. (Note: the so-called 'neon tubes' that are frequently used for advertising are no such thing: they contain other gases).

- Major roads have two main types of lighting. Sodium lights are very orange in colour, and they show a very intense yellow line, which is actually two lines very close together, known as the sodium doublet. Because this line predominates, one gets the impression that these lights are monochromatic, which may give rise to strange effects, especially in road tunnels (surrounding colours disappear; there is a marked stroboscopic effect, etc.).

- Iodine lights are gradually replacing sodium lights. They have a pinkish colour and a spectrum very rich in lines, producing a better chromatic 'balance'.

Note: The fact that the green line of mercury appears brighter than the violet line does not mean that it really is. We should remember that the eye is much more sensitive in the green than in the violet. If you happen to observe sodium lamps as they come alight, the spectrum shows characteristic lines of neon, a gas that is used to initiate the illumination of this type of lamp.

The phenomenon of emission lines may be investigated by using a flame (such as a Bunsen burner), and burning salt in it. If, for example, ordinary cooking salt

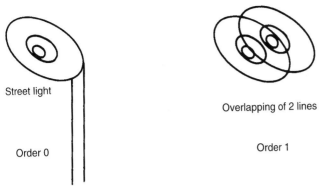

Fig. 18.16. *The appearance of a streetlight with unresolved monochromatic images*

(sodium chloride) is put in the flame, apart from the lines from the gas being used, a brilliant orange line appears, which is obviously the same as that in sodium lights. This line is so bright that the flame (the zero order) will resemble the line. Other chemicals are very suitable for such experiments (copper sulphate, for example).

18.2.3 Improving the spectrum

18.2.3.1 Adding a slit

In observing certain streetlights – in other words the spectra of different elements – it will be noted that some lines overlap, which is because there is insufficient dispersion (Fig. 18.16). The larger the number of lines per mm, or the higher the order, the less important this effect becomes.

If a spectrum of the Sun is examined, all that is seen is a dispersion of the light into a continuous range of colours from red to violet. No absorption lines are seen. This is because the lines have the same shape as the source (i.e., a disk) and are superimposed on one another. Because there are so many of them – more than 2000 may be detected in the visible region with amateur equipment (*see* p. 1038) – this overlap prevents any from being seen.

To understand how the spectrum may be improved, look at that of a mercury lamp at a fair distance. Assuming that the tube is horizontal, if the lines of the grating are vertical the spectrum will consist of white patches where the lines interfere and the colours mix. If, on the other hand, the lines of the grating are parallel to the tube, the spectrum will be resolved, where the lines appear quite narrow, because they have the same shape as the source.

If the source is modified (part of the tube being hidden), the change will alter the appearance of all the lines (Fig. 18.17). From this it is obvious that we can obtain a well-defined spectrum by selecting just a small part of the image, generally with a slit, which therefore prevents interference between the lines.

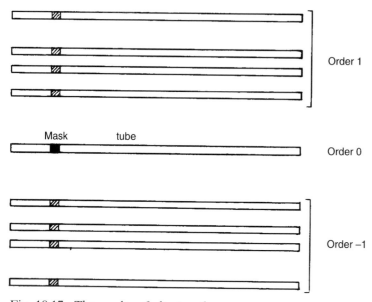

Fig. 18.17. *The results of altering the source*

18.2.3.2 *An experiment with the Sun*

An easy way of obtaining a slit with sunlight is to stand behind the crack of a door or a shutter. We can examine daylight (such as that from clouds or a blue sky). Provided the room is sufficiently dark, we stand far enough away from the 'slit', and the lines of the grating are parallel to it, we should be able to see absorption lines in the spectrum. The narrower and more regular our improvised slit, the sharper these will appear. This simple experiment shows that the quality of this particular element in the design is of paramount importance.

In the section on gratings, we saw that these need to be illuminated by parallel light. The rays of light from a slit are, however, divergent. In our case, the farther away the grating is, the less energy it will intercept (the fainter the spectrum), but the closer the beam of light becomes to parallel (Fig. 18.18). So we need to find some compromise between these two contradictory factors, or else employ the design described a little later.

18.2.3.3 *Trials with stars*

As we have seen, a slit is required for extended objects, and only serves to select a small part of their 'surface area'. Stars, however, are point sources. If the spectrum of a star (such as Sirius, Vega, etc.) is examined with a grating, a narrow band of light will be seen on each side of the star. To detect any lines, this spectrum needs to be spread out sideways. In other words, the star needs to be trailed across an imaginary slit. The spectrum will follow this motion and the lines will appear (Fig. 18.19). Naturally the eye cannot integrate the light received over more than a

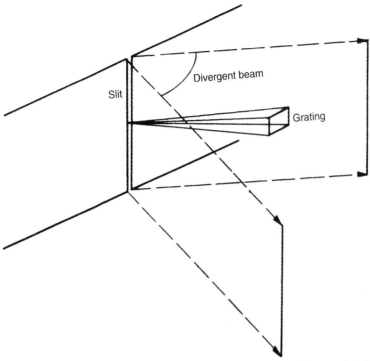

Fig. 18.18. *Effect of distance on the divergence of the beam of light*

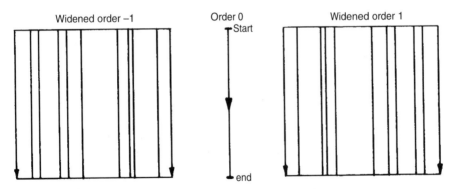

Fig. 18.19. *Broadening a stellar spectrum by trailing the image*

fraction of a second, so another type of detector is required (such as photographic film, for example). We shall see later how to obtain stellar spectra.

18.2.4 *The basic layout*

The various elements used in a spectrograph are:

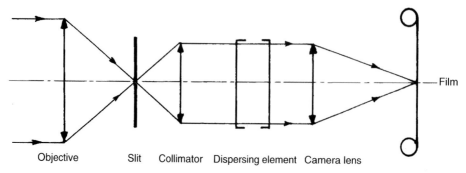

Objective   Slit   Collimator   Dispersing element   Camera lens

Fig. 18.20. *The principal components of a spectrograph*

- a slit,

- a dispersing element,

- a detector.

If the object whose spectrum we require is too faint, we need to add a collecting lens or mirror in front of these three elements. Its role is simply to concentrate more light onto the slit.

There remains the problem of the divergence of the rays of light leaving the slit. These may be turned into a parallel beam by using some form of converging optics, known as a collimator (a simple lens, an objective, or a parabolic mirror), and by placing the slit at its focal point. The basic principle of a spectrograph may be summarized as shown in Fig. 18.20. It will be obvious that optimum conditions will not be obtained by using whatever elements that just happen to be available. We need to optimize the layout and can do so by considering the basic features of each element in turn.

18.2.4.1 The objective

The optics that collect the light do not actually form part of the spectroscope itself. Their role, as indicated by the name, is to collect the maximum amount of light and concentrate it on the slit. This enables us to form an image of an object (such as a nebula, for example) on the slit and obtain spectra of different areas of the object, depending of the portion selected (Fig. 18.21). This introduces the idea of spatial resolution (which is different from the spectral resolution provided by various dispersive elements), and which is a measure of the ability of a particular optical layout to obtain spectra of different (spatial) regions of an object.

The most important properties of an objective are its focal length F and diameter D: just as with any objective used for any other purpose, the larger the diameter, the greater the luminous flux at the entry to the slit; and the longer the focal length, the larger the size of the image of the object on the slit and the better the spatial resolution.

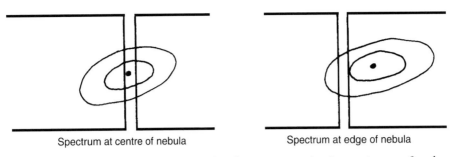

Spectrum at centre of nebula   Spectrum at edge of nebula

Fig. 18.21. *View of a slit onto which the objective is projecting an image of a planetary nebula.*

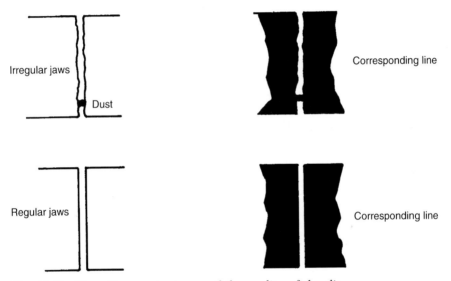

Fig. 18.22. *The effect on the image of the quality of the slit*

18.2.4.2 The slit

With extended objects, the role of the slit is to select part of the source, and thus prevent the lines from overlapping as far as possible. We have seen that the shape of the lines is the same as that of the slit, which is why the quality of the latter is all-important. In general, the jaws of the slit must be even, parallel and clean (Fig. 18.22). If there is any dust on the slit, there will be a corresponding line running perpendicular to the spectral lines over the whole length of the spectrum. This is sometimes known as a 'transversalium line' in spectroscopists' jargon.

On the other hand, all the luminous flux present on one side of the slit must pass through to the other side. In practice, this means that the profile of the jaws of the slit is usually bevelled or brought to a knife-edge. Rectangular-edged jaws are to be avoided, because scattering and reflection may occur along their edges (Fig. 18.23).

The most important property of a slit is its width w. The height h may also be

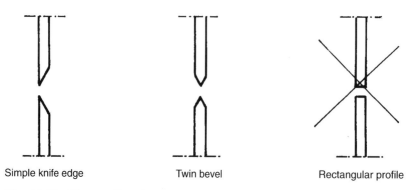

| Simple knife edge | Twin bevel | Rectangular profile |

Fig. 18.23. *The profile of a slit*

important in spatial resolution. We shall describe later how a suitable slit may be made.

18.2.4.3 The collimator
A collimator is used to turn the diverging bundle of rays from the slit into a parallel beam. To do this, the distance between the slit and the collimator must be equal to the latter's focal length. As with the objective, the most important properties of a collimator are its focal length f_1 and its diameter d_1.

18.2.4.4 The dispersing element
This element is obviously of fundamental importance in a spectroscope, because it is the one that disperses the incident light as a function of wavelength. The important properties of a prism are its apex angle A, the length of its base e, the material from which it is made (and thus the variation in the index of refraction with wavelength), and its transparency.

The important properties of a grating are its type (reflection or transmission), the size of usable surface (the ruled area), the number of lines per millimetre, the fact of whether it is blazed or not and, if so, the wavelength for which it was blazed, and the amount of energy within each particular order.

From this information we can calculate the resolution, the efficiency and the true dispersion of the actual component concerned.

18.2.4.5 The camera lens
This serves to create an image of the spectrum on the detector, and it is generally an ordinary photographic lens. The important properties are its focal length f_2 and diameter d_2.

18.2.4.6 The detector
This records the spectrum, and therefore gives the wavelength as a function of distance on the film, at the focal point of the camera lens. The larger its physical size, the larger the region of the spectrum that is recorded, provided the optics in front of it are capable of covering such an area. An important property is its spatial

resolution g, which corresponds to the grain of a photographic emulsion, and to the pixel size of a CCD detector.

The other important factor is the spectral response of the detector being used. We have already discussed the response of the eye, and the consequences in observing a spectrum. The same applies to other detectors, all of which have different response curves, which means that for equal illumination (and thus for equal energy), we do not obtain the same response at different wavelengths. In general, such curves are obtainable from the manufacturer.

18.2.5 Requirements in a spectrograph

It is impossible to design a universal spectrograph that would have a high luminosity for obtaining spectra of faint objects, a high spectral resolution for solar work, and a good spatial resolution for examining extended objects. A spectrograph has to be designed with the object to be studied in mind, within the cost limits that are set, and also with regard to the components that one may already have.

The various elements affect the resolution and efficiency of a spectroscope, and we can consider these in qualitative terms before going on to discuss specific details. Let us just recap on what we mean by resolution and efficiency.

The efficiency of a spectroscope (or spectrograph) is a measure of its ability to obtain spectra of faint objects. Throughout this chapter, we make the assumption that the optics, whether consisting of lenses or mirrors have, respectively, transmission or reflection coefficients that are close to unity, and do not, therefore, have any great effect on the overall efficiency of the spectroscopic system.

The spectral resolution is the ability to separate two wavelengths that are very close together: if $\Delta\lambda$ is the smallest amount that the design can resolve at a wavelength λ, the resolution is given by $R = \lambda/\Delta\lambda$.

18.2.5.1 The characteristics of the dispersing element

The properties of dispersing elements have already been discussed in earlier sections. There is still a problem, however. The resolution of a telescope is limited by its diameter, because diffraction causes a star (a point source) to appear as a succession of rings (the Airy disk) at the instrument's focus. The same applies to any dispersing element illuminated by an infinitely narrow slit. The smaller its size, the larger the diffraction pattern that it forms. The size of the diffraction spot may be obtained from the following equation, which is valid for an infinitely narrow slit, and for monochrome incident light of wavelength λ:

$$d = f_2\,\lambda/a,$$

where d is the diameter of the diffraction spot, f_2 is the focal length of the camera lens that is imaging this spot, and a is the illuminated width of the dispersing element.

18.2.5.2 The effect of slit width

We have just seen that there is an absolute limit on the resolution, set by diffraction in the dispersing element, even if the slit is infinitely narrow. Naturally, this limiting

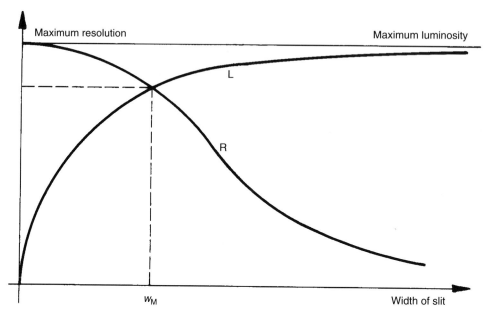

Fig. 18.24. *The effect of increasing the width of the slit*

case can never be achieved. We need to see how the size of the diffraction spot varies as a function of the width of the slit. If the latter is slowly increased, at first the brightness of the spot (and thus the light-grasp of the overall system) increases greatly, without the size (and thus the resolution) changing very much. Later, the illumination remains almost constant, but the size increases. The variation of light-grasp L and resolution R as a function of the width of the slit is shown in Fig. 18.24.

It will be seen that for a width w_M, the resolution and the illumination equal about 80 % of their maximum values. This condition applies when the width of the slit is equal to d/G where d is the diameter of the diffraction spot, when the slit is infinitely small, and G is the system's magnification, i.e., $G = f_2/f_1$, (where f_1 is the focal length of the collimator, and f_2 the focal length of the camera). These values are correct for incoherent illumination, and are approximately true for other cases.

18.2.5.3 Effects caused by the detector and the camera lens
Detectors have limited spatial resolution, and this limitation is determined by the manufacturer. In the case of photographic emulsion, the resolution is given as lines/mm, which is often converted into 'grain size'. This limit is fundamental in calculating the performance of a spectrograph. It is obviously impossible for two images of narrow lines to be separated if they fall on a single grain. Similarly, if the lines are too wide, best use is not being made of the emulsion's potential. It is therefore advisable to use a camera lens that gives a line-width that is approximately equal to the grain size.

The problem is that, in general, the optimum value for the focal length of the

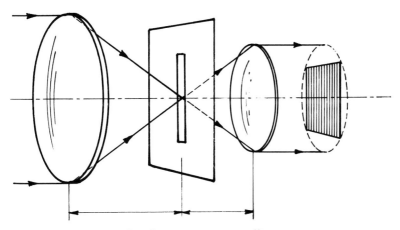

Fig. 18.25. *An example of an appropriate collimator*

lens is quite long, which is a nuisance because the luminous efficiency of a system decreases as the square of the focal length. Once again, we have to adopt a compromise, which will primarily depend on the magnitude of the object whose spectrum we want to obtain. The diameter of the camera lens should be larger than the diagonal of the ruled area, to ensure that it does not cause diffraction itself.

To put some figures to this, the detector's performance will be optimum when $d = g$, where g is the grain size (d has been defined previously). The focal length for optimum resolution is then:

$$f_2 = ga/\lambda.$$

For most arrangements, the efficiency is given by:

$$L = t\,(a/f_2)^2,$$

where t is the system's transmission. L therefore decreases rapidly as the focal length of the camera increases.

18.2.5.4 *The effect of the collimator*
This element must transmit the whole of the flux emerging from the slit to the dispersing element. Similarly, the elements on either side of the slit, the objective and the collimator, should be matched (Fig. 18.25). To meet the first condition, the diameter d_1 of the collimator must be equal to the diagonal of the dispersion element. If the diameter is less, only part of the surface will be illuminated and the resolution will suffer. If the area fully illuminated by the slit is too large, some of the light will be lost, which will lower the overall efficiency. The second condition is met if the objective and collimator have the same focal ratio, i.e., if:

$$F/D = f_1/d_1.$$

It should be noted that the width of the lines is often approximately equal to that of the slit times the magnification f_2/f_1. If the collimator has a long focal length, narrow lines may be obtained without having to resort to very fine slits (which are more expensive or more difficult to obtain in practice).

We shall now discuss some of the major types of spectrograph and the specific problems that these pose, depending on the type of source to be studied.

18.2.6 Slit spectrographs for extended objects

18.2.6.1 Construction of the slit
There are various forms of slit to be considered:

- Needle: with bright sources, the light may be allowed to fall on a needle. The size of the 'slit' varies according to the diameter of the needle.
- Photographic slit: a very fine black line is drawn on white paper and then photographed, using a reprographic-type of emulsion. If the result still lacks contrast, two internegatives may be made using lith film (such as Kodalith). It goes without saying that this type of slit should not be used at the focus of an objective that is to be pointed at the Sun.
- Engraved slit: this is obtained by more or less the same sort of operation as that just described. Using a fine point, a slit is made by removing the coating from a glass plate that has either been painted black or coated with a film of aluminium – the latter being preferable for solar work.

  (NB: In either of the last two forms of slit, the shape does not have to be linear. It might, for example, be circular, which would be very useful for examining the limb of the Sun.)
- Classical, fixed slit: the two jaws of the slit may consist of razor blades, or high-quality, Swedish-steel knife blades. The jaws are refined on a stone using emery, so that they are perfectly straight. It is essential to ensure that they do not subsequently touch, which would damage the edges. One of these jaws is then cemented (using an epoxy glue, for example) onto a rigid support of metal or epoxy resin. Once this is perfectly dry, the second jaw is cemented down with a glue that has a drying time of about 2–3 minutes. The jaws are approximately aligned by examining them against an illuminated background, and then placed in a slide projector set up in a long corridor. The projected image of the slit may then be used to adjust the second jaw until only a thin thread of light is visible. The method also allows the size of the slit to be measured.
- Adjustable slit: Figure 18.26 shows one possible design – there are numerous others. The moveable slit may be used to vary the width of the slit. The most important point is to ensure that the jaws are parallel (using a slip gauge or by projection) before fixing them in place.

18.2.6.2 Mechanical considerations
There are no general rules, although obviously the baseplate for the mounting needs to be as rigid as possible. A metal plate, a stone slab, or – failing anything better – plywood that has been strengthened by longitudinal ribs would serve. As regards the

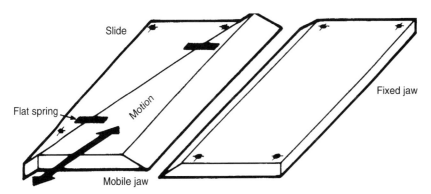

Fig. 18.26. *One design of an adjustable-width slit*

adjustable holders that carry the optics, highly sophisticated versions are available commercially, but are naturally very expensive. Cheaper, home-made forms are likely to be preferred. The important thing is that the supports should allow all the adjustments necessary for correct alignment of the optics. Most of the supports will have to be capable of being adjusted by translation in three axes, and a small amount of adjustment by rotation around two axes is welcome. If reflecting optics are being used, it is worth ensuring that the three adjustment points are of the 'push-pull' type.

18.2.6.3 Alignment and focussing

Optical alignment includes ensuring that the optical axes of the elements lie in the same plane. Because the optical axis is sometimes difficult to locate, the centres of the optical elements and their reflections are adjusted to lie in the same plane. The ideal source for this is a laser (for example, a small helium–neon laser with a power of less than 5 W). Lasers are becoming much more common, and it is not too difficult to obtain (or borrow) one.

The laser is set up in a plane that is parallel to the mounting plane (for example, by aligning these two planes against a third that acts as a reference, and which might be the floor of the room). Once this has been done, the alignment of the optics may begin.

As far as the objective is concerned, if the arrangement allows, the beam is sent through the centre of the optical system, checking to ensure that the spot falls close to the slit, and that the reflection of the spot from the front face of the objective returns to the exit from the laser (Fig. 18.27). The slit is then adjusted so that the spot falls on its centre. If this element is a reflecting one, the image of the spot should fall on the objective's optical centre. The same procedure is followed for the collimator, the grating, and the camera lens (i.e., the optical centres and the reflections of the spot are aligned).

The collimator/slit adjustment is then carried out. To do this, an incoherent source (such as a filament lamp) is placed behind the latter. A plane mirror is placed behind the collimator to receive the rays of light from the slit. One or other of the

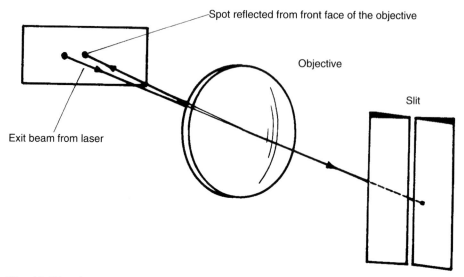

Fig. 18.27. *Aligning the objective and the slit*

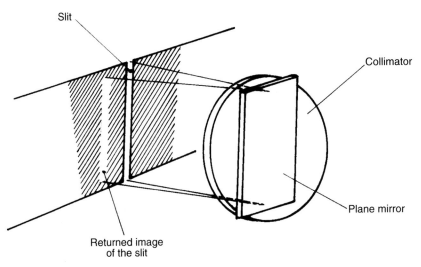

Fig. 18.28. *Aligning the collimator and the slit*

elements is shifted until the return rays form the narrowest possible image of the slit, which is superimposed on the slit itself (Fig. 18.28). This procedure is known as autocollimation.

The rays may be brought to a focus at the slit using light from the Sun. If any suitable means of checking are available (a reflex camera body, a CCD screen, etc.) the correct focus for the detector may be determined using intense light sources (a pigmy 'Pointolite' lamp, or the Sun, for example). If not, it has to be carried out

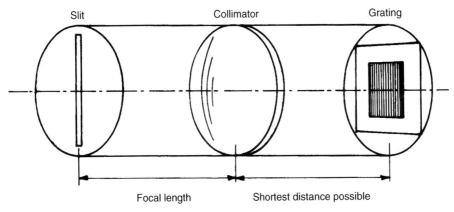

Fig. 18.29. *A simple spectroscope*

by trial and error, as with astrophotography. It should be noted that if all the rest of the system is accurately aligned, and if a photographic lens and ordinary camera body is being used as the camera, accurate focussing will be obtained, in principle, by setting the lens to infinity.

18.2.6.4 A simple mounting

One simple design uses just a slit, a collimator, and a transmission grating. The various elements are mounted in (say) a cardboard tube, with a diameter appropriate to that of the collimator, as shown in Fig. 18.29. We might, for example, use:

- a film grating (usable area about 30 mm per side) with 530 lines/mm;
- a razor-blade slit, 50 μm wide (the blades being cemented onto a slide mount and checked for parallelism by projection);
- a 40-mm diameter collimator – which will thus cover the whole area of the grating – with a focal length of about 200 mm.

This device may be mounted behind a telescope with a focal ratio of f/5 (the same as that of the collimator). Usable spectra may be obtained of bright sources (Sun, Moon, streetlights, etc.) with a resolution that is about a few tenths of a nanometre. A small refracting system (such as a telescope's finder) may be used behind the grating to observe the spectrum, or it may be recorded using a camera and a telephoto lens.

18.2.6.5 A solar spectrograph with good spatial resolution

As examples, we will describe two systems that have been used to obtain spectra of individual regions of the Sun.

The classical mounting Some of the elements are commercially available, but still reasonably priced items. The objective is an ordinary Schmidt–Cassegrain telescope (diameter D about 200 mm, focal length F about 2 m); a reflection grating, 25 mm

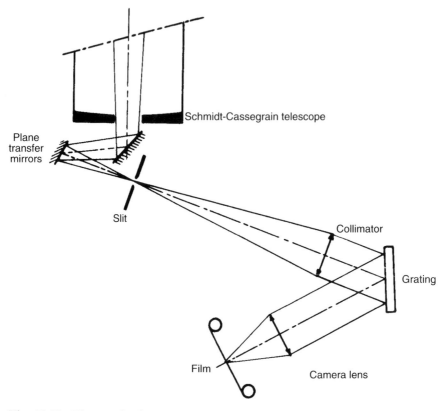

Fig. 18.30. *The standard spectrograph layout described in the text*

square, with 1200 lines/mm. The detector is Kodak TP-2415 film (which has a grain size of approximately 5 μm, or even 3 μm, with accurate development).

The overall design is shown in Figs. 18.30 and 18.31. The remaining elements are chosen with a view to obtaining the best possible spatial resolution, because there is no problem with the amount of light available, given the strength of sunlight.

The camera lens: We have seen that the focal length of the camera's lens should be chosen so that the size of the diffraction spot caused by the grating should be equal to the film's grain size. Using the equations described earlier, and a wavelength of 650 nm, we find $f_2 = 200$ mm. In addition, the diameter of the lens should be greater than the diagonal of the grating, i.e., 36 mm. A 200-mm, f/4 telephoto lens is therefore suitable.

The slit: To obtain the best spatial resolution, consideration needs to be given to turbulence and the image displacements that these cause at the slit. If the value of turbulence is around one second of arc, there is no point in having a slit width w of less than 10 μm. This value is given by:

$$\tan \theta = w/F,$$

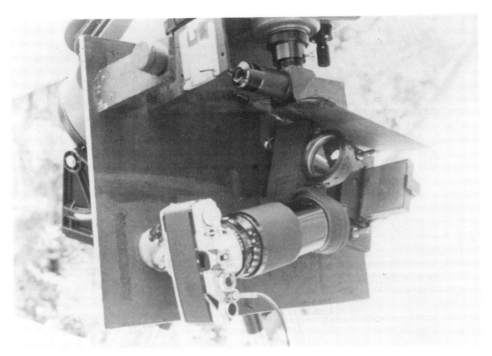

Fig. 18.31. *A photograph of a spectrograph built to the design shown in Fig. 18.30.*

where θ is the angular value of the turbulence and F is the focal length of the objective.

The collimator: If possible, the diameter d_1, should be similar to the diagonal of the grating. In addition (as has been shown in the previous section), if good resolution is required, the width of the slit should be less than, or equal to, the size of the diffraction spot divided by the magnification G of the system. The magnification should therefore be $5 \times 10^{-6}/10 \times 10^{-6}$, i.e., $0.5\times$. The focal length of the collimator should therefore be about 400 mm, or more if the efficiency permits it.

Summing up then, the various elements are as follows:

- objective: Schmidt–Cassegrain telescope, 200 mm in diameter;
- slit: fixed, with a width of 10 μm;
- collimator: 400-mm lens, 60 mm in diameter;
- grating: 25 mm square, 1200 lines/mm, blazed for the first order and red light;
- camera lens: 400-mm, f/4 telephoto lens;
- detector: Kodak TP-2415 film.

The resulting spectrum (Fig. 18.32) has a resolution of approximately 0.06 nm in the first order, and 0.03 nm in the second order. The intrinsic resolution of the grating is about 0.01 nm ($\lambda/\Delta\lambda = 60\,000$), for the second order, so it is obvious that the other elements cause various losses.

Fig. 18.32. *The region around the magnesium triplet in the solar spectrum, obtained with the equipment shown in Figs. 18.30 and 18.31. Resolution: 0.1 nm. Photo: O. Saint-Pé and P. Roth.*

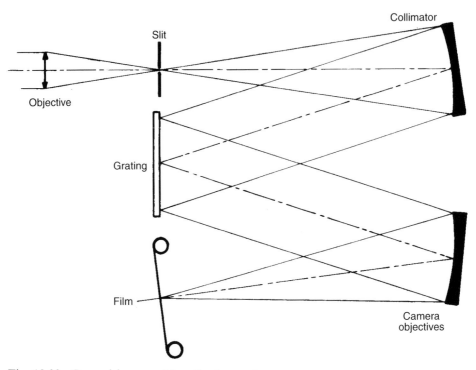

Fig. 18.33. *General layout with reflecting optics*

By moving the image of the solar disk across the slit, spectra of different areas may be obtained. The exposure times, which have to be determined experimentally, are such that guiding is unnecessary.

The Ebert mounting The overall aims of this design are the same as those that applied in the preceding section. Instead of using transmission optics for the collimator and the camera lens, however, this version uses parabolic mirrors. The normal arrangement is as shown in Fig. 18.33.

The principles governing the calculation of the various parameters are the same, but we eliminate the optical errors that are peculiar to lenses. The angles of reflection should be kept as low as possible to limit aberrations.

We can also consider replacing the two parabolic mirrors with a single mirror,

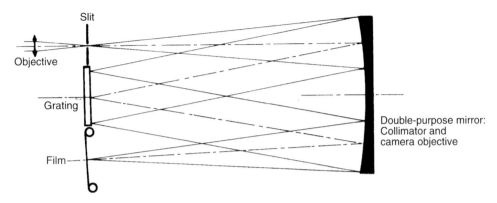

Fig. 18.34. *Ebert mounting*

as shown in Fig.18.34. This provides additional mechanical rigidity, and ease of adjustment. The main inconvenience is that because the focal length of the camera objective is equal to that of the collimator, it is impossible to use a magnification ratio other than unity.

One, fairly typical, example of this sort of mounting incorporated the following elements:

- objective: Zeiss refractor (focal length 840 mm, diameter 63 mm);
- slit: adjustable, width about 10 µm;
- grating: 50 mm square, 1200 lines/mm (not blazed)
- single mirror: diameter 250 mm, focal length 1.15 m;
- detector: photographic emulsion, or linear CCD (pixel size about 13 µm).

Because it is rather unwieldy, the system is stationary, and a coelostat is used to feed the objective. The results obtained show an estimated resolution of about 0.01 nm in the second order. The records shown in Fig.18.35 are spectral scans obtained with this system and a linear CCD.

18.2.7 Spectrographs for point sources: Objective prisms and gratings

18.2.7.1 Basic principles
The technique consists of placing a dispersing element at the entrance aperture of an optical system, which images the spectrum onto a suitable detector (Fig. 18.36). If a prism is used, the angle of incidence is adjusted so that it is approximately equal to the angle of minimum deviation. Because light arriving from a star is parallel, the objective forms a spectrum of the object at the focal plane.

If the apex angle of the prism is small, or if the number of lines per mm on the grating is low, this method is capable (among other things), of photographing the spectra of numerous stars simultaneously. This method may also be used to obtain spectra of monochromatic sources that have small angular diameters (such as gaseous nebulae). Because there is no slit, a spectrum at the focus of the objective consists of a series of monochromatic images of the source. It is therefore impossible

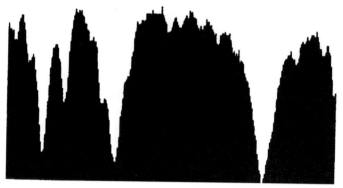

Magnesium triplet

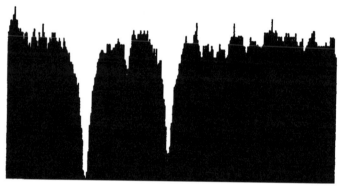

Sodium doublet

Fig. 18.35. *Two regions of the solar spectrum. Observations by: L. Collot, C. Buil and O. Saint-Pé.*

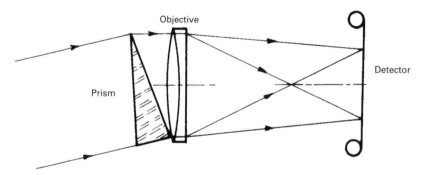

Fig. 18.36. *An objective prism mounting*

to obtain spectra of extended sources that radiate a continuum, or of discrete sources with a large number of lines, because of the problems of interference that have been described previously.

The resolution is obviously dependent on the characteristics of the dispersing element and on the objective's focal length. The most important factor, however, because of the low flux, is the overall efficiency. With point sources, this does not depend on the focal length of the objective. Instead the flux reaching the detector is a function of the transmission of the system – and is thus primarily the percentage of energy transmitted in the spectrum – of the area of the objective, and of a term that describes the luminosity of the source. For an extended source, on the other hand, the focal length does enter into the equation describing the luminosity function.

This type of design may be used just as easily with a simple photographic lens as with a much larger objective, the diameter of which is likely to be limited by the size of dispersing element obtainable.

The dispersing element and camera should be fixed to a common mounting. If a small prism is being used, this must be adjusted to give the angle of minimum deviation. This may be done by using a bright source, but the prism must be capable of being rotated to enable the adjustment to be carried out.

The spectrum of a point source appears as a line. For the details to be visible, this must be spread out perpendicular to the dispersion. There are two possible ways of doing this:

- Diurnal motion (Fig. 18.37): an altazimuth mounting is adequate (such as an ordinary photographic tripod). Selecting a bright star (generally, the magnitude must be above 1.5 for an aperture of f/1.8), the camera is oriented so that the spectrum is perpendicular to the diurnal motion.

  Using the B setting, the rotation of the Earth is used to spread out the spectra. (The lines will appear with a greater curvature the closer the stars are to the pole.) Because the speed of the Earth's rotation is fixed, this technique cannot be used for faint stars.

- Mechanical motion (Fig. 18.38): this time the mounting is an equatorial, and the spectrum is positioned parallel to the lines of right ascension. All that is now required is that the tracking should be fast or slow. If the mounting is motor-driven, a slight adjustment to the speed is all that is needed. Similarly, the spectrum may be spread out in declination.

  The process is supervised by using a cross-hair eyepiece in the guide telescope and arranging the cross-hairs to be parallel to the two axes. It is then sufficient to check that the star follows one or other of the cross-hairs, depending on which direction of trailing is required. This method enables the spectra of much fainter stars to be obtained.

  Examples of the equipment that may be used with the two methods are:

- Diurnal motion: small prism (60° apex angle) mounted in front of a camera with a 50-mm, f/1.4 lens.

- Mechanical motion: reflection grating placed in front of a camera body with a 200-mm, f/3.5 telephoto lens.

Fig. 18.37. *A portion of the spectrum of Jupiter, obtained using a blazed, transmission grating, 600 lines/mm, mounted in front of a 200-mm focal-length, telephoto lens. Diurnal motion was used to widen the spectrum. Note the magnesium triplet in the centre. The resolution is better than 0.6 nm. Photo: O. Saint-Pé and I. Beauvois.*

Fig. 18.38. *A portion of the spectrum of α Centauri, using a blazed grating, 1200 lines/mm, in front of a 300-mm focal length, telephoto lens. The spectrum was broadened by altering the rate of tracking. Photo: Dominique Delmissier and Pascal Goumard.*

18.2.7.2 Large-aperture objective prisms and gratings

The principles are the same as those just mentioned, with the prism or grating being placed in front of an equatorially mounted instrument. The main difference is the large size of the dispersion elements, which means they are correspondingly expensive. There are, however, ways of replacing the prism or grating.

In the case of the prism, an elegant solution is to construct a liquid prism. A solid glass prism, although it may have a smaller apex angle than the prisms used in smaller equipment, is still expensive, because a large piece of optical quality glass free from defects such as bubbles is required. A liquid prism, on the other hand, consists of a suitably shaped glass container filled with liquid. The glass used should be of good quality (for example, the glass used for large photographic plates). The joints need to be made with a suitable cement. The apex angle may be quite small, giving a large field.

With this method, a Schmidt camera is ideal, because it has excellent definition and covers a wide field. The prism is mounted in a suitable holder in front of the entrance aperture. The holder should be adjustable so that the prism may be set to the angle of minimum deviation. Apart from the modest price, the luminous efficiency, and the high quality spectra that are obtained, this method has

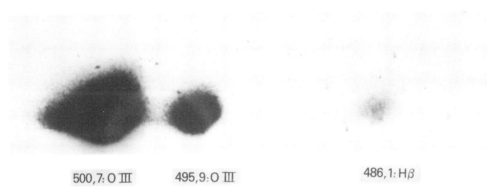

500,7: O III   495,9: O III   486,1: Hβ

Fig. 18.39. *The spectrum of the planetary nebula NGC 6210 in Hercules, obtained by the Club Eclipse, using the 600-mm reflector at the Pic du Midi. Because the image of the nebula was not selected by a slit, its spectrum consists of three monochromatic images of the object, corresponding to Hβ and the two O III lines.*

another advantage in that the liquid in the prism may be changed, giving different dispersions. (Water and benzine are two examples of liquids that may be used.)

In the case of a grating, using one with a low number of lines per millimetre has the advantage of improving the brightness of the system. There are two possible ways of attaining this without spending a fortune. It is possible to buy gratings on gelatine in fairly large sheets rather than mounted in slide mounts. The other possibility is to use printing screens, which consist of large wooden frames carrying two perpendicular sets of wires (several per millimetre). One set may be removed to give a grating.

Whether a prism or a grating is used, the general principles are the same as those described previously. A few points may be noted:

- When the surface area of the dispersing element is smaller than of the objective, the latter should be stopped-down with a suitable diaphragm.

- Because there is no slit, atmospheric seeing has a considerable effect on the quality of the spectra: the lines appear much sharper when turbulence is low.

- The film used should be chosen as if for an astronomical photograph taken through a filter, i.e., with respect to its sensitivity and spectral response curve (TP-2415, 103 aO, 103 aE, etc.).

18.2.7.3 Small-diameter sources with emission spectra
This type of object may be studied with objective prisms or gratings. The only difference is that the spectra do not need to be trailed, and tracking may be at the normal sidereal rate. The purpose in this type of observation is to obtain spectra that consist of monochromatic images of the objects and thus show regions of different abundances of particular elements (Fig. 18.39).

18.2.8 Stellar spectrograph with a slit

The basic design consists of a spectrograph behind an objective. Because the beam of light is divergent beyond the objective's focal plane, collimating optics are used, located at their focal distance from the prime focus, just as in other designs that employ a slit.

In general, although a star is a point source (its image at prime focus is very small, even if it is enlarged slightly by diffraction), a slit (or sometimes a diaphragm) is still placed at the focus of the objective, with the aim of limiting the sky-background flux, and the alterations in the position of the star caused by atmospheric turbulence.

The other elements in the design are chosen to give an appropriate overall luminous efficiency.

18.2.8.1 Individual elements

- The surface area of the objective is the most important factor in the overall efficiency, so the fastest possible instrument should be used.
- The width of the slit depends on the seeing. The smaller the slit, the sharper the definition in the lines; but if the seeing is poor, light will be lost because the image of the star will frequently wander away from the slit. A value of 50 μm is often used.

  It is possible, with care, to make use of the magnification factor (the ratio between the focal length of the camera to that of the collimator) and employ a larger slit than normal, which has the effect of reducing the spectrograph's sensitivity to turbulence. The smaller the magnification, the narrower the lines.
- The collimator should be chosen so that the whole of the flux provided by the objective arrives at the dispersing element. Its focal ratio should therefore be less than that of the telescope.
- The dispersing element should give the brightest possible spectrum. In general a prism will be chosen, or a grating blazed for the wavelength that is of primary interest.
- The aperture of the camera lens determines the overall luminosity of the system. Ideally, the lens may be changed depending on the star being examined. This also allows variations in the system's parameters to be tried experimentally.

If the system has a low luminous efficiency (primarily because of poor performance by the dispersing element), one way of making the best of the situation is for the speed of the optical elements to increase throughout the train, i.e., chose a camera lens that has a lower numerical aperture than that of the collimator. As previously mentioned, the diameter of the latter should be equal to the diagonal of the dispersing element.

18.2.8.2 Implementation

The mounting is generally similar to that of any other spectrograph, either directly behind the prime-focus image (*see* Fig. 18.42), or else carried parallel to the tube of the telescope. (The latter applies particularly to Newtonians, or where the focal lengths of the optical elements are fairly long.)

Adjustment and alignment are carried out as previously described, bearing in mind that for determining the correct position for the dispersing element and checking the focus, a streetlight or a spectral lamp will be perfectly suitable. There are, however, two specific problems:

- Alignment of the slit: there must be a means by which the observer can check if the star is, or is not, on the slit, and also ensure that it remains there. There are three techniques normally adopted:

    (i) A reflecting slit: if the jaws of the slit are reflecting (i.e, consist of polished metal), a sighting telescope is used to examine the reflected image of the star. When this disappears, the light is passing through the slit (Fig. 18.40).

    (ii) Guiding on the zero order: examination of the zero order (assuming a grating is being used) enables an image of the slit to be seen. One is therefore able to check whether the image of the star is falling on it, or not (Fig. 18.41).

    (iii) A beam splitter: using a semi-reflecting plate that diverts just a few percent of the flux provided by the objective, an image of the star is formed, which is observed with a cross-hair eyepiece, correctly oriented so that one of the hairs represents the slit.

- Trailing of the spectrum: as with objective prisms or objective gratings, the spectrum may be set at right-angles to the diurnal motion, and the drive slightly adjusted to deviate from true sidereal rate.

Close to the horizon, the image of a star becomes a short spectrum because of atmospheric dispersion. It is therefore possible, if the slit is very narrow, for part of the spectrum to be lost.

18.2.8.3 *The Littrow mounting*
As mentioned in the section on solar spectroscopy, it is possible to use a single optical element as both the collimator and camera lens. The design discussed was the Ebert type, which uses a parabolic mirror. There is a similar transmission design, the Littrow mounting, which is very suitable for stellar spectroscopy (Fig. 18.43). Its main advantage is that it is compact, but it has the disadvantage of having a fixed magnification.

18.2.9 *A grazing-incidence solar spectrograph*

This technique enables the spectra of extended objects to be obtained with good spectral resolution, and without a slit, hence without a collimator. The major disadvantage is that work requiring good spatial resolution cannot be carried out.

We have already seen that if the source is extended, and the spectrum polychromatic, two different wavelengths may be deviated by the same angle, giving rise to

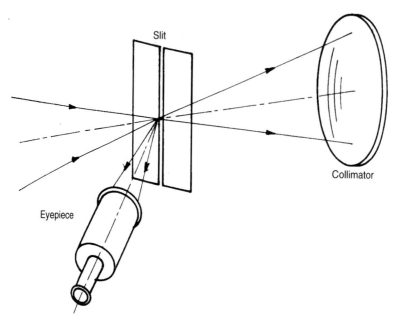

Fig. 18.40. *Viewing arrangement with reflecting jaws*

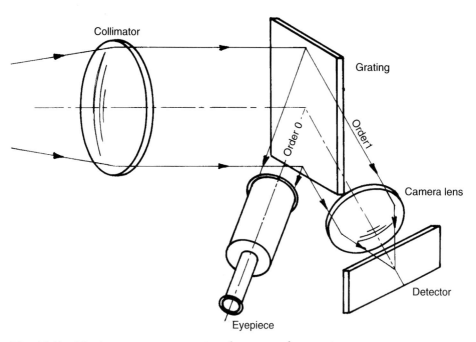

Fig. 18.41. *Viewing arrangement using the zero order spectrum*

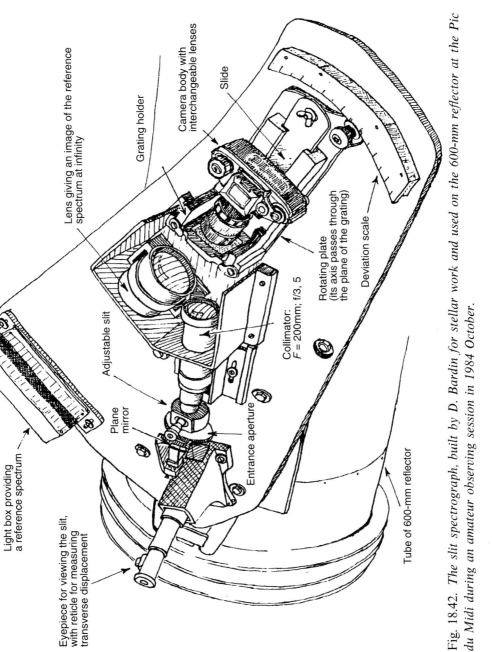

Light box providing
a reference spectrum

Eyepiece for viewing the slit,
with reticle for measuring
transverse displacement

Lens giving an image of the reference
spectrum at infinity

Grating holder

Camera body with
interchangeable lenses

Slide

Adjustable slit

Plane
mirror

Entrance aperture

Collimator:
F = 200mm; f/3, 5

Rotating plate
(its axis passes through
the plane of the grating)

Deviation scale

Tube of 600-mm reflector

Fig. 18.42. *The slit spectrograph, built by D. Bardin for stellar work and used on the 600-mm reflector at the Pic du Midi during an amateur observing session in 1984 October.*

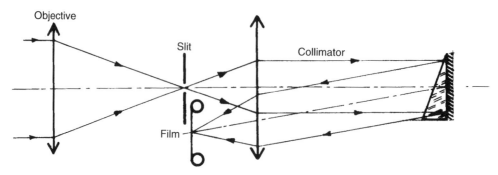

Fig. 18.43. *The Littrow mounting*

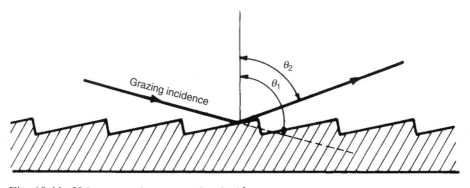

Fig. 18.44. *Using a grating at grazing incidence*

interference. If, on the other hand, the source has a small angular diameter, it is found that the resolution is a function of the angle of incidence of the beam of light.

In fact, if the source has a sufficiently low angular diameter (which is the case with the Sun), the resolution is expressed as follows:

$$\Delta\lambda = \Delta\theta \cos\theta_1/kn,$$

where $\Delta\theta$ is the angular diameter of the source, θ_1 the angle of incidence, k the spectral order, and n the grating's number of lines per millimetre (Fig. 18.44). It will be seen that for a given object and grating, the closer the beam is to grazing incidence, the greater the resolution.

Such a design, consisting of a reflection grating placed in front of a photographic lens, is quite suitable for studying the Sun, given the amount of light available from the latter. In general, a blazed grating is used, with a large number of lines per millimetre (1200 or more). The camera lens should be larger than the diagonal of the grating, and with a focal length that is appropriate to the diffraction spot caused by the grating (for a given wavelength λ):

$$f_2 = ga/\lambda$$

where g is the grain size of the emulsion, and a the length of one side of the grating.

18.2.10 Specific solar designs

Certain designs have been described elsewhere in this book (Chaps. 1 and 2), and are specifically applicable to solar observation:

- The coronagraph, where the interference filter (which is expensive, of limited life and difficult to commission) may be replaced by a high-dispersion grating, adjusted so that the Hα line may be observed and photographed.
- The broad-slit spectrograph, which is based on more-or-less the same principle, where a broad slit (preferably with the same curvature as the projected image of the solar limb) may be used to study solar prominences, enabling the Hα line (for example) to be seen in emission instead of in absorption against the solar disk.
- The spectroheliograph, which enables monochromatic images of the solar disk to be obtained by shifting the position of the detector. Wavelength selection is obtained by rotating the grating.

  It is possible to envisage a form of the last design that has fixed elements, if the photographic emulsion is replaced by a linear CCD, parallel to the lines and centred on one of them (Hα, Ca II, etc.). The diurnal motion would sweep the spectrum across the system's single slit, and the image would only have to be sampled at regular intervals to obtain the information required to reconstitute the image of the solar disk.

18.3 Reducing the spectra

18.3.1 Identifying the lines

Once one has obtained a 'readable' spectrum, i.e., one that is of sufficiently high quality for further work to be carried out on it, we need to know which line corresponds to which element. There are two main possibilities, depending on whether we have a reference lamp. The latter case may be further subdivided into two.

- If known lines appear in the spectrum that has been obtained, then it is possible to determine the system's dispersion curve as a function of wavelength. We have already seen (in the section on dispersing elements) that prisms have non-linear dispersion, and that the lines are much closer together in the red than in the violet. To derive an accurate representation of the curve, we therefore need to be able to recognize several lines.

  With gratings, on the other hand, the dispersion may be considered to be linear over a range of about 100 nm. If at least two lines are known (such as the highly characteristic sodium doublet or the magnesium triplet in the Sun), a simple rule of three calculation enables the wavelengths of all the neighbouring lines to be obtained. The elements that they indicate may then be found from a catalogue of spectral lines.
- If no lines are known, the spectrograph cannot be calibrated. The only solution is to resort to calculation, provided we know the dispersion of the

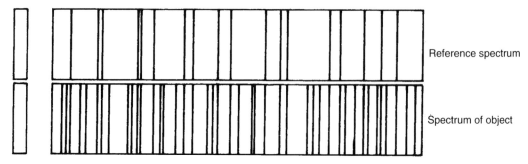

Reference spectrum

Spectrum of object

Fig. 18.45. *Calibration of a spectrum with a reference lamp*

prism or grating that is being used. Fairly accurate results may be obtained, as an example will show.

If a grating with a fairly low dispersion is being used, and if the zero order and part of the first order may be recorded on the same photograph, the wavelengths may be found as follows. Let x be the distance between the zero order and the desired line, and f_2 the focal length of the camera, the deviation angle θ_2 is such that: $\tan \theta_2 = x/f_2$. From this angle, and if the angle of incidence on the grating θ_1 is known (the simplest case is obviously normal incidence), the desired wavelength satisfies the general equation for the grating:

$$\sin \theta_2 - \sin \theta_1 = nk\lambda,$$

where n is the number of lines per millimetre, and k is the order of the spectrum.

If it is not possible to have the two orders on the same photograph, the calculation may still be carried out, provided the angle between the grating and the lens/camera combination is accurately known.

If a reference lamp is available, a slight modification to the design (see, for example, the diagram showing the stellar spectrograph, Fig. 18.42) will enable some of the light from this reference source to illuminate part of the slit. This slit should be capable of being masked so that either the upper or lower portion is exposed. Either before or after the spectrum is obtained using one half of the slit, the other portion of the slit is illuminated (the first being covered). The final result will be as shown schematically in Fig. 18.45. Other arrangements, using more than one lamp, or schemes for placing a reference spectrum above and below the stellar spectrum may also be devised (Fig. 18.46).

18.3.2 Determining the peak wavelength

This value enables us to determine the surface temperature of bodies that resemble black bodies (including stars), by using Wien's law, which was discussed in Sect. 18.1.2.

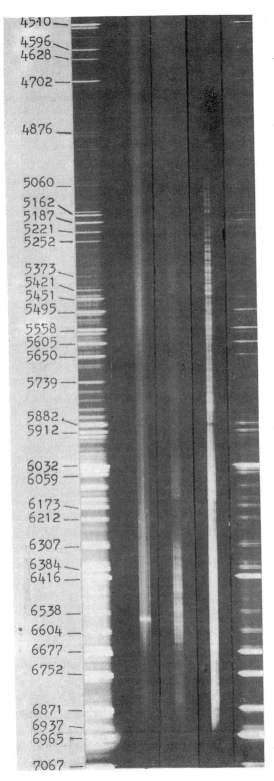

Fig. 18.46. *The three spectra visible in the centre of this photograph are, from left to right, those of ζ Tau (spectral class B1.5 IV), μ Gem (M3 III), and α Ori (M2 Iab). Reference spectra that allow the identification of specific lines are located on both sides of the stellar spectra. The wavelengths are given in Ångström units (0.1 nm). This photograph was obtained by D. Bardin, using the 600-mm reflector at the Pic du Midi and a spectrograph that he had built himself. Film: Kodak TP-2415, hypersensitized for 6 minutes in silver nitrate.*

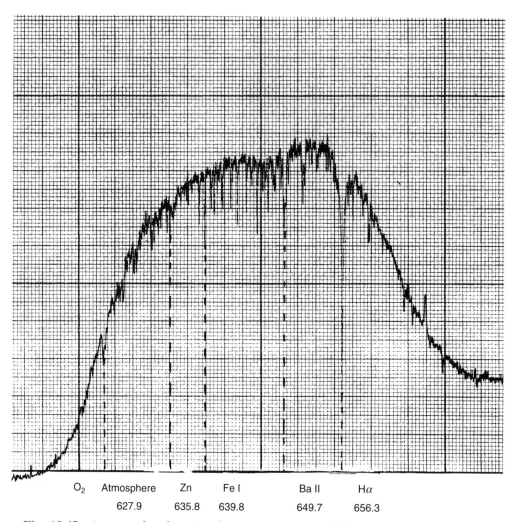

O$_2$	Atmosphere	Zn	Fe I	Ba II	Hα
	627.9	635.8	639.8	649.7	656.3

Fig. 18.47. *An example of a microdensitometer trace of the solar spectrum between 620 and 660 nm. It represents a microdensitometer scan of a photograph obtained with a design like that shown in Fig. 18.30.*

Even if the film is panchromatic, its spectral curve must be taken into account – as must the characteristic curve, if the non-linear portion is used – so that the results obtained may be appropriately corrected. The same applies if the efficiency of the dispersing element varies markedly with wavelength along a single spectrum, as it does with blazed gratings.

This method is of low accuracy, however, and it is preferable, if possible, to calibrate the overall spectrograph system (including the dispersing element and the film) with a standard, black-body reference source, for which the emission curve in known. This enables one to avoid the problems mentioned.

If the zenith distance is fairly large, atmospheric absorption must also be taken

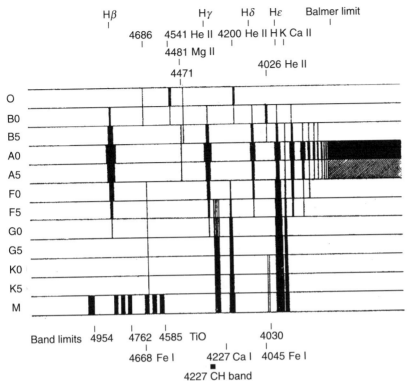

Fig. 18.48. *The definition of stellar spectral classes (wavelengths are given in* Å*(0.1 nm). After E. Schatzman,* Astrophysique, *Masson, Paris, 1963.*

into account. The quantity absorbed varies as a function of zenith distance and wavelength. (See the discussion of Bouguer's extinction law in the chapter on photometry, Sect. 19.4.1.)

In addition, it is distinctly possible that the peak wavelength will not be in the visible. The peak will therefore appear either at the violet end, or at the red. To derive the temperature, other measurements must be taken (such as obtaining the slope of the density curve as a function of wavelength).

When a suitable detector is available, i.e., one that can give an electrical signal as a function of the incident flux (such as a phototransistor, photodiode, etc.), it is possible to determine the darkest region of the spectrum by using an enlarger to project an image onto the detector. The ideal solution, of course, is to have access to a microdensitometer, which enables a complete scan of the spectrum to be obtained, giving a curve corresponding to the differences in density (Fig. 18.47).

18.3.3 Spectral classification

The spectral classification most widely used is the Harvard system. Its classifications depend on the presence or absence of specific lines, and on their relative intensities.

1037

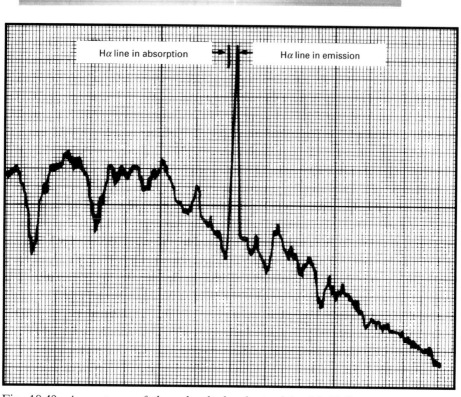

Fig. 18.49. *A spectrum of the solar limb, obtained by M. Tallon, G. Macaisne and I. Bosc with the equipment shown in Fig. 18.30 and a curved slit. In both the photograph and the microdensitometer trace the Hα emission line can be seen, shifted with respect to the Hα absorption line.*

It enables stars to be grouped into classes and sub-classes as a function of their properties (temperature, pressure, diameter, etc.).

The main classes are designated by letters, in order of decreasing temperature:

W, O, B, A, F, G, K, M, R, N, S

where the standard mnemonic is 'Winnie, Oh Be A Fine Girl, Kiss Me Right Now Sweetheart'.

Figure 18.48 shows the main characteristics of the most common spectra in the Harvard system. From its spectrum, it is possible to classify a star, and thus obtain its temperature (as well as other properties).

18.3.4 An example of results attainable: radial velocity measurements

Spectroscopy enables many different physical properties to be deduced, such as surface temperature, pressure (from the line profile), abundance of elements, magnetic field (measured by the Zeeman effect), electrical field (measured by the Stark effect), radial velocity (determined from the Doppler effect), etc. The last of these may be measured in solar prominences.

The Doppler effect is the name given to the shift in wavelength of lines from a moving source. It is the equivalent for light-waves of the phenomenon noticed with sound, such as when a vehicle's horn sounds higher in pitch when approaching the observer, and lower when receding. With light, if a source is moving at velocity v, a line at a wavelength λ will suffer a shift $\Delta\lambda$ such that:

$$\Delta\lambda/\lambda = v/c,$$

where c is the velocity of light. The shift is towards the blue when the source is approaching and towards the red when it is receding.

We know that solar prominences have radial velocities of about 20 km/s, and that they have strong emission in the Hα line (at 656.2 nm), and can thus calculate the spectral resolution required in a spectrograph to be able to detect the effect. We have: $\Delta\lambda = \lambda v/c = 0.04$ nm. This resolution may be fairly easily obtained with a solar spectrograph designed for spatial resolution.

The region close to the solar limb may be examined (*see* Chap. 1), and if a prominence is seen, its spectrum around Hα (which will appear in emission) obtained. This spectrum will also contain absorption lines (caused by other elements), so it will suffice to use one of these as a reference, and compare the distance (in nm) between it and the Hα line here, with the distance as measured on a spectrum of the photosphere (where Hα will be in absorption). The difference between these two values will be caused by the Doppler effect (Fig. 18.49).

18.4 Lists of spectral lines

18.4.1 Some spectral standards

Mercury The principal lines of mercury are shown in Figure 18.50.

Neon The principal lines of neon are at the following wavelengths:
> 650.6 nm
> 640.2 nm
> 638.3 nm
> 633.4 nm
> 614.3 nm

Sodium The principal lines are those in the sodium doublet at 589 and 589.6 nm.

Hydrogen Hydrogen wavelengths are given by the equation:

$$1/\lambda = 109\,677(1/n^2 - 1/m^2)(\text{in reciprical cm, } \text{cm}^{-1}),$$

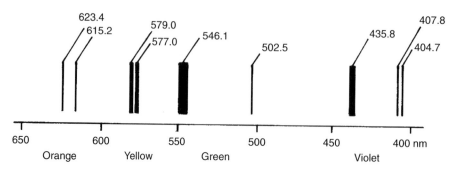

Fig. 18.50. *The spectral lines of mercury*

where n and m are integers and m is greater than n. The Balmer series is obtained when $n = 2$:

$m = 3 \rightarrow 656.2\,\text{nm}$	$m = 7 \rightarrow 397.0\,\text{nm}$
$m = 4 \rightarrow 486.1\,\text{nm}$	$m = 8 \rightarrow 388.9\,\text{nm}$
$m = 5 \rightarrow 434.0\,\text{nm}$	$m = 9 \rightarrow 385.5\,\text{nm}$
$m = 6 \rightarrow 410.1\,\text{nm}$	$m = 10 \rightarrow 379.7\,\text{nm}$

18.4.2 Principal nebular lines

Element	Wavelength (nm)	Relative intensity
N II	658.4	150
Hα	656.3	400
O III	500.7	800
O III	495.9	300
Hβ	486.1	100
He II	468.6	40
Hγ	434.0	40
Ne III	386.9	50

18.4.3 Principal solar lines

Name	Wavelength (nm)	Name	Wavelength (nm)
O₂ (atmospheric)	684.7–694.4	Fe I	426.0
Hα	656.3	Cr I	425.4
Ba II	649.7	Sc II	424.7
Fe I	639.8	Fe I	423.6
Zn	635.8	Fe I	422.7
O₂ (atmospheric)	627.9	Ca I	422.7
Ni	618.0	Sr II	421.5
Si I	594.8	Fe I	420.2
Na I ⎱ sodium	589.6	Fe I	419.8
Na I ⎰ doublet	589.0	Fe I	419.1
Cr	578.6	Fe I	417.3
Ni I	571.2	Mg	416.7
Ni I	570.9*	Fe I	415.4
Fe I	571.2	Fe I	414.4
Fe I	570.7	Fe I	413.2
Fe I	561.4	Ca I	411.9
Mg	552.8	Hg	410.2
Fe I	545.6	Fe I	407.2
Fe I	545.5	Fe I	406.4
Fe I	532.8	Mg I	405.7
Fe I	527.0	Mn I	405.5
Mg ⎱	518.4	Fe I	404.6
Mg ⎰ magnesium	517.3	Mn I	403.6
Mg ⎰ triplet	516.7	Mn I	403.4
Ni	503.6	Mn I	403.3
Fe I	495.6	Mn I	403.1
Ba	493.2	Fe I	400.5
Hβ	486.1	Fe I	399.7
Mn	478.2	Mg I	398.7
Fe I	466.5	Fe I	397.8
Ba II	455.4	Ca II }	396.8
Ti	446.9	Al I	396.1
Hγ	434.0	Fe I ⎱ 2	395.7
CH ⎱	432.4	Fe I ⎰ calcium	395.3
CH ⎰ methydidyne	432.3	Fe I ⎰ bands	395.1
Fe I	430.8	Al I	394.4
Fe I	429.9	Ca II }	393.4
Fe I	427.1	Fe I	392.8

[* There is also a stronger Fe I line at almost the same wavelength that is omitted from the original table. – Trans.]

19 Photoelectric photometry

C. Grégory

19.1 Amateur photoelectric photometry

For a long time, amateur astronomers have made visual magnitude estimates of variable stars, using telescopes, binoculars, or even the naked eye. Many continue to do this. In this field, the use of refractors, and in particular, small reflectors is very widespread. Some observers, still few in number, have taken advantage of commercial production of optics and the pervasive growth of electronics to make a 'quantum jump' from visual estimates to instrumental measurements of magnitudes and determination of colour indices. In the former group, good observers may attain an accuracy of 0.1 magnitude, but in the second, good instruments, used with care and with reliable methods, are able to detect variations of less than 0.01 magnitude. (This is the amount of variation, not the accuracy of the measurements, which will be discussed later.) This jump in accuracy is a result of the adoption of photoelectric photometry (PEP) by amateurs.

Variable-star observers, who are in the majority among photometrists, are thus able to extend their coverage and the physical range of their observations. But PEP does not end there: observation of the occultations of stars by minor planets, eclipses and occultations of the Galilean satellites of Jupiter, measurements of the brightness of minor planets, and contributions to research on the orbital elements of eclipsing binary stars are all within the reach of advanced amateurs.

It should also be noted that the introduction of PEP among amateurs has given rise, in Europe, the United States, and elsewhere, to a new departure: systematic, organised collaboration with astrophysicists on short- or medium-term observational programmes that have been mutually agreed or devised. Nowadays, there are many amateurs who, either because of their training or their jobs, are scientifically on a par with professional astronomers. Because of financial and administrative problems, many professionals have difficulty in gaining access to photoelectric equipment, so such a collaboration was only to be expected.

Photometry demands considerable care and attention, respect for the rules and precautions, an unfailingly critical attitude when it comes to reduction of the data, and experience, which only comes with time and from overcoming the various difficulties that are encountered.

Even the most experienced photoelectric worker is unable to find the solution to every difficulty. Help in arriving at a solution will, however, come from being completely familiar with the instrumentation, by knowing its limits and weaknesses, and by constant observation of the conditions under which it operates. Frequently individual instances and major or minor problems will be encountered that have to be resolved, either during the course of the night (if possible), or later at the desk, to understand what happened and its causes.

When the sky is clear and the equipment functions perfectly, photometry is extremely simple. Unfortunately, in most locations, the sky is subject to considerable change. As we shall see, this type of work is strongly dependent on sky conditions. Hope, persistence, and patience are essential qualities in any photometrist.

It is true that observers with more than a smattering of mathematics will find it easier to undertake more sophisticated calculations and investigations, to devise and write reduction programs, and to develop their own procedures and refinements. Although this will save time and make life easier, their measurements will not be any more accurate than those made by someone who is just about capable of handling elementary algebra and simple logarithms. PEP is not full of mysteries. There are specific methods, of course, and obstacles to overcome, but the latter are the same for everyone and are mainly caused by the vagaries of the sky and instrumental drift. Experience shows that the mathematics, which is not very complicated and becomes routine, is not the most difficult part of the work. It is the observer's experience that makes for good photometry.

The one failing of PEP is that it requires considerable time, and that it is not possible to make many observations during the course of a night, without recourse to one of those wonderful automatic telescopes. Care needs to be taken in preparing for the observations, and they must be reduced with scrupulous attention to detail. One needs to know how to interpret them, avoiding the pitfalls that may be hidden behind apparently satisfactory data, and without jumping to hasty conclusions.

Any amateur who takes up PEP seriously soon finds that it becomes a fascinating interest that leaves little time for other astronomical activities.

There are a number of simple equations in the following sections that are essential to obtaining a full understanding of the subject, and which cannot be avoided in our discussion. A large number of others have been gathered into Sect. 19.5 (Observation and methods). At the end of this chapter, in Sect. 19.6 (The UBV System), the various symbols that are required to understand the equations are defined. Readers who wish to obtain just a review of the subject may ignore most of the equations at this stage.

19.2 Optical and photoelectric detectors

19.2.1 The eye

The observer's raw material is the prime focus image, and this contains a specific amount of intrinsic information that may be extracted in various ways. It is all contained within the diffraction image at the focal plane. The types of detectors used need to be specifically chosen for the spectral region being examined, to the flux that is available, and to the type of information that one wants to obtain. [Flux is, of course, defined technically as the quantity of energy crossing a surface (the detector) per unit time. The flux density is the flux per unit area (*see also* p. 1068).]

The eye handles light 'in bulk'. Like any photosensitive surface, the retina has a response curve that is a function of wavelength over its own particular pass band. But the overall physiological system consisting of the detector and the central nervous system is incapable of determining the various frequencies present and to form quantitative estimates of the luminous flux received. Neither is it capable of retaining a record of the excitation that it has received, except by remembering it qualitatively. It is, on the other hand, capable (with certain reservations), of simultaneously comparing the effect produced by sources of different intensity (estimating relative brightnesses); of determining the frequency region that shows the greatest intensity over the range to which it is sensitive (estimating colour); and of determining quite subtle differences in the surface brightness of extended objects, such as lunar and planetary surfaces. (The eye's response is governed by the law discovered by Fechner in 1850 that the intensity of sensation varies as the logarithm of the stimulus.) But its characteristics vary from one individual to another, and are not always constant in a single person.

An instrument may be calibrated, and thus be capable of producing proper measurements, if it is possible to explain the factors and the relationships that define the measured properties and their fluctuations in a specific reference system. Unfortunately, the eye cannot be calibrated.

19.2.2 Photoelectric photometry

Whereas spectroscopy aims to 'spread out' the signal as a function of wavelength, describe it, and analyze its structure, photometry aims to use filters to select certain pass bands and to measure the luminous flux within these limits. The pass bands are described as being narrow, intermediate, or wide, depending on their spectral extent (from a few nm to more than 200 nm in width.) Photometry offers a means of determining the relative distribution of energy in these pass bands, from which information about the different physical parameters of the objects being observed may be drawn directly. This is why photometry in integrated, visible light, or in a single pass band, is rarely employed, except for accurately determining times or brightnesses in certain phenomena that are found mainly in the field of celestial mechanics (mutual phenomena of Jupiter's satellites, minor-planet occultations, rotation of minor planets, etc.).

Conversion of flux values to magnitudes m is obviously independent of the pass bands. For a given flux F, Pogson's equation gives:

$$m = -2.5 \log F + \text{constant}.$$

The ratio between two fluxes becomes a magnitude difference:

$$m_1 - m_2 = -2.5 \log(F_1/F_2).$$

These are raw magnitudes, applicable to the instrumentation used, but not comparable with others obtained by different instrumentation. Several corrections need to be applied and these will now be considered.

[It is difficult to use the measured flux to derive the theoretical flux that would be received from a source, because the measured flux at a given wavelength is a function of various transmission coefficients: those of the atmosphere, the optical train, and the filter, together with the detector's quantum efficiency. In modern photometry, the term 'flux' is used for the values measured per unit time with a given instrumental system. It is always these values, and the ratios between them, that are used, whatever instrumental units are used. Flux is linked to the radiant energy of the source, and thus to the latter's spectral luminosity.]

The choice of pass band is, of course, controlled by the type of information that it is hoped to obtain. Techniques differ according to the observational methods and the spectral regions that are to be employed. The narrower the pass bands, the smaller the fraction of the flux that the filters transmit, and the larger the collector should be for the received signal to be significant.

The necessity for using large telescopes makes it difficult for amateurs to carry out intermediate-band photometry – diameters of 500 mm at least are required for uvby, or Strömgren, photometry – and the restrictions apply even more forcibly to narrow-band photometry. The latter requires interference filters, which are difficult to make, very costly, fragile, and which degrade fairly rapidly. In addition, the angle of incidence of the rays of light on such filters is critical, requiring perfect collimation, which complicates the optical arrangement.

The most commonly known photometric systems are: the Strömgren uvby (ultraviolet, violet, blue, and yellow) system, perhaps supplemented by photometry of the Hβ line of hydrogen; and the systems devised at the David Dunlap Observatory (DDO); at Vilnius; by Walraven; and at Geneva. These systems are mainly determined by the characteristics of their filters and, as such, it is not always easy, or even possible, to transfer results obtained in one system to another.

Naturally astronomers are quite within their rights in defining their own filters and photometric systems. For results to be compared and reproducible, however, at least one other astronomer somewhere will have to adopt the same system. This has, in fact, occurred for certain narrow-band measurements that are required in very specific investigations.

Unlike the UBV system, the Strömgren system is uniquely defined by its filters' pass bands, which eliminates secondary absorption coefficients (Table 19.1). Its main advantages are that, whatever the detector, it can: provide visual magnitudes; determine effective temperatures; determine the strength of 'metallic' absorption lines and of the Balmer discontinuity. (To astrophysicists, all elements with greater atomic masses than helium are metals. The Balmer series of absorption lines of hydrogen arises from transitions between the second energy level and higher levels. The distance between individual lines in this series becomes smaller and smaller, finally converging at the Balmer limit at 354.6 nm. At the Balmer limit there is a sharp decline in intensity, the Balmer discontinuity, in a continuous spectrum. The Balmer discontinuity is directly linked to spectral type and is therefore a function of stellar temperature.)

The wide-band photometric system universally employed is Johnson and Morgan's UBV (Ultraviolet, Blue, Visible) system, which may be extended to R and I (Red and Infrared), and also to even more remote pass bands (H, J, K, L, M, and N – colloquially known as 'Johnson's alphabet soup'). UBV photometry normally uses Schott or Corning, dyed-in-the-mass filters, 1 or 3 mm thick (Table 19.2).

The UBV system is the one used in most catalogues of stellar magnitudes and photometric colour indices (Fig. 19.1). It is the simplest from the instrumental point of view, and has great potential for amateurs in the spectral range between the near-ultraviolet and the near-infrared. This system does have disadvantages, however,

1045

Table 19.1. *Strömgrem uvby system filters*

Filter	Wavelength of centre of pass band (nm)	Width of the band for filter transmission of 50 %
y	550	20
b	470	10
v	410	20
u	350	40

Table 19.2. *Effective wavelengths defined by the Johnson filters (U, B, V, R, I, etc.)*

Filter	Approximate effective wavelength (nm)
U	360
B	440
V	550
R	700
I	900
J	1250
K	2200
L	3400
M	5000
N	10 200

particularly because the U filter has a very broad pass band, approximately 100 nm wide. Unfortunately, this is not precisely defined on the short-wave side, where the 'cut-off' is determined by atmospheric absorption of ultraviolet radiation.

In addition, the U filter straddles the Balmer discontinuity (*see above*). The U − V colour index is therefore affected by the Balmer discontinuity, and if the latter is intense, the detector will receive very little radiation at wavelengths below the Balmer limit (364.6 nm). It will, however, receive all the radiation at the red end of the pass band, so the effective observational wavelength will not be the nominal centre of the pass band. By contrast, the filter will transmit practically all of the flux from a star that does not show a strong Balmer discontinuity, and the effecive wavelength will be very close to the nominal value of the centre of the pass band. This variation strongly effects a value known as the secondary coefficient of extinction k'' (*see* p. 1075). If the latter is plotted against spectral class from O to M, there are two major deviations from a smooth curve. (When Johnson and Morgan originally established the UBV system, they defined $k''_{ub} = 0$ to avoid long, tedious calculations (*see* p. 1076.)

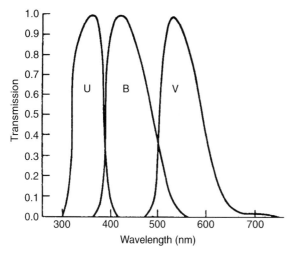

Fig. 19.1. *Normalized transmission curves (maximum transmission = 100%) for the UBV filters*

By using suitable reduction methods, the flux measurements are converted into magnitudes and various colour indices: V, (B−V), (U−B), (V−R), and (R−I), which are referred to as the system's standard magnitudes and indices. These calculations are quite long and tedious, and thus prone to error. They may be carried out easily, quickly, and accurately by the use of a programmable calculator or, better still, a computer, using programs that are readily available, and which may be adapted to one's precise needs.

Quite apart from its own strengths, photometry may be said to complement spectroscopy by being able to make accurate measurements of fluxes.

19.2.3 *The photoelectric effect*

The photoelectric effect was discovered by Hertz in 1887, and later explained by Einstein in 1905, earning him the Nobel Prize. Under certain conditions, a metallic element that is bombarded by light is capable of emitting electrons.

The number of electrons found in an atom is rigorously defined for each chemical element. Schematically, their possible positions are represented by 'shells', corresponding to different energy levels, which are also rigorously defined. The outermost shell of an atom may contain a maximum of eight electrons. In a neutral atom, where the outmost shell (the valency shell) is incomplete, the electrons are bound to the nucleus to a greater or lesser degree. It may capture or lose electrons (becoming ionized). The atom is described as being more or less electropositive depending on the ease with which it will give up an electron or electrons to other atoms with which it enters into molecular combinations. Metal atoms may 'donate' outer electrons as a result of the electrical fields produced by neighbouring atoms (in chemical

1047

reactions), or lose them by gaining enough energy to break the bond: for example, from an energetic photon, a quantum of radiation.

The surface of a metal irradiated by light is subjected to a bombardment by photons and is affected to a depth of less than one micron (1 μm). The structure of the metal is roughly similar to a regular, repetitive lattice. Some photons are simply reflected by the lattice and diffuse into the material. Others are transformed into heat. There is, however, a certain probability that yet others, the least numerous, may cause the emission of electrons. Because of the electrical forces that bind electrons to atomic nuclei, the energy deposited must be enough to enable the outer electrons to escape. If the photons are very energetic, as is the case with high-frequency radiation, electrons may also be expelled from deeper atomic levels. The energy imparted to the emitted electron corresponds to an amount of electromagnetic energy, represented in corpuscular form by the incident photon. The amount of energy represented by the photon, the quantum of energy, is proportional to the frequency of the radiation in accordance with:

$$W = h\nu$$

where ν is the frequency and h is a universal constant (Planck's constant).

The whole of the energy W carried by the photon is transmitted to the emitted electron, but only a fraction of the energy carried by the photon will appear in the form of kinetic energy. Random collisions with other particles within the metal absorb some of the energy. Another fraction is consumed in the work required to free the electron. The third, and remaining fraction appears as the free-electron's kinetic energy.

If we call these three fractions W_1, W_0, and E_K, respectively, we have:

$$W_1 + W_0 + E_K = h\nu \quad , \quad \text{i.e., } E_K = h\nu - W_0 - W_1.$$

Because it involves random interactions, the value of W_1 varies. If it is zero, the kinetic energy carried by the electron is given by:

$$E_K = h\nu - W_0,$$

which is the fundamental, photoelectric-effect equation, or Einstein equation.

For a certain minimum quantum value, equal to the ionization energy W_0, the amount of energy capable of appearing as the electron's kinetic energy is zero. Here:

$$h\nu_0 = W_0 \text{ , and } E_K = 0.$$

This implies that for any frequency ν less than ν_0, no emission will occur. The frequency ν_0 is called the photoelectric threshold, and may also be expressed as a wavelength: $\lambda_0 = c/\nu_0$, (where c is the velocity of light in vacuo, and which by definition $= \lambda\nu$).

For any given metal, there is a threshold frequency, below which radiation is unable to cause the release of photoelectrons, whatever the intensity of the source. This is why a very faint, but high-frequency, source is able to cause the photoelectric effect, and why, on the other hand, powerful radiation that has a frequency below the threshold does not cause any photoemission from the same metal.

To make use of the photoelectric effect, the sheaf of electrons emitted has to be channelled and prevented from dispersing, and also the electron flux, which is only a microcurrent, has to be amplified. Channelling of the electrons is accomplished by using a vacuum tube, where the photoemissive surface is the cathode (photocathode) and the detector is the anode, and applying a sufficiently high potential across the two electrodes. We then have a simple photoelectric cell.

If the voltage applied over the photocathode and the anode is varied, while the photocathode is exposed to constant, monochromatic light, it will be found that as the intensity increases, the voltage increases rapidly until it attains a limiting value, which then remains constant. The point at which this occurs is known as the saturation intensity (Fig. 19.2).

The saturation current I_S, is related to the radiation flux P (the amount of luminous energy received per second at the photocathode) by:

$$I_{S(\lambda)} = \sigma_{(\lambda)}P.$$

The coefficient $\sigma_{(\lambda)}$ is defined as the sensitivity of a given cell to monochromatic radiation of wavelength λ. (It is expressed in amperes/watts.) A plot of the different values of $\sigma_{(\lambda)}$ versus those of λ gives the cell's sensitivity curve as a function of wavelength (Fig. 19.3).

In experiments carried out in 1916 that confirmed Einstein's theory, Millikan showed that if the potential of the photocathode was defined as 0, and a negative potential V (close to 0), was applied to the collector, any photoelectrons having sufficient kinetic energy still reached the collector, producing a current across the device. If, however, the potential of the collector V was made increasingly negative, the photoelectric current ceased for all values of V below a certain value V_0. The electrons were repulsed by the collector. At this potential V_0, the velocity of electrons with the highest kinetic energy as they arrive at the anode is zero. The potential V_0 is the cut-off potential.

This cut-off potential is independent of the intensity of the incident radiation. It depends only on the nature of the photocathode (its threshold) and on wavelength.

If the cathode potential is significantly higher than the cut-off potential, we find (for any given radiation):

(i)   within the limits of error, and for identical conditions, identical measurements of electrical intensity for identical luminous fluxes; and

(ii) measurements of different intensities are proportional to the respective incident fluxes.

At frequencies above the photoelectric threshold, there is a linear relationship between the intensity of the incident radiation and the number of electrons emitted per second. For any given cell and radiation, the ratio of the number of electrons emitted by the cathode to the number of incident photons per unit time is known as the quantum efficiency. This amounts to defining the quantum efficiency as being the percentage of effective photons, i.e., number of photons producing emission / total number of photons incident on the photocathode per unit time.

The quantum efficiency of a cell is therefore not the same at all wavelengths over the spectral range within which it is sensitive. At peak sensitivity, it is very low in classical photoemissive cells: around 3/1000 (i.e., 0.3 %). It is possible to

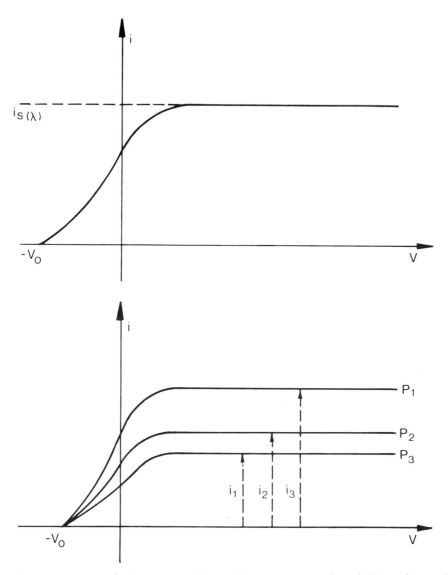

Fig. 19.2. *Current/voltage curves for a cell at a given wavelength.* Top: *for a radiation flux P, the saturation intensity is i_s.* Bottom: *set of curves for three different radiation fluxes at the same wavelength. The value $-V_0$ is the cut-off potential.*

increase this efficiency. Even with much more complex designs, however, efficiencies are rarely higher than about 20 %.

A plot of quantum efficiency versus wavelength gives a cell's response curve. For any specific photometric system, filters should be chosen so that their transmission curves are most suitable for the spectral region to be covered and to the cell's response curve.

The alkaline metals (potassium, sodium, caesium, rubidium, etc.) have a single

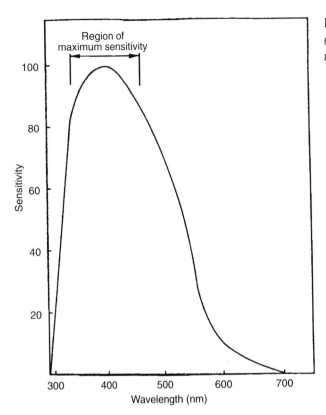

Fig. 19.3. *Sensitivity curve (%) for an RCA 1P21 photomultiplier tube*

valency electron. They are strongly electropositive. The alkaline earths (strontium, calcium, etc.), which have two valency electrons are slightly less electropositive. Generally the alkaline metals, in conjunction with antimony, are used for the photoemissive layer in photomultipliers (Table 19.3).

The photoelectric effect is never perfectly 'clean', as will be seen in the description of detectors. It is accompanied by parasitic and secondary effects, which cause noise to appear in the electrical signal. In photodiodes, light and matter interact in a different way.

19.2.4 Photomultiplier-tube (PMT) photometers

Intermediate metallic plates coated with an emitter (dynodes) are inserted between the photocathode and the anode, each of which is held at a different potential. With 11 dynodes, for example, successive potential differences of $+100$ V could be applied, giving a negative potential of -1200 V at the cathode and 0 V at the anode with respect to earth. Electrons that are emitted from the photocathode with a certain initial energy, caused by the photoelectric effect alone, are then accelerated by the electrical field between the photocathode and the first dynode. Their kinetic energy is thus increased. Their impact on the dynode dislodges other, more numerous

Table 19.3. *Specification and quantum efficiency for different types of PMT*

Photocathode		Win. matl*	Wavelength		Sensitivity at λ_{max} (mA.W^{-1})	q (%)
Type	Material		λ_{min} (nm)	λ_{max} (nm)		
S1(C)	Ag O Cs	1	1100	800	2.3	0.36
S4	Sb Cs	1–3	680	400	50	16
S11 (A)	Sb Cs	1	680	440	60	17
(Super A)	Sb Cs	1	700	440	80	22
S13 (U)	Sb Cs	2	680	440	60	17
S20 (T)	Sb Na KCs	1	850	420	70	20
S20R	Sb Na KCs	1	900	550	35	8
TU	Sb Na KCs	2	850	420	70	20
bialkaline	Sb Rb Cs	1	700	420	85	26
bialkaline D	Sb K Cs	1	630	400	85	26
bialkaline DU	Sb K Cs	2	630	400	85	26
SB	Cs Te	2	340	235	20	10

* Window material – 1: borosilicate or soda-lime glass, 2: cast silica, 3: opaque cathode
λ_{min} is cut-off wavelength; λ_{max} is wavelength of maximum sensitivity; q is quantum efficiency

electrons, which produce a similar effect at the second dynode, and so on. By this means an enormous gain is obtained. For n dynodes with a multiplication factor of k, the internal gain is k^n. This type of design is known as a photomultiplier tube (PMT).

Depending on the number of dynodes and the overall voltage applied, PMTs generally produce more than one million electrons at the anode for every electron dislodged from the cathode.

In theory, for every electron that is emitted by the photocathode, a pulse is obtained at the anode. It is therefore possible to either count the impacts of the photons as they occur, or to consider the series of impacts as an electron flux, turning it into a direct current, variations of which may then be measured.

In photon-counting, consideration has to be given to the height of the pulse and to discriminating such pulse-heights. We will not complicate this discussion by describing these factors in detail, which have to be considered by anyone working on pulse-counting techniques, but which are not required in this introductory discussion.

It would be logical to assume that, contrary to what happens in an amplifier, a photomultiplier device would not introduce noise, because the electrons have not passed through a conductor, or, to put it another way, that the loss of information caused by the process would be almost zero in the high vacuum inside the tube. In fact, this is not strictly the case. For one thing, the gain of the PMT alters over a period of time (months or years), in an irreversible fashion because of aging, and also the dynodes produce variations in the PMT's gain that are perceptible over periods of a few seconds to a few hours. These variations, which cannot yet be prevented, cause the measurements to be subject to errors of up to 1 %. Finally, when the luminous flux is too high (with very bright objects), electrical phenomena occur between the last dynode and the anode that cause a loss at the anode itself, i.e., a decrease in the PMT's gain. This has nothing to do with over-exposure of the photocathode, which is not involved because its response remains linear. It is the effect of fatigue.

To be handled, the output current from the PMT, which is still weak, itself needs to be amplified. Because theoretically the electrical signal is linear if the voltage applied across the PMT is well-regulated, it is obviously essential for the external amplifier to be as stable as possible. Most of the major instrumental errors arise either from fluctuations in the high-voltage supply to the PMT, from unregulated variations in the gain of the external amplifier, or – to a lesser degree – from the variations in the PMT's gain mentioned earlier. In everyday practice, we are forced to ignore this last effect.

The PMTs employed in astronomical photometry are grouped in various classes, corresponding to their response curves (*see* Table 19.3). They are chosen as a function of the spectral regions to be covered and the required sensitivity. There are several methods of mounting the photocathode and the dynodes within the tube: 'side-window' tubes with an opaque lateral cathode and 'squirrel-cage' dynodes; semi-transparent end-on cathode, with 'venetian-blind' or 'box and grid' dynode mounting.

The photocathode's response curve depends on its composition. In S4 and S11 tubes an antimony/caesium (Sb Cs) emitter is used. Other alkaline metals may be incorporated, as in the 'tri-alkali' S20 tube. The S1 tube, which has a very wide spectral range but a very low quantum efficiency (about 1 %), uses a silver/oxygen/caesium emitter.

The classic RCA 1P21 tube (a type S4, side-window PMT), which has an opaque cathode and was used by Johnson and his collaborators to establish the UBV system, is still used in some observatories (Fig. 19.4). However, the more modern S11 and S20, end-on PMTs (Fig. 19.5) have replaced it in many observatories.

The PMT, even if specially selected, is far from being the most expensive component in the overall photometer, because it is absolutely essential not to skimp on the quality of the high-voltage supply or on the amplifier electronics. Both should be designed by fully-qualified specialists, and built professionally. Because 'do-it-yourself' designs are to be avoided, if it is impossible to obtain help from an expert, the best way of ensuring that the electronics are satisfactory is to purchase a complete system from a reputable manufacturer.

19.2.4.1 The objective

The optics required for these detectors simply consist of some form of objective, a 'light-bucket', that is free from chromatic aberration. Refractors are thus to be avoided. Both the magnitude limit of objects that may be measured and the

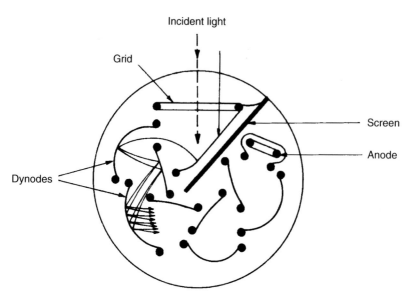

Fig. 19.4. *Cross-section of the classic design of side-window tube, the 1P21. RCA diagram.*

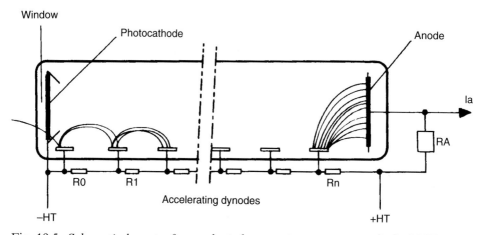

Fig. 19.5. *Schematic layout of an end-window, semi-transparent cathode PMT*

accuracy of the measurements depend on the amount of light gathered, and thus on the diameter of the objective.

Electrons in motion obey Fermi–Dirac statistics, but these do not apply to photons, which are massless. Instead, the probability of their arrival is governed by Bose–Einstein statistics. As a result, any measurement is inherently accompanied by a certain degree of inaccuracy, whatever type of detector is used. The arrival of photons is, by its very nature, a discontinuous process, so the range of measurements follows a Poisson distribution (described in any work on statistics), and the error is approximately equal to the square root of the number of photons recorded. This is the reason for wanting to capture as many photons as possible, because although the absolute error increases, the relative error, known as the photon noise, decreases in inverse proportion to the number of photons:

Absolute error $\sigma = \sqrt{N}$; relative error or photon noise $= \sigma/N = 1/\sqrt{N}$

If 10^6 photons are detected, the error would be 1000, and the relative error $1/1000$, whereas for 10^4 photons, the error would be 100, but the relative error $1/100$.

In practice, the exposure time (integration time) is extended so that the photon noise is low. Using a PMT as a photon-counter, if one wants to obtain a theoretical mean accuracy of 0.001 mag., for example, with a star that has a rate of 50 000 counts per second, a minimum of 10^6 counts needs to be recorded, i.e., the integration time should be 20 seconds.

Amateur reflectors between 200 and 400 mm in diameter are very suitable for stellar photometry. With excellent seeing and a dark sky, a diameter of 200 mm, properly used, is capable of making measurements of 9th magnitude stars. Although small diameters give a relatively high photon noise with faint objects, this disadvantage may be counteracted by extending the integration time to about 1 minute. Cassegrain designs (f/10 or f/12), which are commercially available, are preferable to Newtonians because of their longer focal lengths. This is because it is necessary to use a diaphragm at the focal plane to reduce the amount of sky background around the object being measured. A field diameter of about 1 arc-minute is normally required. The detector does not discriminate between light from the object and light scattered by the Earth's atmosphere. The usable fraction of the overall field is determined by the focal length of the optics and by the diameter of the circular aperture in the diaphragm, which should be held at right angles to the optical axis.

The longer the focal length, the smaller the fraction of the overall field that is given by a given diameter aperture:

$$d = 3600[2\arctan(r/f)]$$

where d is the size of the field given by the diaphragm in seconds of arc on the sky, r is the radius of the circular aperture, and f the focal length, both of the latter being measured in mm. Obviously r is varied as required. Most photometers have a diaphragm wheel or slide with apertures of different sizes.

The sky background gives rise to measurement noise that may be easily removed: the object is measured, together with the sky background passed by the aperture, and then another measurement is made of the sky close to the object. The second reading is subtracted from the first. A bright sky background does, however, raise the limiting magnitude and reduce the accuracy. This is why, without exception, one has to avoid making measurements at Full Moon and during the nights on either side of it, in the area of zodiacal light, etc. Without special, complicated systems that allow separate measurements to be carried out simultaneously on the object and on the nearby sky background, measurements should not be undertaken during astronomical twilight.

The focal length should not be increased by using a negative Barlow doublet, because all glass, with the exception of pure quartz, absorbs ultraviolet radiation.

It is the stopped-down image at the focal plane that is to be used. The circular field provided by the diaphragm must not be too small. This is for two reasons:

a The point-like image of a star as seen in an eyepiece does not contain all the light gathered by a telescope. The light is actually distributed in a diffraction pattern (the Airy disk and rings), the peak of which corresponds to the centre of the image. If the diaphragm is reduced too much, light in the rings and in the outer part of the central peak may be lost. This means losing, in an uncontrollable manner, a significant fraction of the signal, which will have a detrimental effect on the accuracy of the measurements.

b The sidereal rate of the telescope drive can never be absolutely constant. The longer the focal length, the larger any angular displacements of the focal-plane image will be, and the more likely that the image will become decentred. A compromise therefore has to be sought between the two conflicting requirements: reducing the size of the field and retaining a large enough aperture.

Professional instrumentation allows the use of diaphragms 20 arc-seconds in diameter. With amateur equipment, to be on the safe side, it would be better to use rather larger apertures, approaching 1 arc-minute. With a 200-mm, f/10 reflector, a circular aperture 0.48 mm in diameter gives a diaphragm subtending 50 arc-seconds on the sky.

Under optimum observational conditions, a properly centred image in a correctly chosen size of diaphragm transmits more than 95 % of the light incident at the focal plane (Figs. 19.6 and 19.7). The only way of making consistent measurements is to keep the image as closely centred as possible on both the diaphragm and the photocathode itself (Fig 19.8). The former should be checked at each filter change, unless one is completely sure of the accuracy of the drive. Although, for equal diameters, a Cassegrain design is less luminous than a Newtonian, it has two distinct advantages: the balance of the telescope is less affected by the weight of the photometer, and the observer is spared most of the physical effort that would otherwise be required in finding, centring the image, and frequently checking the centring. Short, Schmidt–Cassegrain telescopes are the ones most frequently used by amateur photometrists. The Schmidt plate does absorb a certain amount of UV radiation, but because the plate is thin, this amount is small.

The finest telescope is of no use for photometry unless it has an excellent mounting. This needs to be heavy, stable, smooth in its motion, and to have a sidereal drive that is as accurate and regular as possible, fitted with fast and slow motions in both right ascension and declination. If an ordinary commercial mounting is to be used, it is essential, before contemplating serious photometric work, for a check to be carried out to ensure that a star will remain on the cross-hairs, at resolving magnification, for at least 10 minutes. The mounting itself needs to be fixed to a massive, solid pillar, into the ground and surrounded by a gap to avoid the transmission of vibrations. A pipe filled with concrete would be one way of doing this. Meticulous alignment is essential, and this should be adjusted, if necessary, every few nights.

All this means that equipment that is to be used for accurate, scientific, pho-

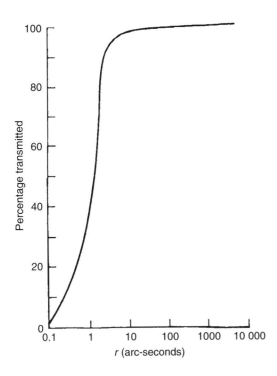

Fig. 19.6. *The effect of the diaphragm's aperture on the percentage of the radiation incident on the PMT. This plot shows the percentage of the light in a stellar diffraction image as a function of the radius of the diaphragm.*

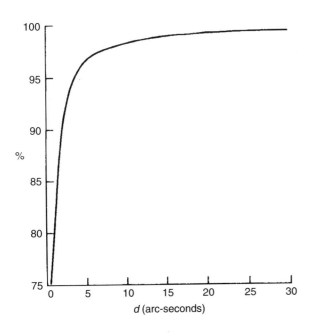

Fig. 19.7. *The percentage of the total amount of energy collected by a 400-mm telescope that is passed by diaphragms of different diameters (in arc-seconds). The curves shown here and in the preceding Figure show that a diaphragm with a clear field of 20 arc-seconds is the safe limit.*

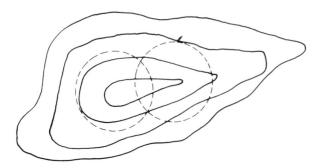

Fig. 19.8. *Isosensitivity curves for a photocathode. Depending on where the spot of light falls, the tube's response will differ. This is why the optics should minimize any angular displacement of the beam of light.*

tometric observations requires a fixed mounting, with a suitable shelter or dome. This is all the more desirable because electronics, even more than optics, suffer from unnecessary adjustments and frequent removal.

If a commercial Schmidt–Cassegrain telescope with a short tube is being used, it is a sensible precaution to guard against stray light and dewing of the corrector plate by fitting a dewcap. This is an essential addition when moonlight is likely to fall on the aperture of the telescope, even when it is faint or coming from the side.

The telescope should be fitted with a finder, which is practically indispensable for ensuring that the object has been correctly identified. It should be a fast refractor, with a diameter of at least 60 mm, and a short focal length. Experience shows that the ideal is to have as large a diameter as possible, a ratio of f/6, and an eyepiece chosen to give a field of 1.5–2° on the sky. This will enable stars down to magnitude 10 to be identified comfortably. Small, 90-mm, Schmidt–Cassegrains or Newtonian 'comet-seekers', which are far less expensive than refractors, may also be used. We shall discuss later why a field of this size is particularly suitable, and helps to save time in differential photometry.

19.2.4.2 The optics of the photometer head

Basically, the head is a light-tight box, which fits the eyepiece holder (Fig. 19.9), and contains:

a A prism or a movable mirror that enables the image of the full field at the focal plane to be sent towards *b*;

b A wide-field eyepiece, which enables the reflected image to be examined, and the object to be identified and centred;

c A diaphragm wheel with a series of carefully prepared, circular apertures of different diameters. Changing the diaphragms should be by mechanical means, and the wheel should be provided with positive stops to ensure that the optical axis passes through the centre of the aperture being used. Once the telescope has been focussed, the prime-focus plane should coincide with that of the diaphragms;

d A viewing eyepiece behind the diaphragm wheel. This consists of an optical train and prism (or flip mirror). The edge of the circular aperture is used to focus the system. After this focussing has been carried out, the telescopic image is focussed in the usual way. Good equipment is fitted with an illumination system, using an LED, preferably red, which enables the edge of the aperture to be seen. (This is quite difficult when the sky background is most favourable, i.e., faintest.) This light should be extinguished automatically when this second prism or mirror is moved to make measurements, or to blank out the sky;

e A converging lens, otherwise known as a Fabry lens. This lens is carefully positioned on the optical axis so that it forms an image of the system's entrance pupil (the objective) on the photocathode. This ensures that the unavoidable slight alterations in the position of the prime-focus image in the plane of the diaphragm produce only imperceptible variations in the image's position on the photocathode. Such shifts would affect the quality of the measurements, because the sensitivity of the surface of a photocathode is far from uniform;

f A filter wheel, with as many positions as filters, plus a blanked-off position that serves to protect the photocathode, which may be irreparably damaged if the illumination is too bright. (Even if it is not permanently affected, the response curve may be affected for several hours.) The coloured filters should have diameters comparable with that of the photocathode. The pass bands are chosen, bearing the photocathode's response curve in mind, to give the best match to the UBV system. In amateur instruments, the aperture and filter wheels, and the mirrors or prisms are operated manually.

19.2.4.3 The detector

The detector consists of the PMT's high-voltage supply, the PMT itself, its pre-amplifier, and the output to the amplifier. The PMT should be electrically shielded. It also needs to be magnetically shielded with a special alloy (mu-metal), so that the tube's electronics, and particularly the photocathode, are protected from the effects of the Earth's magnetic field, which varies according to the position of the telescope. These effects are far from being negligible.

Extreme care should be taken to ensure that the cable carrying the high-voltage supply is properly insulated, because it is carrying a DC voltage of more than 1000 V. It is always advisable to site the amplifier as close to the photometer head as possible, to minimize the danger of losses, which are often difficult to detect, and of parasitic effects in the cable. There is no lack of sources around the telescope that might contribute to this degradation of the signal. This is why most designs incorporate amplification in the photometer head itself, which makes the latter heavier and also more prone to damage.

A considerable improvement to the basic design of photometer heads has been devised comparatively recently by the British astronomer Norman Walker. This form has been specifically designed with amateurs in mind. The head of the 'JEAP' (Joint European Amateur Photometer) consists of just the acquisition optics, i.e.,

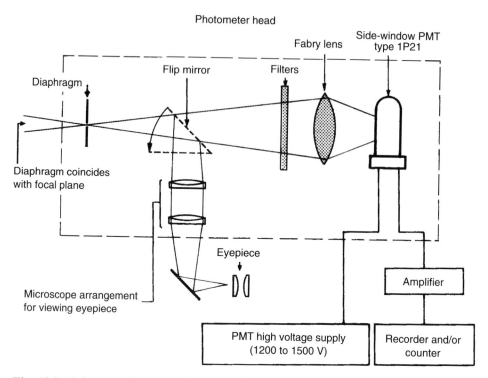

Fig. 19.9. *Schematic diagram of a PMT photometer. PIN-diode photometers are much simpler. No high-voltage supply is required. Amateur models do not have a diaphragm: the optics supplied by the manufacturer determine once and for all the positioning of the spot of light on the detector. As a result, the diaphragm eyepiece is no longer required. All that remains is a wide-field eyepiece that is used to centre the object in the centre of an illuminated reticle with double cross-hairs.*

a simple, robust device for locating and centring the object, using a flip mirror, a set of diaphragms, and an output socket into which the end of a fluid light-guide (3 mm in diameter and 2 m long) is inserted. The fluid light-guide transmits 74 % of the light without any significant differential absorption in the UV (Fig. 19.10). The other end is mounted in front of the filter wheel and the PMT, both of which are in an equipment case located elsewhere in the observatory, together with the amplifier and counter.

Unlike quartz fibres, the liquid light-guide may undergo considerable flexure without having a significant effect on the signal. The anodized-aluminium and brass head weighs only 700 g, and is suitable for any type of amateur telescope as well as larger ones. Amateur observers, working with professionals, have used one with a 5-mm light-guide on the 800-mm reflector at the Observatoire de Haute-Provence. They have also employed a twin version with a double-channel photometer fed by two 300-mm Schmidt–Cassegrain telescopes, mounted on an equatorial table.

Fig. 19.10. *The JEAP photometer head designed by Norman Walker (*$80 \times 70 \times 65\,mm$*, weight 700 g), mounted on a Celestron 11. The fluid light-guide can be seen mounted at the rear. The knob controlling the flip mirror is at the side of the head.*

19.2.5 Amplification

The output from the amplifier may be in digital form, produce a record on a chart recorder or be fed to a computer (or appear in any combination of the three forms). Two types of amplification of the photoelectric signal are used: DC measurement or photon-counting.

19.2.5.1 DC measurement
The packets of electrons arriving at the PMT's anode are treated as a direct current, the average value of which over a specific interval is used as the basis of measurement. The heart of the system is an 'Op-Amp' (operational amplifier), which has a gain of the order of 10^4, and converts the very weak current from the PMT into a voltage. This is actually a very small, cheap component.

Electronic means are used to obtain discrete measurements, similar to those provided by photon-counting. The variations in the PMT's output current, handled as variable voltages, are converted into frequencies. It is these frequencies that are displayed or recorded.

Photometric workers have a tendency, nowadays, to abandon DC measurement in favour of photon-counting, which, despite certain disadvantages (such as the dead time, described later), has considerable advantages.

19.2.5.2 Photon-counting

This method of measurement completely dispenses with heavy and expensive chart recorders, the use of which introduces complications because of the introduction of time constants. It also eliminates calibration problems and periodic checking of the gain of the amplifier. It provides, even before reduction, an idea of the measurements that is easier to interpret immediately than DC measurements. The signal-to-noise ratio is also higher. Variations in the temperature of the tube and fluctuations in the voltage between the dynodes have less effect on the number of pulses than they do on an integrated current. The resolution provided by pulse-counters far exceeds the degree of accuracy that we might require.

Each packet of electrons received at the anode, which corresponds to a photo-electron emitted by the photocathode, becomes a pulse, or counting unit. It is thus possible to count the number of incident photons over a range of a few to several million per second.

An essential part of the electronics consists of the combined amplifier and pulse-height discriminator, which eliminates noise between fixed thresholds that is likely to alter the count-rate. We would refer readers to the literature for details of the construction of suitable amplifiers.

Dead time Photon-counting does, however, involve one problem that has to be solved for every individual PMT and amplifier chain. This is dead time. After a pulse has been detected, a certain amount of time elapses during which both the PMT and certain electronic components in the later amplifiers are unable to register a new electrical pulse.

Every instrument has a resolution time, known as the dead time δ. If the interval t between two pulses is less than δ, the second pulse will not be registered. This is a major disadvantage of photon-counting techniques. If n is the number of recorded pulses per second, and N the actual number per second (n is less than N), the total time during which no counting takes place is $n\delta$. The probability p that a pulse will be counted is directly proportional to the effective counting time, i.e.:

$$p = 1 - n\delta,$$

but $n = pN = N(1 - n\delta)$, whence: $N = n/(1 - n\delta)$.

It should be borne in mind that any calculation made using this equation cannot be guaranteed to be free from error, which is very difficult to determine when the count rate is too high. This is both because of the uncertainty in the value of δ and fatigue in the tube.

The relative error ε of the count rate, caused by dead time is in fact:

$$\varepsilon = (N - n)/n = \delta N = n\delta/(1 - n\delta).$$

Assigning a limiting value to ε, the limiting count rate to be attempted is:

$$n_{\lim} = \varepsilon/\delta(1 - \varepsilon) \approx \varepsilon/\delta, \text{ because } \varepsilon \ll 1.$$

With $\delta = 100$ ns, and ε set at 0.04, the count limit will be: $n_{\text{lim}} \approx 400\,000$ counts per second. This is one of the reasons why it is advisable to set an upper limit on the magnitude (or the maximum flux, which amounts to the same thing), that is observed with any particular telescope. With a 300-mm reflector and a dead time of 130 ns, for example, the counting error increases rapidly above magnitude 5.5. The dead time, which is specific to each system, is stable, except for thermal drift and aging of the components. It may be calculated, either by measurements made in the laboratory, using calibrated artificial sources, or by observations made with appropriate techniques when the equipment is commissioned. The manufacturers of photon-counting photometers sometimes provide an approximate value for the dead time. This should be mistrusted. One of the first things to be done by anyone acquiring a complete, photon-counting system should be to determine its dead time.

Once the dead time δ is known, whether obtained by laboratory tests or by observation, the count rate may be corrected in two different ways:

(i)   We have:

$$n/N = e^{-\delta N} \quad \text{or} \quad \log(N/n) = \delta N$$

where δ (by definition) is equal to $1/N$ for $n = N/e$, (e being the base of natural logarithms). This equation may be solved easily, by iteration, either using a programmable calculator or, more easily still, on a computer.

(ii)   Even more simply, we set:

$$N = n\,(1 - \delta n)$$

which does not sensibly affect the correction.

Although automatic recording of the data by a microcomputer is the fastest and most accurate method, measurements may be made by using a digital read-out. The main disadvantages of this method are the necessity for recording a lot of numbers during the course of an observing session, and the danger of making errors in the transcription, which are difficult to detect at a later stage. So many precautions and checks require the observer's attention that there is every reason to decide on eliminating this time-consuming process by choosing, from the outset, to connect a computer, running a suitable data-acquisition program.

The PMT's dark current   In the section concerning the photoelectric effect, we saw that, in addition to the photon noise, which is a result of the nature of radiation, there are parasitic effects. The most important of these is thermoelectric emission. The noise (or thermoelectric noise) that results is caused by the fact that, even in the absence of any luminous flux, the cathode and the dynodes emit electrons. The sum total of the thermoelectric noise and secondary noise is known as the dark current, and is easily measured by preventing any light from reaching the photocathode. Luckily, this may be eliminated automatically. As we have seen, one measurement of the object is always followed by a measurement of the sky background, and obviously the dark-current noise is subtracted at the same time as the sky background, because it is equally present in both measurements. Despite this, the value of the dark current should be noted down several times during the night, whether DC or photon-counting work is being carried out. If it increases, the signal/noise ratio drops, reducing the accuracy of the measurements and consequently adversely affecting the limiting magnitude that is attainable. With a good-quality PMT, such as an EMI 9924, the dark current is about 200 counts per second at +20°C, 50 per second at 0°C, and about 30 per second at −10°C. It is essentially constant below that temperature. With the extremely expensive Hamamatsu 85 S PMTs, the manufacturer quotes a dark current of 5 counts per second at 20°C. But this fabulous result is achieved by a very significant reduction in the area of the photocathode (which is only a few millimetres square), which is difficult to accommodate in most astronomical photometers. Generally, the dark current has to be reduced to an insignificant level, and this is essential if PMTs with a low quantum efficiency are being used, or if measurements are being made on a very faint object, regardless of the type of tube. The PMT and its pre-amplifier are encased in a chamber that may be cooled by dry ice (solid CO_2), by a circulating liquid refrigerant, or by the thermoelectric effect. In true infrared photometry, isothermal chambers are used, held at a very low temperature by liquid air or nitrogen. Amateur photometers are rarely cooled. They remain capable of excellent measurements, within certain limits, which it would be pointless to exceed. But things are beginning to change: the PMT in Norman Walker's JEAP is cooled thermoelectrically and is held, to an accuracy of about half a degree, at −10°C. The dark current is consequently much less than the sky-background noise. It also appears that cooling reduces variations in the gain of the PMT.

Thermal dark currents are not just a property of PMTs. Photometers based on photodiodes are also affected, and much more severely if they are not cooled. In the United States, the firm of OPTEC manufactures a photodiode photometer, which is cooled by the thermoelectric effect.

19.2.6 Photodiode photometers

Instead of having a surface that emits electrons, the detector may be a photosensitive semiconductor (a PIN diode). The quantum efficiency of this type of detector is far higher than those of PMTs (Fig. 19.11). On the other hand, the signal is strongly affected by a thermal dark current. In amateur photometers silicon photodiodes are generally used.

The physical principles behind a PIN diode are a little more difficult to explain than those in a PMT. It suffices to say that radiation falling on the junction layer (the 'intrinsic' layer) between a p- and an n-type semiconductor creates a migration of electrons and holes across the junction, forming a photovoltaic current. This is then amplified. This type of design uses an Op-Amp electrometer as an amplifier. This has the advantage of being cheap. Its construction is not beyond the capacity of a careful amateur, who is moderately familiar with electronics and capable of carrying out accurate optical alignment. This type functions quite adequately in the visible region. It functions very well in the R and I bands if it is cooled to about −10°C. Blue-sensitive photodiodes are often used.

S. Cortese, at Locarno Monti, in Switzerland, has investigated the construction of a photometer using an avalanche diode, which functions by multiplying the initial

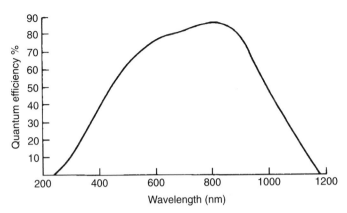

Fig. 19.11. *Quantum efficiency of a PIN-diode sensitized to work in the U and B bands. It is much higher than that of PMTs.*

number of electrons, and which therefore enables a counting method to be used (Fig. 19.12).

Diode photometers are much less expensive, far less fragile, and far simpler to handle than PMT designs. However, the scientific aims and the desired accuracy are at least as important factors as price in determining the choice of equipment. A good PMT photometer enables an experienced observer to obtain measurements accurate to 0.01 mag., or better, with a 200-mm telescope, and down to magnitude 8.5 or 9. A photodiode photometer, used under identical conditions, would not do better than 0.05 mag. In any case, we should not lose sight of the fact that any amateur beginning this sort of work will have to become fully familiar with what is involved, and that reading alone will not suffice. This is why we strongly recommend any amateur astronomer wanting to specialize in PEP work to get in contact with others who can offer advice and encouragement. There are a number of groups around the world who do just this, and there is also the IAPPP (International Amateur–Professional Photoelectric Photometry), which is mentioned later.

19.3 Colour indices and the MK system

The photosphere of a star may generally be taken as approximating a black body (Fig. 19.13), whose radiative energy, or luminosity, may be expressed in terms of its temperature by Stefan's Law:

$$E = \sigma T^4$$

(where σ is Stefan's coefficient, and T is the absolute temperature in Kelvin). A very hot star (50 000 K) is blue, and a cool star (3000 K) is red.

[If we have a source element of area dS, and we consider the radiation emitted in a solid angle $d\Omega$ in direction ω, for a time dt and a wavelength interval $\lambda + d\lambda$, the radiance, perpendicular to the direction ω will be:

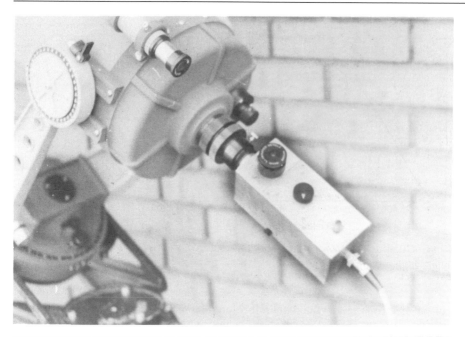

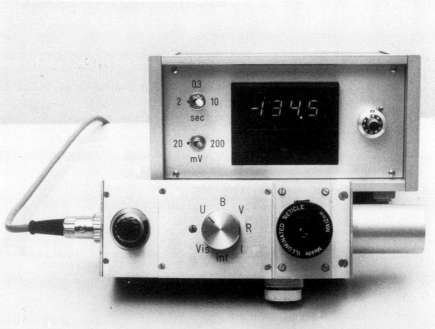

Fig. 19.12. *The PIN-diode photometer designed by S. Cortesi. 1) Mounted on a Celestron 8 telescope. 2) Detail: the knob selects seven filter positions: U, B, V, R, I, integrated light, and 'visible'. The potentiometer, left, controls the degree of illumination of the reticle. Beneath the eyepiece is the knob that controls the flip mirror and switches off the illumination. The switches on the electronics box select the integration time, and the gain in mA. The digital read-out is in mV. The potentiometer, right, adjusts the zero-point.*

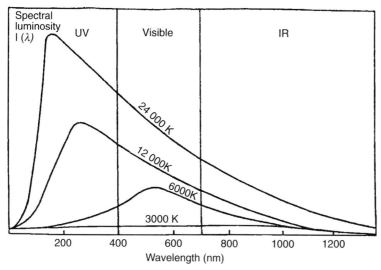

Fig. 19.13. *Radiation curves for a black body. For any black-body temperature there is a specific wavelength λ_m at which the spectral luminosity (or specific intensity) is a maximum. This is Wien's law. The peak luminosity is at a shorter wavelength, the higher the absolute temperature T of the black body.*

$$\mathrm{d}E_\lambda = I_\lambda \cdot \mathrm{d}S \cdot \mathrm{d}\Omega \cdot \mathrm{d}\lambda \cdot \mathrm{d}t,$$

(*see* Fig. 19.14). The direction θ is at an angle Ω to the normal to the surface, so if we prefer we may write:

$$\mathrm{d}E_{\lambda,\omega} = I_{\lambda,\omega} \cdot \mathrm{d}S \cdot \cos\theta \cdot \mathrm{d}\Omega \cdot \mathrm{d}\lambda \cdot \mathrm{d}t,$$

perpendicular to the normal. The factor of proportionality, $I_{\lambda,\omega}$ is known as the radiant intensity of the source. If the source is a black body, or may be treated as one, the radiation will be isotropic, i.e., the same in all directions, and the radiant energy will be given by:

$$\mathrm{d}E_\lambda = I_\lambda \cdot \mathrm{d}S \cdot \cos\theta \cdot \mathrm{d}\lambda \cdot \mathrm{d}t.$$

Planck's formula links a black body's spectral luminance with wavelength and absolute temperature. This is the basis of the curves shown in Fig. 19.13.]

As is well-known, there is a relationship not only between luminosity and temperature, but also between temperature, colour and spectrum. The MK (Morgan–Keenan) spectral system is such that from the luminosity and the spectrum it is possible to derive the temperature or the colour index (Fig. 19.15), and thence the absolute magnitude and, thanks to photometric measurements, the distance modulus.

If U, B, and V are magnitudes obtained from instrumental measurements, simple subtraction provides the difference in magnitude between two pass bands: U − B and B − V. These values are the UBV colour indices of the star. Negative indices (U < B or B < V) mean that the numerical value of the magnitude to the left of

1067

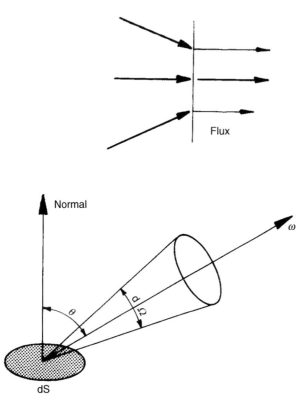

Fig. 19.14. *Intensity and flux.* Top: *flux depends only on the amount of energy crossing a surface per unit time.* Bottom: *the determination of radiance (see text).*

the < sign is less, i.e., that the flux is greater in that pass band. A star with a strongly negative value of (B − V), for example, emits more radiation at the shorter wavelength, and it is therefore hot and blue. On the other hand, a (B − V) value that is strongly positive is characteristic of a red star (*see* Table 19.4 and Figs. 19.16 and 19.17).

If we plot a Hertzsprung–Russell diagram (*see also* Sect. 15.6.3, page 926), where the ordinate indicates stellar luminosity (in absolute magnitude) and the abscissa the effective temperature (or spectra, or colour indices), it will be found that the distribution of points is not random (Fig. 19.18). Most fall near a flattened, shallow, 'S'-shaped curve running across the diagram. (This is known as the Main Sequence, and the stars are of luminosity class V, i.e., they are dwarfs.) Other, less numerous stars form 'horizontal' bands or small, more-or-less isolated groups. It is these various groups that make up the different luminosity classes. They consist of stars having certain physical characteristics in common: supergiants (class I), bright giants (class II), giants (class III), sub-giants (class IV), dwarfs or main-sequence stars (class V), sub-dwarfs, white dwarfs, etc. (Fig. 19.18).

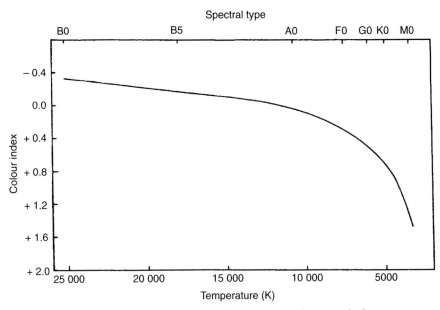

Fig. 19.15. *The relationship between colour index and spectral class*

Table 19.4. *Colour indices and temperature versus Main-Sequence spectral type*

Spectral type	$(B-V)$	$(U-B)$	Effective temperature (K)
O5	−0.32	−1.15	54 000
B0	−0.30	−1.08	29 200
B5	−0.16	−0.56	15 200
A0	0.00	0.00	9600
A5	+0.14	+0.11	8310
F0	0.31	0.06	7350
F5	0.43	0.00	6700
G0	0.59	0.11	6050
G5	0.66	0.20	5660
K0	0.82	0.47	5240
K5	1.15	1.03	4400
M0	1.41	1.26	3750
M5	1.61	1.19	3200

This version of the H–R diagram is of direct interest to those carrying out photometry, because in measuring the flux in the different pass bands, they are determining magnitudes and colour indices.

1069

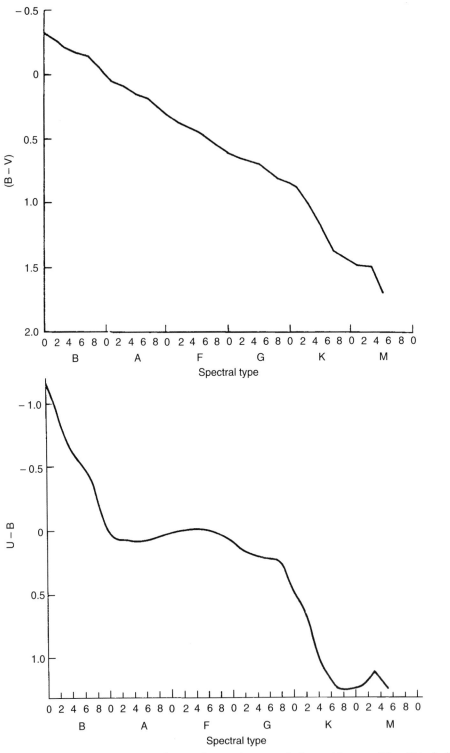

Fig. 19.16. *Main-sequence colour index and spectral class.* Above: *(B − V)*; below: *(U − B)*

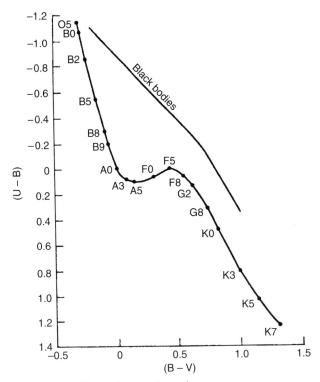

Fig. 19.17. *The colour–colour diagram*

In practice, of course, things are not always so simple. Interstellar clouds are sometimes found between the star and the observer. These clouds absorb some of the light, and this fraction increases towards shorter wavelengths. This causes an apparent reddening, which is described as the colour excess $E(B - V)$ or $E(U - B)$. It is possible to correct the colour indices for colour excess, because each spectral class (in the MK system) corresponds more or less exactly with a theoretical colour index in the absence of interstellar absorption, known as the intrinsic colour. A colour excess is therefore the difference between the observed index and the intrinsic colour, e.g.:

$$E(B - V) = (B - V) - (B - V)_{int}.$$

19.4 Effects of the atmosphere

If the Earth had no atmosphere, the magnitudes and colour indices of objects observed could be simply derived, with only minor corrections, from the measured, instrumental values. Unfortunately for photometrists, the atmosphere behaves like a complex filter, which not only has variable characteristics, but which is, to a great extent, just as unpredictable as the weather itself at any given site.

The atmosphere absorbs some of the light, and scatters another fraction, which

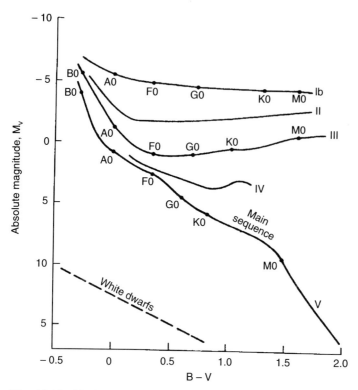

Fig. 19.18. *H–R Diagram*

therefore fails to reach the observer. In addition, it is not physically or chemically homogeneous and contains various substances, either natural or produced by human activities, that are irregularly distributed with respect to height: dust, volcanic ash, water in suspension, etc. The atmosphere does not absorb different frequencies equally in all directions. It undergoes considerable changes because it is a gaseous fluid within which there are various strong energy exchanges, which locally affect its density, and temporarily alter the refractive index and the transmission characteristics of various regions through which radiation has to pass. For photometry, it is essential to allow for these various factors. Overall, the effects that need to be taken into account are:

- (always) *refraction*, which depends on the zenith distance of the object being observed (see Table 19.5); and *extinction*, which depends on zenith distance and on the instantaneous value of transparency for the area of sky being observed;

- (frequently) *turbulence*, which appears as two effects, well-known to observers: agitation of the image, which distorts in a random fashion, and scintillation, which alters the brightness;

- (sometimes) various physical effects that produce interference because of

Table 19.5. *Differential separation of images through refraction**

Zenith distance (°)	Image separation in arc-seconds (between blue and red images)
0	0.00
30	0.35
45	0.60
60	1.04
75	2.24
90	29.00

* Light from a star low on the horizon undergoes significant spectral dispersion. At zenith distances of more than 65°, the red and blue images would not both fall within a small diaphragm.

an increase in sky brightness (astronomical twilight, scattered moonlight, zodiacal light, and airglow emission at certain spectral wavelengths).

19.4.1 Extinction

If we temporarily ignore variations in transparency, we may make the simplifying assumptions (which do not actually hold true) that:

a   the atmosphere has the same physical characteristics, and thus the same average density, in all directions in space. This being so, we may also assume that

b   all the radiation coming from a source outside the atmosphere will be subject to extinction, the amount of which will depend on the thickness of the atmosphere through which it has passed.

The atmosphere consists of a series of layers, approximately concentric with the centre of the Earth, that have different physical properties. This problem may also be eliminated by assuming, for any observing site, that the air is an optically isotropic layer bounded by the local horizon and a plane parallel to it at an arbitrary height above the ground. We can now treat the atmosphere as being an optically homogenous medium, and that our line of sight passes through a cylinder with a diameter equal to that of the telescope. All we require to determine the extinction is the length of this cylinder and the law governing extinction. But the value is unknown. We can get around this difficulty by the following method.

Take the parallel beam of light, with a diameter equal to that of the telescope's entrance pupil, that would arrive from a star at the zenith (i.e., with a zenith distance equal to zero). The distance the light travels through the atmosphere is arbitrarily taken as unity. Because of the assumptions made earlier, we can calculate the length of paths in any other direction, which will simply be the secant of the zenith

Table 19.6. *Error in calculating air mass X if refraction is ignored**

Apparent zenith distance (°)	Apparent X	True X	Error (%)
0	1.000	1.000	0.00
30	1.154	1.154	0.00
60	1.994	1.996	0.10
65	2.356	2.359	0.13
70	2.904	2.910	0.21
75	3.816	3.830	0.37
80	5.598	5.645	0.83
90	10.211	10.468	2.46

* This error becomes significant for zenith distances greater than 40°.

angle of the object being observed, expressed as $\sec Z$. Traditionally, this theoretical path-length X is known as the 'air mass'. Obviously, the value of every air mass will be equal to, or greater than 1 (Table 19.6).

This theory implies that at the horizon the air mass is infinite. This is unimportant, because when carrying out photometry, we never observe close to the horizon, except when carrying out measurements that may be required by calculations of celestial mechanics. Experience shows that use of $\sec Z$ is valid, subject to certain minor corrections, which are included in the standard air-mass equations. The true atmospheric thickness is unimportant. It is the air mass that affects any observations that are made. The diameter of the beam is also unimportant, because we are making flux measurements.

Let F_0 and F_1 be the incident and exit values of the flux for an observation at the zenith, when the length of straight path through the absorbing medium is 1. Bouguer's extinction law is given by:

$$F_1 = F_0 \cdot 10^{-J} \quad \text{or} \quad \log(F_1/F_0) = -J$$

where J is the optical density of the medium.

To convert this to magnitudes we use Pogson's definition, multiplying $\log(F_1/F_0)$ by -2.5. This gives us, for unit air mass, the difference in magnitudes corresponding to the ratio of the flux reaching the telescope to that incident at the top of the atmosphere.

The term on the right-hand side of the second equation is also multiplied by -2.5. Let us define k as the value $2.5J$. This gives us:

$$-2.5 \log(F_1/F_0)_\lambda = k_\lambda \quad \text{or} \quad m_{1\lambda} - m_{0\lambda} = k_\lambda, \quad \text{and} \quad m_{0\lambda} = m_{1\lambda} - k_\lambda.$$

m_0, the magnitude corrected for extinction, is the magnitude observed outside the atmosphere. We could say that k is the value of the extinction for an air mass of 1, or the extinction at the zenith. For an air mass X, and radiation at a wavelength λ, we obtain an instrumental magnitude outside the atmosphere of:

$$m_{0\lambda} = m_{1\lambda} - k_\lambda X.$$

In photometry, k is known as the extinction coefficient. We have defined it in terms of magnitudes. If, for example, the extinction coefficient is 0.20 at a specific wavelength λ, this means that when a telescope in space measured a stellar magnitude of m_0, the same telescope on the ground would measure the magnitude as $m_0 + 0.2$ at the zenith, and $m_0 + 0.2X$ in any other direction. Obviously, some such coefficient is always present, whatever system is used.

The equation given above shows that the coefficient will vary with wavelength. But we are not isolating just a single wavelength. In photometry, therefore, for a given pass band $\lambda + \Delta\lambda$, we use the concept of an equivalent wavelength $\lambda_{eq.}$, which is a weighted mean of all the wavelengths λ in the band, taking account of the sensitivity $S_{(\lambda)}$ of the detector. This amounts to treating the pass band as if it were a single wavelength $\lambda_{eq.}$, and effectively assumes that there is a linear relationship between the actual intensity of the radiation and the wavelength. The previous equation remains valid, provided that the width of the pass band is no greater than a few tens of nanometres (a few hundred ångströms). The UBV pass bands, however, are some 100 nm wide in U and B, and more than 200 nm in V, so this concept cannot be used in practice. We are forced to accept another parameter, and we have to modify the definition of equivalent wavelength and to take into account the actual intensity of the radiation, i.e., the effective temperature of the object. The new parameter is known as the effective wavelength. It serves to relate wide-band photometry to that carried out at a specific wavelength.

In calculating extinction, this translates into a correction that is applied to the colour index. The coefficient k then becomes a combination of a first-order coefficient k' (the principal extinction coefficient), which depends solely on the air mass, and a second-order coefficient k'' (the colour-dependent extinction coefficient), which depends on the air mass and on the instrumental colour of the object (*see* p. 1046), such that $k = k' + k''C$, where C is the $(b - v)$ or $(u - b)$ colour index, relative to the instrumentation used (the sensitivity of the detector being involved). This difficulty, which only applies to the UBV system, is not encountered with medium- and narrow-pass band photometry, which have to allow for just the equivalent wavelength and the single coefficient k.

This would give, for example, in the instrumental v band:

$$V_0 = v_1 - [k'_v + k''_v(b - v)]X$$

or, if expressed as a colour index, setting

$$(b - v)_0 = b_0 - v_0 \quad \text{and} \quad (b - v)_1 = b_1 - v_1 \ :$$

$$(b - v)_0 = (b - v)_1 - [k'_{bv} + k''_{bv}(b - v)]X.$$

These equations are of fundamental importance in UBV photometry. (The subscripts $_{bv}$ indicate that they are subtractive values relative to the colours b and v: $k'_{bv} = k'_b - k'_v$, etc.)

The coefficient k'' changes very slowly for any given site, and it may be calculated three or four times a year. It is always very weak in the v band. The coefficient k''_u is conventionally taken as equal to k''_b, so $k''_{ub} = 0$. (The arbitrary assumption that $k''_{ub} = 0$, leads to errors of 0.02–0.03 in U − B values. There are methods of improving this low accuracy, specifically by using the formulae obtained by Moffat and Vogt.) Photometry in u and in u − b is accompanied by a specific problem regarding the accuracy of results, which may be partially cured by using correction formulae, unless the convention given above is considered adequate. It must be borne in mind, however, that the majority of published data have not been corrected, and that it is often very risky carrying out such a correction when one attempts to compare results with those obtained earlier by other observers.

The first-order extinction coefficient k' does not alter just from night to night, but also during the course of a single night and in different regions of the sky. It will be obvious from this that it is necessary to determine the coefficient k' for each band, i.e., for each colour index, every night, or even several times during a night, and that this must be carried out as carefully as possible. During the course of a night, observers devote 30 % of their time, or sometimes even more, to making measurements destined to be used in calculating these coefficients. On a fairly good night, k' will have a value of 0.20. At the best sites, where the sky remains stable for days on end, average values may be adopted without any appreciable risk.

At the La Silla Observatory in Chile, there are more than 200 photometric nights per year. Elsewhere, such as in Western Europe, sites of exceptional quality can be counted on the fingers of one hand. In France, for example, there are three sites with excellent photometric qualities: the Pic de Chateaurenard (2900 m altitude) in the southern Alps, where there is no photometric installation; the Pic du Midi Observatory, where a k' of 0.10–0.15 is sometimes encountered; and at the Observatory of Haute Provence, where the number of high-quality photometric nights does not, at best, exceed 50 per year. The rest of the time, all that can be done is, whenever possible, to exercise additional care and accept a compromise with the sky conditions.

For reasons that will be explained later, it is rarely possible to calculate k' to an accuracy of greater than 0.01–0.02 mag. If k' is greater than 0.50, significant measurement errors occur, because the inaccuracy with which it can be determined is greater. It would obviously be pointless and disappointing if a PEP instrument were installed, unless one could be sure that the frequency of 'photometric' nights was at least 1 in 7, or 1 in 8, and fairly evenly spread throughout the year in periods of several consecutive nights. It should be noted that a 'photometric' sky is not the more-or-less transparent and unstable sky that can be used easily by visual variable-star observers. Neither is it a 'photographic' sky. It is actually a sky without a trace of cloud that is brilliantly clear from zenith to horizon and free from any light pollution, when 5th-magnitude stars, 45° from the zenith and all round the sky, are visible to the naked eye, and when there is nothing to detract from the star's brightness other than scintillation. Experienced workers have their own 'dodges' that help them judge the quality of the sky before they begin making measurements: the general appearance of the sky at sunset, the clarity of features on the horizon, and the visibility of certain indicative stars or constellations. In any case, one should not expect to obtain measurements to an accuracy of better than about 0.02 mag. at any site lower than about 700–800 m in altitude.

19.4.2 Additional variations in extinction

The extinction caused by the air mass consists of extinction produced by the medium itself, and losses by scattering, dispersion, and absorption (caused by the Rayleigh effect, aerosols, and ozone in the upper atmosphere). [The Rayleigh effect is selective scattering of light by small particles or molecules. It is inversely proportional to the 4th power of the wavelength, and thus particularly affects short wavelengths. Another different effect, Mie scattering, is independent of wavelength and is caused by larger solid particles.]

In addition to the theoretical absorption, there are the effects of water in suspension in the atmosphere. These include mists, thin atmospheric veils, and condensation near the ground. Clouds, even the thinnest ones, such as cirrus invisible to the eye, have the advantage of causing a distinct obstruction that is immediately, and spectacularly, obvious in the output from the photometer. If the cloud-cover is light, homogenous, and stable, it is possible to overcome this under certain observing conditions, by using appropriate techniques.

The two enemies of photometry are atmospheric inhomogeneity and instability, i.e., differences in transparency between different parts of the sky and unpredictable variations in transparency in a specific area, even over a single field and over a short period of time. When the drift in k' is slow but progressive, it is sometime possible to take it into account if enough measurements are to hand. When it is not, any measurements are generally only fit for the wastepaper basket.

Summing up, we can say that photometry, more than any other type of observation, is subject to the whims of the sky. On very poor nights, the photometric observer is simply forced to close the dome, whereas a spectroscopist might well be able to continue, without much difficulty, until dawn. But it may happen that although a night begins with dreadful conditions, these may change and rapidly become excellent. In addition, once having started, many professionals and dedicated amateurs persist at the telescope, hoping for better conditions.

19.4.3 Agitation and scintillation

These two effects are both caused by a greater or lesser degree of turbulence in the atmosphere. These motions may occur at any altitude. Agitation of the image, which is well-known to observers who work with fairly high magnifications, has been defined by Danjon as the angular deviation of a ray of light with respect to the optical axis at the instrument's limit of resolution. It is expressed in seconds of arc. Agitation appears as completely random dilation, deformation, and (in the worst cases) fragmentation of the image. Any observers who are fully familiar with their telescopes make empirical estimates of the degree of agitation present, and whether it is acceptable or not. Under calm conditions it is between 1 and 2 arc-seconds. As an example of the other extreme, in Provence in Southern France (where conditions for photometry are often excellent), when there is a strong Mistral wind, agitation may reach 7–8 arc-seconds. It is dangerous to place any reliability on measurements made when it is above 5 arc-seconds, because the agitation resembles arbitrary changes in the position and diameter of the diaphragm. It has a greater deleterious effect on images produced by large, rather than small, telescopes.

Upper-atmosphere winds cause rapid motion of cells of air with varying density. The effect on light passing through these cells of differing refracting index resembles that produced by lenses of varying curvature. The cells alter the direction of the rays of light and cause changes in the apparent intensity of the source. This is known as scintillation.

The diameter of these cells, perpendicular to the line of sight, is 200–300 mm. Large telescopes are therefore far less sensitive than small ones to their effect. For a given air mass, the effect of scintillation, expressed as a percentage of magnitude, is the same for bright stars as it is for faint ones. The closer the object is to the horizon, the greater is the scintillation.

Agitation corresponds to a shift in the phase of the wave-front, and scintillation to a variation in amplitude. Nothing can be done about agitation. Scintillation may be cured by increasing the integration time.

19.5 Observation and methods

This chapter is intended to serve as an introduction, rather than a detailed description of photometry. We will therefore restrict discussion to the basic principles of observation that are normally applied.

As a general rule, a measurement is only valid if the experimental conditions (the physical parameters and instrumental characteristics) are fully known. For a given night, and for a given photometric chain (optics, detector, electronics, and recorder) it is important to know the amplifier gains (if making DC measurements), the time constants and the recorder gain (if using paper recording), and the dead time (if carrying out photon-counting). In addition, regular measurements should be made of the high-voltage supply to the PMT, and of the dark current. Neither of the latter presents any great difficulty.

If a photodiode photometer with voltage readout is used – the most frequent case – no preliminary regulation needs to be taken into account. Only the zero-point for the readout, which corresponds to zero flux and not to zero on the measurement scale, has to be adjusted occasionally according to the manufacturer's instructions.

Given these differences, the working procedures are strictly the same, whatever form of instrumentation is being used. To reduce the measurements, the first-order extinction coefficients for each pass band are required at the end of each night, together with the zero-point of the scale, if differential photometry is not being used. The second-order extinction coefficients are required and, finally, the transformation coefficients that enable the magnitudes and indices for outside the atmosphere to be linked as accurately as possible to the UBV system.

The times of stellar observations often have to be expressed in heliocentric Julian Days. Depending on the Earth's position in its orbit, the time required for light to reach us from a specific star may vary by several minutes, which complicates the

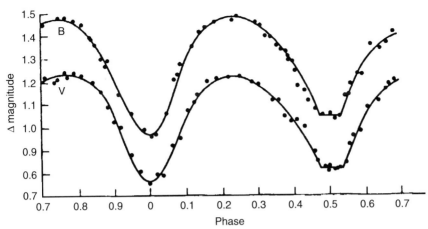

Fig. 19.19. *Light-curves of an eclipsing binary (after Henden and Kaitchuk, 1982)*

comparison of observations in determining accurate periods. There is a standard way of converting geocentric Julian Dates to heliocentric Julian Dates and the appropriate equations may be carried out easily on a programmable calculator or computer.

The types of observation that may be carried out by amateur photometry cover a wide range and present differing degrees of difficulty. A few may be mentioned. Measurements of long-period variables are the simplest. Many of these stars require careful coverage, even if only to determine their maxima, which are often poorly known. The work entails the production of light-curves, identical to those produced by 'visual' observers. All the variable-star organisations (whether they have photometric programmes or not), have programmes devoted to long-period stars (including Mira objects): these include the AAVSO, the BAA VSS, the AFOEV, the IAPPP, etc., in which one may participate.

Measurements of eclipsing binaries are particularly rewarding for workers with a little experience, because they allow light-curves to be drawn from observations made over fairly short periods of time (i.e., extending over just a few days). The difficulty, and the interest, in such observations lies in determining the minima, or in obtaining limits within which they must occur (Fig. 19.19). Longer-period stars (with periods between 25 and about 100 days) of the RV Tauri type are particularly interesting because of the variations in the shape, amplitude and period of their light-curves.

Short-period variables, such as the δ Scuti stars, have periods below 8 hours. Periods of the true Cepheids are clustered around 50 days. Those of the RR-Lyrae stars do not exceed 1 day. This means that, under favourable conditions, the latter may be followed over a short interval of time. The main difficulty, which is an insuperable one for small instruments, is that they are generally faint, which means that measurements are inherently inaccurate and of lesser value.

Recently, solar-type stars (both eclipsing and non-eclipsing) of the RS CVn and BY Dra classes have attracted considerable attention by amateurs. These G- and K-type stars, like the Sun, exhibit sunspots that vary in size and position, causing changes in brightness. Douglas Hall of Vanderbilt Observatory coordinates observations of RS-CVn stars for the IAPPP.

More details about the different classes of stars may be found in the chapter dealing with variable stars. A judicious choice from the many objects available will provide the basis for a PEP programme.

Occultations of stars by minor planets are difficult to carry out visually, but quite easy with PEP. The most important point here is that the time of occultation should be determined as accurately as possible. Such determinations may be carried out to a fraction of a second visually, but to hundredths of a second, or even less, using a PMT.

The eye is subject to fatigue if it has to concentrate on an object over a period of some 20 minutes in waiting for an occultation, the exact time of which is unknown. It may miss the event or even record a 'false' occultation when it did not actually take place. A PMT is never fooled. The photometry of minor-planet occultations does, however, involve the use of an accurate time-base, which is easily obtained over a period of about an hour (some computers have a quartz clock that will serve for this purpose), and accurate calibration against Universal Time. This involves interfacing the data-acquisition microcomputer with a receiver that will pick up the time signals, a decoder, and an encoder, written in machine language that will take account of the distance of the transmitter. This complicates the instrumentation. It may be simplified by recording the pips from a speaking clock. However, an accurate receiver to pick up these pips is still required. A similar arrangement was used by J. E. Arlot (from the Bureau des Longitudes, Paris), working at the Haute-Provence Observatory, for the PHEMU 86 programme (Jovian satellite mutual phenomenon programme).

Finally, we should note that many minor planets may be observed just like variable stars, because they show variations in brightness caused by the axial rotation. Some of them have irregular shapes and therefore reflect a varying amount of light from the Sun as they rotate. They also show varying albedo, arising from differences in their surface composition. Others, which are quasi-spherical, still show variations in albedo, as does any planetary surface. (See Chap. 7 on Minor Planets.)

Some amateurs, working with professional astronomers, have mounted specific programmes: e.g. the study of shells of dust surrounding certain cool stars (a collaboration between F. Querci, of the Pic du Midi and Toulouse Observatories, and C. Grégory and B. Fontaine of GEOS). The description of this sort of programme is beyond the scope of this introductory survey, because it involves the use of fairly large telescopes. In addition, some of the programmes involve other, simultaneous investigations, in particular, spectroscopy.

19.5.1 Methods

Whatever type of work is undertaken, PEP methods all involve:

- measuring the flux in different pass bands;

- the conversion of these fluxes into magnitudes and indices that apply to the instrumental system used, taking account of the various corrections that have already been mentioned;
- finally, linking the values obtained to the universal system, defined by that system's pass bands, by the use of a number of reference objects or standards, observed under the same conditions.

During the course of a night, the procedure may either be to take measurements of the programme stars interspersed with measurements of standard stars, or to alternate several measurements of each programme star and those of one or more, suitably chosen, comparison (reference) stars. Apart from the case where a star and a single comparison are followed for a long time, however, it is impossible to dispense with the periodic measurements of standard stars that are required to calculate extinction. Both methods, whatever their respective advantages and disadvantages, always involve referring the values obtained to specific standards.

Even if the measurements are not specifically linked to any system, and consist of just atmosphere-free, differential magnitudes and indices between a star and a calibrated comparison, the measurements are being implicitly referred to standards, because the comparison star, which is considered to be non-variable, would not have been chosen without direct or indirect reference to fixed standards.

The UBV system, established by Johnson, Morgan and collaborators, consists of a small number of standard stars, measured some 40 years ago, with specific instrumentation and PMTs (RCA's 1P21, with a type S4 photocathode, which was the only one available for a long time, but which is more or less obsolete today). These are the primary standards. Secondary standards, referred to them, were added later, followed by several thousand other stars (in the Arizona–Tonantzintla Catalogue, Landolt, Purgathoffer, etc.), more or less accurately related to the standard stars. Some photometrists use strict criteria and employ only stars, measurements of which have been made over a number of years and show standard deviations of less than 0.01 mag. But nothing can guarantee such stability over a very long period of time, because variability is an inherent characteristic in the evolution of every star. In addition, some of the standards established by Johnson, and which are still included as such in some catalogues, ought to have been discarded as having been shown to be variable.

From all this we can see that to obtain some protection against such unknown errors occurring in the calculations that are required, we need to use a sufficiently large number of standard stars, which will decrease the errors.

19.5.2 Reduction methods

Sections 19.5.3 and 19.6 deal with photoelectric observations and the fundamental equations that enable measurements to be reduced. As we have seen, however, no reductions are possible without knowing the extinction coefficients, by means of which the raw data may be transformed into instrumental magnitudes outside the atmosphere. The transformation coefficients also enable the measurements to be linked to the UBV system. The quality of the results depends directly on the degree of care taken to determine these coefficients, and on their accuracy. The finest measurements are useless if satisfactory coefficients are not available.

19.5.2.1 First-order extinction coefficients

These coefficients will have a greater or lesser accuracy, depending on the method used and on the observing conditions. The conditions guide the choice of method. It may be too late by the time one notices that the choice was the incorrect one. It might be said that for photometrists the sky turns more rapidly than for other observers.

Bouguer comparison-star coefficients When a variable is followed for the greater part of a night, and quite a large number of measurements over a fairly wide range of air-masses are available, it is, in principle, sufficient to calculate the Bouguer coefficients k for the comparison star, k'' being known. This procedure is theoretically advantageous because the variable and the comparison are close together.

We derive a linear regression of the air masses (X) against the raw instrumental magnitudes (Y). The coefficient of the first-order equation is k, and the constant is the instrumental magnitude outside the atmosphere: $m = kX + m_0$. If the sky is unstable, however, ridiculous values will be obtained. At most mid-latitude sites, even when the sky appears to be favourable, stability cannot be guaranteed. This is not the only problem associated with this method. If the correlation coefficient does not exceed 90 % it is always advisable to plot Bouguer's straight-line relationship, which will show the scatter, and enable one to determine whether to eliminate some points and if so, which ones. Provided the scatter is not abnormal, the Bouguer straight-line relationship laid through the points by eye may be improved by triangulating consecutive points and plotting a new straight line through the intersections of the medians.

The amount of absorption may vary considerably. At ground level, under very humid, low-pressure conditions, there is a strong break in the linear Bouguer's law relationship when the temperature reaches the dewpoint. At height, invisible cirrus and aircraft contrails, which have an annoying tendency to initiate the slow formation of very broad bands of cloud, may unpredictably alter the transparency.

Observation of standard stars (known coefficients of k'', ε, μ, Ψ) If several variables are to be observed, which is usually the case, the k' coefficients for the night are determined from standard stars, once again on the assumption that it is a good photometric night.

Roughly a dozen standard stars of different colours are chosen that are bright enough to minimize measurement errors, and these are observed over the greatest possible range of air-masses. Ideally, each one would be measured twice, once before, and once after meridian passage, but this is a luxury that is rarely possible. It would take the whole night, to the detriment of the programme stars. It is, however, a good idea to repeat, in pairs, the measurement of standard stars of different colours.

Measurements of the programme stars have to be interspersed between measurements of the standards so that the reference data are spread as evenly as possible throughout the observing session.

The coefficients k'' and the transformation coefficients are assumed to be known. We have:

$$v - V - k_v''X(b - v) + \varepsilon_v(B - V) = k_v'X - \zeta_v.$$

Using rectangular graph paper, plot the values of the left-hand term on the y-axis, and those of X on the x-axis. Each star provides one point. A straight line is then laid through the points and its slope is the value of k'. The intercept at the origin gives ζ. A linear-regression method may be used if there are considered to be enough good points, but attention needs to be paid to any wide discrepancies.

Regressions are made of:

$$\begin{bmatrix} v - V - k_v''X(b - v) + \varepsilon_v(B - V) \\ b - B - k_b''X(b - v) + \varepsilon_b(B - V) \\ u - U - k_u''X(u - b) + \varepsilon_u(U - B) \end{bmatrix} \qquad \text{in } y$$

or (k_{ub}'' being considered as zero),

$$\begin{bmatrix} (b - v) - k_{bv}''X(b - v) - (B - V)/\mu \\ (u - b) - (U - B)/\Psi \end{bmatrix} \qquad \text{in } y$$

and versus X in x.

Apart from the coefficients k', one obtains, for information, the values of ζ_v, ζ_b, and ζ_u (or the values of ζ_{bv}/μ and ζ_{ub}/Ψ) as constants, i.e., as 'zero-points' for the night, but these are not required for single-channel differential photometry. Regression may be used, even if the transformation coefficients are unknown, but a regression with two variables has to be used, starting with:

$$v - V - k_v''X(b - v) = k_v'X - \varepsilon_v(B - V) - \zeta_v,$$

as the first term, versus X and $(B-V)$ respectively as the first and second independent variables.

These procedures take considerable time.

Observation of stars with A0 and A1 spectra (unknown coefficients k'', ε, μ, and Ψ)
Certain A0 or A1 stars, fairly well distributed on the sky, are chosen each night. These stars should have colour indices that are zero or very small values (a few hundredths). This means that the secondary extinction term and the transformation term may be ignored. For V, we will have:

$$v - V = k_v'X - \zeta_v.$$

For a value of $(B - V)$ of 0.05, a relatively high $X = 2$, and $k' = -0.03$, the secondary extinction term in b reduces to 0.003 mag. If an A0 star with a $(B - V)$ of 0.02 is observed with an air-mass of 1.5, the value of secondary extinction falls to 0.001, a value below which the standard deviation of the measurements themselves has little significance, and the colour extinction even less, because first, the determination of k'' does not reach this degree of accuracy, and second, the $(B - V)$ values for the standard and comparable stars rarely attain such a level.

On the other hand, if ε_V is −0.09, the transformation term will be less than −0.002. The error introduced by these two simplifications will be around 0.002 mag., which is not significant as regards the accuracy of k'.

A linear regression of the first term versus X will give k' and the night's zero-point as constants. A similar procedure is carried out for B and U.

The same reservation applies to this method, which is of interest because it simplifies the calculations required, as to the use of the comparison star and Bouguer's law: namely the stability of the sky.

The Bouguer method   This method cannot claim to be fast, but it has the merit of being very accurate, and giving instantaneous values. Two standard stars with the same or very similar spectra are chosen, one high, and the other low, in the region of sky where the programme star is situated. If the greatest possible degree of accuracy is sought, the pair of standard stars are measured once before and once after the programme star. Because, the values of $\Delta(B-V)$ are zero or very small, the k'' term may be ignored. The transformation term disappears in a similar manner:

$$\Delta V = \Delta v - k' \Delta X$$

whence,

$$k' = (\Delta v - \Delta V)/\Delta X.$$

The difficulty lies in satisfying the requirement that the colour indices should be very close, especially for bright standard stars, which are always to be preferred, because of the accuracy of measurement that they allow. If there are no standard stars available, it is possible to make do by taking stars from the *Bright Star Catalogue*, provided two precautions are observed:

(i)   check to ensure that they are not variables, by consulting the *General Catalogue of Variable Stars*, and recent literature;

(ii)   interrogate the *UBV Catalogue* held by the Centre de Données Stellaires (Stellar Data Centre) at Strasbourg (using SIMBAD) to find out recent measurements of the V magnitudes and colour indices for these stars. If there are several, sufficiently consistent measurements, the average values and standard deviations should be used.

The absolute values of the standard deviations are dV_1 and dV_2, so in the worst case we will have, in V:

$$k' = \left(\Delta v - \Delta V \pm \sqrt{dV_1^2 + dV_2^2} \right) / \Delta X.$$

Let $dV_1 = 0.01$, $dV_2 = 0.01$, and $\Delta X = 1$ (a high value of X is required to minimize the error). The difference in the result of calculating k' will only depend on:

$$\left(\sqrt{dV_1^2 + dV_2^2} \right) / \Delta X.$$

(Verification of these equations is given in Sect. 19.7.)

It will be approximately ± 0.014, which is all that can be tolerated under the circumstances. This is a method that should be used with caution, and it should be borne in mind that the calibration of the standard stars themselves is far from being impeccable.

The MD method When the transparency of the whole sky changes only slowly over the course of a night, this is the most practical method to use. It has the advantage, like the Bouguer method, of requiring only two standards, and these are retained throughout the observing session. There is another advantage: two standards of different colours are used, which are easier to find than two standards with the same colour indices.

One of the chosen standards should be rising, M, and the other setting, D. ['Montante' and 'Descendante' in French, whence the accepted term for the method. – Trans.] They are measured as a pair, some 10 or 12 times, alternating with the programme stars that are in their vicinity. We thus have N pairs of measurements, at times t_i, where $i = 1, 2, 3, \ldots, N$.

We assume that, because the pairs of measurements of the two stars are taken as quickly as possible each time, the respective values of X vary by a negligible amount while the pair of measurements are being made. Because the two stars were chosen to have very different colour indices, the values of k_M and k_D are not identical. In UBV we would have:

$$k_{v[M]} = k'_v + k''_v (b-v)_{[M]} \qquad k_{v[D]} = k'_v + k''_v (b-v)_{[D]}$$
$$k_{b[M]} = k'_b + k''_b (b-v)_{[M]} \qquad k_{b[D]} = k'_b + k''_b (b-v)_{[D]}$$
$$k_{u[M]} = k'_u + k''_b (u-b)_{[M]} \qquad k_{u[D]} = k'_u + k''_b (u-b)_{[D]}.$$

The coefficients $k_{v[M]}$ and $k_{v[D]}$, for example, differ by:

$$\Delta k_{v[MD]} = [k'_v + k''_v (b-v)_{[M]}] - [k'_v + k''_v (b-v)_{[D]}] = k''_v \Delta (b-v)_{[MD]}.$$

The two principal equations required in calculation, are, for time t_i:

$$v_{[M]i} = k_{v[M]} X_{[M]i} + v_{0[M]}$$

and

$$v_{[D]i} = k_{v[D]} X_{[D]i} + v_{0[D]},$$

or:

$$\Delta v_{[MD]i} = k_{v[M]} X_{[M]i} - k_{v[D]} X_{[D]i} + \Delta v_{0[MD]}.$$

A linear regression for two independent variables, with values of $\Delta v_{[MD]i}$ in y_i, $X_{[M]i}$ in x_{1i}, and $-X_{[D]i}$ in x_{2i} will give $k_{v[M]}$, $k_{v[D]}$ and $\Delta v_{0[MD]}$ as constants.

The values of k that are obtained are mean values. They may be used to check that the standard deviation of the estimates from the linear-regression line is acceptable. It would be better, however, in every case, if all the values of $v_{0[M]i}$ and $v_{0[D]i}$ corresponding to the pairs i were calculated from these values of $k_{v[M]}$ and $k_{v[D]}$.

Theoretically, all the values of v_0 for a constant star are identical for a given night and instrumentation, and for zero drift. Obviously, this never applies, even if only because of random errors. Pairs of measurements are discarded when the v_0 of one or other star shows a deviation from the mean that is greater than the standard deviation. It is advisable to have at least a dozen pairs, so that dropping a pair will not create too large a gap in the distribution of measurements over the course of the night.

The various values of k_v' are calculated for each of the values of $v_{[M]i}$ that are retained from:

$$(v_{[M]i} - v_{0[M]})/X_{[M]i} - k_v''(b - v)_{[M]i} = k_{v[M]i}' \tag{19.1}$$

$$(v_{[D]i} - v_{0[D]})/X_{[D]i} - k_v''(b - v)_{[D]i} = k_{v[D]i}' \tag{19.2}$$

k_v'' being known.

The regression is repeated. In the cases of k_v' and k_b', a standard deviation of less than 0.01 is excellent. At low altitudes, the standard deviation for k_u', will, perforce, be worse.

The condition:

$$(v_{[M]i} - v_{0[M]})/X_{[M]i} - k_v''(b - v)_{[M]i}(v_{[D]i} - v_{0[D]})/X_{[D]i} = k_{v[MD]i}'$$

cannot be satisfied exactly. In other words:

$k_{v[M]i}'$ being $\neq k_{v[D]i}'$,

$k_{v[M]i} - k_{v[D]i} \neq k_v''\Delta(b - v)_{[MD]i}.$

For the present, it suffices if the difference between the k_v values for the two stars does not differ from $k_v''\Delta(b - v)_{[MD]i}$ by more than a few thousandths. When measurements of a star are reduced with reference to a relatively close comparison star, a difference in air mass of 0.02 will give rise to an error of less than 0.002 mag. in the differential extinction term. The values of $k_{v[MD]i}'$ are obtained by taking the mean of $k_{v[M]i}'$ and $k_{v[D]i}'$. To reduce programme stars, the values of $k_{v[MD]i}'$ are interpolated from the times at which the estimates were made.

The method just described is a simplified version that is quite adequate for amateurs concerned with quality. When the MD method is used, Rufener (1984) suggests an algorithm that takes into account the pass bands of the filters, as determined in the laboratory. During the course of the observation, each sample is sub-divided into hundredths of a second. This algorithm, which is used by Geneva Observatory, not only allows the coefficients of extinction to be calculated, but also, thanks to appropriate instrumentation and computer software, enables measurements that have been affected by scintillation to be checked and eliminated during the course of sampling. It continues to take samples until it obtains the best standard deviations possible for the reigning conditions.

19.5.2.2 Second-order extinction coefficients

Second-order extinction coefficients, like the magnitude and UBV colour-index transformation coefficients (*see* Sect. 19.5.2.3), should be calculated three or four times a year, given the slow speed at which they change. It is not advisable to use the values of k'' obtained at the same site with different instruments, because the instrumental parameters and the measurement errors will differ, as will the filters, which directly affect the colour indices observed.

Measurements are made over several hours of two neighbouring stars whose magnitudes should be well-calibrated and as similar as possible, but with very different colours. Δv_0, the difference in the magnitude outside the atmosphere, is a constant. ΔX, the difference in the air masses, is also practically constant, given that the stars are close together.

For N pairs of measurements, with $i = 1, 2, 3, \ldots, N$, the equation:

$$\Delta v = k_v'' \bar{X} \Delta(b - v) + k_v' \Delta X + \Delta v_0$$

becomes:

$$\Delta v_i = k_v'' \bar{X}_i \Delta(b - v)_i + k_v' \Delta X + \Delta v_0$$

The sum $k_v' \Delta X + \Delta v_0$ is the ordinate at which the equation originates.

A linear regression for N pairs of measurements for two stars with Δv_i in y_i and $\bar{X}_i \Delta(b - v)_i$ in x_i gives k_v'' as the coefficient of the straight line.

Regression with Δb_i in y_i and $\bar{X}_i(b - v)$ in x_i gives k_b''. If we accept UBV system conventions, k_u'' is equal to k_b'', by definition.

The difference $k_b'' - k_v''$ gives k_{bv}''. It may also be calculated directly, because:

$$\Delta(b - v) = k_{bv}'' \bar{X} \Delta(b - v) + [\Delta(b - v)_0 + k_{bv}' \Delta X].$$

By regression with $\Delta(b - v)_i$ in y_i and $\bar{X}_i \Delta(b - v)$ in x_i, we obtain the coefficient k_{bv}''; by definition, k_{ub}'' is zero (see above).

19.5.2.3 UBV transformation coefficients

The UBV transformation coefficients are calculated from observations of one or more pairs of stars (red/blue), the same as above, measured between ten and twenty times per night, over a period of one or more nights.

The principle consists of making a series of measurements like those described earlier. We then carry out a linear regression with Δv_i in y_i and ΔX in x_i. This gives Δv_0 as a constant.

Because

$$\Delta V \quad = \quad \Delta v_0 + \varepsilon_v(B - V) \quad (cf. \text{ later}),$$
$$\varepsilon_v \quad = \quad (\Delta V - \Delta v_0)/\Delta(B - V)$$

similarly,

$$\varepsilon_b \quad = \quad (\Delta B - \Delta b_0)/\Delta(B - V) \quad \text{and}$$
$$\varepsilon_u \quad = \quad (\Delta U - \Delta u_0)/\Delta(U - B).$$

The advantage of this method is that the same measurements are used to derive both the k'' coefficients for the prevailing conditions, and the transformation coefficients. A point that must be borne in mind, however, is that if the transformation is not linear, a single pair of stars will not reveal the discrepancy.

19.5.2.4 Using standard stars or members of clusters

1. A dozen standard stars are chosen, with different colours and air-masses. To minimize the values of X, these stars are observed within 2 hours of culmination. Values of k'' should have been calculated previously.

For each star, we have:

$$V - v + k''_v X(b - v) = \varepsilon_v(B - V) - k'_v X + \zeta_v.$$

Carrying out a linear regression for two independent variables with the first term in y, $(B - V)$ in x_1 and X in x_2, etc., gives ε_v. The same is done for ε_b and ε_u. At the same time the values of k'_v, k'_b and k'_u for the night are obtained, as well as those for ζ, which are the ordinate values at the origin. It must be borne in mind, however, that with neighbouring stars (and thus similar values of X) regression will give good transformation coefficients, but low accuracy for the k' values. The opposite holds true if there are considerable differences in air-mass.

2. A good method is to refer all the observations to the standard that lies close to the average right ascension. The sequence is:

$$R - A - B - C - R - D - E - F - R - G - H - I - R \ldots$$

This should be repeated, if there is time to do so.

Linear regression is used to calculate the values of Bouguer's k for star R. If the standard deviation and correlation obtained are satisfactory, the values of $\Delta v_{(\text{St}-R)}$ are obtained for each standard, after having related the measurements of R to those of each star by interpolation between the air-masses and the measurements:

$$\Delta V_{(\text{St}-R)} - \Delta v_{(\text{St}-R)} + k''_v \bar{X} \Delta(b - v) = \varepsilon_v \Delta(B - V) - k'_v \Delta X \quad, \text{etc.}$$

We then carry out a linear regression for two independent variables, as before.

It is best, if possible, to use standards that are close together and well-calibrated. This applies to stars in the Pleiades, Praesepe and IC 4665, which were calibrated when the UBV system was established. One disadvantage of these clusters is that they include few red stars, and these are fainter than the others. In addition, there are slight differences in calibration between one cluster and another. To do things properly, a large number of measurements should be made for each cluster, increasing the number obtained for red stars. The weighted means of the coefficients may then be calculated, using the estimates' standard deviations as given by the regressions (*cf.* later).

The coefficients μ and Ψ are calculated from the values of $(b - v)_0$ and $(u - b)_0$:

$$(B - V) = \mu(b - v)_0 + \zeta_{bv} \tag{19.3}$$

$$(b - v)_0 = [(B - V) - \zeta_{bv}]/\mu \tag{19.4}$$

Subtracting $(b - v)_0$ from each side of (19.3):

$$(B - V) - (b - v)_0 = \mu(b - v)_0 - (b - v)_0 + \zeta_{bv} = (\mu - 1)(b - v)_0 + \zeta_{bv}.$$

Substituting (19.4) into the third term, and dividing by μ, we have:

$$(B - V) - (b - v)_0 = (1 - 1/\mu)(B - V) + \zeta_{bv}/\mu.$$

Similarly, we have:

$$(U - B) = \Psi(u - b)_0 + \zeta_{ub}$$

$$(u - b)_0 = [(U - B) - \zeta_{ub}]/\Psi$$

$$(U - B) - (u - b)_0 = \Psi(u - b)_0 - (u - b)_0 + \zeta_{ub} = (\Psi - 1)(u - b)_0 + \zeta_{ub}.$$

Substituting and dividing as previously, we have:

$$(U - B) - (u - b)_0 = (1 - 1/\Psi)(U - B) + \zeta_{ub}/\Psi.$$

A regression with

$[(B - V) - (b - v)_0]$	or	$[(U - B) - (u - b)_0]$	in y, and
$(B - V)$	or	$(U - B)$	in x

gives the angular coefficients $(1 - 1/\mu) = a$ and $(1 - 1/\Psi) = b$. Then $\mu = 1/(1 - a)$ and $\Psi = 1/(1 - b)$.

The intercepts of the straight regression lines are the 'zero-points' ζ_{bv} and ζ_{ub}.

Although μ may be obtained directly from $1/\mu = 1 - \varepsilon_b - \varepsilon_v$, the same does not hold for Ψ, because Ψ and ε_u are coefficients of $(U - B)$, not of $(B - V)$. In differential photometry, therefore, it is essential to know μ and Ψ before commencing reductions.

In carrying out the correlations, it will in fact be seen – even if the reductions are in magnitudes and not in colour index – that $\Delta(B - V)$ appears in the calculation of V and B via the $\varepsilon\Delta(B-V)$ term, and $\Delta(U-B)$ in that for U via $\varepsilon\Delta(U-B)$. Although the values of $(B - V)$ and $(U - B)$ may be known for the comparison stars, they are not known for the programme stars, and are in fact what we want to obtain from the equations:

$(B - V) = \mu(b - v)_0 + \zeta_{bv}$	$\Delta(B - V) = \mu\Delta(b - v)_0$
$(U - B) = \Psi(b - v)_0 + \zeta_{ub}$	$\Delta(U - B) = \Psi\Delta(u - b)_0.$

19.5.3 Observations

A typical photon-counting measurement consists of measuring, through each of the three filters, the number of effective photons incident over a period of at least 20 seconds, reduced to the number of counts per second. If possible, it is a good idea to sub-divide each sample into smaller units, which allows the standard deviation of each unit to be determined, and its quality to be assessed. Changes in the sky may be monitored in this way. Careful observers, concerned with quality, adopt longer integration times – up to about 1 minute with small instruments. Measurements of faint stars requires such integration times to reduce the photon-noise, caused by the very nature of the signal. The level of this noise is approximately inversely proportional to the Poisson standard error: $1/\sqrt{n}$ (where n is the number of photons received during the measurement interval). If measurements accurate to 0.01 mag. are required, the signal/noise ratio should never be less than 100. This precludes any measurements with a total of less than 10 000 counts. In general, a much larger safety margin is allowed. To avoid overwhelming, and also fatiguing the photocathode, measurements are not made of stars giving more than 10^6 counts per second. In exceptional cases, when bright stars have to be measured, the telescope is stopped-down, or a neutral-density filter is used. (Neutral-density filters must be of high quality to ensure that they absorb equally at all wavelengths.) It is then essential to remember that such modifications alter the instrumental parameters and that it would be difficult, if not impossible, to relate any measurements made under such circumstances to those made with the normal instrumentation.

It is particularly desirable to increase the total integration time to 1 minute when there is strong scintillation, which may be considerable in small instruments. Even with the 800-mm reflector at the Observatoire de Haute Provence, this precaution is taken when there is appreciable turbulence. The increase in the total time reduces the effect of the variations in flux to a reasonable level.

In principle, each measurement through each of the filters is preceded or followed by a measurement of the sky background. In practice, the three sky-background measurements are made immediately following those of the star. With bright stars, the sky-background measurements are very faint relative to the stars. When the variable and the comparison are very close to one another, so that moving from one to the other may be achieved very rapidly, the sky background is measured immediately after the comparison, and the same sky-background value is assumed for the variable.

Measurements are discontinued at moonrise. In UBV work, observations are avoided at First and Last Quarter. Photometric observations of stars made during astronomical twilight are always suspect, because it is very difficult to estimate the variations in sky background over several minutes at the beginning and end of the night. In summer particularly, during the time when the altitude of the Sun is higher than $-18°$, it is essential to check the ephemerides, otherwise work may be undertaken in vain.

The sky background is subtracted from each of the raw measurements to give the actual measurement. When photon-counting is being undertaken, the computer is used to obtain the true count-rate, correcting the measured value for the instrumental, electronic dead time. In amateur instruments, this dead time generally amounts to about 100×10^{-9} seconds.

One argument that is as old as photometry itself is that between the advocates of working with specific standards and those who favour differential photometry.

19.5.3.1 The standard-star method
This method requires two conditions to be met:

(i) Absolute stability of the sky throughout the night. (This is a prerequisite with this method that is seldom achieved.)

(ii) Observation of the greatest possible number of standard stars, with different colours and air-masses. Rufener (at Geneva) recommends dividing the night into two halves, so that two sets of standards may be assigned to two parts of the programme. The programme stars are measured in between measurements of the standards.

The quality of the correlation with the system and the accuracy of reductions obviously depend on the number of high-quality measurements of the standard stars that are available. If such measurements are to hand, it is possible to obtain good k' coefficients, and in particular, better values for ε, μ, and Ψ than if (as in differential photometry) seasonal transformation coefficients are employed that do not take account of instrumental drift.

Correlation by a least-squares fit (or regression) yields coefficients for the night. If the correlation coefficient is high ($> 95\%$), even if this means rejecting some aberrant measurements for certain of the standards, it is possible to be reasonably confident about reductions of the programme stars. The latter will have been adequately bracketed in both air-mass and colour, because this type of photometry, sometimes wrongly called 'absolute photometry', is directly linked to the photometric reference system.

This, naturally, only applies if the PMT-filter response curves do not differ appreciably from the Johnson system. It is also obvious that using the correlation principle, it is also possible (and perhaps better) to use the Bouguer method or the MD method.

19.5.3.2 The differential method

The basis of the differential method is simple: measurements of the programme star are compared with those of a constant star, which has colour indices that are as close to those of the variable as possible, and which is as bright or slightly brighter. If colour indices are not available for the variable, we have to make do with spectra.

For $(B - V)$ we have:

$$\Delta(B - V) = \mu\Delta(b - v)_0\,,$$

$\Delta(b - v)_0$ being equal to $\Delta(b - v)(1 - k_v''\bar{X}) - k_{bv}'\Delta X\,,$

and in V:

$$\Delta V = \Delta v - k_v'\Delta X - k_v''\bar{X}\Delta(b - v) + \varepsilon\Delta(B - V).$$

$\Delta(B - V)$ should be calculated first, so that it can be used in the equation for ΔV, because the colour index of the programme star is unknown at the time of observation.

At exceptional sites, using highly accurate and fully automated telescopes (at the European Southern Observatory at La Silla, in Chile, for example), one measurement of the variable, preceded and followed by one of the comparison, is frequently regarded as adequate: $C \rightarrow \mathbf{V} \rightarrow C$.

Under other conditions, a series of at least three measurements of the variable is obtained: $C \rightarrow \mathbf{V} \rightarrow C \rightarrow \mathbf{V} \rightarrow C \rightarrow \mathbf{V} \rightarrow C$. An experienced observer, using a non-automated telescope, would take about half-an-hour to carry out these estimates.

Times should be recorded in UT. The air-masses and instrumental magnitudes of the comparison are interpolated for the times of the measurements of the variable. These values are then used to calculate the Δ mags. It is customary to subtract the magnitudes and to give the results in the form $\mathbf{V} - C$.

When two comparison stars are available, the set of measurements should be: $C_1 \rightarrow V \rightarrow C_1 \rightarrow C_2 \rightarrow V \rightarrow C_2 \rightarrow C_1 \rightarrow V \rightarrow C_1 \rightarrow C_2 \rightarrow V \rightarrow C_2$.

This sequence has the double advantage that separate reductions may be made for each of the comparisons, and also that it allows the differential magnitudes between the comparisons themselves to be checked. The latter is made all the easier by the fact that on three occasions, measurements of the comparisons are made one after the other. The more frequently these sets are repeated during the course of the night, the better the mean of the measurements will be. It is always best to have series of measurements that consist of symmetrical sets $C_n \rightarrow V \rightarrow C_n$.

When only one comparison is considered reliable, which is the normal situation, an additional degree of safety may be obtained by using a second comparison, known as a 'check star' (**Ch**), which is measured twice. The sequence is: $C \rightarrow \mathbf{Ch} \rightarrow C \rightarrow V \rightarrow C \rightarrow V \rightarrow C \rightarrow V \rightarrow C \rightarrow \mathbf{Ch} \rightarrow C$. This is a luxury for which there is not always time. Most observers simply measure **Ch** at the beginning of the sequence only, thus saving the last two measurements.

In the absence of unstable conditions or any temporary problems with the sky, any differences between the reduced values for the comparison might indicate that, contrary to expectations, either the check-star or the comparison is not constant. The observations would not enable one to decide which. The observations of the programme star will probably have to be rejected, but it is possible that the comparison might prove to be a new variable. In which case it would be as well to concentrate on checking the comparison and, if there is a serious indication of variability, to inform the IAU Central Bureau for Astronomical Telegrams, and the International Bulletin of Variable Stars. Discoveries of new variables are by no means rare, even among bright stars. However, beginners should not 'jump the gun'. It would be wise to let a more experienced colleague or a professional astronomer examine the data first.

When observing a rapid variable such as an RR-Lyrae type, or an eclipsing binary, the sequence of measurements is normally: $C \rightarrow \mathbf{Ch} \rightarrow C \rightarrow V \rightarrow V \rightarrow V \rightarrow V \rightarrow V \rightarrow C \rightarrow V \ldots \rightarrow V \rightarrow C$, or even longer if the variable fluctuates very rapidly. This intrinsically allows the variation to be followed better, because gaps in the series of measurements are avoided, at the risk of lesser accuracy because of the separation in time of measurements of the comparison.

After the Δ mags have been calculated, the UBV values can be derived from the Δ mags and from the catalogue value for the comparison:

$$\mathrm{Mag_{var}} = \mathrm{Mag_{comp}} + \Delta\mathrm{Mag_{var.comp}}.$$

The differential method is undoubtedly the most popular, because it decidedly increases the accuracy of the measurements. However, for the reasons already described as regards the comparison stars, it is often difficult to be certain of the reliability of measurements made with just a single comparison. It remains true, however, that differential measurements of a star may attain an accuracy of a few millimags., provided that:

   (i)   there are enough of them;

   (ii)   the low standard deviation of the corrected measurements of the comparison indicate (if necessary) that it is not variable and, more importantly, confirms that the sky was stable during the period(s) when the measurements were obtained.

19.5.3.3 *The comparison stars*

In measuring a variable in UBV, the V magnitude and the colour indices of the comparison star are the relative reference points for the system. The final accuracy of differential measurements therefore depends directly on the quality, known or unknown, of the values of V, $(B - V)$, and $(U - B)$ for the comparison. These values are only rarely completely definite, for several reasons:

(i)   In general, the comparison star is not one of the system's standards. However, even the Johnson standards have a mean error of 0.02 mag.

(ii) In most cases, the quality of the calibration of the comparison stars that have to be adopted is completely unknown. Some catalogues give it on a scale of 1–3, or 1–4, or of a–d, and a very few authors give standard deviations or mean errors for their calibrations. Most papers that give details of the comparison stars used for making estimates of variables, however, do not include the magnitudes and colour indices that were adopted for the comparisons. It is therefore possible for various observers to observe a star that remains constant throughout the period, but still obtain different results, because their reductions did not use identical calibrations of the single comparison star that was available. The differential measurements might be essentially the same, but not the final UBV results.

The only ways of overcoming this fault are: either minimizing the unknown errors by using two or three comparison stars – which is not always possible with bright stars – or carefully calibrate the comparison star oneself. This requires sufficient observations, spread over several nights of excellent photometric quality. Such a calibration is advisable, nevertheless, if one intends an extended campaign of observation of the variable.

(iii) In general, most UBV measurements are given in the various catalogues (such as the SIMBAD database held at the Centre de Données Stellaires at Strasbourg, amongst others) as magnitudes rounded to hundredths, which reduces their significance, not for measurements made to a higher degree of precision, but for any transformations for which they are used. This is why, when publishing or describing photometric work, it is essential to state the magnitudes and colour indices of the comparison that served as a reference.

19.5.3.4 *Carrying out the observations*

A typical night would be one when just a few stars and their comparisons are measured, preceded and followed by measurements of several standards. The quality of the results may be judged by making independent reductions using the various methods – provided the seeing has been suitable – and provided the work has been carried out properly. However, to prevent disappointment it should be borne in mind that only a few privileged (and very rare) sites are capable of providing closely correlated observations of standard stars over a period of some hours. The method most frequently used by the UBV stellar-photometry group at the Observatoire de Haute Provence is to use two of the methods described earlier, whenever possible, depending on the circumstances and the type of observations that are envisaged.

Sets of differential measurements are obtained, using a comparison star, a check star and the programme star. These observations are bracketed by measurements of standard stars using the Bouguer method, or the MD method on very good nights, which provide extinction coefficients for the various objects being measured. In reducing and calibrating the data, the secondary-extinction and transformation coefficients that are employed are already known and have been checked.

The reductions and calibration to the UBV system are carried out using the various equations that are set out in the appropriate sections. There is no need to go into details of the calculations here.

It is very important to make proper preparations before observing, and this may require some hours: checking the order in which objects will be available for observation, selecting the type of measurements to be made, and choosing standard and comparison stars. This last point is sometimes difficult when it comes to bright stars. It may be impossible to find a red, 6th-magnitude comparison within a field that is 1, 2 or even 3 degrees in diameter. Charts of the fields concerned should be prepared from atlases that cover stars at least 2 magnitudes fainter than the magnitudes that will be measured. In this connection, when working in very crowded fields, such as those in Cygnus, care should be taken not to be deceived by the sparse number of stars shown in atlases with limiting magnitudes of about 9, and which might lead one to choose a diaphragm that is too large in diameter. The same precautions are needed to determine star-free areas of the field, where the sky background may be measured most easily without the risk of stray light from faint stars.

(i) Whatever type of measurements are being made, it is essential to have the coordinates, corrected for precession, for each star that is to be measured; the Universal Time and hour angle, and thus the sidereal time, when filters are changed; and, from the star's position and the hour angle, the air mass.

(ii) When pulse-counting, the equipment's dead time must be established. When DC measurements are being made, the amplifier's gain setting should be noted, and, when the observations are being plotted on a chart recorder, the time constant must also be known. When using a photodiode, the zero point should be adjusted and checked.

(iii) The same diaphragm and the same integration time should be used throughout a single series of measurements.

(iv) If a PMT is being used, the high-voltage supply and the amplifier should be switched on at least an hour before observations start. Many professional astronomers leave the HT supply and the amplifier permanently switched on. No measurements should be made until the HT supply has stabilized. Between sets of measurements a voltmeter or, better still, a digital meter should be used to check the stability of the high-voltage supply. If the HT supply varies more than 1 % from one set of measurements to the next, it will do the same in the middle of a single series and no accurate measurements are possible. As a matter of principle, the supply should not be altered during the course of a night. This is absolutely essential when standard stars are being measured over a period of several hours. From time to time, the stability of the amplifier should be checked by following a constant star at different air masses. Abnormal discrepancies serve to alert the observer: aberrant measurements show that the discrimination threshold is varying and that the instrumentation needs to be checked. During the course of a night the HT supply and the rest of the electronics will drift, with fatal results, if the ambient temperature drops significantly, unless the equipment is installed in a temperature-controlled environment, which is the ideal. The long, very cold, dry nights in winter, with no appreciable changes in temperature are best for the electronics and, generally, for sky quality. They are the most satisfying and also the hardest for astronomers. Dampness is the other enemy of the HT supply. If the voltage collapses, it is a sign that condensation on the PMT's pins is causing a discharge. There is nothing to do but stop, take the PMT housing into a warm room, and wait for conditions to improve. If a PMT is not perfectly enclosed, it is advisable to interrupt measurements when the humidity reaches 90 %. Packs of dessicant may cause more trouble than they are worth. When full of moisture, they may maintain the humidity of their surroundings rather than decreasing it. When humidity is above 90 %, saturation point is soon reached, especially if the rise from 70 to 90 % has taken less than an hour. This is when the optics start to be covered, and then to run, with dew, unless specific steps are taken, which would in any case not help photometry, because atmospheric absorption makes measurements impossible.

(v) Unless the system records the data automatically, it is essential to note down, for the star and for the sky background, each of the integration times that form a single measurement through each filter (6 units each of 10 seconds, for example); the time in UT, and any details that may be useful in reducing and interpreting the measurements. The dark current should be recorded from time to time.

It is always important to keep an observing book in which details of each star observed are entered, together with any remarks and brief details of events. Experience proves how essential this record is. Loose-leaf books should never be used, because the sheets may be misplaced or even lost, just when they are most needed.

(vi) It is a good idea to draw up, once and for all, a check-list of all the individual operations that have to be carried at the beginning and end of the night. This is invaluable in ensuring both that the equipment is fully operational (and safe) and that measurements proceed smoothly.

At the Observatoire de Haute Provence, a program known as 'Astro-Ack', written for a Macintosh computer by Bruno Fontaine, controls a secondary, buffer microprocessor which stores each measurement for a short while. The program reads data about each star from a catalogue file, calculates the corrections for precession, hour angle and air mass every second, and records the three-colour measurements passed by the interface, which are then corrected for dead time and sky background. It also calculates the least-squares discrepancies. The data are written to the screen or sent to a printer (or both). On request, it plots on the screen the light-curve from any set of measurements consisting of n integrations (for example, 200 integrations each lasting 100 ms). Times, measurements and air-mass data are written to a file for each night. The catalogue file, which is handled by a separate program, contains fundamental data for all the stars, whether they are standards, comparisons, or variables that have been measured or are to be measured.

We have a set of calibration and reduction programs, and others for calculating the various coefficients, depending on the methods and options that are chosen. The 'Astro-Ack' program was written to be used with the EPP-1 photometer, which uses a liquid light-guide, and which has been used since January 1985, sometimes on the 800-mm reflector (f/15) and sometimes on a 300-mm reflector (f/10) mounted on the equatorial table. The two-channel EPP-2 photometer, which is mounted on twin 300-mm reflectors, was finished in the winter of 1986–7 (Figs. 19.20 and 19.21). This new equipment raised some problems that cannot be discussed in detail here. Its primary interest lies in the fact that simultaneous measurements may be made of the variable and the comparison, thus suppressing errors caused by variations in sky conditions. Instead of long series of measurements with intermittent calibration, it will give a series of pairs of simultaneous measurements which may be reduced individually and with a considerably increased accuracy. One other advantage is that it will be possible to operate under mediocre sky conditions.

19.6 UBV conventions and relationships

The standard symbols and relationships are as follows:

- V, B, U: magnitudes in the standard system.
- v, b, u: raw instrumental magnitudes. The indices $_{bv}$ and $_{ub}$ refer to $(b - v)$ and $(u - b)$.
- v_0, b_0, u_0: magnitudes outside the atmosphere in the observation's instrumental system.
- X: air mass. A variable that enters into the calculation of atmospheric extinction, approximately equal to the secant of the zenith distance Z at which an observation is carried out. Various empirical equations enable X to be calculated to a close approximation from the secant of the apparent zenith distance.
- $\bar{X}$: mean air-mass (generally of 2 air masses).
- k: global extinction coefficients (k_v, k_b, k_u, k_{bv}, k_{ub}) The coefficients k may be expanded to give:

  $$k_v = k'_v + k''_v(b - v), \quad k_{bv} = k'_{bv} + k''_{bv}(b - v).$$

- k': principal extinction coefficient (k'_v, k'_b, k'_u, k'_{bv}, k'_{ub})

  $$k'_{bv} = k'_b - k'_v \text{ and } k'_{ub} = k'_u - k'_b.$$

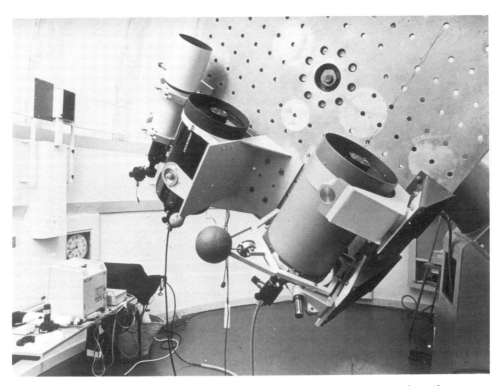

Fig. 19.20. *The two-channel, twin-telescope, EPP-2 photometer, mounted on the eastern equatorial table at the Haute-Provence Observatory. The top telescope is fixed. The bottom telescope may be adjusted, thanks to an adjustable mounting designed by R. Chanal. Its movements are controlled by fine-thread screw jacks, and monitored by dial-gauges. The guide telescope at the top is 125-mm aperture, f/6, fitted with a filar micrometer. The two light-guides run to the 'black box' (on the table) which contains a filter wheel, the PMT and the electronics. The computer, alongside it, is a Macintosh 512 kb.*

- k'': second-order extinction coefficient (k''_v, k''_b, k''_{bv}) applied to the colour index. By definition in the UBV system, $k''_u = k''_b$, whence $k''_{ub} = 0$, and $k_{ub} = k'_{ub}$; but it may be useful to reject this convention, which increases the error in U − B. In this case, the correction to be applied is slightly complicated, and uses effective values of $k''_{ub} \neq 0$, obtained by the methods described.

  k''_v, which is always small, is assumed to be negligible by some observers. This leads to appreciable errors when the variable being observed and the comparison star have very different colour indices, and are measured far from the zenith or when the comparison star is not close to the variable (or both).

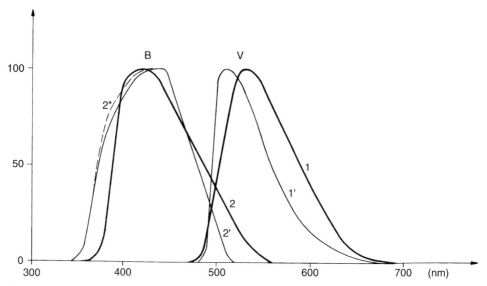

Fig. 19.21. *Response curves and filters chosen for an XP 2010 RTC PMT, fitted to the EPP-1 and EPP-2 photometers built by B. Fontaine. 1 and 2, in bold, are the standard Johnson system curves for V and B. 1′: cell plus Schott GG 495 filter. 2′: cell plus GG 385, BG 12, and BG 23 filters. 2″: with the BG 23 filter removed.*

- ε_v: transformation coefficient linking the magnitudes outside the atmosphere v_0 to the standard system (V magnitudes). ε_b and ε_u are used if the correlation is made in B and U magnitudes.

- μ: transformation coefficient for direct correlation in (B − V) index.

- Ψ: transformation coefficient for direct correlation in (U − B) index.

- $\zeta_{v,\ b,\ bv,\ ub}$: zero-points of the measurement scale in each band or colour index, considered to be constant throughout the night. They are eliminated algebraically in single-channel, differential photometry.

In general terms, all the standard transformations for a given wavelength are derived from two fundamental relationships:

$$M = m_{0\lambda} + \beta_\lambda C + \gamma_\lambda$$

where C is the star's standard colour index, β is the coefficient of instrumental colour, or the transformation coefficient, and γ is the zero-point; and

$$C = \delta C_0 + \gamma_C$$

where C_0 is the instrumental colour index outside the atmosphere, δ is a colour coefficient, and γ_C is the zero-point for the colour.

In the UBV system conventions, the reduction equations are:

I $\quad V = v_0 + \varepsilon_v(B - V) + \zeta_v$
$\quad\quad B = b_0 + \varepsilon_b(B - V) + \zeta_b$
$\quad\quad U = u_0 + \varepsilon_u(U - B) + \zeta_u$
$\quad\quad (B - V) = \mu(b - v)_0 + \zeta_{bv}$
$\quad\quad (U - B) = \Psi(u - b)_0 + \zeta_{ub}$

In differential reductions (Programme Star–Comparison Star), we have:

II $\quad \Delta V = \Delta v_0 + \varepsilon_v\Delta(B - V)$
$\quad\quad \Delta B = \Delta b_0 + \varepsilon_b\Delta(B - V)$
$\quad\quad \Delta U = \Delta u_0 + \varepsilon_u\Delta(U - B)$
$\quad\quad \Delta(B - V) = \mu\Delta(b - v)_0$
$\quad\quad \Delta(U - B) = \Psi\Delta(u - b)_0$

Magnitudes outside the atmosphere are:

III $\quad v_0 = v - [k'_v + k''_v(b - v)]X$
$\quad\quad b_0 = b - [k'_b + k''_b(b - v)]X$
$\quad\quad u_0 = u - [k'_u + k''_b(u - b)]X$
$\quad\quad (b - v)_0 = (b - v) - [k'_{bv} + k''_{bv}(b - v)]X$
$\quad\quad (u - b)_0 = (u - b) - k'_{ub}X$

and for differential reductions:

IV $\quad \Delta v_0 = \Delta v - k'_v\Delta X + k''_v\Delta(b - v)\bar{X}$
$\quad\quad \Delta b_0 = \Delta b - k'_b\Delta X + k''_b\Delta(b - v)\bar{X}$
$\quad\quad \Delta u_0 = \Delta u - k'_u\Delta X + k''_b\Delta(u - b)\bar{X}$
$\quad\quad \Delta(b - v)_0 = \Delta(b - v) - k'_{bv}\Delta X - k''_{bv}\Delta(b - v)\bar{X}$
$\quad\quad \Delta(u - b)_0 = \Delta(u - b) - k'_{ub}\Delta X$

The relationships given above under I, II, III, and IV constitute the set of fundamental equations that are used for all reductions in the UBV system.

19.7 Statistics

Any set of measurements should be accompanied by details of the standard deviation. Here we mean the standard deviation of a set of measurements, not the individual standard errors of each measurement, which we have discussed earlier. In any sample, the standard deviation hardly varies with the number of measurements, unless these are extremely numerous. The standard deviation of the mean, however, decreases as the number of measurements increases. The standard deviations are calculated as follows.

Let us assume that we have n reduced measurements of a slow variable star, obtained on the same night. For $i = 1, 2, 3, \ldots, N$, we have a set of values x_i and a mean:

$$\bar{x} = \frac{1}{n}\sum_{i=1}^{n} x_i$$

The differences $(x_i - \bar{x})$ are known as residuals, and we next obtain the sum of the squares of the residuals. The value

$$\frac{1}{n-1} \sum_{i=1}^{n} (x_i - \bar{x})^2$$

is known as the variance of the sample, or σ^2.

In general, the variance is not used because it is a quadratic expression, its positive square root, the standard deviation being employed instead, and which is expressed in the same units as the measurements themselves:

$$\sigma = \sqrt{\frac{1}{n-1} \sum_{i=1}^{n} (x_i - \bar{x})^2}.$$

To be accurate, and avoid compounding errors, the standard deviation, σ should not be calculated from just the reduced values, but using the simple equation:

$$\sigma = \sqrt{\sigma_{\text{comp.}}^2 + \sigma_{\text{var.}}^2}.$$

Let us assume that we have three measurements of the magnitude of the variable and three of a comparison star: $C_1 = 6.732$; $C_2 = 6.739$; $C_3 = 6.727$; and $V_1 = 6.124$; $V_2 = 6.132$; $V_3 = 6.119$. The standard deviation of the comparison is 0.006; that of the variable is 0.007. The standard deviation of the measurements is therefore 0.009, the difference Δ between the means being -0.608. The observed magnitude of the variable is $6.7326 - 0.608 = 6.125 \pm 0.009$. Calculating the values of Δ individually, we have: $\Delta_1 = -0.608$, $\Delta_2 = -0.607$, $\Delta_3 = -0.608$, giving a mean Δ of -0.608 as before, but with a standard deviation of the means of only 0.0006. This would give a value of 6.125 ± 0.0006, which is wonderful, but wrong. In the first case, the error is approximately 0.01, and in the second, 0.001. The difference is considerable. Many observers have a tendency to use the second procedure, which apparently improves their results, but which is not scientifically correct.

By indicating the scatter found among the measurements, the standard deviation gives an indication of their accuracy. A σ of 0.01 mag. would indicate that each measurement of the variable may be considered accurate to within ± 0.01 mag.

The Gaussian (or normal) distribution, which applies to most experimental data of this sort, indicates that for any given sample of measurements, 68 % may be expected to fall within the range $\bar{x} - \sigma$ to $\bar{x} + \sigma$ and 95 % within the range $\bar{x} - 2\sigma$ to $\bar{x} + 2\sigma$.

The standard deviation of the mean, σ_m is:

$$\sqrt{\frac{\sum_{i=1}^{n} (x_i - \bar{x})^2}{n(n-1)}} = \frac{\sigma}{\sqrt{n}}$$

This indicates that for 12 measurements the standard deviation of the mean will be less than 0.003 mag. for $\sigma \pm 0.01$.

Some authors give the probable error, which is $\pm 0.675\sigma$. It indicates the range, on either side of the mean, within which there is a 50 % probability that the true value will lie. For the values just given, the probable error of any one measurement is ± 0.007.

If a very slow variable or a non-variable star – for example, a comparison star that is to be calibrated – is measured over several nights using the same method, a weighting may be given to the measurements obtained on the various nights. A similar weighting may be given to measurements of the same star observed by different observers on the same night.

The statistical weight of a sample of measurements is inversely proportional to the square of the standard deviation: $\bar{\omega} = 1/\sigma_m^2$.

The weighted mean of n sets of measurements x is the ratio of the two sums:

$$\frac{\sum_{i=1}^{n}(\bar{\omega}_i x_i)}{\sum_{i=1}^{n}\bar{\omega}_i}$$

If over three nights, the measurements are 4.725; 4.748 and 4.723 and the values of σ_m are ± 0.009; ± 0.014 and ± 0.006, we obtain: $\bar{\omega}_1 = 12346$; $\bar{\omega}_2 = 5102$; $\bar{\omega}_3 = 27778$. The calculation is simplified, of course, by taking the weights as 12, 5 and 28:

Mag.$_{\bar{\omega}} = [(12 \times 4.725) + (5 \times 4.748) + (28 \times 4.723)]/(12 + 5 + 28)$

i.e.,

Mag.$_{\bar{\omega}} = 4.726$.

The weight is just a quality coefficient defined by the standard deviation of the measurements and by the number of the latter included in each sample. The weighted mean is a statistical estimate of the result of the combined observations.

If we eliminate the second sample, which has a low weighting, we then have:

Mag.$_{\bar{\omega}} = [(12 \times 4.725) + (28 \times 4.723)]/40 = 4.724,$

giving a result very little different from the previous one. There are problems with such a procedure, however: short of making a detailed examination of each measurement and the conditions under which it was obtained, there is nothing to indicate that the rejected sample was in error, other than its larger standard deviation. But the latter might indicate that the object displayed unexpected microvariability on one of the nights, at a time that was not covered by the 'good' observers. We would be discarding the sample that actually represents a discovery. Any standard deviation that shows a major discrepancy could be an indication of either a significant error, or of some other circumstances that need to be investigated. Because weighting effectively determines a 'reliability' factor for each sample, it must be borne in mind that rejection of any sample actually contravenes the principle on which the procedure is based, and that such a step should therefore be taken only with the greatest care. Any weighting also assumes that none of the means have been affected by systematic errors in measurement.

It goes almost without saying that standard deviations of the type just discussed have no significance whatsoever with rapid variables, because any means would themselves be meaningless. In drawing up the light-curve, however, the standard errors of the various estimates should be taken into account.

20 Image-intensifiers and CCDs

C. Buil & T. Midavaine

Electronic imaging is coming to play a larger and larger role in professional astronomy, to the detriment of conventional photography. There are few amateurs, however, who mount opto-electronic devices on their telescopes, but it is safe to assume that they will become more and more numerous, and that the few people obtaining digital or intensified images nowadays are merely the pioneers, who are foreshadowing amateur astronomy in years to come.

We therefore thought it advisable to end this book by including a short chapter of information about these modern techniques, which are beginning to make inroads into amateur astronomy, not with the aim of providing an exhaustive description of their use (for which a whole book would be necessary), but rather with the view of giving the reader a summary of their very considerable possibilities.

The present discussion is restricted to the two technologies that appear to give the best astronomical performance: image-intensifiers, whose significant gain enables spectacularly reduced exposure times for faint objects; and charge-coupled device (CCD) cameras, whose quantum efficiency is superior to that of any other detectors, and which, by means of an analogue-to-digital conversion, provide a digitized image that may be manipulated with a computer.

20.1 Image-intensifiers

Image-intensifiers are new to most amateur astronomers. They have been used extensively (and with reason) in observatories for a number of years. They have been mainly developed, however, for military use in night-time combat. This is why series production has been established, resulting in lower costs, and the appearance on the market of surplus or sub-standard devices that are within the amateurs' means.

20.1.1 Principles

Image-intensifiers have been the subject of a series of developments and three generations of devices exist. They consist of various elements:

- a photocathode,
- electrostatic optics,
- an electron amplifier,
- a phosphorescent screen,
- a fibre-optic window.

Not all these elements need to be used simultaneously; on the other hand, the same elements may be present several times in a single device. The second generation is

A

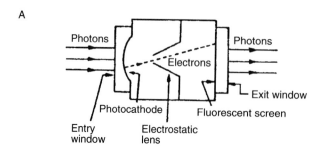

B

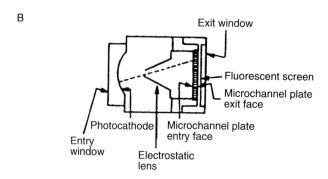

C

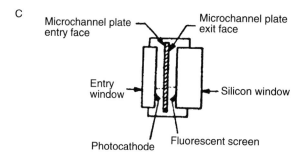

Fig. 20.1. *The design of different types of image-intensifier (from RTC data-sheets). A: first-generation, single-stage intensifier tube. B: second-generation intensifier with an electrostatic lens. C: second-generation double-focus intensifier.*

easily identified by the presence of a microchannel plate as an electron amplifier. The third generation incorporates new gallium arsenide photocathodes, which have a higher quantum efficiency. We will briefly describe the roles of the various components before covering the typical characteristics of a complete intensifier. Figure 20.1 shows the three different types of design in use.

20.1.1.1 Photocathodes

The photocathodes are identical to those used in photomultipliers and consist of caesium-based material that exhibits the photoelectric effect. When a sufficiently

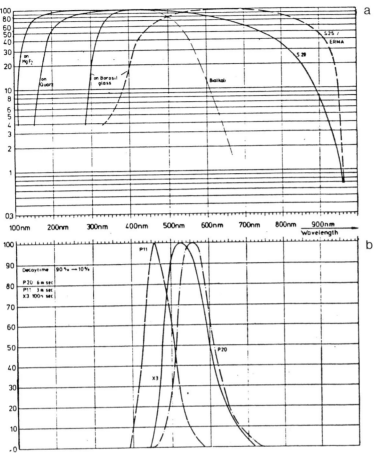

Fig. 20.2. *a) Relative spectral sensitivities of bialkali S20 and S25/ERMA photocath-odes, with MgF₂, quartz, and borosilicate windows. b) Relative spectral luminance of P11, X3, and P20 phosphors. (From Proxitronic data-sheets.)*

energetic photon hits the material, an electron is ejected from the surface. This takes place with an efficiency of some 10–20 %. Literature on photomultipliers will give more information about the efficiency of the various photocathodes as a function of wavelength. The most common photocathodes are the S20R or S25, the spectral response curves of which are shown in Fig. 20.2a. It should be noted from this diagram that the sensitivity drops sharply beyond 950 nm at long wavelengths, because the energy carried by each photon is not sufficiently high to liberate an electron from the photocathode. At short wavelengths, the sensitivity is limited by the transmission of the photocathode window. For glass, the cut-off is at 300 nm. The spectral region therefore extends beyond the visible region and, relative to standard photographic emulsions, offers a new window between 700 nm and 950 nm.

20.1.1.2 Electrostatic lenses

As in an electron microscope, the beams of electrons ejected by the photocathode may be manipulated by electrostatic lenses. These consist either of a truncated, hollow metal cone that is connected to the anode, or of a solenoid or permanent magnet that produces an axial magnetic field. The lens has two effects. First, it accelerates the electrons: each of these becomes more energetic, so the lens contributes to the tube's overall amplification. Second, like an optical lens, it has an imaging function: electrons arising from a single point on the photocathode are directed to a single point on the screen. Depending on the geometry and overall arrangement of the lenses, the image may be inverted or reversed (Figs. 20.1a and 20.1b).

20.1.1.3 Microchannel plates

Microchannel plates may be simply described as consisting of a sheet of glass pierced by a large number of tiny channels arranged in a honeycomb pattern (Fig. 20.3). The walls of the channels are treated to give a coating that produces secondary emissions: if an electron hits a wall it ejects several other electrons. A potential difference applied across the two, metallic, surface coatings accelerates the electrons towards the positive pole, producing an amplification cascade. The gain is a complex function of the thickness of the plate, the voltage applied, and the shape of the channels. Typical gains that may be obtained lie within the range 10^1 to 10^6. The number of electrons that may emerge from a channel is limited, because the electrons' mutual repulsion tends to push them towards the walls of the channels, causing a saturation effect. The response time of the plate depends on the number of electrons and active channels: a working figure is around 10 ms. The dark current, resulting from any electrons that encounter the plate in the absence of illumination, is generally very low, around 1 electron per second per cm^2 of the plate. This is one of the great advantages of this type of amplifier. Current manufacturing techniques are able to achieve channels 10 μm across and plates 125 mm in diameter.

20.1.1.4 Screens

In an image-intensifier, the electrons are destined to be transformed back into light, except in certain cases where the detector is able to detect and store the electrical charges directly. In general, the electrons hit the surface of a phosphor screen. The phosphor – a generic term that does not correspond to the element with a similar sounding name – is a material with luminescent properties. When bombarded by the electrons that have been accelerated by the electrostatic field operating within the tube, the material emits visible radiation. There are various compositions, which may emit in different spectral regions, ranging from the ultraviolet to the infrared, with varying response times and persistence (decay time). These phosphors are identical to those found in the screens of television sets, computer VDUs and oscilloscopes. The most widely used material is type P20, which emits in the green, and therefore has the advantage of matching the sensitivity of the eye (*cf.* Fig. 20.2b). Under certain circumstances this may be preferable to screens that emit in the blue (P11) or in the violet, which give a better match with the sensitivity of photographic emulsions, and have a shorter persistence.

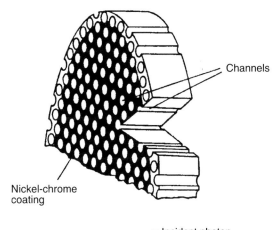

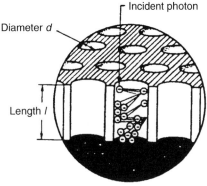

Fig. 20.3. *Schematic representation and principles of a microchannel plate (from RTC literature).*

20.1.1.5 Windows

The entrance and exit windows generally consist of optical-quality glass. In fact, all the active components are held under vacuum so that the electrons can propagate freely. The photocathode and the phosphor are deposited on the internal faces of the entrance and exit windows respectively. Light therefore penetrates the former and illuminates the photocathode, and the emission from the phosphor is visible through the second window. In some cases, the entrance window consists of a different material, such as silica or monocrystalline magnesium fluoride, which enable the photocathode's sensitivity in the ultraviolet to be utilized (*cf.* Fig. 20.2a). In addition, fibre-optic windows may be used. These consists of numerous parallel fibres, the axes of which are perpendicular to the faces of the window. They have the property of transferring the image plane from one face of the window to the other. This type of component enables the intensifier to be coupled to the optical system in a particularly elegant manner. For example, the faces of the fibre-optic windows may be finished with specific curvatures to match the concave curvature of the objective that is being used, and also to match the curvature of the equipotentials generated by the electrostatic lens. We shall see how it may be very desirable to use

1107

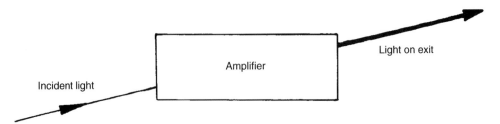

Fig. 20.4. *A schematic representation of the way in which a laser operates: the emergent wave has a preferred direction, which depends on that of the incident wave.*

such a component for the exit window, because it enables it to be placed in direct contact with a photographic emulsion without any intervening optical system.

20.1.2 Characteristics of the tube

We shall simply consider the tube as a 'black box', which, for a certain input signal at the photocathode, provides a certain output signal on the screen. We shall discuss the main factors that determine the intensifier's performance and give typical operating values, and then go on to consider the way in which intensifiers are integrated into an overall system, with an objective, a detector and, perhaps, an image-processing system.

20.1.2.1 Gain

Before considering this parameter, there is one important point, relevant to understanding this section, that needs to be described. Some stimulated-emission devices produce a straightforward amplification of the electromagnetic field. (This is the effect that is utilized in lasers.) They may be understood in general terms as multiplying the amplitude of the incident radiation; they are true (dimensionless) amplifiers in that they give a gain without altering the direction of propagation (Fig. 20.4).

An image-intensifier, however, does not operate in this way. It does provide considerable gain, as we shall see, but it is not dimensionless and does affect the directional component. The multiplication in the number of electrons produces an amplification of the luminous flux, but the latter's geometrical properties are considerably modified between arriving at the photocathode and leaving the screen (Fig. 20.5).

The photocathode receives a certain illumination (expressed in $W\,m^{-2}$), which results in a certain luminance (expressed in $W\,m^{-2}\,sr^{-1}$) at the screen. The intensifier's performance may therefore be measured in terms of the ratio: exit luminance/entrance illumination, which is consequently expressed in sr^{-1}.

Other definitions exist; in particular, manufacturers often express the gain in the form of the ratio: exit flux (lumens/m^2)/entrance illumination (lux). Sometimes radiometric units are used: emittance (W/m^2)/irradiance (W/m^2). In the last two cases, the gain is dimensionless, but cannot be exploited, because it would never be possible to collect all the light emitted by the screen. Such gains are measured with

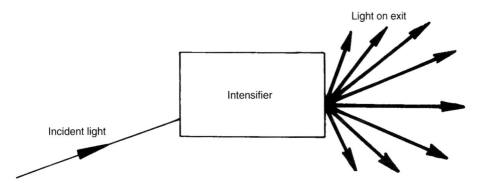

Fig. 20.5. *A schematic representation of the way in which an image-intensifier operates: the light from the output screen is radiated in all directions, quite independent of the geometry of the incident light.*

a lamp having a tungsten filament at 2856 K, or else with a monochromatic source of specified wavelength.

An illustration may clarify the effect of an intensifier's gain. Only a single eye can observe a star directly through a telescope. But if an intensifier is placed at the focus of the same telescope, the star will not only appear much brighter, but it may be seen by several pairs of eyes simultaneously, just as if it were on a television screen. The art of using intensifiers lies in capturing the maximum amount of light, which is emitted over a solid angle of 2π steradians. Typical gains offered by microchannel intensifiers are $1000\,\text{sr}^{-1}$ to $1\,000\,000\,\text{sr}^{-1}$ in the visible for a source at 2856 K.

Finally, with visual use, another effect of spectral origin means that in certain cases the visual gain is increased. As we have seen, the photocathode has a wider spectral sensitivity than the eye and that, on the other hand, the spectral emission of the screen is significantly narrower than the eye's spectral bandwidth, both of which are centred near 550 nm. As a result there are considerable variations in the visual gain, depending on the spectra of the actual sources. This may significantly alter the appearance of the fields being observed. The apparent relative brightness of stars may be upset. Sources that mainly radiate outside the visible region may be detected through the intensifier.

This phenomenon is particularly striking with nebulae, where the eye is mainly sensitive to just the O III lines at 500 nm. With an intensifier, the Hα line emission becomes very strong, and even predominates, completely altering the appearance of the object.

20.1.2.2 Resolution

The detail in the image obtained through the tube is generally limited by the amplification process. With an entrance image consisting of an infinitely small, bright point, the screen shows a spot of light with a distinct diameter, whose luminosity distribution follows a specific law (Fig. 20.6). It may also be described in terms of the modulation transfer function, MTF (Fig. 20.7), or by the number of pairs of lines that may be resolved, as with photographic emulsions.

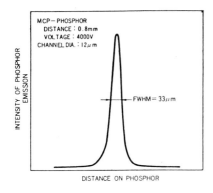

Fig. 20.6. *Typical response profile for an image-intensifier (from Hamamatsu literature)*

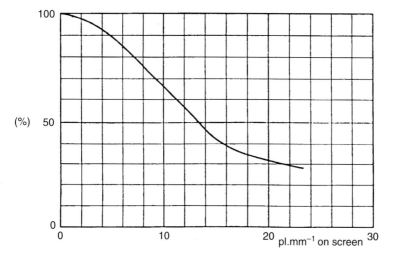

Fig. 20.7. *Typical modulation transfer function (from RTC data-sheets)*

20.1.2.3 Quantum efficiencies

Independently of the gain, the performance of an intensifier is also a result of its quantum efficiency. In general, this is fixed by the quantum efficiency of the photocathode. The potential for detecting a very weak source is limited by the quantum nature of light: i.e., by the fact that it consists of photons. When a device has an inherently low noise level (which is the case here), and its response time is fast relative to the rate at which photons arrive, the device is able to detect the latter one by one. A photocathode does not systematically convert a photon into an electron; there is a specific probability of the effect occurring. This is known as the quantum efficiency. Typically, quantum efficiencies that may be attained are around 10 to 20 % – just as in the case of photomultipliers.

20.1.2.4 Dark noise

The noise introduced by the device is very low, as we have just noted; nevertheless it does exist. It may be determined easily by occulting the photocathode. While this is subject to zero illuminance, the luminance of the exit window is measured. There are two principal causes of this dark noise:

- Thermal noise: thermal agitation may spontaneously release electrons from the various components of the device, which strike the screen and give rise to a luminous signal.
- High-energy radiation: everything is subjected to bombardment by high-energy cosmic radiation, which may penetrate varying thicknesses of material and thus reach the photocathode and other parts of the device, releasing a number of electrons.

20.1.2.5 Typical characteristics

By way of example, we give the characteristics of RTC's XX1390 intensifier, a second-generation device that has been used to obtain the illustrations shown in this chapter. Mounting this sort of device is simple, because of its small dimensions: 22 mm deep, 45 mm in diameter, with a mass of just 35 g.

Photocathode:
    Spectral response   S20R
    Window   glass, plane
    Diameter   20 mm
Screen:
    Phosphor   P20
    Window   glass, plane
    Diameter   20 mm
Gain   10 000
Resolution   30 pairs of lines mm^{-1}
Illumination equivalent of max. noise   0.5 µlux

20.1.3 Mountings

The overall system in which an image-intensifier is to be used should be designed to make the best use of the device, depending on the source to be examined and the desired result, with the proviso that the system is limited by specific factors and the laws of physics. An outline of a typical system is shown in Fig. 20.8.

As we shall see, the use of an intensifier can only be justified because the detector is imperfect. A perfect detector would have a quantum efficiency of unity, and thus introduce no noise and have an infinitely fine resolution. When an intensifier is used, the detector's quantum efficiency may be raised to 20 %, with low noise, and a resolution that is dependent on that of the intensifier tube.

20.1.3.1 Objective

The component that collects the light (the objective) may be defined by two parameters: its area and its focal length. The first is fixed by the reflector or refractor

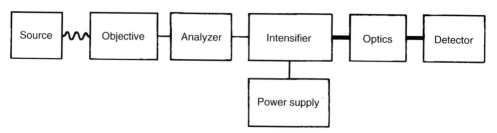

Fig. 20.8. *Block diagram of the overall imaging chain*

and, naturally, the larger the better. The focal length may be optimized by using one of the normal optical methods that will be familiar to any amateur who has taken astronomical photographs. Here, we have two choices, because we can decide to favour resolution or luminosity.

For the former, the resolution of the intensifier is matched to the best angular resolution ϵ that may be attained. If the exposure time envisaged is very short, the resolution used may be that dictated by diffraction ($\epsilon = 1.22\lambda/D$). Turbulence will limit the resolution for instruments larger than 150 mm in diameter. Typically, one can expect to reach 1 arc-second, i.e., where $\epsilon = 5 \times 10^{-6}$ radians. With an intensifier that resolves 30 pairs of lines per mm, i.e., 33 µm, the minimum focal length F that may be used is: $F = 33 \times 10^{-6}/5 \times 10^{-6} = 6.6$ m!

In the second case, when we want the maximum luminosity to reduce the exposure time, we can shorten the focal length. This may be appropriate for extended objects that have very small images at the photocathode.

20.1.3.2 *The analyzer*
We shall not discuss this element in detail. One typical device is a spectroscope, the subject of Chap. 18. Intensifiers are particularly valuable when used with any equipment that is selective in its operation and thus requires extended exposure times. An analyzer generally reduces the light available at the focal plane, and its effect is described by a transmission coefficient, which is a factor that is applied to the flux of light that is received.

20.1.3.3 *The intensifier*
An image-intensifier is light and physically small; it is therefore easy to integrate into an overall system that is destined to be used on a small telescope. Nevertheless, the greatest problem arises from the high-voltage supply that is required (several thousand volts). The material on which the intensifier is mounted should preferably be an insulator (such as wood or PVC), to avoid any stray electrostatic effects. Checks should be made to ensure that no metal is in its vicinity, especially if it has sharp edges. The electrical supply therefore needs to be carefully designed; high voltages can always be a source of spectacular flash-overs. It is becoming more common, however, for modern tubes to be supplied with a high-voltage converter fitted within the housing, so the constructor has to supply only a few volts. When

this is not the case, the tube has to be connected to a high-voltage supply containing a resistor bridge that will provide the various voltages required.

An XX1390 intensifier, for example, requires three voltages (100, 645 and 5700 V), to different parts of the tube. The current required is extremely low, being just a few mW. To avoid any hazards, it is also advisable to limit the current that may be drawn: if a short-circuit occurs the voltage applied will drop.

20.1.3.4 Visual use

Because of their diameter and resolution, intensifiers are not very suitable for visual work, and a large fraction of the light is lost. Nevertheless, as we shall see in Sect. 20.1.4.1, results are spectacular. The image formed by the telescope is projected onto the photocathode. The same image appears, brighter and now coloured green by the phosphor, on the screen. This final image may simply be observed, like one that appears on a television screen, by several observers. Because both the screen and the pixels (from 'picture elements') are small, however, it is best to bring one's eye close to the screen. A fine image may be obtained by using an eyepiece, and the latter's focus should coincide with the plane of the screen. The type of eyepiece therefore needs to be one that has its focal plane ahead of the field lens. In general, a focal length greater than 20 mm is required to ensure that the whole of the screen is covered. An alternative is to use a 50 mm objective fixed to the tube. The eye may then view an aerial image, and in many cases this may be more comfortable.

20.1.3.5 Photographic use

The problem consists of projecting the image of the screen onto the plane of the film. Several solutions are possible. The first consists of choosing a tube that has a plane, fibre-optic output screen. All that has to be done is to press the film against the free surface of the exit window. This will give contact images on the emulsion. This solution is perhaps the most elegant, and is, in any case, the most compact. In addition, fibre optics have a large numerical aperture (low F/D ratio), but it remains limited, implying that some of the light is lost. Mechanically, this solution is more difficult to attain, because it requires a camera body that will allow the film to be placed in contact with the exit window, and also because it means finding a way of avoiding scratches on the emulsion when the film is advanced. The main inconvenience is that the image is not accessible visually.

The second solution is to use a photographic lens, for example a standard 50-mm f/1.4 lens, fitted with macro extension tubes. This type of mounting has the advantage in that there is no vignetting, but it is slow, because extension rings reduce the apparent aperture to 2.8 for a magnification of 1×. On the other hand, the image may be seen easily through the reflex finder.

The third solution consists of using two lenses (for example, 50-mm f/1.4) held face-to-face by an adaptor ring screwed into the filter threads. The first lens is fixed behind the screen, and the second forms an image inside the camera body (Fig. 20.9). Here, we retain the full aperture ratio of 1.4, and gain in exposure times by a factor of 4 when compared with the previous scheme. On the other hand, vignetting becomes significant. Anyone with plenty of money could use lenses with apertures of 1.2 or even less.

Fig. 20.9. *Equipment mounted on a Celestron 8 telescope. In the left foreground is the box containing the resistance bridges, providing the high-voltage supply via the three connectors. From left to right, we have the off-axis guider, the intensifier between two PVC rings, two 50-mm, f/1.4 lenses, and the SLR body. Typical exposure times for deep-sky objects are a few seconds with Tri-X developed for 4 minutes in D19.*

The fourth solution is to use an optical system specifically designed for the task in hand. In fact, the first three solutions mentioned all include a certain redundancy: the resolution of photographic lenses is superior to the resolution of the tube; in addition the lenses are corrected for chromatic aberration, whereas the output from the screen is limited to a narrow band in the green. A combination of four or six lenses should ensure that the focal planes are flat and the resolution is adequate, preserving, at the same time, a fast overall system.

For choice, the film to be used should be a black-and-white emulsion with a maximum sensitivity at around 550 nm, with a spatial resolution higher than that of the tube. Since 1988, one film has been pre-eminent: TMAX P 3200. Its gain is a factor of 8 greater than older films of ISO 800. Calculations show that on a 200-mm Schmidt–Cassesgrain, with the third type of arrangement described earlier, it is possible to reach magnitude 19 in a 10-second exposure!

20.1.3.6 Video use

Most video cameras do not allow the lens to be removed, so the third solution suggested above has to be employed. In addition, the weight of professional equipment does not allow it to be rigidly fixed to a small telescope. The camera

1114

therefore has to be mounted on a separate tripod, with the optical axes approximately aligned, the required angular tolerance being provided by the fields of the lenses used.

20.1.4 Astronomical applications

20.1.4.1 Visual use

It is difficult to describe visual impressions, but nevertheless we may give a few, arbitrary details. Deep-sky objects that are normally visible only in long-exposure photographs may be seen directly. M57 or the Sombrero Galaxy are brilliant, and a sweep through the clusters in Leo or Coma Berenices is a joy! With stellar fields, however, the images are perhaps somewhat disappointing because of their lack of sharpness. The first remark made by an observer using this type of tube for the first time usually concerns the significant amount of noise in the image, and which consists of random flickering points of light. In fact this noise is our signal! With faint objects we are actually seeing the photons arriving one by one; it is only persistence of vision (about 100 ms) that enables us to detect them. With brighter sources, on the other hand, the flux is higher, and the image remains visible all the time.

20.1.4.2 Limiting magnitude

The image of M57 shown in Fig. 20.10 was obtained with a 600-mm reflector having an effective focal length of 20 m and an exposure time of just 2 minutes! The meteorological condition were far from perfect, because the sky was covered by a layer of cirrus. The limiting magnitude was still around 18. This may be taken as only a rough guide, because as previously stated, the spectral sensitivity of the eye and of the tube are very different. The appearance of the nebula is different from that shown in ordinary photographs or visual observations. This is because of Hα emission, which increases the brightness of the central region in the red, at a range where photographic emulsions and the eye are less sensitive. The film used was Tri-X, pushed to about ISO 1000.

20.1.4.3 Spatial resolution

Figure 20.11 shows the central region of M13. As will be seen, stars are resolved on the photograph, which was obtained with the same equipment as that used for Fig. 20.10. The stellar resolution may be estimated as 1 arc-second (width at half-height). This result was obtained thanks to the long focal length and the short exposure time, which limited random tracking errors. This points to another advantage of intensifiers: the possibility of dispensing with an accurate equatorial mounting and having to guide using cross-hairs throughout long exposures. With stars, it is conceivable that turbulence could be fixed by photographing the 'speckle' image with a longer effective focal length that matches the tube's resolution to the diffraction disk of the main telescope's optics.

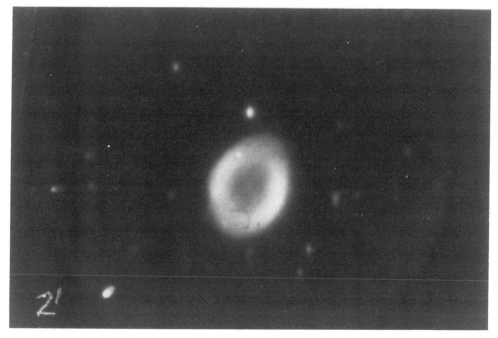

Fig. 20.10. *M57, taken with the 600-mm reflector at the Pic du Midi. Exposure: 2 minutes, effective focal length 20 m. Note the number of stars in the field. Photo: CARPE and T. Midavaine; T60 Association.*

20.1.4.4 Temporal resolution

The capability of being able to make short exposures opens up fields that are not available with standard techniques. We have, for example, been able to obtain images of Comet Halley, through a C8 telescope, in only 2 seconds! Figure 20.12 shows two different prints made from negatives obtained at an interval of just 2 minutes. Figure 20.13 shows examples of images produced by masking negatives obtained 2 minutes apart.

A video camera enables an even finer resolution in time to be achieved. However, this type of camera has a lower sensitivity than the eye and only partially reproduces the visual impression.

20.1.4.5 For 'sight-seeing'

Those who collect photographs can use an intensifier to set up a virtual production line for images of deep-sky objects. An hour's juggling with a telescope and the Messier catalogue may produce 36 exposures on film. Figures 20.14 and 20.15 bear witness to the quality of the images produced by a 200-mm telescope. Figure 20.14 may be compared with one (published elsewhere) that shows identical results but which took an exposure of 2 hours as against 4 seconds. This is a measure of the amount of progress that has been made in just a few years!

Fig. 20.11. *The central region of M13; 4-minute exposure with the same equipment as Fig. 20.10. Photo: CARPE and T. Midavaine.*

20.1.5 The future of intensifiers

Image-intensifiers should become increasingly popular among amateur astronomers. Apart from the provision of a high-voltage supply to tubes that are not fitted with an integral supply, and intensifiers' intolerance to stray light, they are extremely easy to use. Apart from Schmidt cameras, they may be used on any form of telescope, increasing their potential for visual, photographic and video use. Microchannel-plate intensifiers, in particular, because of their low mass, insignificant bulk, and high gain, greatly simplify the photography of deep-sky objects, reducing exposure times to a few seconds or tens of seconds at the most. This type of device opens up new horizons in astrophysical applications, with its excellent temporal resolution and also because it allows sophisticated analyzers (e.g., a Fabry-Perot grating) to be used on faint objects – as in spectroscopic work on galaxies. When employed in conjunction with a video camera, it may find applications as diverse as the production of films to popularize astronomy, or photon-counting close to a telescope's quantum-efficiency limit.

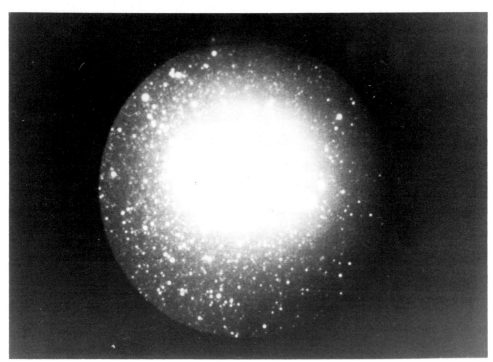

Fig. 20.12. *The central region of Comet Halley (photographed in 1986 April from the island of La Réunion in the Indian Ocean), on a C8 at f/5, with a 2-second exposure for both plates. Only the exposure time during printing is different, illustrating the great dynamic range of the image tube. Photos: Club Eclipse and T. Midavaine.*

1118

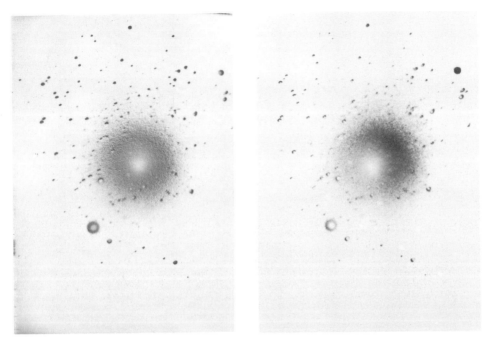

Fig. 20.13. *Photographs of Comet Halley taken with an image intensifier. a: Negative masked with a positive taken from the same negative. b: Negative masked with a positive derived from a negative exposed 2 minutes later. The motion of the coma is easily visible.*
Photos: Club Eclipse and T. Midavaine.

20.2 Charge-coupled devices

20.2.1 Principles

Charge-coupled devices (CCDs) form a family of solid-state, semiconductor detectors based on the photoelectric effect. The first types appeared in 1970, and they were immediately adopted by professional astronomers. Today the cost of CCDs has become reasonable, and it is quite feasible for amateur astronomers to use them for their own observations.

The CCDs used in astronomical observation are arrays of small MOS (metal–oxide–silicon) capacitors capable of storing an electrical charge induced by light. Such a unit is shown schematically in Fig. 20.16: a doped silicon semiconductor substrate is covered by an insulating layer of silicon oxide (SiO_2) on which a metallic electrode, known as the gate, is deposited. When this gate is appropriately polarized, it acts to create a series of potential wells, which have the property of being able to store electrical charge.

The gate is semi-transparent, and photons are able to pass through it. When a sufficiently energetic photon penetrates the silicon, there is a certain probability that it will displace an electron, which will be captured by the nearest potential well.

Fig. 20.14. *The Sombrero; 4-second exposure, 200-mm reflector, f/10. Photo: Club Eclipse and T. Midavaine.*

If individual MOS capacitors are arranged in a regular pattern and an image is formed on the surface of this array, the image will be broken up into a number of individual elements corresponding to the number of sites, and which are known as pixels. The size of pixels in current CCD detectors lies between about ten microns (10 µm) and several tens of microns across.

The number of units of charge created is proportional to the amount of light received. The longer the device is exposed to light, the more electrons accumulate, similar to the effect found in the more familiar process of photography. The exposure time is also known as the integration time. It is important to note, however, that some electrons, generated by thermal noise within the silicon, are captured by the potential wells. This noise is also proportional to the integration time. In astronomical applications that require long exposures, this parasitic effect is usually reduced by cooling the CCD.

After a greater or shorter exposure time, the individual electrons that have accumulated in each site have to be transferred to an output stage. This is achieved by altering the polarity of adjacent MOS capacitance sites in sequence. When the potential wells of two adjacent sites intercommunicate, the electrons accumulate where the gate potential is highest. By applying appropriate voltages to adjacent sites, charges may be transferred from one site to the next. A number of electrodes may be connected together in a periodic pattern, making it possible to transfer several

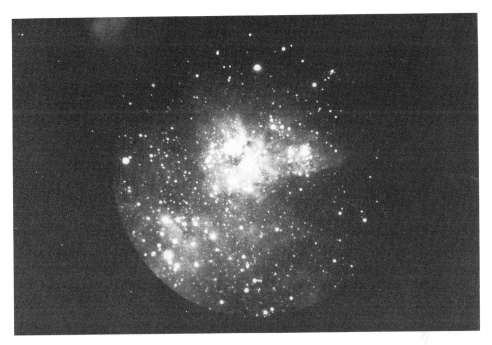

Fig. 20.15. *η Carinae; 8-second exposure, 200-mm reflector, f/5. Photo: Club Eclipse;
T. Midavaine and A. T. Savoye.*

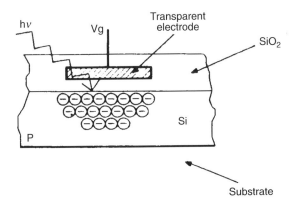

Fig. 20.16. *Schematic representation of a photosite*

charge packets along the chain simultaneously. The interconnected electrodes form a phase.

Figure 20.17 shows how a three-phase charge transfer takes place. There are a number of other methods of transfer: two-phase, four-phase, virtual phase, etc. For users they differ in the ease with which they may be applied, in the speed of transfer that may be attained, and in their efficiency. The last parameter is a measure of the amount of residual charge left behind in transferring a packet to an adjacent site.

1121

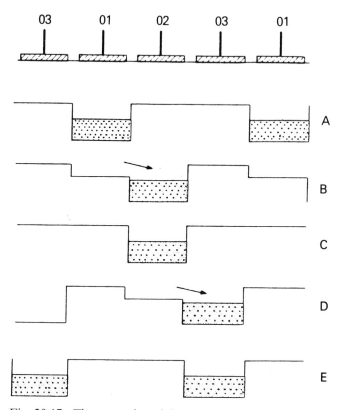

Fig. 20.17. *The principles of three-phase transfer. Initially the potential well is created beneath electrode 1. Then the potential of electrode 2 is gradually increased while that of 1 is decreased. The electrons move from 1 to 2. The process continues with electrodes 2 and 3. It will be seen that at least one electrode separates two successive charge-packets.*

With modern CCDs, the transfer efficiency is about 0.999 990, which is well-nigh perfect. A CCD's output stage converts the charges into a voltage, i.e., it converts the individual packets of charge into a video signal.

There are two main ways in which the pixels are arranged:

- As a linear CCD, with the pixels (as many as 2048 in current devices) in a single line, which is typically 25 mm long. The width of this linear array is about 10 μm.
- As a rectangular matrix of rows and columns. Currently matrices of 512 × 512 pixels are used. Larger arrays and mosaics are made, but the price of a detector array consisting of 1024 × 1024 pixels (or larger) is ...astronomical!

Matrices often operate on the frame transfer principle (Fig. 20.18). The detector is divided into three parts:

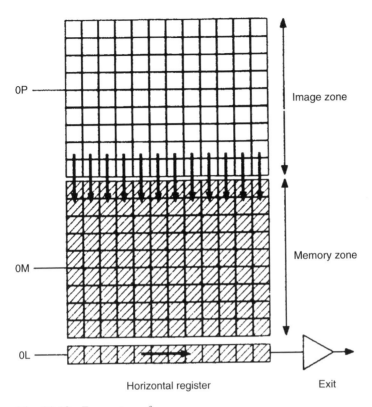

Fig. 20.18. *Frame transfer*

- the image zone, which is where the electronic image is produced;
- the memory zone, which differs from the image zone only in that it is protected by an opaque mask, which prevents photons from reaching the silicon;
- a horizontal shift register, which receives the information from successive lines of the matrix.

[It may also be noted that some amateurs have successfully used a different type of CCD, the interline transfer CCD, for astronomical work. This differs from the type just described in that every other line of pixels is masked and serves as a memory zone to the immediately adjacent line of pixels. The obvious disadvantage is the lower resolution in the image area, which means that it is not highly regarded by the astronomical purists. Its particular advantage, however, is that this type is used in video cameras – where fast read-out is important – and is thus readily available at a reasonable cost. – Trans.]

At the end of the integration time, the charges are transferred from the image zone into the memory zone as rapidly as possible so that the image being recorded is not affected by light incident during the read-out. The information held in the memory zone is then transferred, line by line, towards the horizontal register, which transfers the packets of charge to the output one at a time. This stage is a bottleneck, which means that the time taken to read the whole matrix is relatively long, and this is why the electronic image has to be protected by being shifted into the memory zone. When the first image has been read out, another is shifted into the memory zone, and so on.

CCDs fabricated specifically for astronomical work do not have a memory zone, because the read-out time is negligible in comparison with the integration times required, given the low value of the incident flux.

20.2.2 *The performance of CCDs*

CCDs respond to light in the same way as all other silicon-based detectors.

- The quantum efficiency (the ratio of the number of photoelectrons created for a given number of incident photons) may attain 80 % in components that have been carefully optimized. The most common CCDs have a quantum efficiency of about 40 %. These figures may be compared with those for the best, hypersensitized, photographic emulsions, which are estimated at 3–4 %.

- The spectral sensitivity extends from the blue into the near-infrared (from 400–1000 nm) as shown in Fig. 20.19. This range, which is roughly twice that of common photographic emulsions, offers a number of possibilities. It is also worth noting that with specific surface treatment, some CCDs have an extended response into the far-ultraviolet (100 nm).

- The linearity is excellent over practically the whole dynamic range, which enables accurate photometric measurements to be determined from the image. The dynamic range is the ratio between the maximum output signal and a CCD's dark noise. A saturated potential-well may contain, typically, 5×10^5 electrons, and the noise in the best professional-quality CCD cameras is about 10 electrons (the effective noise). Under these conditions, the dynamic range is 50 000, which far exceeds the performance of other detectors commonly used in astronomy.

- CCDs are extremely robust in comparison with more traditional detectors, such as image tubes. They withstand excessive illumination and suffer no residual effects. The mechanical rigidity of the silicon means that they may be applied to astrometric work.

Nevertheless there are some disadvantages to CCDs, the foremost of which is that the area of the detector is very small. When used at the prime focus of a telescope, modern CCDs are far from being capable of exploiting the whole of the available field. This limit to the available surface area is caused by technological restrictions as well as by the problems that arise in handling large, digitized images (see Sect. 20.2.3.2). This situation will undoubtedly change in the future, and already CCD matrices with 4096 × 4096 pixels are being tested, the overall detector size being 50 mm square.

Any housing is relatively large, mainly because of the necessity for cooling the detector when imaging deep-sky objects.

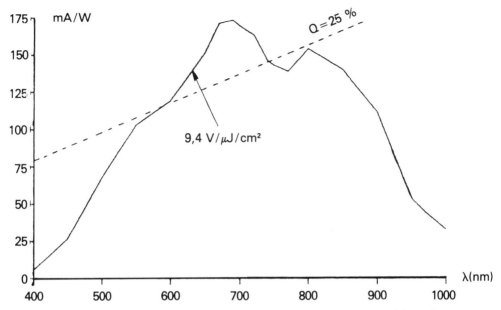

Fig. 20.19. *The spectral response of a Thomson TH7852 matrix. The oblique line joins points that have a quantum efficiency of 25 %.*

Problems may arise with the efficiency of transfer at weak fluxes, as well as reduced spatial resolution in the infrared when compared with that recorded in the visible. But these points are of minor significance in comparison with CCDs' intrinsic qualities. It could be said that they are practically perfect detectors, capable of exploiting to the full the low fluxes collected by telescopes.

20.2.3 CCD applications

20.2.3.1 Electronics

A video signal is particularly suitable for numerical treatment. In addition, a computer is the ideal adjunct to any CCD camera. Once the image has been digitized and stored in the computer's memory, it is possible to manipulate it to extract the maximum amount of information.

A preamp is placed as close to the detector as possible, boosting the signal enough for it to be passed to the next stage, perhaps 0.5 m away, where an analogue-to-digital converter changes the electrical voltages into appropriate binary numbers. The converter is generally preceded by a sample-and-hold (or more than one) that locks the voltage measured during the conversion time. In professional cameras, 16-bit digitization is often used (giving 65 536 levels). The binary output from the converter is then amplified and transmitted to a computer, which may be some metres away from the camera.

The charge transfer within the CCD requires a number of clock pulses, pro-vided by wired logic, which ensures perfect synchronism between the phases. This

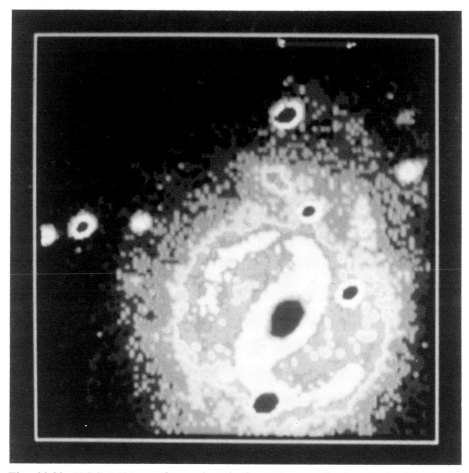

Fig. 20.20. *NGC 5435 as observed with the 600-mm reflector at the Pic du Midi Observatory in 1986 June by C. Buil and E. Thouvenot. The exposure was 600 seconds with a CCD cooled to −32°C. Image: T60 Association.*

logic produces what are known as TTL [transistor/transistor logic – Trans.] signals (switching between 0 and 5 V) and they should be amplified by TTL/CMOS translators before being passed to the CCD.

This electronics assembly is a fairly complicated piece of circuitry, which, by definition, is difficult for an amateur to construct. However, as we shall see later, the electronics may be simplified considerably by making the most of the possibilities offered by computerization.

20.2.3.2 The computer

We have just said that a computer is essential in conjunction with any CCD camera being used for observation. The correct choice of computer is imperative, so it should meet the following criteria:

- it should be easy to interface, which means both having access to the microprocessor bus, and also having access to the appropriate information needed to construct an interface;
- the graphics should be capable of handling the images required. A resolution of 320×200 in 16 colours could be considered the minimum, but experience shows that such a screen resolution is soon found to be inadequate. As a general rule, it is better to have a larger number of colours or a more extended grey scale than to increase the number of pixels on the screen;
- even a small image of just 100×100 points with 8-bit digitization requires 10k octets. Imagery is extremely greedy for memory, especially as most of the image-processing algorithms require at least two images to be held in active memory at any given time;
- a floppy-disk drive is essential for manipulating and storing the images.

The earlier generation of 8-bit computers are not particularly suitable, because they are unable to address very much memory. Sixteen-bit machines avoid this problem, and usually have suitable graphics capacity. Among the simplest, widespread machines is the Apple IIgs, which falls within this category.

IBM-compatibles have become much cheaper in recent years, and they have the advantage of being well-documented and easy to interface. The old XT-style machines are now technically outdated, and an AT-compatible offers a very substantial gain in speed, very important in image-processing. There are many different graphics cards available, but an EGA (Enhanced Graphics Adaptor), with a resolution of 640×350 in 16 colours from a palette of 64, should be considered the minimum. If finance is no consideration, cards are available giving much higher resolutions and as many as 256 simultaneous colours.

The basic version of Commodore's Amiga 2000 has exceptional graphics and uses the powerful Motorola 68000 microprocessor. Two buses are available, one Commodore's own, and the other IBM-compatible. This is a machine to be seriously considered.

Computing equipment and techniques are evolving rapidly, and it is sometimes difficult to choose a machine without regretting it just a few months later. Current trends are towards machines with a large amount of memory and extensive graphics possibilities (such as some of the Macintosh models, for example). This is all to the good in obtaining the CCD camera of one's dreams. [It should be noted, however, that it will probably be necessary to develop one's own image-processing software (or adapt programs written elsewhere), so it is essential to ensure that the appropriate language(s) are available for any specific machine, and that full use may be made of the graphics capacity available. Windowing-type of environments may not be the best for this purpose. – Trans.]

20.2.3.3 Cooling

At ambient temperatures, a CCD is completely saturated by thermal noise after just 2 second's integration. Imagery of faint objects requires exposures of several minutes (and sometimes fully an hour), so it is essential to cool the CCD. Professional astronomers use liquid nitrogen, which maintains the detector at temperatures

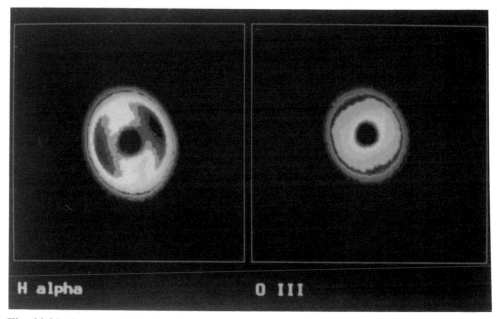

Fig. 20.21. *Images of the planetary nebula IC 418 obtained with the 1-m reflector at the Pic du Midi Observatory in 1986 December. Left: with a filter isolating the Hα line (656.3 nm). Right: in the forbidden oxygen line (500.7 nm). Note the difference in structure between the two images. On this scale, one pixel represents an angle of just 0.38 arc-second. These images were obtained with a TH7852A matrix, controlled by an Apple IIe, and then processed on an IBM PC fitted with an EGA card.*

around $-100\ ^{\circ}$C. (There is no point, and it is even deleterious to use still lower temperatures.) Under such circumstances, integration times of more than 1 hour pose no problems. The CCD is mounted inside an enclosure, capable of maintaining a high vacuum, which prevents frost and image degradation from air convection. Heat is extracted from the detector by a copper conductor, one end of which is placed in the cryogen.

The liquid-nitrogen method, although ideal, is extremely difficult for an amateur to utilize, with all the problems of obtaining liquid nitrogen, obtaining a high vacuum, etc. It is more sensible to use dry ice (solid carbon dioxide), which gives a temperature of around $-50\ ^{\circ}$C, which is more than sufficient for most applications. (The integration time under reasonable conditions may be extended to as long as 30 minutes.) Dry ice may be stored relatively well in a freezer. It should also be noted that dry ice may be produced by discharging a cylinder of CO_2 into a suitable container.

The refrigeration circuit from an old refrigerator may be used, but this method requires a certain knowledge of compressors and regulators. To minimize problems, a circuit using an ammonia coolant (a camping refrigerator) should be used. The degree of cooling to be expected is about $-20\ ^{\circ}$C.

The most practical method for an amateur CCD camera is to use thermoelectric cooling. A Peltier module is a block several millimetres thick and a few centimetres square. The thermoelectric couples are mounted between the two faces of the block, which are generally ceramic. When an electrical current is passed through the couples, the module acts as a heat-pump: it extracts heat from one face and rejects it from the other. The latter needs to be provided with an efficient method of removing the heat, which includes a contribution caused by the Joule effect (i.e., the resistance) of the module itself.

The two fundamental characteristics of a Peltier unit are the heat transfer potential, which is a maximum when the temperature difference between the two plates is zero, and the overall temperature difference, which is a maximum when the heat transfer potential is zero. A temperature difference of 70 °C may be achieved when the module is free from thermal loading. This is a theoretical case, and in practice performance is distinctly poorer. To obtain a greater degree of cooling, several modules are often mounted in cascade. A well-designed, three-stage system is able to attain a temperature difference of about 75 °C.

The optimization of a thermoelectric system is an art. The most important point is trying to cool the detector and nothing else. This means that the electrical connections to the CCD must be poor conductors of heat (e.g., very fine platinum wire). The most difficult problem is that of removing the heat from the heat-rejection plate. This is generally achieved by using a finned radiator over which air is blown by a fan. Heat may also be extracted by a liquid cooling system that circulates a suitable fluid, which may or may not be independently cooled. Despite these problems, use of thermoelectric modules is very attractive, because they enable a very compact design to be devised (offering the possibility of mounting a CCD camera on a small telescope), which is self-contained and also cheap. (The price of a thermoelectric module is about $100.) [Compact, commercial CCD systems with prices beginning at around $1100 are now available in North America. – Trans.]

20.2.4 Applications

The remarkable optical and electronic performance of CCDs enables them to detect objects under extreme conditions. The high quantum efficiency and the low read-out noise mean that their performance is often limited by the quality of seeing at the observing site, and by problems of radiometric corrections to the images. A CCD achieves the same limiting magnitude as a photographic emulsion with exposures that are only one-tenth as long. In addition, there is no question of a CCD being affected by reciprocity failure, whatever the length of the exposure: integrate for twice as long and effectively objects only half as bright are detected.

Using a CCD detector on an amateur-sized telescope offers the prospect of being able to carry out true scientific work. Magnitude 20 may be reached with a 200-mm reflector and a properly cooled matrix with an exposure of about 20 minutes. Superimposing several images is a common technique with electronic imaging, and this enables even fainter objects to be detected without much difficulty.

A CCD array may be used, for example, to monitor the fields of galaxies in a search for supernovae (Fig. 20.20). The limiting magnitude attainable enables a

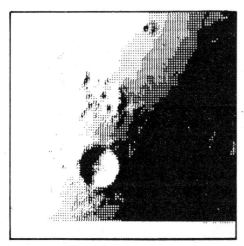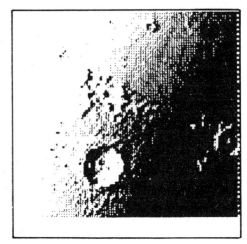

Fig. 20.22. *Image of the Moon obtained at prime focus of a Schmidt–Cassegrain 200-mm telescope. The left-hand image is unsharp, because of strong turbulence. On the right, the image has been processed to emphasize detail.*

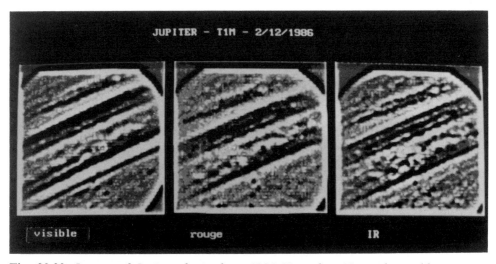

Fig. 20.23. *Images of Jupiter obtained on 1986 November 30, in the visible region (left), the red (centre), and in the near infrared (right). The strong colours of the planet's cloud-cover are easily seen. The CCD matrix was placed at the Cassegrain focus of the 1-m reflector at the Pic du Midi Observatory.*

thorough investigation to be carried out, thus increasing the chances of success. In addition, image-processing offers the possibility of comparing the image with older reference frames, which in some cases may establish the priority of a discovery.

Observations made within a CCD's wide, overall spectral range enable the distri-

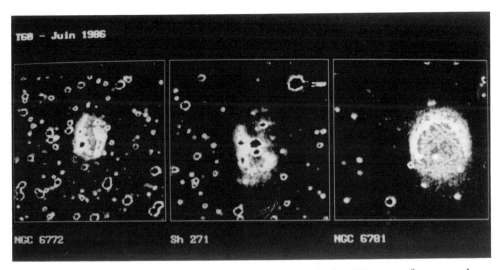

Fig. 20.24. *Observations of planetary nebulae made with the 600-mm reflector at the Pic du Midi Observatory. From left to right: NGC 6772, Sh 271, and NGC 6781. The last image was obtained through an Hα filter.*

bution of stars within a galaxy to be determined, or the precise state of ionization within a nebula (Figs. 20.21 and 20.24). Again, image-processing means that different images may be superimposed and then subtracted from one another (Fig. 20.25) or otherwise manipulated to show morphological differences (Fig. 20.26).

CCDs may also be used with bright objects (Fig. 20.22). They equal the best photographic emulsions (such as TP 2415) as regards contrast, and the highly accurate geometrical properties of the array of pixels enable planetary details to be measured very accurately (Fig. 20.23). It even appears to be possible to observe the surface of the Sun effectively, thanks to the extremely short integration times that are feasible, which freeze turbulence.

Linear CCDs should not be forgotten. They are easier to use than matrices, and are a more-or-less essential first stage in developing CCD techniques. Linear CCDs are very attractive for spectroscopic work. By aligning the spectrum and the device, it is possible to record the whole range simultaneously. Two-dimensional imaging is also possible by shifting the image of the object at right-angles to the linear array and taking regular readings. The image is reconstructed by aligning the digitized scan lines side-by-side on the screen. This is the same technique as that used in a scanner, and may be used to digitize photographs, by using an enlarger to project an image of the negative onto the linear CCD. The latter may be shifted sideways using a micrometer stage and a stepper motor controlled by the computer.

20.2.5 An amateur CCD camera

We will briefly describe a CCD camera built by the author and Eric Thouvenot. This camera was the first of its type to be constructed in France. The heart of the

1131

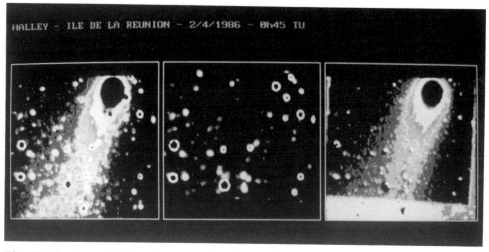

HALLEY - ILE DE LA REUNION - 2/4/1986 - 0h45 TU

Fig. 20.25. *Comet Halley observed on 1986 April 2, 00:45 UT, from the island of La Réunion in the Indian Ocean. The left-hand image was obtained by using a 135-mm focal-length, f/2.8, lens. The comet was then in the galactic plane and the large number of background stars interfered with interpretation of the image. An image of the same field, without the comet, was obtained several days later (centre). The right-hand image is the result of subtracting these two images, once they had been correctly aligned with one another. The process suppressed most of the stars.*

camera is a Thomson TH7852A array. This matrix is 144×204 pixels, each $30\,\mu\text{m}$ square. It uses frame transfer and requires four clock signals to operate:

- one controlling the transfer of the image into the memory area;
- one driving the transfer of individual lines from the memory area into the horizontal register;
- one for the horizontal register's read-out;
- one resetting the CCD's output stage after each packet is read.

A particularly interesting feature of this device is it is designed to prevent charges from spilling over from one highly over-exposed pixel into neighbouring ones. This means that it may be used to observe the areas immediately adjacent to bright objects (such as galactic nuclei or shell stars).

The CCD is a small, ceramic, 24-pin, DIL chip. A glass window protects the sensitive surface while allowing light to reach it. The Thomson TH7852 is sold in several grades, depending on the number of defects on the sensitive surface. The best class (class A) costs about $75. It is quite easy to use and of very high quality.

The computer used to store the images is an Apple IIe. One of the camera's unique features is the provision of clock pulses generated by a programme written in assembler. This has enabled the electronics to be simplified considerably, because there are practically no other logic circuits. It is also very easy to use, because

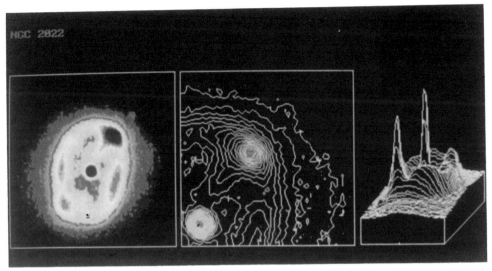

Fig. 20.26. *The use of high-resolution graphics enables various aspects of the same object to be depicted. This is the planetary nebula NGC 2022.* Left: *a representation in false colours – here reproduced in black-and-white – reveals the distribution of light in the object.* Centre: *isophotes depicting a portion of the nebula.* Right: *a three-dimensional representation gives a subjective impression of the ring of nebulosity that surrounds the central star.*

altering a few instructions in the program completely changes the camera's method of operation.

After amplification, the video signal is passed to an LF398 sample-and-hold, then to a 10-bit AD571 converter. An interface circuit (PIA MC6851) handles the computer input/output. An equivalent circuit generates the different CCD phases.

Because of memory and storage restrictions, the digitized images are only 96×96 pixels. A new version of the camera using an IBM AT machine enables 12-bit digitization of the whole matrix. It takes about 1.5 seconds to digitize an image of 96×96 pixels (the speed is partly dictated by the analogue-to-digital converter which has a relatively long conversion time of 25 μs).

Apart from the acquisition software itself, display (grey-scale and false-colour), and image-manipulation programs have been written. For speed reasons, all these programs are in assembler. The image-processing software is particularly comprehensive. It has various features that enable the enhancement of detail (high-pass filtration, for example); removal of noise (low-pass and medium-range filtration, etc.); image-amplification (histogram equalization and translation tables); contour plotting; arithmetical operations on more than one image (addition, subtraction, and division); alignment of one image with respect to another (registration); etc. Writing this progam alone took about half the time required to develop the camera. Such a program is essential to make full use of a CCD camera.

Cooling is achieved by two Peltier stages. The heat is carried to a large, finned radiator, with forced air cooling (Fig. 20.27). A thermistor is used to monitor the

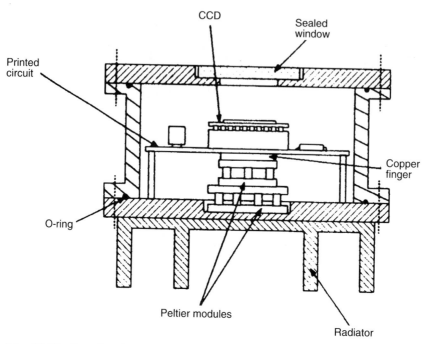

Fig. 20.27. *Details of one method of cooling a CCD*

temperature of the CCD itself. The latter and the cooling modules are mounted in a hermetically sealed chamber, which is evacuated to 10^{-3} Torr with a small vane pump. At an ambient temperature of 15°C, the CCD operates at −40°C.

A typical image-acquisition session using this camera is as follows:

- the electronics are switched on at least half an hour before any image is acquired, to ensure that they are fully stabilized;
- at the same time, the detector is progressively cooled (the life of a CCD is prolonged by avoiding thermal shocks);
- the object is then identified by taking a series of short exposures that may be displayed immediately on the screen. (A 5-second exposure will reach magnitude 13 with a 200-mm telescope.) A so-called 'wide-field' acquisition mode is also used, in which the whole of the matrix is employed, but with a reduced resolution;
- the proper image is acquired. The result may be assessed immediately after the integration has finished. If there is any problem, a further integration is carried out. This facility for interpreting an image in what amounts almost to real-time is a decisive advantage when compared with photography;
- the image is stored on floppy disk for later processing.

Apart from the image itself, integrations are also carried out in absolute darkness. These will later be subtracted from the raw image to remove the non-uniform effects between pixels of thermal noise. (This is required because of the insufficiently low temperature of the detector.)

Uniform-field images are also acquired by imaging the twilight sky background. The dark current signal is first subtracted from the raw image, which is then divided by the uniform-field result, thus allowing the response at different sites to be normalized. The image obtained in this way may then be used for photometric work.

20.2.6 *The future of CCDs*

The price of charge-coupled devices militated against their use by amateurs for a long time. This situation is changing, thanks to large-scale fabrication of these components, which has reduced prices. CCDs are now found in most video cameras. The price of a linear CCD is now around $50 (or less), and the price of a matrix suitabe for astronomical use is close to that of a good camera body. In addition, the continuing decline in the price of computers and the rise in their performance mean that CCD cameras are likely to become even more widespread among amateurs.

Installation requires some knowledge of electronics, computing, cryogenics, etc. Although it is fairly difficult for an individual amateur to construct a CCD camera from scratch, it is not too difficult for a group of astronomers to build one by pooling their expertise. For those who like practical projects, building a CCD camera is extremely worthwhile, because the field is almost completely unexplored. Image-processing is a fascinating occupation, which will sometimes enable a computer – perhaps purchased without any clear idea of its potential – to be used to the utmost of its capacity.

Given their performance and declining prices, CCDs are obviously likely to slowly replace photographic material in amateur telescopes. This change will completely alter past habits: scientific exploitation of observations will become much easier, and even the rule. This does not mean to say that the visual aspect of images will be neglected, because a digitized image may be very beautiful. A revolution in amateur astronomy is in progress.

Appendix 1   Time-scales

For certain astronomical observations, it is of the utmost importance that measurements should be specified on an appropriate time-scale. Failing this, the observations may be unusable. This is the case for events such as the mutual phenomena of Jupiter's satellites, for example, or for lunar occultations.

The problem of time-measurement and timing an event has become very complicated because of the accuracies required, and also obtained. The theory of relativity, which states that time is not identical for observers who are in relative motion, does not simplify the question – so much so that even the specialists themselves, the astronomers and physicists, are not always in agreement of the precise interpretation of the various time-scales that they have been forced to define.

We shall discuss here just the essential points, without trying to give a series of theoretical justifications, so that observers may know what to use, and why.

The notion of time gives rise to the concept of duration. It would be possible to define a standard duration as given by (say) a specific sand-glass. Instead, the preferred, standard duration is the Système Internationale second (often known as the 'SI second'). This is defined in terms of events related to the caesium-133 atom. This is not crucial, however, and all we need to know is that, using a good atomic clock in the laboratory, it is possible to measure duration, expressed in SI seconds, to a high degree of accuracy.

If we want to specify in an unequivocal manner when an event occurs, however, it needs to be located in what is termed a time-scale. Such a scale may be obtained by a succession of seconds, one after the other, numbered in a specific manner. This is where our troubles begin.

One important property of a time-scale is its uniformity. Let us assume that we know that, for theoretical reasons, a certain physical event should have an absolutely constant duration. Let us measure that duration as the difference between the date of the end of the event and that of its beginning, in the time-scale under consideration. We can say that the time-scale is uniform if this duration is found to be the same, at whatever epoch the measurement is made. This property of uniformity is obviously one that we expect to find in a time-scale, but one, unfortunately, that cannot be perfectly obtained.

The best current implementation of a uniform time-scale is given by International Atomic Time (TAI). Previously, Ephemeris Time (ET), defined by the Earth's motion around the Sun, was used. In practice, this could be regarded as TAI + 32.184 seconds. ET was abandoned because it suffered from certain inconveniences, which we will not discuss here. Instead of adopting TAI for the calculation of ephemerides, astronomers defined another time-scale, called Terrestrial Dynamic Time (TDT). This is the time-scale used for the tables of geocentric planetary ephemerides given in the *Astronomical Almanac*. [Note that as far as amateur astronomers are concerned, TDT may be taken as equivalent to the old Ephemeris Time. – Trans.]

In practice, therefore, we have:

ET = TDT = TAI + 32.184 seconds

We now need to discuss another set of time-scales, which all have in common the fact that they are related to the duration of the Earth's rotation about its axis. Again, we will simplify the discussion as much as possible.

The periodic return of the Sun to a specific meridian (the Greenwich Meridian, for example) naturally leads to a time-scale known as True Solar Time. Because the Earth orbits the Sun in an ellipse, rather than a circle, and its motion therefore obeys the law of equal areas rather than being uniform, and also because the orbit is inclined at an angle of approximately 23.5° to that of the equator, it has been known for a long time that this time-scale is not uniform. By calculating a quantity (known as the Equation of Time), it may, however, be corrected, enabling us to define a time-scale known as Universal Time (UT), which, for a long time, was accepted as being uniform. We now know that it is not: the Earth's rotation is subject to various changes, in particular to a gradual slowing down. For various reasons it is desirable to retain this time-scale. In fact, a compromise was adopted: broadcast time signals give what is known as Coordinated Universal Time (UTC), which is a succession of SI seconds, like TAI, but where the numbering of the seconds is occasionally altered, such that UTC does not differ from UT by more than 0.9 second. Such accuracy is sufficient for the vast majority of users (navigators, for example). There are tables showing the difference UT−UTC, and thus the dates on which leap seconds were inserted, but the important thing to know is the difference TDT−UTC (or TAI + 32.184 s − UTC – or even ET−UTC – which, as far as we are concerned, are equivalent). From 1992 June 30, this difference was 59 seconds, but it is increasing, currently by about 1 second a year. This difference is only known *a posteriori*; so it has to be extrapolated at intermediate times. For the convenience of observers, a number of ephemerides, particularly those intended for amateurs, give the coordinates of celestial objects, using the UT time scale. This enables immediate comparisons to be made with an observation recorded in UTC. But it is not actually correct, unless the ephemeris is of low accuracy, because the difference TDT−UTC is unpredictable.

[For the calculation of minor-planet, cometary, and similar ephemerides, from 1991 December 24, a slightly different time-scale has been adopted. This is Terrestrial Time (TT), which for practical purposes may be considered equivalent to TDT (or, if you wish, to the old Ephemeris Time). For the present and future:

TT − UT = (32 + N) seconds,

where N is the integral number of leap seconds that have been inserted. – Trans.]

To summarize, let us assume that I want to make an observation of an event, such as an occultation.

- The prediction is for a time *t* in TT. I make my preparations for the time

$t + (TT - UTC)$. If the prediction is given in UT, no correction needs to be applied.

- I then listen to, or record, suitable time signals, such as a speaking clock, or broadcast radio signals. NB – Speaking clocks and most broadcast signals give local standard time or, in summer, local summer time. Either of these may be shifted a specific (usually integral) number of hours from UT. This must not be forgotten. Such an error is (usually) easily detectable, however. It is rare for there to be an error of as much as an hour in the prediction of any event.

- In reducing the observation, it will be converted from UTC to TT, and compared with the predicted time. If one is observing the occultation of stars by the Moon, what one does, in effect, is determine the old Ephemeris Time. Now that this has disappeared, we actually obtain the position of the Moon at a given instant on the TDT time-scale (or TAI + 32.184 s).

Finally, a word about sidereal time. Like UT, it depends on the rotation of the Earth, but is determined when the vernal equinox, not the Sun, crosses the meridian. It is used in calculating the rising and setting times of celestial objects, and details may be found in the ephemerides. It is no longer used for recording observations: sidereal clocks deserve to be relegated to museums. [This is not, strictly speaking, correct, because many observers find them extremely useful as an indication of the RA that is on the meridian, and for use with setting circles. – Trans.]

Appendix 2   The T60 Association

In 1982, the Pic du Midi Observatory (Fig. A2.1) decided to lend one of its telescopes, the T60 – i.e., the 60-cm Reflector – to amateur astronomers. This is a Newtonian telescope, whose primary has a diameter of 600 mm and a focal length of 2100 mm. Its short focal ratio of f/3.5 means that it is ideal for the observation of faint objects (Fig. A2.1). In fact, the brightness at the image plane means that it is an excellent light-collector, although various aberrations, primarily caused by the short focal length prevent any high-resolution work.

The equatorial mounting is very stable and is motor-driven on both axes. The setting circles allow the telescope to be aligned to an accuracy of about 1 minute of arc. The telescope may be moved manually, however, and a finder allows it to be set by identifying star fields.

The drive is very accurate and the slow motions are very smooth. A guide telescope with a diameter of 130 mm and a focal length of 1900 mm, enables guided exposures to be made.

At the Newtonian focus there is a metal plate that can be fitted with an eyepiece mounting, photographic equipment, or any heavy equipment. The telescope will probably be fitted with a CCD camera and a spectrograph.

A number of results obtained with this telescope are given in this book. To give an idea of its performance, it is possible to reach 20th magnitude with a 1-hour exposure on hypersensitized Kodak TP-2415.

Usage of the telescope is controlled by the 'T60 Users' Association' (AT60), an amateur association, located at the headquarters of the Pic du Midi and Toulouse Observatories (OPMT, 14 avenue Edouard Belin, 31400 Toulouse). This association's executive groups are a Programme Committee and a Technical Support Group.

The Programme Committee meets thrice yearly, in February, June and October, to consider requests for telescope time, and to allocate usage during the following four months. Appropriate details and request forms are issued by the AT60. Any amateur astronomer may request telescope time, but knowledge of the telescope is required, and this may be obtained by helping with a previous observing run, for example. The criteria for the selection of specific projects are the telescope's suitability for the proposed work, and the scientific merit of the proposed experiment. The maximum number of persons in an observing team is four.

The Technical Support Group is responsible for the telescope's maintenance, and also for carrying out alterations to improve the telescope's performance. It consists of volunteer, amateur astronomers, and makes several technical trips to the observatory every year.

The T60 is a powerful tool for amateur astronomers; it enables a large number of observations to be made in a quasi-professional manner. We can only encourage amateurs to use it, and to use it to the full.

Fig. A2.1. *An aerial view of the Pic du Midi Observatory; the T60's dome is indicated by the arrow.*

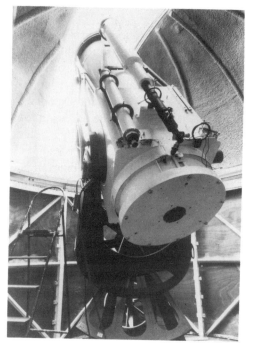

Fig. A2.2. *The T60 reflector itself*

Notes, references and bibliography – Volume 2

Chapter 11: Aurorae

Addresses

British Astronomical Association (BAA), Burlington House, Piccadilly, London W1V 0NL, United Kingdom

Norsk Astronomisk Selskap, Postboks 677, 4001 Stavanger, Norway

Royal Astronomical Society of Canada (RASC), 136 Dupont Street, Toronto, Ontario M5R 1v2, Canada.

For Canadian and North-American noctilucent-cloud observations only: Mark Zelik, #2 14225 - 82 Street, Edmonton, Alberta T5E 2vy, Canada

Royal Astronomical Society of New Zealand, PO Box 3181, Wellington, New Zealand

Svenska Astronomska Sallskapet, Stockholms Observatorium, 13300 Saltsjobaden, Sweden

Ursa Astronomical Association: Laivanvarustajankatu 3, 00410 Helsinki, Finland. Tel:+358 0 174 048; Fax:+358 0 657 728. Internet: ursa@f861.n220.z2.fidonet.org

References

Giraud, A. & Petit, M., 1978, *Ionospheric Techniques and Phenomena*, Geophysics and Astrophysics Monographs, D. Reidel, Dordrecht

Livesey, R. J., 1982, 'A "jamjar" magnetometer', *J. Brit. Astron. Assoc.*, **93**, pp. 17–19

Livesey, R. J., 1989, 'A jam-jar magnetometer as an "aurora detector"', *Sky & Telescope*, **78**, pp. 426–32

Pettitt, D. O., 1984, 'A fluxgate magnetometer', *J. Brit. Astron. Assoc.*, **94**, pp. 55–61

Simon, P. & Legrand, J.-P., 1990, *l'Astronomie*, **104**, pp. 363–72 (November–December)

Simmons, D. A. R., 1985a, 'An introduction to the aurora', *Weather*, **40**, pp. 147-55

Simmons, D. A. R., 1985b, 'A study of auroral emissions by interference filter photography', *J. Brit. Astron. Assoc.*, **95**, pp. 252–6

Simmons, D.A.R., 1988, 'Auroral photography at high latitudes', *J. Brit. Astron. Assoc.*, **98**, pp. 93–7

Smillie, D.J., 1992, 'Magnetic and radio detection of aurorae', *J. Brit. Astron. Assoc.*, **102**, pp. 16–20

Bibliography

Akasofu, S.-I., 1979 *Alaska Geographic*, **6** (2), special edition

Bone, N., 1991, *The Aurora*, Ellis Horwood, Chichester
Brekke, A. & Egeland, A., 1983, *The Northern Light*, Springer-Verlag, Heidelberg
Eather, R., 1979, *Majestic Lights*, American Geophysical Union, Washington, D.C.
Gadsden, M. & Schröder, W., 1989, *Noctilucent Clouds*, Springer-Verlag, Heidelberg
International Union of Geodesy and Geophysics, 1963, *International Auroral Atlas*, Edinburgh University Press, Edinburgh
Paton, J., 1973, 'The Aurora', *in* Moore, P. (ed.), *Practical Amateur Astronomy*, Lutterworth

Chapter 12: Meteors

References

Betlem, H. & Mostert, H. E., 1982, 'How to automate a 35-mm all-sky camera for an automatic fireball network', *J. Brit. Astron. Assoc.*, **93**, pp.11–16
Cook, A. F., 1973, *Evolutionary and Physical Properties of Meteoroids*, (NASA SP-319), Washington, DC. [Despite its age, this remains the best source for reliable information about meteor streams. – Trans.]
Kresakova, M., *Contr. Astr. Obs. Skalnate Pleso*, **3**, pp. 75–109
Öpik, E., 1922, 'A statistical method of counting shooting stars and its application to the Perseid shower of 1920', *Tartu Publ.*, **25**, pt 1, pp. 1–56
Sarma, T. & Jones, J., 1986, *Bull. Astron. Inst. Czechosl.*, **36**, pp. 9–24
Witze, A. M., 1992, 'A brilliant flash', *Sky & Telescope*, **83**, p. 585, (May)
Zvolankova, J., 1983, *Bull. Astron. Inst. Czechosl.*, **34**, pp. 122–8

Bibliography

Hawkins, G. S., 1964, *Meteors, Comets and Meteorites*, McGraw-Hill, N.Y.
Hughes, D. W., 1978, 'Meteors' in *Cosmic Dust*, ed. Macdonnell, J.A.M., Macmillan
Kronk, G. W., 1988, *Meteor Showers: a descriptive catalogue*, Enslow, N.J.
Levy, D. H. & Edberg, S. J., 1986, *Observe Meteors* (The ALPO Meteor Observer's Guide), Astronomical League, Washington, D.C.

Chapter 13: Double stars

References

Baize, P., 1982a, 'Sur quelques causes d'erreur dans les observations d'étoiles doubles vraies et fausses étoiles doubles', *l'Astronomie*, **96**, p. 7 (January)
Baize, P., 1982b, *l'Astronomie* [Not in April issue given in original reference – Trans.]
Baize, P., 1982c, 'La mesure des étoiles doubles par interférometrie "speckle"', *l'Astronomie* **96**, p. 389 (September)
Baize, P., 1984, 'Méthode simple pour calculer les éléments orbitaux d'une étoile double', *l'Astronomie*, **98**, p. 67 (February)
Couteau, P., 1981a, *Observing visual double stars*, MIT Press, Cambridge, Mass., pp. 116–20

Danjon, A., 1980, *Astronomie Générale*, Paris, 2nd edn, p. 432

Durand, P., 'Pour la mesure des étoiles doubles: construction d'un micromètre pour amateurs, adaptable sur un telescope', *l'Astronomie*, **93**, p. 375 (September)

de Froment, G., 1987, 'Note sur ADS 8231 (BD +28°2022)', *l'Astronomie*, **101**, p. 83, February

Ingalls, R., (ed.), 1946, 'Micrometers – a composite chapter', in *Amateur Telescope Making Advanced*, (i.e., Vol. 2), pp. 447–59

Kitchin, C. R., 1984, *Astrophysical Techniques*, Adam Hilger, Bristol, pp. 204–7

Labeyrie, A., 1982, 'Stellar interferometry: a widening frontier', *Sky & Telescope*, **63**, pp. 334–8 (April)

Minois, J., 1984, 'Présentation du micromètre à diffraction de M. Duruy', *l'Astronomie*, **98**, p. 231 (May)

Muller, P., 1939, 'Construction et mode d'emploi d'un nouveau micromètre astronomique', *la Revue d'Optique*, **18**, pp. 172–196

Muller, P., 1972, 'Les mesures par double image', *l'Astronomie*, **86**, p. 472 (October–November)

Muller, P., 1978, 'Mouvements relatifs de 17 étoiles doubles visuelles', *Astronomy and Astrophysics Suppl.*, **32**, pp. 165–72

Peterson, H.H., 1954, 'Double stars in a 3-inch refractor', *Sky & Telescope*, **13**, p. 396 (September)

Pither, C., 1980, 'Measuring double stars with a grating micrometer', *Sky & Telescope*, **59**, pp. 519–23 (June)

Proust, D., Ochsenbein, F. & Petterson, B. R., 1981, 'A catalogue of variable visual binary stars', *Astronomy & Astrophysics Suppl.*, pp. 179–85 (May)

Robertson, T. J., 1985, *Sky & Telescope*, **69**, pp. 361–2 (April)

Soulié, E., 1986, 'L'amélioration de l'orbite d'une étoile double visuelle', *Astronomy and Astrophysics*, **164**, p. 408–14

Bibliography

Aitken, R. G., 1932, *New General Catalogue of Double Stars Within* 120° *of the North Pole*, Carnegie Inst. of Washington Publ. No.417, Washington, DC

Couteau, P., 1981, *Observing Visual Double Stars*, MIT Press, Cambridge, Mass. [Contains many additional, useful references – Trans.]

Couteau, P., Morel, P. J. and Fulconis, M., 1986, *Cinquième Catalogue d'Ephémérides d'Etoiles Doubles*, Paris

Danjon, A., & Couder, A., 1935, *Lunettes et Télescopes*, Ed. de la Revue d'Optique, Paris

Dommanget, J. & Nys, O., 1964, *Catalogue de Trajectoires Rectilignes Relatives d'Étoiles Doubles Visuelles*, Annales de l'Observatoire Royal de Belgique, Serie 3, **IX**, Fasc. 6

Heintz, W. D., 1978, *Double Stars*, Reidel, Dordrecht

Hirschfeld, A. & Sinnot, R. W., 1985, *Sky Catalogue 2000.0*, Cambridge University Press, Cambridge and Sky Publishing Corp., Cambridge, Mass. [Also contains useful references – Trans.]

Jeffers, H. M., Van den Bos, W. H. & Greeby, F. M., 1963, *Index Catalogue of Visual Double Stars, 1961.0*, Lick Observatory, University of California

Minois, J., 1978, *Catalogue de 2500 Étoiles Doubles avec Éphemérides et Élements Orbitaux*, privately published, Paris

Société Astronomique de France, *Observations et Travaux*, **2** & **3**

Worley, C. E. and Heintz, W. D., 1983, 'Fourth catalog of orbits of visual binary stars', *Publications U.S. Naval Obsy*, 2nd Series, **24**, Part 7, Washington, D.C.

Chapter 14: Variable stars

Addresses

AAVSO: 25 Birch Street, Cambridge, Mass. 02138, U.S.A.

AFOEV: c/o E. Schweitzer, 16 rue de Plobsheim, 61700 Strasbourg, France

BAA VSS: British Astronomical Association, Burlington House, Piccadilly, London, W1V 0NL, United Kingdom

VSOLJ: c/o Dr Sei-ichi Sakuma, 2-21-9, Kami-Aso, Kawasaki, 215 Japan

RASNZ VSS: P.O. Box 3039, Greerton, Tauranga, New Zealand

References

Ells, J. W. & Ells, P. E., 1989 & 1990, 'A simple automatic photoelectric telescope', *Journal Brit. Astron. Assoc.*, **99**, p. 299; **100**, p. 24 & p.30

Kholopov, P. N. *et al.*, 1982, *New Catalogue of Suspected Variable Stars*, Nauka, Moscow

Kholopov, P. N. *et al.*, 1985–7, *General Catalogue of Variable Stars*, Nauka, Moscow

Lecacheux, J., (1970), 'Analyse des observations de Nova Delphini 1967', *Ciel et Terre*, **86**, (1), pp. 46–60

Bibliography

Burnham, R., 1966, *Burnham's Celestial Handbook*, Dover, New York

Hoffmeister, C., Richter, G. & Wenzel, W., 1985, *Variable Stars*, Springer-Verlag, Heidelberg

Jones, G. (ed.), 1990, *Webb Society Deep-Sky Observer's Handbook, Vol. 8: Variable Stars*, Enslow, Hillside N.J.

Levy, D. H., 1989, *Observing Variable Stars*, Cambridge University Press, Cambridge

Madore, B. F. (ed.), 1985, *Cepheids: Theory and Observations*, Cambridge University Press, Cambridge.

Meeus, J., 1982, *Astronomical Formulae for Calculators*, 2nd edition, Willmann-Bell, Richmond, VA

Percy, J. R. (ed.), 1986, *The Study of Variable Stars using Small Telescopes*, Cambridge University Press, Cambridge

Percy, J. R., Mattei, J. & Sterken, C., 1992, *Variable Star Research: An International Perspective*, Cambridge University Press, Cambridge

Petit, M., 1987, *Variable Stars*, John Wiley & Sons, Chichester

Chapter 15: Deep sky, novae and supernovae

Address

UK Nova/Supernova Patrol: Guy Hurst, *The Astronomer*, 16 Westminster Close, Kempshott Rise, Basingstoke, Hants. RG22 4PP, United Kingdom. Tel. & fax: +44 256 471074;
Email: GMH@ASTRO1.BNSC.RL.AC.UK

References

di Ciccio, D., 1984, 'A conversation with Robert Evans', *Sky & Telescope*, **67**, pp. 94–6

di Ciccio, D., 1986, 'Excerpts from an Australian journal', *Sky & Telescope*, **72**, p. 347 (October)

Doggett, J. B., and Branch, D., 1985, 'A comparative study of supernova curves', *The Astronomical Journal*, **90**, (11), pp. 2303–11

Harrington, P., 1985, 'A Messier marathon', *Sky & Telescope*, **69**, pp. 81–2 (January)

Heidmann, J., 1982, 'Clumpy irregular galaxies: an astronomical adventure', *Mercury*, **11**, pp. 170–7 (November–December)

Heidmann, J., 1986, 'Hyperactive star-bursts in clumpy irregular galaxies' in *Star Forming Regions* (IAU Symposium No. 115, Tokyo, 1985 November 11–15), Reidel

Hirschfield, A. & Sinnott, R. W., 1985, *Sky Catalogue 2000.0*, Cambridge University Press & Sky Publishing Corp.

Houston, W. S., 1979, 'Deep-sky wonders', *Sky & Telescope*, **57**, p. 315

Iburg, W., 1987, 'Adventures in gas hypering', *Sky & Telescope*, **73**, p. 110–12 (January)

Kukarkin, B. V., *et al.*, 1982, *New catalogue of suspected variable stars* (NSV), Moscow

Lagerkvist, C.-I., 1975, 'Photographic photometry of small asteroids', *Minor Planet Bulletin*, **3**, pp. 11–19.

Marsden, B. G., 1982, *IAU Circular* No. 3746, 1982 December 29

Mayer, B. & Liller, W., 1985, *The Cambridge Astronomy Guide* Cambridge University Press, Cambridge

Quirk, S., 1991, 'Testing Kodak's new films for deep-sky photography', *Sky & Telescope*, **81**, p. 328–30 (March)

Sagot, R. and Texereau, J., 1963, *La Revue des Constellations*, Société Astronomique de France

Schur, C., 1982, 'Experiments with all-sky phototography', *Sky & Telescope*, **63**, pp. 621–4 (June)

Schwartz, R., 1986, 'A hunt for flashing stars', *Sky & Telescope*, p. 560 (December)

de Vaucouleurs, G., 1985, 'The supernova of 1885 in Messier 31', *Sky & Telescope*, **70**, pp. 115–18 (August)

Woolf, N., 1985, 'Will Mount Graham have an observatory?', *Sky & Telescope*, **70**, pp. 424–5 (November)

Chapter 16: Plate comparators

Notes

1 This is not correct: stars that have varied in brightness (including novae, of course) may be detected quite easily, especially if the plates (or the head) may be moved slightly. The different-sized images do not merge into a single image, and thus give the immediate impression that something is 'wrong'. Some of the most successful professional and amateur discoverers of variables – such as Cuno Hoffmeister and Mike Collins, respectively – have used and still use the stereoscopic method. – Trans.

References

Bourge, P., Dragesco, J. & Dargery, Y., 1977, *La Photographie Astronomique d'Amateur*, Paul Montel, Paris

Godillon, D., 1971, *Atlas de l'Astronomie Amateur*, Douin, Paris

Mayer, B. & Liller, W., 1985, *The Cambridge Astronomy Guide*, Cambridge University Press, Cambridge

Texereau, J., 1947, 'Un blink microscope d'amateur', *L'Astronomie*, **61**, p. 355 (December)

Chapter 17: Astrometry

References

Acker, A. & Jaschek, C., 1986, *Astronomical Methods and Calculations*, Wiley, Chichester, pp. 24–6

Everhart, E., 1982, 'Constructing a Measuring Engine', *Sky & Telescope*, **64**, pp. 279–82 (September)

Marche, J. D., 1990, 'Measuring positions on a photograph', *Sky & Telescope*, **80**, pp. 71–5 (July)

Marsden, B. G., 1982a, in *Comets*, ed. Wilkening, L. L., Univ. Arizona Press, pp. 713–15

Marsden, B. G., 1982b, 'How to reduce plate measurements', *Sky & Telescope*, **64**, p. 284 (September)

Bibliography

Edberg, S., 1983, *IHW Amateur Observer's Manual for Scientific Comet Studies*, Sky Publishing Corp.

Gordon, B., 1985, *Astrophotography*, Willmann-Bell, Richmond, VA

Meeus, J., 1982, *Astronomical Formulae for Calculators*, 2nd edition, Willmann-Bell, Richmond, VA

Montenbruck, O. & Pfleger, T., 1990, *Astronomy on the PC*, Springer-Verlag, Heidelberg and New York

van de Kamp, P., 1967, *Principles of Astrometry*, W. H. Freeman, New York

Wallis, B.D. & Provin, R.W., 1988, *A Manual of Advanced Celestial Photography*, Cambridge University Press, Cambridge [This contains an excellent, comprehensive list of references. – Trans.]

Yeomans, D., West, R. M., Harrington, R. S. & Marsden, B. G., (eds), 1984, *Cometary Astrometry*, Proceedings of a Workshop held June 18–19, 1984 at the European Southern Observatory. JPL publication 84-82, November 15, 1984

Chapter 18: Spectroscopy

Bibliography

Dufay, J., 1964, *Introduction to Astrophysics: The Stars*, Dover, New York

Dufay, J., 1968, *Galactic Nebulae and Interstellar Matter*, Dover, New York

Jaschek, C. & Jaschek, M., 1990, *The Classification of Stars*, corrected edn, Cambridge University Press, Cambridge

Kaler, J. B., 1989, *Stars and their Spectra*, Cambridge University Press, Cambridge

Kitchin, C. R., 1984, *Astrophysical Techniques*, Adam Hilger, Bristol

Thackeray, A. D., 1961, *Astronomical Spectroscopy*, Eyre & Spottiswood, London

Chapter 19: Photoelectric photometry

Reference

Rufener, F., 1984, 'Reduction to outside the atmosphere and statistical tests used in Geneva photometry', in Borucki, W. J. & Young, A., eds, *Proceedings of the workshop on Improvements to Photometry*, San Diego, 18–19 June 1984. NASA Conf. Publ., NASA CP-2350, pp. 108–23

Address

International Amateur-Professional Photoelectric Photometry (IAPPP), Dyer Observatory, Vanderbilt University, Nashville, Tennessee 37235

Bibliography

Cooper, W. A. & Walker, E. N., 1989, *Getting the Measure of the Stars*, Adam Hilger, Bristol

Golay, M., 1974, *Introduction to Astronomical Photometry*, D. Reidel, Dordrecht

Hall, D. S. & Genet, R. M., 1982, *Photoelectric Photometry of Variable Stars*, IAPPP

Henden, A. & Kaitchuk, R., 1982, *Astronomical Photometry*, Van Nostrand Reinhold, New York

Sterken, C. & Manfroid, J., 1992, *Astronomical Photometry, A Guide*, Kluwer, Dordrecht

Young, A. T., 1974, *Methods in Experimental Physics, Vol. 12A, Astrophysics*, Academic Press, New York

Zelik, M. & Smith, E. v. P., 1987, *Introductory Astronomy & Astrophysics*, Saunders, Philadelphia

Chapter 20: Image Intensifiers and CCDs

References

Buil, C., 1985, 'A charge-coupled device for amateurs', *Sky & Telescope*, **69**, pp. 71–83 (January)

Palermiti, M., 1980, 'Photography with an intensifier camera', *Sky & Telescope*, **59**, p. 530 (June)

Tichenor, C.L., 1982, 'Observing with a TV camera', *Sky & Telescope*, **63**, pp. 533–5 (May)

Bibliography

Barbe, D. F., 1980, *Charge-Coupled Devices*, Springer-Verlag, Heidelberg

Buil, C., 1991, *CCD Astronomy*, Willman-Bell, Richmond, VA

Eccles, M. J., Sin, M. E. & Tritton, K. P., 1983, *Low light-level Detectors in Astronomy*, Cambridge University Press, Cambridge

Howes, M. & Morgan, D., 1979, *Charge-Coupled Devices and Systems*, John Wiley and Sons, New York

Melen, R. and Buss, D., 1977, *Charge-Coupled Devices: Technology and Applications*, IEEE Press

Platt, T., 1992, *Practical CCD Astronomy*, Cambridge University Press, Cambridge